NONLINEAR SYSTEMS IN AVIATION, AEROSPACE, AERONAUTICS, AND ASTRONAUTICS

ADVANCES IN DYNAMICS AND CONTROL

Nonlinear Systems in Aviation, Aerospace, Aeronautics, and Astronautics

A series edited by:
S. Sivasundaram
Embry-Riddle Aeronautical University, Daytona Beach, FL USA

Volume 1
Stability Domains
L. Gruyitch, J.-P. Richard, P. Borne, and J.-C. Gentina

Volume 2
Advances in Dynamics and Control
S. Sivasundaram

Volume 3
Optimal Control of Turbulence
S.S. Sritharan

NONLINEAR SYSTEMS IN AVIATION, AEROSPACE, AERONAUTICS, AND ASTRONAUTICS

ADVANCES IN DYNAMICS AND CONTROL

S. SIVASUNDARAM

CHAPMAN & HALL/CRC

A CRC Press Company

Boca Raton London New York Washington, D.C.

Library of Congress Cataloging-in-Publication Data

Advances in dynamics and control / edited by S. Sivasundaram.
p. cm. -- (Nonlinear systems in aviation, aerospace, aeronautics, and astronautics ; 2)
Includes bibliographical references and index.
ISBN 0-415-30852-6 (alk. paper)
1. Flight control. 2. Aerodynamics. I. Sivasundaram, S. II. Series.

TL589.4.A33 2004
629.135--dc22 2003070019

Direct all inquiries to CRC Press LLC, 2000 N.W. Corporate Blvd., Boca Raton, Florida 33431.

Visit the CRC Press Web site at www.crcpress.com

International Standard Book Number 0-415-30852-6
Library of Congress Card Number 2003070019
1 2 3 4 5 6 7 8 9 0

CONTENTS

CONTRIBUTORS

Alessandro Astolfi, Department of Electrical and Electronic Engineering, Imperial College, Exhibition Road, SW7 2BT London, England

George Avalos, Department of Mathematics and Statistics, University of Nebraska–Lincoln, Lincoln, NE 68588-0323, USA

A.M. Balint, Department of Physics, University of West Timisoara, Blv. V. Pârvan No. 4, 1900 Timisoara, Romania

St. Balint, Department of Mathematics, University of West Timisoara, Blv. V. Pârvan No. 4, 1900 Timisoara, Romania

Oscar Castillo, Department of Computer Science, Tijuana Institute of Technology, P.O. Box 4207, Chula Vista CA 91909, USA

Hsin-Yuan Chen, Department of Automatic Control Engineering, Feng Chia University, Taichung, Taiwan

M.N. Demenkov, Department of Computing Sciences and Control, Bauman Moscow State Technical University, Moscow, Russia

V.M. Glumov, Trapeznikov Institute of Control Sciences, Russian Academy of Sciences, Moscow, Russia

M.G. Goman, De Montfort University, Faculty of Computing Sciences and Engineering, Hawthorn Building, Leicester LE1 9BH, UK

Norihiro Goto, Kyushu University, Fukuoka 812-8581, Japan

Lyubomir T. Gruyitch, University of Technology Belfort–Montbeliard, Site Belfort, 90010 Belfort Cedex, France

Sen-Wei Huang, Department of Applied Mathematics, National Chung Hsing University, Taichung 402, Taiwan

Vikram Kapila, Department of Mechanical, Aerospace, and Manufacturing Engineering, Polytechnic University, Brooklyn, NY 11201, USA

Takashi Kawakita, Kyushu University, Fukuoka 812-8581, Japan

Jinlu Kuang, Satellite Engineering Center, The School of Electrical and Electronic Engineering, Nanyang Technological University, 639798, Singapore

Rei-Min Lai, Institute of Automatic Control Engineering, Feng Chia University, Taichung 40724, Taiwan

V. Lakshmikantham, Florida Institute of Technology, Department of Mathematical Sciences, Melbourne, FL 32901, USA

Irena Lasiecka, Department of Mathematics, Kerchof Hall, University of Virginia, Charlottesville, VA 22904, USA

Chun-Liang Lin, Department of Electrical Engineering, National Chung Hsing University, Taichung 402, Taiwan

Marco Lovera, Dipartimento di Elettronica e Informazione, Politecnico di Milano, Piazza Leonardo da Vinci 32, 20133 Milano, Italy

Steffen Marburg, Institut für Festkörpermechanik, Technische Universität, 01062 Dresden, Germany

V.M. Matrosov, Stability and Nonlinear Dynamics Research Center, Mechanical Engineering Research Institute, Russian Academy of Sciences, 5 Dmitry Ul'yanov Street, Moscow, 117333 Russia; Moscow Aviation Institute (State Technical University), 4 Volokolamskoye Avenue, Moscow, 125871, Russia

Raman K. Mehra, Scientific Systems Company Inc., 500 West Cummings Park, Suite 3000, Woburn, MA 01801, USA

Patricia Melin, Department of Computer Science, Tijuana Institute of Technology, P.O. Box 4207, Chula Vista, CA 91909, USA

Haizhou Pan, Department of Mechanical, Aerospace, and Manufacturing Engineering, Polytechnic University, Brooklyn, NY 11201, USA

Ravi K. Prasanth, Scientific Systems Company Inc., 500 West Cummings Park, Suite 3000, Woburn, MA 01801, USA

Marcio S. de Queiroz, Department of Mechanical Engineering, Louisiana State University, Baton Rouge, LA 70803, USA

V.Yu. Rutkovsky, Trapeznikov Institute of Control Sciences, Russian Academy of Sciences, Moscow, Russia

Marianna A. Shubov, Department of Mathematics and Statistics, Texas Tech University, Lubbock, TX 79409, USA

S. Sivasundaram, Embry-Riddle Aeronautical University, Department of Mathematics, Daytona Beach, FL 32114, USA

Ye.I. Somov, Stability and Nonlinear Dynamics Research Center, Mechanical Engineering Research Institute, Russian Academy of Sciences, 5 Dmitry Ul'yanov Street, Moscow, 117333 Russia; Research Institute of Mechanical Systems Reliability, Samara State Technical University, 244 Molodogvardyeskaya Street, Samara, 443100 Russia

V.M. Sukhanov, Trapeznikov Institute of Control Sciences, Russian Academy of Sciences, Moscow, Russia

Soonhie Tan, Satellite Engineering Center, The School of Electrical and Electronic Engineering, Nanyang Technological University, 639798, Singapore

Roberto Triggiani, Department of Mathematics, Kerchof Hall, University of Virginia, Charlottesville, VA 22904, USA

Firdaus E. Udwadia, Aerospace and Mechanical Engineering, Civil Engineering, Mathematics, and Information and Operations Management, 430K Olin Hall, University of Southern California, Los Angeles, CA 90089-1453, USA

Hong Wong, Department of Mechanical, Aerospace, and Manufacturing Engineering, Polytechnic University, Brooklyn, NY 11201, USA

Chi-Ching Yang, Department of Electrical Engineering, Hsiuping Institute of Technology, Taichung, Taiwan

S.D. Zemlyakov, Trapeznikov Institute of Control Sciences, Russian Academy of Sciences, Moscow, Russia

PREFACE

Nonlinear phenomena in aviation and aerospace have stimulated cooperation among engineers and scientists from a range of disciplines. Developments in computer technology have allowed for solutions of nonlinear problems, while industrial recognition of the use and applications of nonlinear mathematical models in solving technological problems is increasing.

Advances in Dynamics and Control comprises research papers contributed by expert authors in dynamics and control and is dedicated to Professor A.V. Balakrishnan, University of California, Los Angeles, USA in recognition of his significant contribution to this field of research.

Professor A.V. Balakrishnan earned his Ph.D. in mathematics from the University of Southern California in 1954. Professor Balakrishnan has been with the University of California, Los Angeles as a professor of engineering since 1962 and a professor of mathematics since 1965. He was chair of the Department of System Science in the School of Engineering from 1969–1975. Since 1985, he has served as the director of the NASA–UCLA Flight Systems Research Center. He was chairman of the Technical Committee on System Modelling and Optimization, International Federation of Information Processing, 1970–1980, and is currently the president of the Com Con Conference Board. He has received honors and awards including: Fellow IEEE (1966), Silver Core IFIP (1977), the Guillemin Prize (1980), the NASA Public Service Medal (1996), and the AACC Richard E. Bellman Control Heritage Award (2001).

The work of Professor A.V. Balakrishnan has been an inspiration for generations of engineers and applied mathematicians. During more than 40 years of his outstanding scientific career, he has made significant contributions to the analysis and design of control systems. His contributions range from the theory of optimal control (in the 1960s, he developed a celebrated method — the epsilon technique — for the computation of optimal controls for distributed parameter systems) to filtering and identification theory, to a number of difficult engineering applications that include the control of aircraft under wing turbulence, the control of flexible structures in space, and the aeroelastic modeling of aircraft wings. Professor Balakrishnan has demonstrated the ability to master difficult engineering problems in a rigorous mathematical way and has produced effective engineering solutions.

Advances in Dynamics and Control explores many new ideas, results, and directions in the rapidly growing field of aviation and aerospace. It encompasses a wide range of topics including rotorcraft dynamics, stabilization of unstable aircraft, spacecraft and satellite dynamics and control, missile auto-pilot and guidance design, hybrid systems dynamics and control, and structural and acoustic modeling.

The contributors to *Advances in Dynamics and Control* are to be highly commended for their excellent chapters. The volume will provide a useful source of reference for graduate and postgraduate students, and researchers working in areas of applied engineering and applied mathematics, in university and in industry.

S. Sivasundaram

1

Global Spacecraft Attitude Control Using Magnetic Actuators

Marco Lovera† and Alessandro Astolfi†*

†*Dipartimento di Elettronica e Informazione, Politecnico di Milano, Milano, Italy*

**Department of Electrical and Electronic Engineering, Imperial College, London, England*

The problem of inertial pointing for a spacecraft with magnetic actuators is addressed. It is shown that a global solution to the problem can be obtained by means of (static) attitude and rate feedback and (dynamic) attitude feedback. Simulation results demonstrate the feasibility of the proposed approach.

1 INTRODUCTION

The problem of (global) regulation of the attitude of rigid spacecraft, i.e., spacecraft modeled by the Euler's equations and by a suitable parameterization of the attitude, has been widely studied in the recent years.

If the spacecraft is equipped with three independent actuators a complete solution to the set point and tracking control problems is available. In [5, 11] these problems have been solved by means of PD-like control laws, i.e., control laws that make use of the angular velocities and of the attitude, whereas [1], building on the general results developed in [3], has solved the same problems using dynamic output feedback control laws. It is worth noting that, if only two independent actuators are available, as discussed in detail in [4], the problem of attitude regulation is not solvable by means of continuous (static or dynamic) time-invariant control laws, whereas a time-varying control law, achieving local asymptotic (nonexponential) stability, has been proposed in [10].

The above results, however, are not directly applicable if the spacecraft is equipped with magnetic coils as attitude actuators. As a matter of fact, it is not possible by means of magnetic actuators to provide three independent torques at each time instant, yet as the control mechanism hinges on the variations of the Earth's magnetic field along the spacecraft orbit, *on average* the system possesses strong controllability properties. In [2, 7, 8, 13] the regulation problem has been addressed exploiting the (almost) periodic behavior of the system, hence resorting to classical tools of linear periodic systems, if local results are sought after, or to standard passivity arguments, if (global) asymptotic stabilization of open-loop stable equilibria is considered.

However, several problems remain open. In particular, if only inertial pointing is considered, the global stabilization problems by means of full- (or partial-) state feedback is still theoretically unsolved. Note, however, that from a practical point of view these

0-415-30852-6/04/$0.00+$1.50

problems have an engineering solution, as demonstrated by the increasing number of applications of this approach to attitude control.

The aim of this chapter is to show how control laws achieving global[1] inertial pointing for magnetically actuated spacecraft can be designed by means of arguments similar to those in [1, 11], provided that time-varying feedback laws are used and that the control gains satisfy certain scaling properties. In particular, while previous work ([9]) dealt with the case of state feedback control for a magnetically actuated, isoinertial spacecraft, this chapter deals with the more general problems of full (attitude and rate) and partial (attitude only) state feedback for a generic magnetically actuated satellite.

The chapter is organized as follows. In Section 2 the model of the system is presented, while in Section 3 the model of the geomagnetic field used in this study is described. In Section 4 a general result on the stabilization of magnetically actuated spacecraft is presented. Namely, using the theory of generalized averaging, it is shown how stabilizing control laws designed for spacecraft with three independent control torques have to be modified to construct stabilizing laws in the presence of magnetic actuators. In Section 5 the general theory is used to design control laws using only attitude information, so avoiding the need for rate measurements in the control system. Finally, Sections 6 and 7 present some simulation results and concluding remarks.

2 THE MODEL

The model of a rigid spacecraft with magnetic actuation can be described in various reference frames [12]. For the purpose of the present analysis, the following reference systems are adopted.

- Earth Centered Inertial reference axes (ECI). The origin of these axes is in the Earth's center. The X-axis is parallel to the line of nodes, that is the intersection between the Earth's equatorial plane and the plane of the ecliptic, and is positive in the Vernal equinox direction (Aries point). The Z-axis is defined as being parallel to the Earth's geographic north–south axis and pointing north. The Y-axis completes the right-handed orthogonal triad.

- Pitch–Roll–Yaw axes. The origin of these axes is in the satellite center of mass. The X-axis is defined as being parallel to the vector joining the actual satellite center of gravity to the Earth's center and positive in the same direction. The Y-axis points in the direction of the orbital velocity vector. The Z-axis is normal to the satellite orbit plane and completes the right-handed orthogonal triad.

- Satellite body axes. The origin of these axes is in the satellite center of mass; the axes are assumed to coincide with the body's principal inertia axes.

The attitude dynamics can be expressed by the well-known Euler's equations: [12]:

$$I\dot{\omega} = S(\omega)I\omega + T_{\text{coils}} + T_{\text{dist}}, \tag{1}$$

[1] To be precise, the control laws guarantee that almost all trajectories of the closed-loop system converge to the desired equilibrium.

where $\omega \in \mathbb{R}^3$ is the vector of spacecraft angular rates, expressed in body frame, $I \in \mathbb{R}^{3\times 3}$ is the inertia matrix, $S(\omega)$ is given by

$$S(\omega) = \begin{bmatrix} 0 & \omega_z & -\omega_y \\ -\omega_z & 0 & \omega_x \\ \omega_y & -\omega_x & 0 \end{bmatrix}, \tag{2}$$

$T_{\text{coils}} \in \mathbb{R}^3$ is the vector of external torques induced by the magnetic coils and $T_{\text{dist}} \in \mathbb{R}^3$ is the vector of external disturbance torques, which will be neglected in what follows.

In turn, the attitude kinematics can be described by means of a number of possible parameterizations (see, e.g., [12]). The most common parameterization is given by the four Euler parameters (or quaternions), which lead to the following representation for the attitude kinematics:

$$\dot{q} = W(\omega)q, \tag{3}$$

where $q = \begin{bmatrix} q_1 & q_2 & q_3 & q_4 \end{bmatrix}^T = \begin{bmatrix} q_r^T & q_4 \end{bmatrix}^T$ is the vector of unit norm Euler parameters and

$$W(\omega) = \frac{1}{2}\begin{bmatrix} 0 & \omega_z & -\omega_y & \omega_x \\ -\omega_z & 0 & \omega_x & \omega_y \\ \omega_y & -\omega_x & 0 & \omega_z \\ -\omega_x & -\omega_y & -\omega_z & 0 \end{bmatrix}. \tag{4}$$

It is useful to point out that Eq. (3) can be equivalently written as

$$\dot{q} = \tilde{W}(q)\omega, \tag{5}$$

where

$$\tilde{W}(q) = \frac{1}{2}\begin{bmatrix} q_4 & -q_3 & q_2 \\ q_3 & q_4 & -q_1 \\ -q_2 & q_1 & q_4 \\ -q_1 & -q_2 & -q_3 \end{bmatrix}. \tag{6}$$

Note that the attitude of inertially pointing spacecraft is usually referred to the ECI reference frame.

The magnetic attitude control torques are generated by a set of three magnetic coils, aligned with the spacecraft principal inertia axes, which generate torques according to the law:

$$T_{\text{coils}} = m_{\text{coils}} \wedge b(t),$$

where $m_{\text{coils}} \in \mathbb{R}^3$ is the vector of magnetic dipoles for the three coils (which represent the actual control variables for the coils) and $b(t) \in \mathbb{R}^3$ is the vector formed with the components of the Earth's magnetic field in the body frame of reference. Note that the vector $b(t)$ can be expressed in terms of the attitude matrix $A(q)$ (see [12] for details) and of the magnetic field vector expressed in the ECI coordinates, namely $b_0(t)$, as

$$b(t) = A(q)b_0(t).$$

The dynamics of the magnetic coils reduce to a very short electrical transient and can be neglected. The cross-product in the above equation can be expressed more simply as a matrix-vector product as

$$T_{\text{coils}} = B(b(t))m_{\text{coils}}, \tag{7}$$

where

$$B(b(t)) = \begin{bmatrix} 0 & b_z(t) & -b_y(t) \\ -b_z(t) & 0 & b_x(t) \\ b_y(t) & -b_x(t) & 0 \end{bmatrix} \tag{8}$$

is a skew symmetric matrix, the elements of which are constituted by instantaneous measurements of the magnetic field vector.

As a result, the overall dynamics, after application of the preliminary feedback,

$$m_{\text{coils}} = B^T(b(t))u,$$

can be written as

$$\begin{aligned} \dot{q} &= \tilde{W}(q)\omega \\ I\dot{\omega} &= S(\omega)I\omega + \Gamma(t)u, \end{aligned} \tag{9}$$

where $\Gamma(t) = B(b(t))B^T(b(t)) \geq 0$.

3 MAGNETIC FIELD MODEL

A time history of the International Geomagnetic Reference Field (IGRF) model for the Earth's magnetic field [12] along five orbits in Pitch–Roll–Yaw coordinates for a near-polar orbit (87° inclination) is shown in Fig. 1.

As can be seen, $b_x(t)$, $b_y(t)$ have a very regular and almost periodic behavior, while the $b_z(t)$ component is much less regular. This behavior can be easily interpreted by noticing that the x and y axes of the Pitch-Roll-Yaw coordinate frame lie in the orbit plane while the z axis is normal to it. As a consequence, the x and y components of $b(t)$ are affected only by the variation of the magnetic field due to the orbital motion of the coordinate frame (period equal to the orbit period) while the z component is affected by the variation of $b(t)$ due to the rotation of the Earth (period of 24 h).

When one deals with the problem of inertial pointing, however, it is more appropriate to consider a representation of the magnetic field vector in Earth centered inertial coordinates to be more convenient, as shown in Fig. 2.

4 STATE FEEDBACK STABILIZATION

In this section, a general stabilization result for a spacecraft with magnetic actuators is given, in the case of full-state feedback (attitude and rate). For, let $\bar{q} = \begin{bmatrix} 0 & 0 & 0 & 1 \end{bmatrix}^T$ and consider the system

$$\begin{aligned} \dot{q} &= \tilde{W}(q)\omega \\ I\dot{\omega} &= S(\omega)I\omega + \tau \end{aligned} \tag{10}$$

and the control law

$$\tau = k_p q_r - k_v \omega. \tag{11}$$

In the light of Theorem 1 in [11], the control law (11) guarantees that $q_r \to 0$ and $\omega \to 0$ as $t \to \infty$ for the closed-loop system (10) and (11). Also, an analysis of the Lyapunov function used in the same reference shows that the equilibrium $(\bar{q}, 0)$ of the closed-loop system (10) and (11) is asymptotically stable, while the other possible equilibrium $(-\bar{q}, 0)$ is unstable.

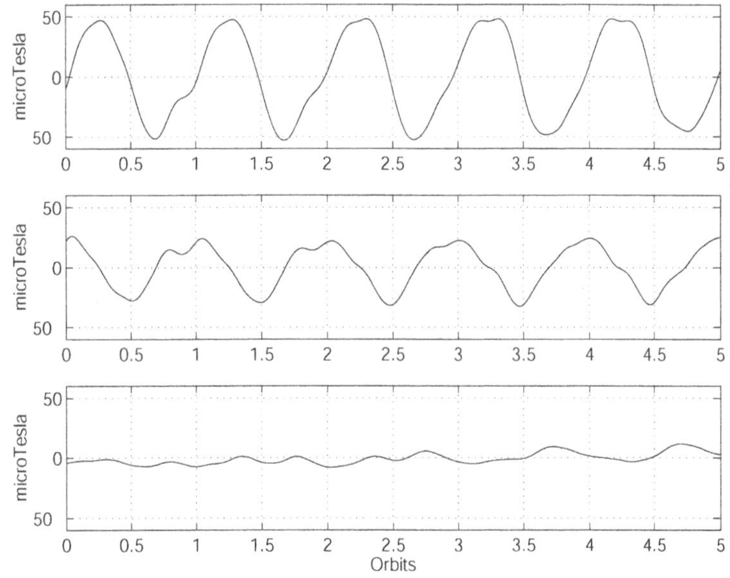

Figure 1 Geomagnetic field in Pitch–Roll–Yaw coordinates, 87° inclination orbit, 450 km altitude.

Proposition 1 *Consider the system (9) and the control law*

$$u = \varepsilon^2 k_p q_r - \varepsilon k_v \omega. \tag{12}$$

Then, there exists $\varepsilon^\star > 0$ *such that for any* $0 < \varepsilon < \varepsilon^\star$ *the control law (12) ensures that* $(\bar{q}, 0)$ *is a locally exponentially stable equilibrium of the closed-loop system (9–12). Moreover, almost all trajectories of (9–12) converge to* $(\bar{q}, 0)$.

Proof. In order to prove the first claim, introduce the coordinates transformation

$$z_1 = q \qquad z_2 = \frac{\omega}{\varepsilon}. \tag{13}$$

In the new coordinates, the system (9) is described by the equations:

$$\begin{array}{rcl} \dot{z}_1 & = & \varepsilon \tilde{W}(z_1) z_2 \\ I\dot{z}_2 & = & \varepsilon S(z_2) I z_2 + \varepsilon \Gamma(t)(k_p z_{1r} - k_v z_2). \end{array} \tag{14}$$

System (14) satisfies all the hypotheses for the application of generalized averaging theory ([6, Theorem 7.5]). Moreover, using the Lyapunov function of Theorem 1 in [11] one can conclude that the system obtained applying the generalized averaging procedure has $(\bar{q}, 0)$ as locally asymptotically stable equilibrium provided that

$$\Gamma = \lim_{T \to \infty} \frac{1}{T} \int_0^T B(t) B^T(t) dt > 0.$$

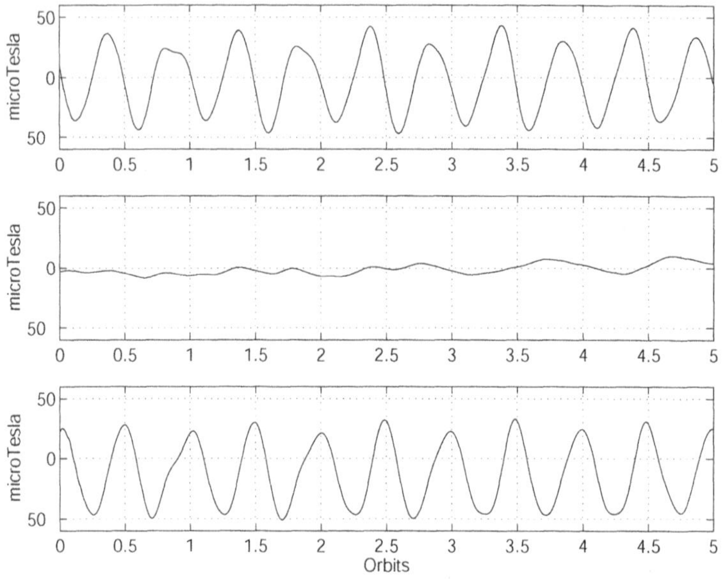

Figure 2 Geomagnetic field in Earth-centered inertial coordinates, 87° inclination orbit.

To conclude the proof of the second claim it is necessary to prove that the matrix Γ is generically positive definite. For, note that the matrix Γ is obtained by integration of a three-by-three square (symmetric) matrix of rank two, namely the matrix $B(t)B^T(t)$. However, the kernel of the matrix $B(t)B^T(t)$ is not generically a constant vector, which implies $\Gamma > 0$ generically. The set of *bad* trajectories, i.e., the trajectories for which the matrix Γ is singular, is described by the simple relation

$$\mathcal{K} = \mathrm{Ker}(B(t)) = \mathrm{Im}(\bar{b}),$$

for some constant vector $\bar{b}$. However, by a trivial property of the vector product one has

$$\mathcal{K} = \mathrm{Im}(b(t)) = \mathrm{Im}(A(q)b_0(t)),$$

hence all *bad* trajectories are such that, for all t,

$$A(q)b_0(t) = \lambda(t)\bar{b},$$

for some scalar function $\lambda(t)$, which is obviously a nongeneric condition. $\triangleleft$

5 STABILIZATION WITHOUT RATE FEEDBACK

The ability of ensuring attitude tracking without rate feedback is of great importance from a practical point of view. The problem of attitude stabilization without rate feedback has been recently given an interesting solution in [1] for the case of a fully actuated spacecraft. In this section a similar approach is followed in the development of a dynamic control law that solves the problem for a magnetically actuated satellite. First, notice that the system (10) and the control law (which is similar in spirit to the one proposed in [1]):

$$\begin{aligned} \dot{z} &= q - \lambda z \\ \tau &= -k_p q_r - \alpha \tilde{W}^T(q)(q - \lambda z), \end{aligned} \tag{15}$$

(where $\lambda > 0$ and $\alpha > 0$) give rise to a closed-loop system having $(\bar{q}, 0, \bar{q}/\lambda)$ as a locally asymptotically stable equilibrium and $q_r \to 0$ and $\omega \to 0$ as $t \to \infty$. On the basis of this consideration, which can be proved by means of the Lyapunov function

$$V = k_p[(q_4 - 1)^2 + q_r^T q_r] + \frac{1}{2}\omega^T I \omega + \frac{1}{2}(q - \lambda z)^T \alpha (q - \lambda z), \tag{16}$$

it is possible to give a solution to the magnetic attitude control problem without rate feedback.

Proposition 2 *Consider the system (9) and the control law*

$$\begin{aligned} \dot{z} &= q - \varepsilon \lambda z \\ u &= -\varepsilon^2 (k_p q_r + \alpha \tilde{W}^T(q)(q - \varepsilon \lambda z)). \end{aligned} \tag{17}$$

Then there exists $\varepsilon^\star > 0$ such that for any $0 < \varepsilon < \varepsilon^\star$ the control law renders the equilibrium $(\bar{q}, 0, \bar{q}/\varepsilon\lambda)$ of the closed-loop system (9–17) locally asymptotically stable. Moreover, almost all trajectories of the closed-loop system converge to this equilibrium.

Proof. As in the case of the state feedback control law we now introduce the coordinates transformation

$$\eta_1 = q \qquad \eta_2 = \frac{\omega}{\varepsilon} \qquad \eta_3 = \varepsilon z \, . \tag{18}$$

In the new coordinates, the system (9) is described by the equations:

$$\begin{aligned} \dot{\eta}_1 &= \varepsilon \tilde{W}(\eta_1)\eta_2 \\ I\dot{\eta}_2 &= \varepsilon S(\eta_2) I \eta_2 - \varepsilon \Gamma(t)(k_p \eta_{1r} + \alpha \tilde{W}^T(\eta_1)(\eta_1 - \lambda \eta_3)) \\ \dot{\eta}_3 &= \varepsilon(\eta_1 - \lambda \eta_3). \end{aligned} \tag{19}$$

Again, system (19) satisfies all the hypotheses for the application of [6, Theorem 7.5] and using the Lyapunov function given in Eq. (16) one can conclude that the system obtained applying the generalized averaging procedure has $(\bar{q}, 0, \bar{q}/\varepsilon\lambda)$ as a locally asymptotically stable equilibrium provided that

$$\Gamma = \lim_{T \to \infty} \frac{1}{T} \int_0^T B(t) B^T(t) dt > 0,$$

and this holds nongenerically as demonstrated in the proof of Proposition 1. ◁

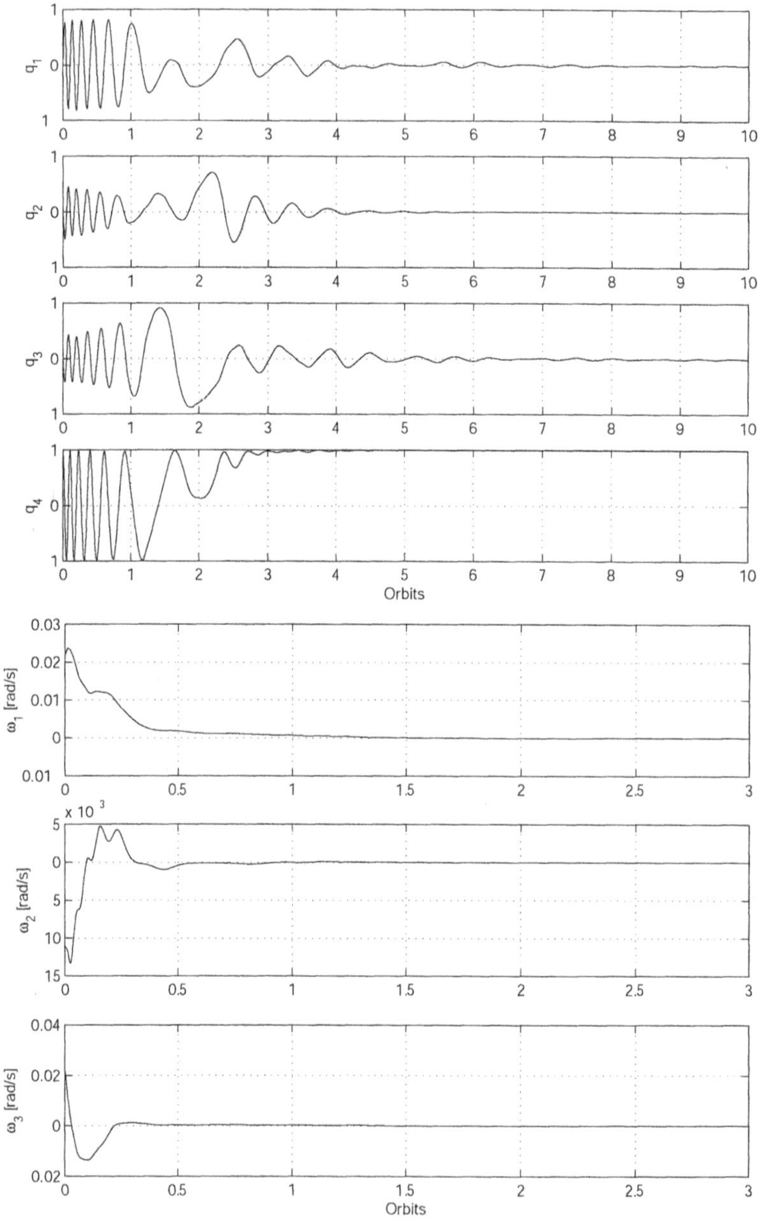

Figure 3 Quaternion and angular rates for the attitude acquisition: state feedback controller.

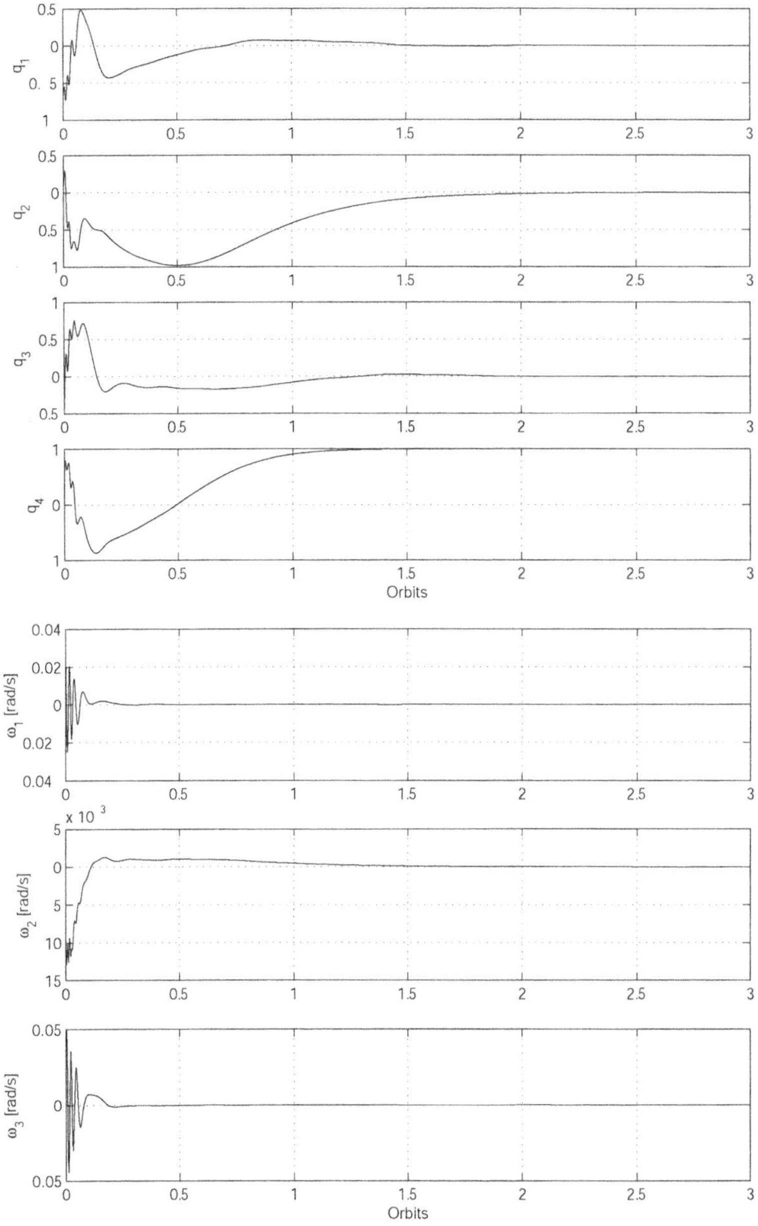

Figure 4 Quaternion and angular rates for the attitude acquisition: output feedback controller.

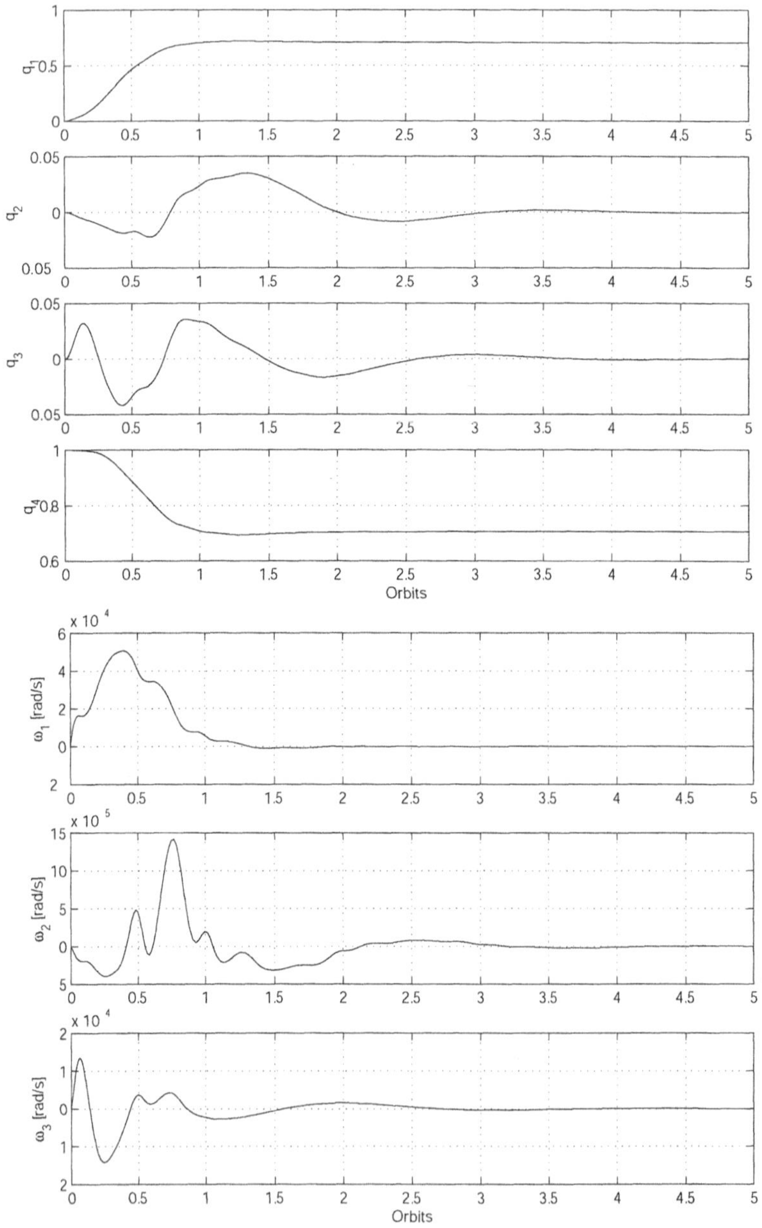

Figure 5 Quaternion and angular rates for the attitude maneuver: state feedback controller.

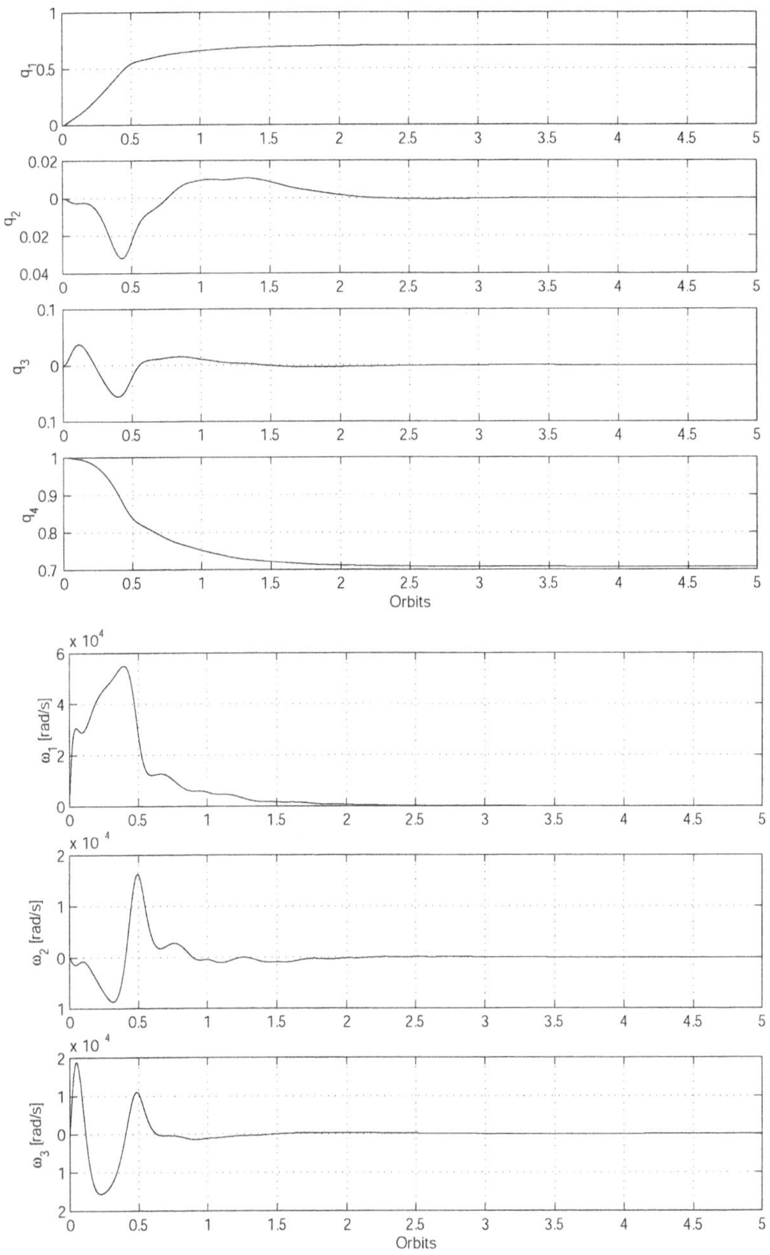

Figure 6 Quaternion and angular rates for the attitude maneuver: output feedback controller.

6 SIMULATION RESULTS

To assess the performance of the proposed control laws the following simulation test case has been analyzed. The considered spacecraft has an inertia matrix given by $I = \text{diag}[27, 17, 25]$ kg m^2 and it is operating in a near-polar (87° inclination) orbit with an altitude of 450 km and a corresponding orbit period of about 5600 s. For such a spacecraft, two sets of simulations have been carried out; the first is related to the acquisition of the target attitude $\bar{q} = \begin{bmatrix} 0 & 0 & 0 & 1 \end{bmatrix}^T$ from an initial condition characterized by a high initial angular rate; the second is related to a point-to-point attitude maneuver from the initial attitude given by $q_0 = \begin{bmatrix} 0 & 0 & 0 & 1 \end{bmatrix}^T$ to the target attitude $\bar{q} = \frac{1}{\sqrt{2}} \begin{bmatrix} 1 & 0 & 0 & 1 \end{bmatrix}^T$.

In all cases, both the full-state feedback control law and the control law without rate feedback have been applied.

The results of the simulations are displayed in Figs. 3 and 4 for the attitude acquisition, and Figs. 5 and 6 for the attitude maneuver, from which the good performance of the proposed control laws can be seen.

7 CONCLUDING REMARKS

The problem of inertial attitude regulation for a small spacecraft using only magnetic coils as actuators has been analyzed and it has been shown that a nonlinear low-gain PD-like control law yields (almost) global asymptotic attitude regulation even in the absence of additional active or passive attitude control actuators such as momentum wheels or gravity gradient booms.

Acknowledgments

The work for this chapter was partially supported by the European network "Nonlinear and Adaptive Control" and by the MURST project, "Identification and Control of Industrial Systems."

REFERENCES

1. M.R. Akella. Rigid body attitude tracking without angular velocity feedback. *Systems and Control Letters*, 42:321–326, 2001.
2. C. Arduini and P. Baiocco. Active magnetic damping attitude control for gravity gradient stabilised spacecraft. *Journal of Guidance and Control*, 20(1):117–122, 1997.
3. S. Battilotti. Global output regulation and disturbance attenuation with global stability via measurement feedback for a class of nonlinear systems. *IEEE Transactions on Automatic Control*, 41(3):315–327, 1996.
4. C. I. Byrnes and A. Isidori. On the attitude stabilization of rigid spacecraft. *Automatica*, 27(1):87–95, 1991.
5. O.E. Fjellstad and T.I. Fossen. Comments on 'the attitude control problem.' *IEEE Transactions on Automatic Control*, 39(3):699–700, 1994.
6. H.K. Khalil. *Nonlinear Systems*. Macmillan, New York, 1992.

7. M. Lovera. Periodic H_∞ attitude control for satellites with magnetic actuators. In *3rd IFAC Symposium on Robust Control Design*, Prague, Czech Republic, 2000.
8. M. Lovera. Optimal magnetic momentum control for inertially pointing spacecraft. *European Journal of Control*, 7(1):30–39, 2001.
9. M. Lovera and A. Astolfi. Global attitude regulation using magnetic control. In *IEEE Conference on Decision and Control*, Orlando, Florida, 2001.
10. P. Morin, C. Samson, J.-B. Pomet, and Z.-P. Jiang. Time-varying feedback stabilization of the attitude of a rigid spacecraft with two controls. *Systems and Control Letters*, 25:375–385, 1995.
11. J. T.-Y. Wen and K. Kreutz-Delgado. The attitude control problem. *IEEE Transactions on Automatic Control*, 36(10):1148–1162, 1991.
12. J. Wertz. *Spacecraft Attitude Determination and Control.* D. Reidel Publishing Company, Dordrecht, 1978.
13. R. Wisniewski and M. Blanke. Fully magnetic attitude control for spacecraft subject to gravity gradient. *Automatica*, 35(7):1201–1214, 1999.

2

Adaptive Learning Control for Spacecraft Formation Flying

Hong Wong,† Haizhou Pan,† Marcio S. de Queiroz,* and Vikram Kapila†

†*Department of Mechanical, Aerospace, and Manufacturing Engineering, Polytechnic University, Brooklyn, NY*

**Department of Mechanical Engineering, Louisiana State University, Baton Rouge, LA*

This chapter considers the problem of spacecraft formation flying in the presence of periodic disturbances. In particular, the nonlinear position dynamics of a follower spacecraft relative to a leader spacecraft are utilized to develop a learning controller that accounts for the periodic disturbances entering the system model. Using a Lyapunov-based approach, a full-state feedback control law, a parameter update algorithm, and a disturbance estimate rule are designed that facilitate the tracking of given reference trajectories in the presence of unknown spacecraft masses. Illustrative simulations are included to demonstrate the efficacy of the proposed controller.

1 INTRODUCTION

Spacecraft formation flying is an enabling technology that distributes mission tasks among many spacecraft. Practical applications of spacecraft formation flying include surveillance, passive radiometry, terrain mapping, navigation, and communication, where using groups of spacecraft permits reconfiguration and optimization of formation geometries for single or multiple missions.

Distributed spacecraft performing space-based sensing, imaging, and communication provide larger aperture areas at the cost of maintaining a meaningful formation geometry with minimal error. The ability to enlarge aperture areas beyond conventional single spacecraft capabilities improves slow target detection for interferometric radar and allows for enhanced resolution in terms of geolocation [1]. Current spacecraft formation control methodologies provide good tracking of relative position trajectories between a leader and follower spacecraft pair. However, in the presence of disruptive disturbances, designing a control law that compensates for unknown, time-varying, disturbance forces is a challenging problem.

The current spacecraft formation flying literature largely addresses the problem of control of spacecraft relative positions using linear and nonlinear controllers. For example, linear and nonlinear formation dynamic models have been developed for formation maintenance and a variety of control designs have been proposed to guarantee the desired

0-415-30852-6/04/$0.00+$1.50

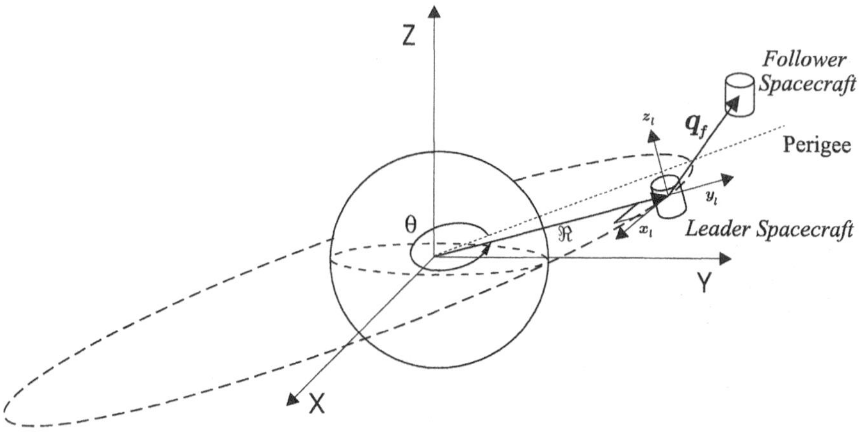

Figure 1 Schematic representation of the spacecraft formation flight system.

formation performance [2–10]. Specifically, a linear formation dynamic model known as Hill's equation is given in [8, 11], which constitutes the foundation for the application of various linear control techniques to the distributed spacecraft formation maintenance problem [3, 4, 6, 8, 10]. Model-based and adaptive nonlinear controllers for the leader-follower spacecraft configuration are given in [2, 7, 9]. However, control design to track desired trajectories under the influence of exogenous disturbances within the formation dynamic model has not been fully explored.

In this chapter, we consider the problem of spacecraft formation flying in the presence of periodic disturbances. In particular, the nonlinear position dynamics of a follower spacecraft relative to a leader spacecraft are utilized to develop a learning controller [12], which accounts for the periodic disturbances entering the system model. First, in Section 2, the Euler–Lagrange method is used to develop a model for the spacecraft formation flying system. Next, a trajectory tracking control problem is formulated in Section 3. In Section 4, using a Lyapunov-based approach, a full-state feedback control law, a parameter update algorithm, and a disturbance estimate rule are designed that facilitate the tracking of given reference trajectories in the presence of unknown spacecraft masses. Illustrative simulations are included in Section 5 to demonstrate the efficacy of the proposed controller. Finally, some concluding remarks are given in Section 6.

2 SYSTEM MODEL

In this section, referring to Fig. 1, we develop the nonlinear model characterizing the position dynamics of follower spacecraft relative to a leader spacecraft using the Euler–Lagrange methodology. We assume that the leader spacecraft exhibits planar dynamics in a closed elliptical orbit with the Earth at its prime focus. In addition, we consider that the inertial coordinate system $\{X, Y, Z\}$ is attached to the center of the Earth. Next, let $\Re(t) \in \mathbb{R}^3$ denote the position vector from the origin of the inertial coordinate frame to the leader spacecraft. Furthermore, we assume a right-hand coordinate frame $\{x_\ell, y_\ell, z_\ell\}$ is attached to the leader spacecraft with the y_ℓ-axis pointing along the direction of the vector $\Re$, the z_ℓ-axis pointing along the orbital angular momentum of the leader spacecraft, and x_ℓ-axis being mutually perpendicular to the y_ℓ-axis and z_ℓ-axis, and pointing in the direction that completes the right-handed coordinate frame.

Modeling the dynamics of a spacecraft relative to the Earth utilizes the fact that the energy of the spacecraft moving under the gravitational influence is conserved. Next, let the Lagrangian function L be defined as the difference between the specific-kinetic energy T and the specific-potential energy V, i.e., $L \triangleq T - V$. Then, the Lagrangian function for the leader spacecraft $L_\ell(t) \in \mathbb{R}$ is given by

$$L_\ell \triangleq 0.5v_\ell^2 + \frac{\mu}{||\Re||}, \tag{2.1}$$

where $||a|| \triangleq \sqrt{a^T a}$, for an arbitrary n-dimensional vector a, $v_\ell(t) \triangleq ||\mathbf{v}_\ell(t)||$, with $\mathbf{v}_\ell(t)$ as defined in (2.4) below, μ is the Earth gravity constant, and $-\frac{\mu}{||\Re||}$ is the specific-potential energy of the leader spacecraft. Similarly, the Lagrangian function for the follower spacecraft $L_f(t) \in \mathbb{R}$ is given by

$$L_f \triangleq 0.5v_f^2 + \frac{\mu}{||\Re + q_f||}, \tag{2.2}$$

where $q_f(t) \in \mathbb{R}^3$ is the position vector of the follower spacecraft relative to the leader spacecraft, expressed in the coordinate frame $\{x_\ell, y_\ell, z_\ell\}$ as $q_f \triangleq x_f\hat{\imath}_\ell + y_f\hat{\jmath}_\ell + z_f\hat{k}_\ell$ and $v_f(t) \triangleq ||\mathbf{v}_f(t)||$, with $\mathbf{v}_f(t)$ as defined in (2.3) below.

Next, let $\mathbf{v}_\ell(t)$ and $\mathbf{v}_f(t)$ denote the absolute velocities (i.e., velocity relative to the inertial coordinate system) of the leader spacecraft and the follower spacecraft, respectively, expressed in the moving coordinate frame $\{x_\ell, y_\ell, z_\ell\}$. In addition, let $\mathbf{v}_{\mathrm{rel}_f}(t)$ denote the velocity of the follower spacecraft relative to the leader spacecraft, measured in the coordinate frame $\{x_\ell, y_\ell, z_\ell\}$. Then, it follows that

$$\mathbf{v}_f = \mathbf{v}_\ell + \mathbf{v}_{\mathrm{rel}_f} + \omega\hat{k}_\ell \times q_f, \tag{2.3}$$

where

$$\mathbf{v}_\ell = -r\omega\hat{\imath}_\ell + \dot{r}\hat{\jmath}_\ell, \tag{2.4}$$

$$\mathbf{v}_{\mathrm{rel}_f} = \dot{x}_f\hat{\imath}_\ell + \dot{y}_f\hat{\jmath}_\ell + \dot{z}_f\hat{k}_\ell, \tag{2.5}$$

$r(t) \in \mathbb{R}$ denotes the instantaneous distance of the leader spacecraft from the center of the earth and $\omega(t) \in \mathbb{R}$ is the orbital angular speed of the leader spacecraft. In addition, $\Re(t)$ in the moving coordinate frame $\{x_\ell, y_\ell, z_\ell\}$ can be expressed as $\Re = r\hat{\jmath}_\ell$.

To obtain the leader spacecraft dynamics relative to the Earth, we use the conservative form of the Lagrange's equations [13] on the leader spacecraft given by

$$\frac{d}{dt}\left(\frac{\partial L_\ell}{\partial \dot{\alpha}_\ell}\right) - \frac{\partial L_\ell}{\partial \alpha_\ell} = 0, \tag{2.6}$$

where $\alpha_\ell \in \{r, \omega\}$, which denotes the set of polar orbital elements describing the motion of the leader spacecraft. This yields a set of differential equations of motion for the leader spacecraft in the chosen coordinate system. Specifically, substituting the magnitude of the velocity vector of (2.4) into (2.1) yields

$$L_\ell = 0.5r^2\omega^2 + 0.5\dot{r}^2 + \frac{\mu}{r}. \tag{2.7}$$

For L_ℓ given by (2.7), an application of (2.6) results in a set of planar dynamics describing the elliptical motion of the leader spacecraft given as

$$\ddot{r} - r\omega^2 + \frac{\mu}{r} = 0, \tag{2.8}$$

$$r^2\omega = 0. \tag{2.9}$$

Taking the time derivative of (2.9) produces a dynamic relationship coupling ω and r as

$$r\dot{\omega} + 2\dot{r}\omega = 0. \tag{2.10}$$

To obtain the follower spacecraft dynamics relative to the Earth, we utilize the same structure of (2.6) in the form

$$\frac{d}{dt}\left(\frac{\partial L_f}{\partial \dot{\alpha}_f}\right) - \frac{\partial L_f}{\partial \alpha_f} = 0, \tag{2.11}$$

where $\alpha_f \in \{x_f, y_f, z_f\}$, which denotes the set of cartesian elements describing the motion of the follower spacecraft relative to the leader spacecraft. Substituting the magnitude of the velocity vector of (2.3) into (2.2) yields

$$L_f = 0.5(\dot{x}_f - r\omega - \omega y_f)^2 + 0.5(\dot{r} + \dot{y}_f + \omega x_f)^2 + 0.5\dot{z}_f^2 + \frac{\mu}{\|\Re + q_f\|}. \tag{2.12}$$

For L_f given by (2.12), an application of (2.11) results in

$$\begin{aligned} (\dot{r}\omega + 2r\dot{\omega}) + \ddot{x}_f - 2\omega\dot{y}_f - (\dot{\omega}y_f + \omega^2 x_f) + \tfrac{\mu x_f}{\|\Re+q_f\|^3} &= 0, \\ (\ddot{r} - r\omega^2) + \ddot{y}_f + 2\omega\dot{x}_f + (\dot{\omega}x_f - \omega^2 y_f) + \tfrac{\mu(y_f+r)}{\|\Re+q_f\|^3} &= 0, \\ \ddot{z}_f + \tfrac{\mu z_f}{\|\Re+q_f\|^3} &= 0. \end{aligned}$$

Substituting the leader planar dynamics of (2.8) and (2.10) into the above homogeneous dynamics results in the thrust free dynamics of the follower spacecraft relative to the leader spacecraft, expressed in the coordinate frame $\{x_\ell, y_\ell, z_\ell\}$, given by

$$\begin{aligned} \ddot{x}_f - 2\omega\dot{y}_f - \dot{\omega}y_f - \omega^2 x_f + \tfrac{\mu x_f}{\|\Re+q_f\|^3} &= 0, \\ \ddot{y}_f + 2\omega\dot{x}_f + \dot{\omega}x_f - \omega^2 y_f + \tfrac{\mu(y_f+r)}{\|\Re+q_f\|^3} - \tfrac{\mu r}{\|\Re\|^3} &= 0, \\ \ddot{z}_f + \tfrac{\mu z_f}{\|\Re+q_f\|^3} &= 0. \end{aligned} \tag{2.13}$$

After premultiplying the nonlinear dynamics of (2.13) with the follower spacecraft mass m_f and using the method of virtual work for the insertion of external forcing terms [13], e.g., disturbance and thrusting forces for the leader spacecraft and the follower spacecraft, the nonlinear position dynamics of the follower spacecraft relative to the leader spacecraft can be arranged in the following form [14, 15]:

$$m_f\ddot{q}_f + C(\omega)\dot{q}_f + N(q_f, \dot{\omega}, \omega, r, u_\ell) + F_d = u_f, \tag{2.14}$$

where C is a Coriolis-like matrix defined as

$$C \triangleq 2m_f\omega \begin{bmatrix} 0 & -1 & 0 \\ 1 & 0 & 0 \\ 0 & 0 & 0 \end{bmatrix}, \tag{2.15}$$

N is a nonlinear term consisting of gravitational effects and inertial forces

$$N \triangleq m_f \begin{bmatrix} -\omega^2 x_f - \dot{\omega} y_f + \frac{\mu x_f}{\|\Re + q_f\|^3} + \frac{u_{\ell x}}{m_\ell} \\ -\omega^2 y_f + \dot{\omega} x_f + \frac{\mu(y_f + r)}{\|\Re + q_f\|^3} - \frac{\mu r}{\|\Re\|^3} + \frac{u_{\ell y}}{m_\ell} \\ \frac{\mu z_f}{\|\Re + q_f\|^3} + \frac{u_{\ell z}}{m_\ell} \end{bmatrix}, \tag{2.16}$$

$F_d(t) \in \mathbb{R}^3$ is a composite disturbance force vector given by

$$F_d \triangleq F_{d_f} - \frac{m_f}{m_\ell} F_{d_\ell}, \tag{2.17}$$

where $u_\ell(t) \in \mathbb{R}^3$ and $u_f(t) \in \mathbb{R}^3$ are the control inputs to the leader spacecraft and the follower spacecraft, respectively, m_ℓ is the mass of the leader spacecraft, $u_{\ell x}, u_{\ell y}, u_{\ell z}$ are components of the vector u_ℓ, and $F_{d_f}(t) \in \mathbb{R}^3$ and $F_{d_\ell}(t) \in \mathbb{R}^3$ are disturbance vectors for the follower spacecraft and the leader spacecraft, respectively. In this chapter, we assume that F_d is a periodic disturbance with a known period $\tau > 0$ such that $F_d(t + \tau) = F_d(t)$, $t \geq 0$. Note that periodic disturbances in formation dynamics may arise due to, e.g., solar pressure disturbance and gravitational disturbance, among others.

The following remarks further facilitate the subsequent control methodology and stability analysis.

Remark 1 The Coriolis matrix C of (2.15) satisfies the skew symmetric property of

$$x^T C x = 0, \quad \forall x \in \mathbb{R}^3. \tag{2.18}$$

Remark 2 The left-hand side of (2.14) produces an affine parameterization

$$m_f \xi + C\dot{q}_f + N = Y(\xi, \dot{q}_f, q_f, \dot{\omega}, \omega, r, u_\ell)\phi, \tag{2.19}$$

where $\xi(t) \in \mathbb{R}^3$ is a dummy variable with components ξ_x, ξ_y, and ξ_z, $Y \in \mathbb{R}^{3\times 2}$ is a regression matrix, composed of known functions, defined as

$$Y \triangleq \begin{bmatrix} \xi_x - 2\omega\dot{y}_f - \dot{\omega} y_f - \omega^2 x_f + \frac{\mu x_f}{\|\Re + q_f\|^3} & u_{\ell_x} \\ \xi_y + 2\omega\dot{x}_f - \omega^2 y_f + \dot{\omega} x_f + \frac{\mu(y_f + r)}{\|\Re + q_f\|^3} - \frac{\mu r}{\|\Re\|^3} & u_{\ell_y} \\ \xi_z + \frac{\mu z_f}{\|\Re + q_f\|^3} & u_{\ell_z} \end{bmatrix}, \tag{2.20}$$

$\phi \in \mathbb{R}^2$ is the unknown, constant system parameter vector defined as

$$\phi \triangleq \begin{bmatrix} m_f & \frac{m_f}{m_\ell} \end{bmatrix}^T. \tag{2.21}$$

Remark 3 In this chapter, we consider that the composite disturbance vector in the position dynamics of the follower spacecraft relative to the leader spacecraft can be upper bounded as follows

$$\|F_d(t)\|_\infty \leq \beta, \quad t \geq 0, \tag{2.22}$$

where β is a positive constant and $\|\cdot\|_\infty$ denotes the usual infinity norm.

3 PROBLEM FORMULATION

In this section, we formulate a control design problem such that the relative position q_f tracks a desired relative trajectory $q_{df}(t) \in \mathbb{R}^3$, i.e., $\lim_{t\to\infty} q_f(t) = q_{df}(t)$. The effectiveness of this control objective is quantified through the definition of a relative position error $e(t) \in \mathbb{R}^3$ as

$$e \triangleq q_{df} - q_f. \tag{3.1}$$

The control design methodology is to construct a control algorithm that obtains the aforementioned tracking result in the presence of the unknown composite disturbance vector defined in (2.17) and the unknown constant parameter vector ϕ of (2.21). We assume that the relative position and velocity measurements (i.e., q_f and $\dot{q}_f$) of the follower spacecraft relative to the leader spacecraft are available for feedback.

To facilitate the control development, we assume that the desired trajectory q_{df} and its first two time derivatives are bounded functions of time. Next, we define the parameter estimation error $\tilde{\phi}(t) \in \mathbb{R}^2$ as the difference between the actual parameter vector ϕ and the parameter estimate $\hat{\phi}(t) \in \mathbb{R}^2$, i.e.,

$$\tilde{\phi} \triangleq \phi - \hat{\phi}. \tag{3.2}$$

In addition, we define the composite disturbance error $\tilde{F}_d(t) \in \mathbb{R}^3$ as the difference between the composite disturbance vector F_d and the disturbance estimate $\hat{F}_d(t) \in \mathbb{R}^3$, i.e.,

$$\tilde{F}_d \triangleq F_d - \hat{F}_d. \tag{3.3}$$

Finally, we define the components of a saturation function $\mathrm{sat}_\beta(\cdot) \in \mathbb{R}^3$ as

$$\mathrm{sat}_\beta(s)_i \triangleq \begin{cases} s_i & \text{for } |s_i| \le \beta \\ \mathrm{sgn}(s_i)\beta & \text{for } |s_i| > \beta \end{cases} \quad \forall s \in \mathbb{R}^3, \tag{3.4}$$

where $i \in \{x, y, z\}$ and s_x, s_y, s_z are the components of the vector s.

To facilitate the subsequent stability analysis, we define the saturated disturbance estimation error variable $\varphi \in \mathbb{R}^3$ as $\varphi(\sigma) \triangleq \mathrm{sat}_\beta(F_d(\sigma)) - \mathrm{sat}_\beta(\hat{F}_d(\sigma))$.

Remark 4 The saturation function (3.4) satisfies the following useful property:

$$\left(\mathrm{sat}_\beta(a) - \mathrm{sat}_\beta(b)\right)^T \left(\mathrm{sat}_\beta(a) - \mathrm{sat}_\beta(b)\right) \le (a-b)^T(a-b), \quad a, b \in \mathbb{R}^3. \tag{3.5}$$

4 ADAPTIVE LEARNING CONTROLLER

In this section we develop an adaptive learning controller based on the system model of (2.14) such that the tracking error variable e exhibits asymptotic stability. Before we begin the control design, we define an auxiliary filter tracking error variable $\eta(t) \in \mathbb{R}^3$ as

$$\eta \triangleq \dot{e} + \alpha e, \tag{4.1}$$

where $\alpha \in \mathbb{R}^{3\times 3}$ is a constant, diagonal, positive definite, control gain matrix.

4.1 Controller Design

To initiate the control design, we take the time derivative of (4.1) and premultiply the resulting equation by m_f to obtain

$$m_f \dot{\eta} = m_f (\ddot{q}_{df} - \ddot{q}_f) + m_f \alpha \dot{e}, \tag{4.2}$$

where the second time derivative of (3.1) has been used. Substituting (2.14) into (4.2) results in

$$m_f \dot{\eta} = m_f (\ddot{q}_{df} + \alpha \dot{e}) + C\, \dot{q}_f + N + F_d - u_f. \tag{4.3}$$

Simplifying (4.3) into a more advantageous form, we obtain

$$m_f \dot{\eta} = Y(\ddot{q}_{df} + \alpha \dot{e}, \dot{q}_f, q_f, \dot{\omega}, \omega, r, u_\ell)\phi + F_d - u_f, \tag{4.4}$$

where (2.19) has been used with $\xi = \ddot{q}_{df} + \alpha \dot{e}$ in the definition of (2.20). Equation (4.4) characterizes the open-loop dynamics of η and is used as the foundation for the synthesis of the adaptive learning controller.

Based on the form of the open-loop dynamics of (4.4), we design the control law u_f as

$$u_f = Y\hat{\phi} + K\eta + \hat{F}_d, \tag{4.5}$$

where $K \in \mathbb{R}^{3\times3}$ is a constant, diagonal, positive definite, control gain matrix. Guided by the subsequent Lyapunov stability analysis, the parameter update law for $\hat{\phi}$ in (4.5) is selected as

$$\dot{\hat{\phi}} = \Gamma Y^T \eta, \tag{4.6}$$

where $\Gamma \in \mathbb{R}^{2\times2}$ is a constant, diagonal, positive definite, adaptation gain matrix. Finally, the disturbance estimate vector $\hat{F}_d$ is updated according to

$$\hat{F}_d(t) = \text{sat}_\beta\big(\hat{F}_d(t-\tau)\big) + k_L \eta(t), \tag{4.7}$$

where $k_L \in \mathbb{R}$ is a constant, positive, learning gain.

Using the control law of (4.5) in the open-loop error dynamics of (4.4) results in the following closed-loop dynamics for η:

$$m_f \dot{\eta} = Y\tilde{\phi} - K\eta + \tilde{F}_d. \tag{4.8}$$

In addition, computing the time derivative of (3.2), using the fact that the parameter vector ϕ is constant, and the parameter update law of (4.6), the closed-loop dynamics for the parameter estimation error is given by

$$\dot{\tilde{\phi}} = -\Gamma Y^T \eta. \tag{4.9}$$

4.2 Stability Analysis

The proposed control law of (4.5)–(4.7) provides a stability result for the position and velocity tracking errors as illustrated by the following theorem. In order to state the main result of this section, we define

$$\lambda \triangleq \lambda_{\min}\left(K + \frac{1}{2}k_L I_3\right), \tag{4.10}$$

where I_3 is the 3×3 identity matrix and $\lambda_{\min}(\cdot)$ denotes the smallest eigenvalue of a matrix.

The adaptive learning control law described by (4.5)–(4.7) ensures asymptotic convergence of the position and velocity tracking errors as delineated by

$$\lim_{t\to\infty} e(t), \dot{e}(t) = 0. \tag{4.11}$$

Proof. We define a nonnegative function as follows:

$$V(t) \triangleq \frac{1}{2} m_f \eta^T \eta + \frac{1}{2} \tilde{\phi}^T \Gamma^{-1} \tilde{\phi} + \frac{1}{2k_L} \int_{t-\tau}^{t} \varphi(\sigma)^T \varphi(\sigma) d\sigma. \tag{4.12}$$

Taking the time derivative of (4.12) along the closed-loop dynamics of (4.8) and (4.9) results in

$$\dot{V}(t) = \eta^T \left(\tilde{F}_d(t) - K\eta \right) + \frac{1}{2k_L} \varphi^T(t)\varphi(t) - \frac{1}{2k_L} \varphi^T(t-\tau)\varphi(t-\tau). \tag{4.13}$$

Substituting the composite disturbance estimate of (4.7), noting that $\text{sat}_\beta(F_d(t-\tau)) = F_d(t)$ due to (2.22), and utilizing the definition of (3.3) into (4.13) produces

$$\begin{aligned}\dot{V}(t) &= \eta^T \left(\tilde{F}_d(t) - K\eta \right) + \tfrac{1}{2k_L} \varphi^T(t)\varphi(t) \\ &- \tfrac{1}{2k_L} \left(\tilde{F}_d(t) + k_L \eta \right)^T \left(\tilde{F}_d(t) + k_L \eta \right).\end{aligned} \tag{4.14}$$

Expanding (4.14) and combining like terms produces

$$\dot{V}(t) = -\eta^T \left(K + \frac{1}{2} k_L I_3 \right) \eta - \frac{1}{2k_L} \left(\tilde{F}_d^T(t) \tilde{F}_d(t) - \varphi^T(t)\varphi(t) \right). \tag{4.15}$$

Utilizing the inequality of (3.5) and definitions of $\tilde{F}_d$ and φ, we can upper-bound (4.15) as follows:

$$\dot{V}(t) \;\leq\; -\eta^T \left(K + \frac{1}{2} k_L I_3 \right) \eta. \tag{4.16}$$

Taking the norm of the right-hand side of (4.16) and using the definition of (4.10) results in

$$\dot{V}(t) \;\leq\; -\lambda \|\eta\|^2. \tag{4.17}$$

Since V is a nonnegative function and $\dot{V}$ is a negative semidefinite function, V is a nonincreasing function. Thus $V(t) \in \mathcal{L}_\infty$ as described by

$$V(\eta(t), \tilde{\phi}(t), \varphi(t)) \leq V(\eta(0), \tilde{\phi}(0), \varphi(0)),\ t \geq 0. \tag{4.18}$$

Using standard signal chasing arguments, all signals in the closed-loop system can now be shown to be bounded. Using (4.8) along with the boundedness of all signals in the closed-loop system, we now conclude that $\dot{\eta}(t) \in \mathcal{L}_\infty$. Solving the differential inequality of (4.17) results in

$$V(0) - V(\infty) \;\geq\; \lambda \int_0^\infty \|\eta(t)\|^2 dt. \tag{4.19}$$

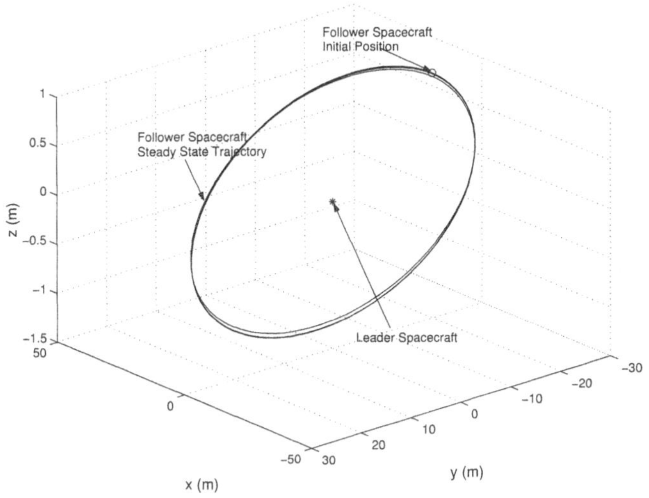

Figure 2 Actual trajectory of follower spacecraft relative to leader spacecraft.

Since $V(t)$ is bounded, $t \geq 0$, we conclude that $\eta(t) \in \mathcal{L}_\infty \bigcap \mathcal{L}_2$, $t \geq 0$. Finally, using Barbalat's Lemma [16, 17], we conclude that

$$\lim_{t \to \infty} \eta(t) = 0. \tag{4.20}$$

Using the definition of η in (4.1), the limit statement of (4.20), and Lemma 1.6 of [16], yield the result of (4.11). □

5 SIMULATION RESULTS

The illustrative numerical example considered here utilizes orbital elements to propagate the leader spacecraft in a low-altitude orbit similar to the TechSat 21 mission specifications [1]. The adaptive learning control law described in (4.5) is simulated for the dynamics of the follower spacecraft relative to the leader spacecraft. The leader spacecraft is assumed to have the following orbital parameters $\hat{a} = 7200$ km, $\hat{e} = 0.001$, $\theta(0) = 0$ rad, $\hat{T} = \tau = 1.6889$ hr, where $\hat{a}$ is the semimajor axis of the elliptical orbit of the leader spacecraft, $\hat{e}$ is the orbital eccentricity of the leader spacecraft, $\theta(t) \in \mathbb{R}$ is the time-varying true anomaly of the planar dynamics of the leader spacecraft, and $\hat{T}$ is the orbital period of the leader spacecraft. The relative trajectory between the follower spacecraft and the leader spacecraft was generated by numerically solving the thrust-free dynamics given by (2.13). The initial conditions for the relative position and velocity between the follower spacecraft and the leader spacecraft were obtained in the same manner as in [15] and are given by

$$q_{df}(0) = [0 \ -20 \ 1]\,\mathrm{m}, \quad \dot{q}_{df}(0) = [-149.1374 \ 0 \ 0]\,\frac{\mathrm{m}}{\mathrm{hr}}. \tag{5.1}$$

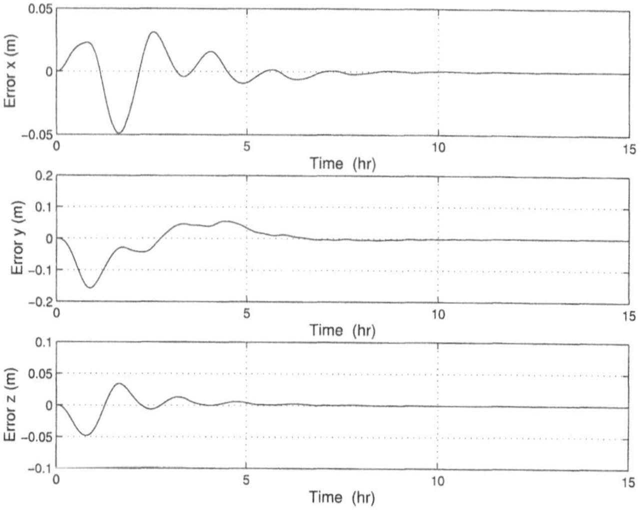

Figure 3 Tracking error of follower spacecraft relative to leader spacecraft.

Additional parameters used for simulation within the spacecraft formation flying system are as follows $\mu = 5.16586248 \times 10^{21}\ \frac{\text{m}^3}{\text{hr}^2}$, $m_\ell = 100\,\text{kg}$, $m_f = 100\,\text{kg}$, $u_\ell = [0\ 0\ 0]^T$ N, $F_d = [1.9106\ \ -1.906\ \ -1.517]^T \cdot \sin\left(\frac{2\pi}{T}t\right) \times 10^{-5}$N. Finally, in the following simulation, the parameter and disturbance estimates were all initialized to zero.

The control, adaptation, and learning gains, in the control law of (4.5)–(4.7), are obtained through trial and error in order to obtain good performance for the tracking error response. The following resulting gains were used in this simulation $K = \text{diag}\,(3,3,3) \times 10^3$, $\alpha = \text{diag}\,(1,1,1)$, $\Gamma = \text{diag}\,(1,1) \times 10^2$, and $k_L = 1000$. The actual trajectory q_f, shown in Fig. 2, is initialized to be the same as the desired trajectory in (5.1). Figures 3 and 4 show the tracking error e and velocity tracking error $\dot{e}$, respectively. The control input u_f and the disturbance estimate $\hat{F}_d$ are shown in Figs. 5 and 6, respectively. Finally, since we assume the leader is in a thrust free orbit about the earth, the second component in the parameter estimate is neglected while the first component of the parameter estimate is shown in Fig. 7.

6 CONCLUSION

In this paper, we designed an adaptive learning control algorithm for the position dynamics of the follower spacecraft relative to the leader spacecraft. A Lyapunov-type design was used to construct a full-state feedback control law and parameter and disturbance estimates that facilitate the tracking of reference trajectories with global asymptotic convergence. Simulation results were given to illustrate the efficacy of the control design in the presence of unknown periodic disturbance forces.

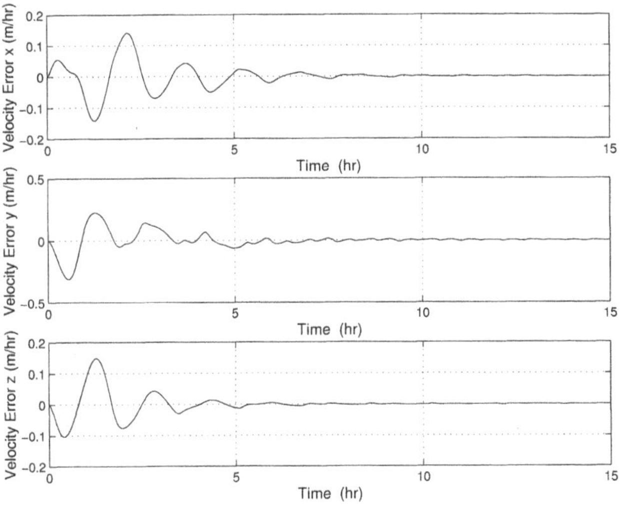

Figure 4 Velocity tracking error of follower spacecraft relative to leader spacecraft.

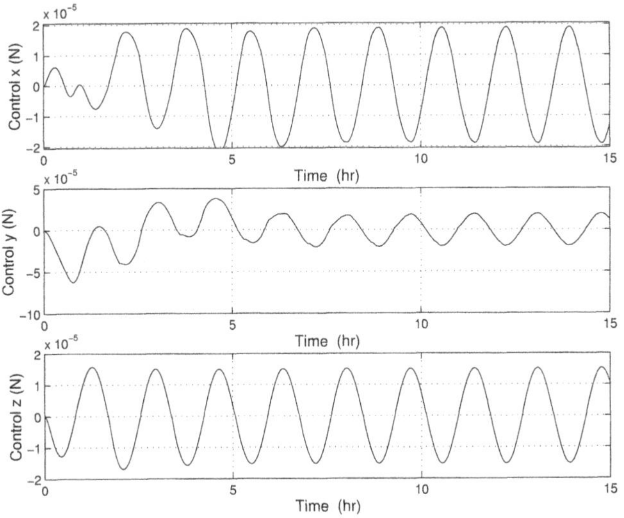

Figure 5 Control effort for follower spacecraft.

Acknowledgments

Research was supported in part by the National Aeronautics and Space Administration–Goddard Space Flight Center under Grant NAG5-11365; AFRL/VACA, WPAFB, OH; and the New York Space Grant Consortium under Grant 39555-6519.

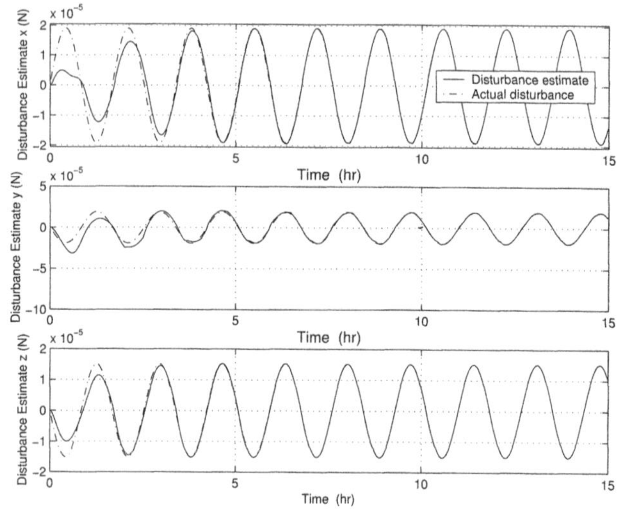

Figure 6 Disturbance estimate for follower spacecraft.

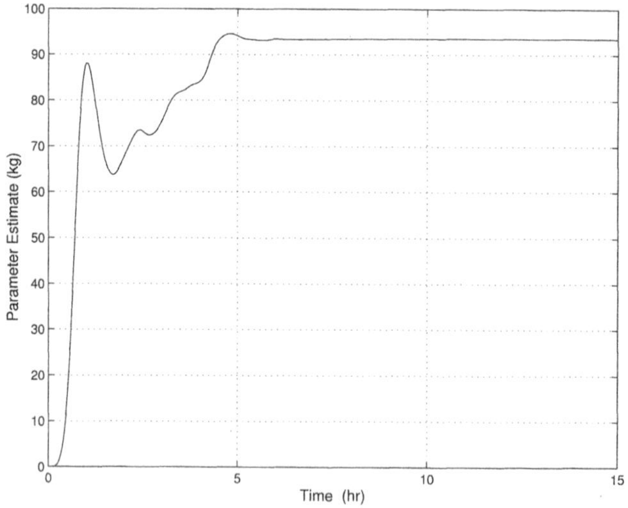

Figure 7 Parameter estimate for follower spacecraft.

H.W. and V.K. are grateful to the Air Force Research Laboratory/VACA, Wright-Patterson Air Force Base, OH, for their hospitality during the summer of 2001.

REFERENCES

1. http://www.vs.afrl.af.mil/factsheets/TechSat21.html.
2. F. Y. Hadaegh, W. M. Lu, and P. C. Wang, "Adaptive control of formation flying spacecraft for interferometry," *IFAC Conference on Large Scale Systems*, pp. 97–102, 1998.
3. V. Kapila, A. G. Sparks, J. Buffington, and Q. Yan, "Spacecraft formation flying: Dynamics and control," *AIAA J. GCD.*, vol. 23, pp. 561–564, 2000.
4. C. Sabol, R. Burns, and C. McLaughlin, "Formation flying design and evolution," *Proc. of the AAS/AIAA Space Flight Mechanics Meeting*, Paper No. AAS–99–121, 1999.
5. H. Schaub, S. R. Vadali, J. L. Junkins, and K. T. Alfriend, "Spacecraft formation flying control using mean orbit elements," *AAS G. Contr. Conf.*, pp. 97–102, 1999.
6. R. J. Sedwick, E. M. C. Kong, and D. W. Miller, "Exploiting orbital dynamics and micropropulsion for aperture synthesis using distributed satellite systems: Applications to Techsat 21," *D.C.P. Conf.*, AIAA Paper No. 98-5289, 1998.
7. M. S. de Queiroz, V. Kapila, and Q. Yan, "Adaptive nonlinear control of multiple spacecraft formation flying," *AIAA J. GCD.*, vol. 23, pp. 385–390, 2000.
8. R. H. Vassar and R. B. Sherwood, "Formation keeping for a pair of satellites in a circular orbit," *AIAA J. GCD.*, vol. 8, pp. 235–242, 1985.
9. H. Wong, V. Kapila, and A. G. Sparks, "Adaptive output feedback tracking control of multiple spacecraft," *Proc. ACC*, 2001.
10. H.-H. Yeh and A. G. Sparks, "Geometry and control of satellite formations," *Proc. ACC.*, pp. 384–388, 2000.
11. R. R. Bate, D. D. Mueller, and J. E. White, *Fundamentals of Astrodynamics.* New York: Dover, 1971.
12. B. Costic, M. de Queiroz, and D. Dawson, "A new learning control approach to the magnetic bearing benchmark system," *Proc. ACC*, pp. 2639–2643, 2000.
13. H. L. Langhaar, *Energy Methods in Applied Mechanics.* New York: John Wiley and Sons, 1962.
14. M. de Queiroz, V. Kapila, , and Q. Yan, "Adaptive nonlinear control of multiple spacecraft formation flying," *J. GCD.*, vol. 23, pp. 385–390, 2000.
15. Q. Yan, G. Yang, V. Kapila, and M. S. de Queiroz, "Nonlinear dynamics, trajectory generation, and adaptive control of multiple spacecraft in periodic relative orbits," *AAS G. Contr. Conf.*, Paper No. 00-013, 2000.
16. D. M. Dawson, J. Hu, and T. C. Burg, *Nonlinear Control of Electric Machinery.* New York: Marcel Dekker, 1998.
17. J.-J. E. Slotine and W. Li, *Applied Nonlinear Control.* Englewood Cliffs, NJ: Prentice-Hall, 1991.

3

Spectral Properties of the Generalized Resolvent Operator for an Aircraft Wing Model in Subsonic Airflow

Marianna A. Shubov
Department of Mathematics and Statistics, Texas Tech University, Lubbock, TX

In this chapter we announce a series of results on the asymptotic and spectral analysis of an aircraft wing model in a subsonic air flow. This model has been developed in the Flight Systems Research Center of UCLA and is presented in the works by A.V. Balakrishnan. The model is governed by a system of two coupled integro-differential equations and a two-parameter family of boundary conditions modeling action of the self-straining actuators. The unknown functions (the bending and torsion angle) depend on time and one spatial variable. The differential parts of the above equations form a coupled linear hyperbolic system; the integral parts are of the convolution type. The system of equations of motion is equivalent to a single operator evolution-convolution equation in the state space of the system equipped with the energy metric. The Laplace transform of the solution of this equation can be represented in terms of the so-called generalized resolvent operator. The generalized resolvent operator is a finite-meromorphic function on the complex plane having the branch cut along the negative real semi-axis. The poles of the generalized resolvent are precisely the aeroelastic modes and the residues at these poles are the projectors on the generalized eigenspaces. In this paper, we present the following results: (i) spectral properties of an analytic operator-valued function associated with the generalized resolvent operator; (ii) asymptotic distribution of the aeroelastic modes; (iii) the Riesz basis property of the system of mode shapes in the energy space.

1 INTRODUCTION

In this chapter, we formulate a series of results on the asymptotic and spectral analysis of an aircraft wing model. The model has been developed in the Flight Systems Research Center of the University of California at Los Angeles. The mathematical formulation of the problem can be found in the works by A.V. Balakrishnan [2–4]. This model has been designed to find an approach to control the flutter phenomenon in an aircraft wing in a surrounding airflow by using the so-called self-straining actuators.

The model, which is used in [2–4], is the 2-D strip model that applies to bare wings of high-aspect ratio [12, 15]. The structure is modeled by a uniform cantilever beam that bends

0-415-30852-6/04/$0.00+$1.50

and twists. The aerodynamics is assumed to be subsonic, incompressible, and inviscid. In addition, the author of [2–4] has added self-straining actuators using a currently accepted model (e.g., [4–10, 13, 18, 20, 23, 25, 26]).

Flutter, which is known as a very dangerous aeroelastic development, is the onset, beyond some speed–altitude combinations, of unstable and destructive vibrations of a lifting surface in an airstream. Flutter is most commonly encountered on bodies subjected to large lateral aerodynamic loads of the lift type, such as aircraft wings, tails, and control surfaces. The only air forces necessary to produce it are those due to the deflection of the elastic structures from the undeformed state. The flutter or critical speed u_f and frequency ω_f are defined as the lowest airspeed and corresponding circular frequency at which a given structure flying at given atmospheric density and temperature will exhibit sustained, simple harmonic oscillations. Flight at u_f represents a borderline condition or neutral stability boundary, because all small motions must be stable at speeds below u_f, whereas divergent oscillations can ordinarily occur in a range of speeds (or at all speeds) above u_f.

Probably, the most dangerous type of aircraft flutter results from coupling between the bending and torsional motions of a relatively large aspect–ratio wing and tail. The model presented in [2–4] has been designed to treat flutter caused by this type of coupling. We also mention paper [14] by P. Dierolf et al., in which the authors treat the flutter problem as a perturbation problem for semigroups.

Our main objective is to find the time–domain solution of the initial–boundary value problem from [2–4]. This objective requires very detailed mathematical analysis of the properties of the system. Now, we describe the content of this chapter. In Section 2, we give a precise mathematical formulation of the problem. The model is governed by a system of two coupled partial integro-differential equations subject to a two-parameter family of boundary conditions. The parameters are introduced in order to model the action of self-straining actuators as known in current engineering and mathematical literature. In Section 3, we reformulate the problem and set it into an operator format. We also show that the dynamics are defined by two matrix operators in the energy space. One of the aforementioned operators is a matrix differential operator, and the second one is a matrix integral convolution-type operator. *The aeroelastic modes* (or the discrete spectrum of the problem) are closely related to the discrete spectrum of the matrix differential operator while the continuous spectrum is completely defined by the matrix integral operator. We note that, if the speed of an airstream $u = 0$, then the integral operators vanish and the appropriate purely structural problem will have only a discrete spectrum. In Section 4, we formulate the asymptotic and spectral results related to the matrix differential operator. In Section 5, we present asymptotics for aeroelastic modes and formulate and discuss the importance of Balakrishnan's Theorem (see Theorem 5.2). In Section 6, we discuss the properties of the Laplace transform of the matrix integral operator. In Section 7, we formulate the results concerning the properties of the adjoint operator-valued function, the properties of which are important for the spectral decomposition of the resolvent operator generated by the aircraft wing model.

In the conclusion of the Introduction, we describe what kind of a control problem will be considered in connection with the flutter suppression. In the specific wing model considered in the current chapter, both the matrix differential operator and the matrix integral operator contain entries depending on the speed u of the surrounding air flow. Therefore, the aeroelastic modes are functions of $u : \lambda_k = \lambda_k(u)$, $(k \in \mathbb{Z})$. The wing is stable if $Re\, \lambda_k(u) < 0$ for all k. However, if u is increasing, some of the modes move to the right half-plane. The flutter speed u_k^f for the k-th mode is defined by the relation $Re\,\lambda_k(u_k^f) = 0$. To understand the flutter phenomenon, it is not sufficient to trace the motion of aeroelastic modes as functions of a speed of airflow. It is necessary to have efficient representations for the solutions of our boundary-value problem, containing the contributions from both the discrete and continuous parts of the spectrum. Such a representation will provide a precise description of the solution behavior. It is known that flutter cannot be eliminated completely. *To successfully suppress flutter, one should design self-straining actuators (i.e., in the mathematical language, to select parameters in the boundary conditions that are the control gains β and δ in formulas (2.10) and (2.11) of Section 2) in such a way that flutter does not occur in the desired speed range. This is a highly nontrivial boundary control problem.*

2 STATEMENT OF THE PROBLEM

In this section, we give a precise formulation of the initial-boundary value problem.

Following [2–4], let us introduce the dynamical variables

$$X(x,t) = \begin{pmatrix} h(x,t) \\ \alpha(x,t) \end{pmatrix}, \qquad -L \le x \le 0,\ t \ge 0, \tag{2.1}$$

where $h(x,t)$ is the bending and $\alpha(x,t)$ the torsion angle. The model can be described by a linear system

$$(M_s - M_a)\ddot{X}(x,t) + (D_s - uD_a)\dot{X}(x,t) + (K_s - u^2K_a)X = \begin{bmatrix} f_1(x,t) \\ f_2(x,t) \end{bmatrix}. \tag{2.2}$$

We will use the notation "." (dot) to denote differentiation with respect to t. We use the subscripts "s" and "a" to distinguish the structural and aerodynamical parameters, respectively. All 2×2 matrices in Eq. (2.2) are given by the following formulae:

$$M_s = \begin{bmatrix} m & S \\ S & I \end{bmatrix}, \qquad M_a = (-\pi\rho)\begin{bmatrix} 1 & -a \\ -a & (a^2 + 1/8) \end{bmatrix}, \tag{2.3}$$

where m is the density of the flexible structure (mass per unit length), S is the mass moment, I is the moment of inertia, ρ is the density of air, and a is linear parameter of the structure $(-1 \le a \le 1)$.

$$D_s = \begin{bmatrix} 0 & 0 \\ 0 & 0 \end{bmatrix} \qquad D_a = (-\pi\rho)\begin{bmatrix} 0 & 1 \\ -1 & 0 \end{bmatrix}, \tag{2.4}$$

$$K_s = \begin{bmatrix} E\frac{\partial^4}{\partial x^4} & 0 \\ 0 & -G\frac{\partial^2}{\partial x^2} \end{bmatrix} \qquad K_a = (-\pi\rho)\begin{bmatrix} 0 & 0 \\ 0 & -1 \end{bmatrix}, \tag{2.5}$$

where E is the bending stiffness, and G is the torsion stiffness. The parameter u in Eq. (2.2) denotes the stream velocity. The right-hand side of system (2.2) can be represented as the following system of two convolution-type integral operations:

$$f_1(x,t) = -2\pi\rho \int_0^t \left[uC_2(t-\sigma) - \dot{C}_3(t-\sigma)\right] g(x,\sigma)d\sigma, \tag{2.6}$$

$$f_2(x,t) = -2\pi\rho \int_0^t \Big[1/2C_1(t-\sigma) - auC_2(t-\sigma) + a\dot{C}_3(t-\sigma)$$

$$+uC_4(t-\sigma) + 1/2\dot{C}_5(t-\sigma)\Big] g(x,\sigma)d\sigma, \tag{2.7}$$

$$g(x,t) = u\dot{\alpha}(x,t) + \ddot{h}(x,t) + (1/2 - a)\ddot{\alpha}(x,t). \tag{2.8}$$

The aerodynamical functions C_i, $i = 1\dots5$, are defined in the following ways [11]:

$$\begin{aligned}
\hat{C}_1(\lambda) &= \int_0^\infty e^{-\lambda t}C_1(t)dt = \frac{u}{\lambda}\frac{e^{-\lambda/u}}{K_0(\lambda/u) + K_1(\lambda/u)}, \qquad Re\lambda > 0,\\
C_2(t) &= \int_0^t C_1(\sigma)d\sigma, \qquad C_4(t) = C_2(t) + C_3(t),\\
C_3(t) &= \int_0^t C_1(t-\sigma)(u\sigma - \sqrt{u^2\sigma^2 + 2u\sigma})d\sigma,\\
C_5(t) &= \int_0^t C_1(t-\sigma)((1+u\sigma)\sqrt{u^2\sigma^2+2u\sigma} - (1+u\sigma)^2)d\sigma,
\end{aligned} \tag{2.9}$$

where K_0 and K_1 are the modified Bessel functions of the zero and first orders [1, 19]. The self-straining control actuator action can be modeled by the following boundary conditions:

$$Eh''(0,t) + \beta\dot{h}'(0,t) = 0, \qquad h'''(0,t) = 0, \tag{2.10}$$

$$G\alpha'(0,t) + \delta\dot{\alpha}(0,t) = 0, \qquad \beta,\delta \in \mathfrak{C}^+ \cup \{\infty\}, \tag{2.11}$$

where $\mathfrak{C}^+$ is the closed right half-plane. The boundary conditions at $x = -L$ are

$$h(-L,t) = h'(-L,t) = \alpha(-L,t) = 0. \tag{2.12}$$

Let the initial state of the system be given as follows:

$$h(x,0) = h_0(x), \quad \dot{h}(x,0) = h_1(x), \quad \alpha(x,0) = \alpha_0(x), \quad \dot{\alpha}(x,0) = \alpha_1(x). \tag{2.13}$$

We will consider the solution of the problem given by Eq. (2.2) and conditions (2.10)–(2.13) in the energy space $\mathcal{H}$. To introduce the metric of $\mathcal{H}$, we assume that the parameters satisfy the following two conditions:

$$det\begin{bmatrix} m & S \\ S & I \end{bmatrix} > 0, \qquad 0 < u \le \frac{\sqrt{2G}}{L\sqrt{\pi\rho}}. \tag{2.14}$$

Let $\{\tilde{C}_i\}_{i=1}^2$ be the kernels in the convolution operations in (2.6), (2.7), i.e.,

$$\tilde{C}_1(t) = -2\pi\rho(uC_2(t) - \dot{C}_3(t)),$$
$$\tilde{C}_2(t) = -2\pi\rho\,(1/2C_1(t) - auC_2(t) + a\dot{C}_3(t) + uC_4(t) + 1/2\dot{C}_5(t)), \tag{2.15}$$

and let M, D, K be the following matrices:

$$M = M_s - M_a, \;\; D = D_s - uD_a, \;\; K = K_s - u^2 K_a \,. \tag{2.16}$$

Then Eq. (2.2) can be written in the form

$$M\ddot{X}(x,t) + D\dot{X}(x,t) + KX(x,t) = \; (\mathcal{F}\dot{X})(x,t)), \;\; t \geq 0, \tag{2.17}$$

where the matrix integral operator $\mathcal{F}$ can be given by the formula

$$\mathcal{F} \;\; = \;\; \begin{bmatrix} \int_0^t \; \tilde{C}_1(t-\sigma) \left(\frac{d}{d\sigma}\cdot\right) \, d\sigma & \int_0^t \; \tilde{C}_1(t-\sigma)[u \cdot + (1/2 - a) \left(\frac{d}{d\sigma}\cdot\right)] d\sigma \\ \int_0^t \; \tilde{C}_2(t-\sigma) \left(\frac{d}{d\sigma}\cdot\right) \, d\sigma & \int_0^t \; \tilde{C}_2(t-\sigma)[u \cdot + (1/2 - a) \left(\frac{d}{d\sigma}\cdot\right)] d\sigma \end{bmatrix}$$

$$= \;\; \begin{bmatrix} \tilde{C}_1 * \left(\frac{d}{d\sigma}\cdot\right) & \tilde{C}_2 * (u \cdot + (1/2 - a)\frac{d}{d\sigma}\cdot) \\ \tilde{C}_2 * \left(\frac{d}{d\sigma}\cdot\right) & \tilde{C}_1 * (u \cdot + (1/2 - a)\frac{d}{d\sigma}\cdot) \end{bmatrix}, \tag{2.18}$$

where the notation "$*$" has been used for the convolution.

Remark 2.1. The model described by Eq. (2.17) in the case $u = 0$ occurs actually in aeroelastic problems (see classic textbooks [12, 15]) if one ignores aeroelastic forces. However, the boundary conditions in [12, 15] that complemented the system of Eq. (2.17) are totally different. To the best of our knowledge, the whole problem consisting of Eq. (2.17) ($u = 0$) and boundary conditions (2.10) and (2.11) has been considered only in papers [5, 9].

3 OPERATOR REFORMULATION OF THE PROBLEM

We start with the state space of the system (the energy space). Let $\mathcal{H}$ be the set of 4-component vector-valued functions $\Psi = (h, \dot{h}, \alpha, \dot{\alpha})^T \equiv (\psi_0, \psi_1, \psi_2, \psi_3)^T$ (the superscript "T" means the transposition) obtained as a closure of smooth functions satisfying the conditions

$$\psi_0(-L) = \psi_0'(-L) = \psi_2(-L) = 0 \tag{3.1}$$

in the following energy norm:

$$||\Psi||^2_{\mathcal{H}} = 1/2 \int\limits_{-L}^{0} \Big[E|\psi_0''(x)|^2 + G|\psi_2'(x)|^2 + \tilde{m}|\psi_1(x)|^2 + \tilde{I}|\psi_3(x)|^2$$
$$+ \tilde{S}(\psi_3(x)\bar{\psi}_1(x) + \bar{\psi}_3(x)\psi_1(x)) \, -\pi\rho u^2 |\psi_2(x)|^2 \Big] \, dx, \tag{3.2}$$

where

$$\tilde{m} = m + \pi\rho, \;\; \tilde{S} = S - a\pi\rho, \;\; \tilde{I} = I + \pi\rho(a^2 + 1/8). \tag{3.3}$$

Note that under the first condition in (2.14), formula (3.2) defines a positively definitive metric. Our goal is to rewrite Eq. (2.17) as the first order in time evolution–convolution equation in the energy space. As the first step, we will represent Eq. (2.17) in the form

$$\ddot{X} + M^{-1}D\dot{X} + M^{-1}KX = M^{-1}\mathcal{F}\dot{X}. \tag{3.4}$$

Note that, due to the first condition in (2.14), M^{-1} exists.

Using formula (3.1), one can easily see that the initial-boundary value problem defined by Eq. (3.4) and conditions (2.10)–(2.13) can be represented in the form

$$\dot{\Psi} = i\mathcal{L}_{\beta\delta}\Psi + \tilde{\mathcal{F}}\dot{\Psi}, \quad \Psi = (\psi_0, \psi_1\psi_2, \psi_3)^T, \quad \Psi|_{t=0} = \Psi_0. \tag{3.5}$$

$\mathcal{L}_{\beta\delta}$ is the following matrix differential operator in $\mathcal{H}$:

$$\mathcal{L}_{\beta\delta} = -i \begin{bmatrix} 0 & 1 & 0 & 0 \\ -\frac{E\tilde{I}}{\Delta}\frac{d^4}{dx^4} & -\frac{\pi\rho u\tilde{S}}{\Delta} & -\frac{\tilde{S}}{\Delta}\left(G\frac{d^2}{dx^2} + \pi\rho u^2\right) & -\frac{\pi\rho u\tilde{I}}{\Delta} \\ 0 & 0 & 0 & 1 \\ \frac{E\tilde{S}}{\Delta}\frac{d^4}{dx^4} & \frac{\pi\rho u\tilde{m}}{\Delta} & \frac{\tilde{m}}{\Delta}\left(G\frac{d^2}{dx^2} + \pi\rho u^2\right) & \frac{\pi\rho u\tilde{S}}{\Delta} \end{bmatrix} \tag{3.6}$$

defined on the domain

$$\begin{aligned} \mathcal{D}(\mathcal{L}_{\beta\delta}) = \{\Psi \in \mathcal{H} : \ \psi_0 \ \in \ H^4(-L,0), \ \psi_1 \ \in H^2(-L,0), \ \psi_2 \ \in \ H^4(-L,0), \\ \psi_3 \ \in H^1(-L,0); \quad \psi_1(-L) = \psi_1'(-L) = \psi_3(-L) = 0; \ \ \psi_0'''(0) = 0; \\ E\psi_0''(0) + \beta\psi_1'(0) = 0, \ \ G\psi_2'(0) + \delta\psi_3(0) = 0\}, \end{aligned} \tag{3.7}$$

where $H^i, i = 1, 2, 4$, are the standard Sobolev spaces. $\tilde{\mathcal{F}}$ is a linear integral operator in $\mathcal{H}$ given by the formula

$$\tilde{\mathcal{F}} = \begin{bmatrix} 1 & 0 & 0 & 0 \\ 0 & [\tilde{I}(\tilde{C}_1*) - \tilde{S}(\tilde{C}_2*)] & 0 & 0 \\ 0 & 0 & 1 & 0 \\ 0 & 0 & 0 & [-\tilde{S}(\tilde{C}_1*) + \tilde{m}(\tilde{C}_2*)] \end{bmatrix} \begin{bmatrix} 0 & 0 & 0 & 0 \\ 0 & 1 & u & (1/2-a) \\ 0 & 0 & 0 & 0 \\ 0 & 1 & u & (1/2-a) \end{bmatrix}. \tag{3.8}$$

It turns out that the spectral properties of both the differential operator $\mathcal{L}_{\beta\delta}$ and the integral operator $\tilde{\mathcal{F}}$ are of crucial importance for the representation of the solution.

Important Remark. At this moment, we would like to emphasize the difference between the aircraft wing model considered in this chapter and models related to other flexible structures that have been studied by the author in recent years. Each of the aforementioned models (a spatially nonhomogeneous damped string, a 3-dimensional damped-wave equation with spatially nonhomogeneous coefficients, Timoshenko beam model, and couple Euler–Bernoulli and Timoshenko beam model) can be described by an abstract evolution equation in a Hilbert space $\mathcal{H}$:

$$\dot{\Psi}(t) = iA\Psi(t), \ \ \Psi(t) \in \mathcal{H}, \ t \geq 0. \tag{3.9}$$

The dynamics generator A is an unbounded nonself-adjoint operator in $\mathcal{H}$. This operator is our main object of interest in the aforementioned works. In each of the above examples, A is a specific differential or matrix differential operator in $\mathcal{H}$. The domain of A is defined by the differential expression and the corresponding boundary conditions. A has a compact resolvent and, therefore, has a purely discrete spectrum. The main results concerning the dynamics generator A established in the author's works can be split into the following three parts: (i) the explicit asymptotic formulae for the spectrum; (ii) asymptotic representations for the generalized eigenvectors; (iii) the Riesz basis property of the generalized eigenvectors in $\mathcal{H}$.

In contrast with the evolution equation (3.9), the aircraft wing model is described by an evolution–convolution equations of the form

$$\dot{\Psi}(t) = i\,A\,\Psi(t) + \int_0^t F(t-\tau)\dot{\Psi}(\tau)d\tau. \tag{3.10}$$

Here $\Psi(\cdot) \in \mathcal{H}$ – the energy space of the system, Ψ is a 4-component vector function, A ($A = \mathcal{L}_{\beta\delta}$) is a matrix differential operator, and $F(t)$ is a matrix-valued function.

Equation (3.10) does not define an evolution semigroup and does not have a dynamics generator. So it is necessary to explain what is understood as the spectral analysis of Eq. (3.10).

Let us take the Laplace transformation of both parts of Eq. (3.5). Formal solution in the Laplace representation can be given by the formula

$$\hat{\Psi}(\lambda) = \left(\lambda I - iA - \lambda\hat{F}(\lambda)\right)^{-1}\left(I - \hat{F}(\lambda)\right)\Psi_0, \tag{3.11}$$

where Ψ_0 is the initial state, i.e., $\Psi(0) = \Psi_0$, and the symbol "$\hat{}$" is used to denote the Laplace transform. In order to have representation of the solution in the space–time domain, we have to calculate the inverse Laplace transform of Eq. (3.11). To this end, it is necessary to investigate the generalized resolvent operator

$$R(\lambda) = \left(\lambda I - iA - \lambda\hat{F}(\lambda)\right)^{-1}. \tag{3.12}$$

In the case of the 1-dimensional wing model, $R(\lambda)$ is an operator-valued meromorphic function on the complex plane with a branch cut along the negative real semi-axis. The poles of $R(\lambda)$ are called the eigenvalues, or *the aeroelastic modes*. The residues of $R(\lambda)$ at the poles are precisely the projectors on the corresponding generalized eigenspaces. The branch cut corresponds to the continuous spectrum.

Definition 3.1. Those values of λ for which the equation

$$[\lambda I - i\mathcal{L}_{\beta\delta} - \lambda\hat{\mathfrak{F}}(\lambda)]\Phi = 0 \tag{3.13}$$

has nontrivial solutions are called *aeroelasitc modes*. The corresponding solutions are called *mode shapes*.

4 ASYMPTOTIC AND SPECTRAL PROPERTIES OF THE MATRIX DIFFERENTIAL OPERATOR $\mathcal{L}_{\beta\delta}$

In this section, we describe the main properties of the matrix differential operator $\mathcal{L}_{\beta\delta}$ [20–23].

Theorem 4.1. *a) $\mathcal{L}_{\beta\delta}$ is a closed linear operator in $\mathcal{H}$ whose resolvent is compact, and therefore, the spectrum is discrete [16, 17].*
b) Operator $\mathcal{L}_{\beta\delta}$ is nonselfadjoint unless $\Re\,\beta = \Re\,\delta = 0$. If $\Re\,\beta \geq 0$ and $\Re\,\delta \geq 0$, then this operator is dissipative, i.e., $\Im\,(\mathcal{L}_{\beta\delta}\Psi, \Psi) \geq 0$ for $\Psi \in \mathcal{D}(\mathcal{L}_{\beta\delta})$ [16, 17]. The adjoint operator $\mathcal{L}^*_{\beta\delta}$ is given by the same matrix differential expression (3.6) on the domain obtained from (3.7) by replacing the parameters β and δ with $(-\bar{\beta})$ and $(-\bar{\delta})$, respectively.

c) When $\mathcal{L}_{\beta\delta}$ is dissipative, then it is maximal, i.e., it does not admit dissipative extensions [24].

In the next theorem, we provide asymptotics of the spectrum for the operator $\mathcal{L}_{\beta\delta}$. It is convenient now to recall the notion of an *associate vector* [16].

Definition 4.1. A vector Φ in a Hilbert space $\mathcal{H}$ is an *associate vector* of a nonselfadjoint operator A of order m corresponding to an eigenvalue λ if $\Phi \neq 0$ and

$$(A - \lambda I)^m \Phi \neq 0 \quad \text{and} \quad (A - \lambda I)^{m+1} \Phi = 0. \tag{4.1}$$

If $m = 0$, then Φ is an eigenvector. The set of all associate vectors and eigenvectors together will be called the set of *root vectors.* The following results from [21, 22] hold.

Theorem 4.2. *a) The operator $\mathcal{L}_{\beta\delta}$ has a countable set of complex eigenvalues. If*

$$\delta \neq \sqrt{G\tilde{I}}, \tag{4.2}$$

then the set of eigenvalues is located in a strip parallel to the real axis.
b) The entire set of eigenvalues asymptotically splits into two different subsets. We call them the β-branch and the δ-branch and denote these branches by $\{\lambda_n^\beta\}_{n\in\mathbb{Z}}$ and $\{\lambda_n^\delta\}_{n\in\mathbb{Z}}$, respectively. If $\Re\,\beta \geq 0$ and $\Re\,\delta > 0$, then each branch is asymptotically close to its own horizontal line in the closed upper half-plane. If $\Re\beta > 0$ and $\Re\delta = 0$, then both horizontal lines coincide with the real axis. If $\Re\,\beta = \Re\,\delta = 0$, then the operator $\mathcal{L}_{\beta\delta}$ is self-adjoint and, thus, its spectrum is real. The entire set of eigenvalues may have only two points of accumulation: $+\infty$ and $-\infty$ in the sense that $\Re\,\lambda_n^{\beta(\delta)} \longrightarrow \pm\infty$ and $|\Im\,\lambda_n^{\beta(\delta)}| \leq const$ as $n \longrightarrow \pm\infty$ (see formulae (4.3) and (4.4) below).
c) The following asymptotics are valid for the β-branch of the spectrum as $|n| \longrightarrow \infty$:

$$\lambda_n^\beta = (sgn\ n)(\pi^2/L^2)\sqrt{E\tilde{I}/\Delta}\ (n - 1/4)^2 + \kappa_n(\omega), \quad \omega = |\delta|^{-1} + |\beta|^{-1}, \tag{4.3}$$

with Δ being defined by $\Delta = \tilde{m}\tilde{I} - \tilde{S}^2$. A complex-valued sequence $\{\kappa_n\}$ is bounded above in the following sense: $\sup_{n\in\mathbb{Z}}\{|\kappa_n(\omega)|\} = C(\omega)$, $C(\omega) \longrightarrow 0$ as $\omega \longrightarrow 0$.
d) The following asymptotics are valid for the δ-branch of the spectrum:

$$\lambda_n^\delta = \frac{\pi n}{L\sqrt{\tilde{I}/G}} + \frac{i}{2L\sqrt{\tilde{I}/G}}\ ln\ \frac{\delta + \sqrt{G\tilde{I}}}{\delta - \sqrt{G\tilde{I}}} + O(|n|^{-1/2}), \ |n| \longrightarrow \infty. \tag{4.4}$$

In (4.4), "ln" means the principal value of the logarithm. If β and δ stay away from zero, i.e., $|\beta| \geq \beta_0 > 0$ and $|\delta| \geq \delta_0 > 0$, then the estimate $O(|n|^{-1/2})$ in (4.4) is uniform with respect to both parameters. There may be only a finite number of multiple eigenvalues of a finite multiplicity each. Therefore, only a finite number of the associate vectors may exist.

The next result is concerned with the properties of the root vectors of the operator $\mathcal{L}_{\beta\delta}$. Before we formulate this result, we recall the definition of the *biorthogonal vectors* [16].

Definition 4.2. Two sequences of vectors $\{\phi_n\}$ and $\{\chi_n\}$ in a Hilbert space $\mathcal{H}$ are said to be *biorthogonal* if for every m and n,we have

$$(\phi_m, \chi_n)_{\mathcal{H}} = \delta_{mn}. \tag{4.5}$$

In the case when an operator has a simple spectrum (i.e., there are no associate vectors), the biorthogonal set consists of the eigenvectors of the adjoint operator. However, in general, the assumption that the spectrum is simple is quite artificial. Numerical simulations show that minor changes of the parameters of the problem often result in appearance of numerous multiple eigenvalues. So, one cannot disregard the existence of associate vectors. In that case the relationship between the root vectors of the operator and its adjoint becomes less obvious. (For the corresponding results, see [22].) It is also shown in [22] that the set of the root vectors of the operator $\mathcal{L}_{\beta\delta}$ is *complete* in the state space. (Recall the set of vectors is complete in a Hilbert space if finite linear combinations of the vectors from the set are dense in this space [16, 17].)

Definition 4.3. A basis in a Hilbert space is a *Riesz basis* if it is linearly isomorphic to an orthonormal basis, that is, if it is obtained from an orthonormal basis by means of a bounded and boundedly invertible operator.

The next result from [22] describes a very important property of the root vectors of the differential operator $\mathcal{L}_{\beta\delta}$.

Theorem 4.3. *The set of root vectors of the operator $\mathcal{L}_{\beta\delta}$ forms a Riesz basis in the energy space $\mathcal{H}$.*

The Riesz basis property of the root vectors has been proven in [22] based on the *functional model* for nonself-adjoint operators by Sz. Nagy–C. Foias [24], the main elements of which have been reproduced in Section 7 of [22].

5 ASYMPTOTICAL DISTRIBUTION OF AEROELASTIC MODES

Our first result in this section is the following statement.

Theorem 5.1. *a) The set of all aeroelastic modes (which are the poles of the generalized resolvent operator) is countable and does not have accumulation points on the complex plane $\mathbb{C}$. There might be only a finite number of multiple poles each of a finite multiplicity. There exists a sufficiently large $R > 0$ such that all aeroelastic modes, whose distance from the origin is greater than R, are simple poles of the generalized resolvent. The value of R depends on the speed u of an airstream, i.e., $R = R(u)$.*
b) The set of the aeroelastic modes splits asymptotically into two series, which we call the β-branch and the δ-branch. Asymptotical distribution of the β and the δ-branches of the aeroelastic modes can be obtained from asymptotical distribution of the spectrum of the operator $\mathcal{L}_{\beta\delta}$. Namely if $\{\mu_n^\beta\}_{n\in\mathbb{Z}}$ is the β-branch of the aeroelastic modes, then $\mu_n^\beta = i\hat{\mu}_n^\beta$ and the asymptotics of the set $\{\hat{\mu}_n^\beta\}_{n\in\mathbb{Z}}$ are given by the right-hand side of formula (4.3). Similarly, if $\{\mu_n^\delta = i\hat{\mu}_n^\delta\}_{n\in\mathbb{Z}}$ is the δ-branch of the aeroelastic modes, then the asymptotical distribution of the set $\{\hat{\mu}_n^\delta\}_{n\in\mathbb{Z}}$ is given by the right-hand side of formula (4.4).

The next result is of particular importance for us. It does not follow from the asymptotic representations (4.3) and (4.4). This result has been formulated in paper [4]. The author of [4] has suggested the proof using the general ideas of the Semigroup Theory. We are able to prove the same fact by using a totally different approach. We divide our proof into two parts, and the first part is formulated as Theorem 6.3 below.

Theorem 5.2. (A.V. Balakrishnan, [4]). *For any $u > 0$, there might exist only a finite number of the aeroelastic modes having nonnegative real parts.*

Now we provide an explanation why Theorem 5.2 cannot be considered as a corollary of Theorem 5.1. We discuss each branch individually. First we note that the δ-branch of the aeroelastic modes may have only a finite number of modes with nonnegative real parts. Indeed, since the points $\hat{\mu}_n^\delta$ are asymptotically close to the eigenvalues $\{\lambda_n^\delta\}_{n\in\mathbb{Z}}$, and the latter set has a horizontal asymptote in the upper half-plane (see (4.4)), the δ-branch of the aeroelastic modes $\{\mu_n^\delta = i\hat{\mu}_n^\delta\}_{n\in\mathbb{Z}}$ has a vertical asymptote in the open left half-plane. This means that for $|n| \geq N >> 1$, all modes $\{\mu_n^\delta\}_{|n|\geq N}$ are located in the left half-plane. However, this is not the case for the β-branch of the aeroelastic modes. Indeed, the set $\{\hat{\mu}_n^\beta\}_{n\in\mathbb{Z}}$ is asymptotically close to the set $\{\lambda_n^\beta\}_{n\in\mathbb{Z}}$, and the latter set is close to the real axis. Therefore, the β-branch of the aeroelastic modes $\{\mu_n^\beta = i\hat{\mu}_n^\beta\}_{n\in\mathbb{Z}}$ is close to the imaginary axis. Moreover, we know that $\Im\, \lambda_n^\beta > 0$ (i.e., $\Re\,(i\lambda_n^\beta) < 0$) for all $n \in \mathbb{Z}$ due to the fact that the operator $\mathcal{L}_{\beta\delta}$ is dissipative. However, the points $\{\hat{\mu}_n^\beta\}_{n\in\mathbb{Z}}$ are not the eigenvalues of any operator, let alone a dissipative operator. So we cannot be sure that $\Im\, \hat{\mu}_n^\beta > 0$ (i.e., $\Re\, \mu_n^\beta < 0$) for any value of n. Thus, though the fact that there may be only a finite number of modes of the β-branch with $\Im\, \mu_n^\beta > 0$ is valid, it requires quite involved proof.

6 STRUCTURE AND PROPERTIES OF MATRIX INTEGRAL OPERATOR

In this section, we describe the properties of the Laplace transform of the convolution-type matrix integral operator given in (3.8).

Lemma 6.1. *Let $\hat{\mathfrak{F}}$ be the Laplace transform of the kernel of matrix integral operator from (3.8). The following formula is valid for $\hat{\mathfrak{F}}$:*

$$\hat{\mathfrak{F}}(\lambda) = \begin{bmatrix} 0 & 0 & 0 & 0 \\ 0 & \mathcal{L}(\lambda) & u\mathcal{L} & (1/2 - a)\mathcal{L}(\lambda) \\ 0 & 0 & 0 & 0 \\ 0 & \mathcal{N}(\lambda) & u\mathcal{N}(\lambda) & (1/2 - a)\mathcal{N}(\lambda) \end{bmatrix},$$

where

$$\mathcal{L}(\lambda) = -\frac{2\pi\rho\, u}{\Delta\ \lambda}\left\{-\frac{\tilde{S}}{2} + \left[\tilde{I} + (a + 1/2)\,\tilde{S}\right] T(\lambda/u)\right\}, \tag{6.1}$$

$$\mathcal{N}(\lambda) = -\frac{2\pi\rho\, u}{\Delta\ \lambda}\left\{\frac{\tilde{m}}{2} - \left[\tilde{S} + (a + 1/2)\,\tilde{m}\right] T(\lambda/u)\right\}, \tag{6.2}$$

and T is the Theodorsen function defined by the formula

$$T(z) = \frac{K_1(z)}{K_0(z) + K_1(z)}, \tag{6.3}$$

K_0 and K_1 are the modified Bessel functions [1, 19].

We note that the Theodorsen function is defined as a single-valued analytic function on the complex plane having a branch cut along the negative real semi-axis. The following asymptotic approximation is valid:

$$T(z) = 1/2 + V(z), \qquad V(z) = 1/(16z) + O(z^{-2}), \qquad |z| \to \infty. \tag{6.4}$$

Taking into account representation (6.5) and the fact that $z = \lambda/u$, we can write $\lambda\hat{\mathfrak{F}}(\lambda)$ as the following sum:

$$\lambda\hat{\mathfrak{F}}(\lambda) = \mathfrak{M} + \mathfrak{N}(\lambda). \tag{6.5}$$

The matrix $\mathfrak{M}$ in (6.6) is defined by the formula

$$\mathfrak{M} = \begin{bmatrix} 0 & 0 & 0 & 0 \\ 0 & A & uA & (1/2-a)A \\ 0 & 0 & 0 & 0 \\ 0 & B & uB & (1/2-a)B \end{bmatrix}, \tag{6.6}$$

where according to (6.2) and (6.3), we have for A and B

$$A = -\pi\rho u\Delta^{-1}[\tilde{I} + (a-1/2)\tilde{S}], \qquad B = \pi\rho u\Delta^{-1}[\tilde{S} + (a-1/2)\tilde{m}]. \tag{6.7}$$

The matrix-valued function $\mathfrak{N}(\lambda)$ is defined by the formula

$$\mathfrak{N}(\lambda) = \begin{bmatrix} 0 & 0 & 0 & 0 \\ 0 & A_1(\lambda) & uA_1(\lambda) & (1/2-a)A_1(\lambda) \\ 0 & 0 & 0 & 0 \\ 0 & B_1(\lambda) & uB_1(\lambda) & (1/2-a)B_1(\lambda) \end{bmatrix}, \tag{6.8}$$

where, according to (6.2) and (6.3), we have for $A_1(\lambda)$ and $B_1(\lambda)$

$$A_1(\lambda) = -2\pi\rho u\Delta^{-1}V(z)[\tilde{I} + (a+1/2)\tilde{S}] \equiv -2\pi\rho u\Delta^{-1}V(z)\, d_1,$$
$$B_1(\lambda) = 2\pi\rho u\Delta^{-1}V(z)[\tilde{S} + (a+1/2)\tilde{m}] \equiv 2\pi\rho u\Delta^{-1}V(z)\, d_2, \quad z = \lambda/u. \tag{6.9}$$

Therefore, the generalized resolvent (see (3.12) and (3.13)) can be written in the form

$$\mathcal{R}(\lambda) = (\lambda I - i\mathcal{L}_{\beta\delta} - \mathfrak{M} - \mathfrak{N}(\lambda))^{-1}, \tag{6.10}$$

where the matrix $\mathfrak{M}$ and the matrix-valued function $\mathfrak{N}(\lambda)$ are defined by (6.7)–(6.10).

Theorem 6.1. *$\mathfrak{M}$ is a bounded linear operator in $\mathcal{H}$. The operator $\mathcal{K}_{\beta\delta}$ introduced by the formula*

$$\mathcal{K}_{\beta\delta} = \mathcal{L}_{\beta\delta} - i\,\mathfrak{M} + i\|\mathfrak{M}\|\,\mathbb{I}, \tag{6.11}$$

where $\mathbb{I}$ is the identity operator, is an unbounded dissipative operator in $\mathcal{H}$.

Remark 6.1. The fact that $\mathcal{K}_{\beta\delta}$ is dissipative is very important for the proof that the root vectors of this operator form a Riesz basis in $\mathcal{H}$.

Theorem 6.2. *$\mathfrak{N}(\lambda)$ is an analytic matrix-valued function on the complex plane with the branch-cut along the negative real semi–axis. For each λ, $\mathfrak{N}(\lambda)$ is a bounded operator in $\mathcal{H}$ with the following estimate for its norm:*

$$\|\mathfrak{N}\|_{\mathcal{H}} \le C(1+|\lambda|)^{-1}, \tag{6.12}$$

where C is an absolute constant the precise value of which is immaterial for us.

We recall that Balakrishnan's Theorem means that for a given speed u, there may be only a finite number of unstable mode shapes. In other words, only a finite number of modes can be subject to flutter. Let us return to Eq. (3.13) for the aeroelastic modes. We know that $\lambda\hat{\mathfrak{F}}(\lambda)$ is a sum of the 4×4 matrix $\mathfrak{M}$ with constant entries and the matrix-valued function $\mathfrak{N}(\lambda)$, whose norm goes to zero as $|\lambda| \to \infty$. We show that the result similar to Balakrishnam's Theorem is valid for the equation obtained from Eq. (3.13) in which the term containing $\mathfrak{N}(\lambda)$ has been omitted, i.e., that an addition of $\mathfrak{N}(\lambda)$ cannot destroy the result obtained for the equation

$$i\mathcal{L}_{\beta\delta}\Psi + \mathfrak{M}\Psi = \lambda\Psi. \tag{6.13}$$

Theorem 6.3. *For a given value of u, there may exist only a finite number of eigenvalues of the operator $i\mathcal{L}_{\beta\delta} + \mathfrak{M}$ having nonnegative real parts.*

7 PROPERTIES OF THE ADJOINT OPERATORS AND OPERATOR-VALUED FUNCTIONS

To write the spectral decomposition for the generalized resolvent operator, we need detailed information about the set of functions that is biorthogonal to the set of the mode shapes. It turns out that this biorthogonal set is closely related to the set of nontrivial solutions of the following equation:

$$\mathfrak{S}^*(\lambda)\Phi \equiv [-i\mathcal{L}^*_{\beta\delta} + \mathfrak{M}^* + \mathfrak{N}^*(\lambda)]\Phi = \lambda\Phi. \tag{7.1}$$

The set of complex points λ for which Eq. (7.1) has nontrivial solutions will be called the set of the *adjoint aeroelastic modes* and the set of corresponding solutions will be called the set of *adjoint mode shapes*.

Now we describe the structure of the adjoint analytic operator-valued function $\mathfrak{S}^*(\lambda)$. We already know that $\mathcal{L}^*_{\beta\delta}$ is given by the same matrix differential expression (3.6) and the only difference is in the description of the domain of $\mathcal{L}^*_{\beta\delta}$, i.e., the parameters β and δ from (3.7) should be replaced with $(-\bar{\beta})$ and $(-\bar{\delta})$ respectively.

Theorem 7.1. *The operator-valued function* $\mathfrak{M}^* + \mathfrak{N}^*(\lambda)$ *can be represented in the following form:*

$\mathfrak{M}^* + \mathfrak{N}^*(\lambda) =$

$$\pi\rho u\begin{bmatrix} 0 & 0 & 0 & 0 \\ 0 & -\left(\tilde{I}-b\tilde{S}\right)\left(1+2\bar{V}(\lambda/u)\right)/\Delta & 0 & \left(\tilde{I}-b\tilde{S}\right)\left(b+2\hat{b}\bar{V}(\lambda/u)\right)/\Delta \\ 0 & -u\left(1+2\bar{V}(\lambda/u)\right)\int_{-L}^{0}\mathcal{K}(x,\xi)\cdot d\xi & 0 & u\left(1+2\bar{V}(\lambda/u)\right)\int_{-L}^{0}\mathcal{K}(x,\xi)\cdot d\xi \\ 0 & \left(\tilde{S}-b\tilde{m}\right)\left(1+2\bar{V}(\lambda/u)\right)/\Delta & 0 & -\left(\tilde{S}-b\tilde{m}\right)\left(b+2\hat{b}\bar{V}(\lambda/u)\right)/\Delta \end{bmatrix},$$

where $b = 1/2 - a$, $\hat{b} = 1/2 + a$. *The kernel of the integral operator in (7.2) is given by the following formulae:*

$$\mathcal{K}(x,\xi) = \frac{1}{W(\omega)}\begin{cases} \sin\omega(x+L) & \cos\omega\xi, \quad x \le \xi, \\ \sin\omega(\xi+L) & \cos\omega x, \quad x \ge \xi, \end{cases} \tag{7.2}$$

and $W(\omega) = \omega\cos\omega L$, $\omega = u\sqrt{\pi\rho/G}$. *If* u *satisfies the second condition from (2.14), then* $W(\omega) \neq 0$.

Theorem 7.2. *The set of adjoint aeroelastic modes is countable and does not have accumulation points on the complex place* $\mathbb{C}$. *This set splits asymptotically into two series, which we call the* $\beta*$*-branch and the* $\delta*$*-branch. Asymptotical distribution of the* $\beta*$ *and the* $\delta*$*-branches of the adjoint aeroelastic modes can be obtained from the asymptotical distribution of the spectrum of the operator* $\mathcal{L}_{\beta\delta}$. *Namely, if* $\{\mu_n^{\beta*}\}_n \in \mathbb{Z}$ *is the* $\beta*$*-branch of the adjoint aeroelastic modes, then* $\mu_n^{\beta*} = i\hat{\mu}_n^{\beta*}$, *and the asymptotics of the set* $\{\mu_n^{\beta*}\}_n \in \mathbb{Z}$ *are given by the right-hand side of formula (4.3). Similarly, if* $\{\mu_n^{\delta*} = i\hat{\mu}_n^{\delta*}\}_n \in \mathbb{Z}$ *is the* $\delta*$*-branch of the adjoint aeroelastic modes, then the asymptotical distribution of the set* $\{\mu_n^{\delta*}\}$ *can be obtained from the right-hand side of the formula (4.4) by replacing the sign "+" with the sign "−" before the logarithmic term.*

Note that due to the latter theorem, one can see that the set of the adjoint aeroelastic modes is closely related to the spectrum of the operator $\mathcal{L}^*_{\beta\delta}$.

Finally, we formulate an important result about the operator

$$\mathcal{K}_{\beta\delta} = \mathcal{L}_{\beta\delta} + i\mathfrak{M}. \tag{7.3}$$

Let us denote by $\{\Phi_n^\beta\}_{n\in\mathbb{Z}}$ and $\{\Phi_n^\delta\}_{n\in\mathbb{Z}}$ the β and δ-branches of root vectors of the operator $\mathcal{K}_{\beta\delta}$, respectively.

Theorem 7.3. *a) The entire set of the root vectors* $\{\Phi_n^\beta\}_{n\in\mathbb{Z}} \bigcup \{\Phi_n^\delta\}_{n\in\mathbb{Z}}$ *of the operator* $\mathcal{K}^*_{\beta\delta}$ *is complete in* $\mathcal{H}$. *b) This set forms a Riesz basis in* $\mathcal{H}$. *The properties (a) and (b) imply that the operator* $\mathcal{K}_{\beta\delta}$, *is a Riesz spectral operator in the sense of Dunford [22]. c) The operator* $\mathcal{K}^*_{\beta\delta}$, *which is adjoint to the operator* $\mathcal{K}_{\beta\delta}$, *is also Riesz spectral.*

To prove Theorem 7.3, we have used the same approach as for the case of the operator $\mathcal{L}_{\beta\delta}$, i.e., the Sz. Nagy–C. Foias functional model for nonself-adjoint operators. To formulate the result on the mode shapes, let us use the following notations: $\{F_n^\beta\}_{n\in\mathbb{Z}}$ are the β-branch mode shapes, $\{F_n^\delta\}_{n\in\mathbb{Z}}$ are the δ-branch mode shapes.

Theorem 7.4. *a) The entire set of the mode shapes* $\{F_n^\beta\}_{n\in\mathbb{Z}} \bigcup \{F_n^\delta\}_{n\in\mathbb{Z}}$ *is complete in the energy space* $\mathcal{H}$. *b) The set of mode shapes is quadratically close to the set of root vectors of the operator* $\mathcal{K}_{\beta\delta}$, *i.e.,*

$$\sum_{n\in\mathbb{Z}} ||\Phi_n^\beta - F_n^\beta||^2_{\mathcal{H}} + \sum_{n\in\mathbb{Z}} ||\Phi_n^\delta - F_n^\delta||^2_{\mathcal{H}} < \infty. \tag{7.4}$$

By combining (a) and (b) of Theorem 7.4 with the fact that $\mathcal{K}^*_{\beta\delta}$ is a Riesz spectral operator, we show that the set of the mode shapes forms a Riesz basis in $\mathcal{H}$. A similar fact on the Riesz basis property of the set of adjoint mode shapes can be shown as well.

The Riesz basis property of the mode shapes is crucially important for the spectral decomposition of the generalized resolvent operator.

Acknowledgments

Partial support by the National Science Foundation Grants ECS #0080441, DMS#0072247, DMS-9972748, and the Advanced Research Program-97 of Texas Grant 0036-44-045 is highly appreciated.

REFERENCES

1. Abramowitz M., Stegun I., Ed., *Handbook of Mathematical Functions*, Dover, New York (1972).
2. Balakrishan A.V., Aeroelastic control with self-straining actuators: continuum models. In: *Smart Structures and Materials, Mathematics Control in Smart Structures,* Vasundaran V., Ed., *Proceedings of SPIE*, **3323** (1998), pp. 44–54.
3. Balakrishnan A.V., Subsonic flutter suppression using self-straining acutators. To appear in: *Journal of the Franklin Institute*, Special Issue on Control, Udwadia F., Ed.
4. Balakrishnan A.V., Aeroelastic control with self–straining actuators: unsteady aerodynamics solution. Unpublished manuscript.

5. Balakrishnan A.V., Shubov M.A., Peterson C.A., Spectral analysis of Euler–Bernoulli and Timoshenko beam model. Submitted to: *Integral and Differential Equations* (2001).
6. Balakrishnan A.V., Damping performance of strain actuated beams, *Comput. Appl. Math.* **18** (1) (1999), pp. 31–86.
7. Balakrishnan A.V., Vibrating systems with singular mass–inertia matrices, *First Intl. Conf. Nonlinear Problems in Aviation and Aerospace*, Sivasundaram S., Ed., pp. 23–32 (1997), Embry-Riddle Aeronautical University Press.
8. Balakrishnan A.V., Theoretical limits of damping attainable by smart beams with rate feedback. In: *Smart Structures and Materials, 1997; Mathematics and Control in Smart Structures*, Vasundaran V., Jagdish C., Eds., *Proc. SPIE*, **3039** (1997), pp. 204–215.
9. Balakrishnan A.V., Control of structures with self-straining actuators: coupled Euler/Timoshenko model, *Nonlinear Problems in Aviation and Aerospace*, Gordon and Breach Sci. Publ., Reading, United Kingdom (1998).
10. Balakrishnan A.V., Dynamics and control of articulated anizotropic Timoshenko beams. In: *Dynamics and Control of Distributed Systems*, Tzou H.S. and Bergman L.A., Eds., Cambridge University Press (1998), pp. 121–201.
11. Balakrishnan A.V. and Edwards J.W., Calculation of the transient motion of elastic airfoils forced by control surface motion and gusts, *NASA TM* 81351 (1980).
12. Bisplinghoff R.L., Ashley H., and Halfman R.L., *Aeroelasticity*, Dover Publ. Inc., New York (1996).
13. Chen G., Krantz S.G., Ma D.W., Wayne C.E., and West H.H., The Euler–Bernoulli beam equations with boundary energy dissipation. *Operator Methods for Optimal Control Problems*, Lecture Notes in Math., **108**, pp. 67–96, Marcel Dekker (1987).
14. Dierolf P., Schröder D., and Voight J., Flutter as a perturbation problem for semigroups. In: *Semigroups of Operators: Theory and Applications*, Balakrishnan A.V., Ed., Birkhauser (2000), pp. 89–95.
15. Fung Y.C., *An Introduction to the Theory of Aeroelasticity*, Dover Publ. Inc., New York (1993).
16. Gohberg I.Ts. and Krein M.G., Introduction to the theory of linear nonselfadjoint operators, *Trans. of Math. Monogr.*, **18**, AMS, Providence, RI (1996).
17. Istratescu V.I., *Introduction to Linear Operator Theory*, Pure Appl. Math Series of Monog., Marcel Dekker Inc., New York (1981).
18. Lee C.K., Chiang W.W., and O'Sullivan T.C., Piezoelectric modal sensor/actuator pairs for critical active damping vibration control, *J. Acoust. Soc. Am.*, **90**, pp. 384–394 (1991).
19. Magnus W., Oberhettinger F., and Soni R.P., *Formulas and Theorems for the Special Functions of Mathematical Physics*, 3rd Ed., Springer Verlag, New York (1966).
20. Shubov M.A., Mathematical analysis of aircraft wing model in subsonuc airflow, *IMA Journal of Applied Math.*, **66** (2001), pp. 319–356.
21. Shubov M.A., Asymptotic representations for root vectors of nonselfadjoint operators generated by aircraft wing model in subsonic air flow, *J. Math. Anal. Appl.*, **260** (2001), pp. 341–366.
22. Shubov M.A., Riesz basis property of root vectors of nonselfadjoint operators generated by aircraft wing model in subsonic air flow, *Math. Methods in Applied Sciences*, **23** (2000), pp. 1585–1615.

23. Shubov M.A., Asymptotics of the aeroelastic modes and basis property of mode shapes for aircraft wing model, *J. of the Fraklin Institute.*, **338** (2/3) (2001), pp. 171–185.
24. Sz.–Nagy B. and Foias C., *Harmonic Analysis of Operators on Hilbert Space*, North Holland Publ. Co. (1970).
25. Tzou H.S. and Gadre M., Theoretical analysis of a multi-layered thin shell coupled with piezoelectric shell actuators for distributed vibration controls, *J. Sound Vibr.*, **132**, pp. 433–450 (1989).
26. Yang S.M. and Lee Y.J., Modal analysis of stepped beams with piezoelectric materials, *J. Sound Vibr.*, **176**, pp. 289–300.

4

Bifurcation Analysis for the Inertial Coupling Problem of a Reentry Vehicle

Norihiro Goto and Takashi Kawakita
Kyushu University, Fukuoka, Japan

This chapter conducts a bifurcation analysis for the inertial coupling problem that may occur in the case of automatic flight control system failures of a reentry vehicle. Equilibrium paths are first obtained with the use of the continuation method in the planes of the vehicle's motion variables vs. control surface angles. Over all the ranges of the equilibrium paths, stability analyses are then undertaken by means of Lyapunov's first method. Special emphasis is put on the convergence nature of the motion variables' temporal trajectories for the case where more than one stable equilibrium point exists for a combination of control surface angles. Based on the equilibrium paths with stability information, the control sequences of control surface angles are finally sought that may bring the vehicle in a high roll rate motion back to a zero roll rate steady-state flight. It is pointed out that a steady flight with no roll-rate can be recovered only if the basic solution branch is stable.

1 INTRODUCTION

Stable flights of a reentry vehicle rely heavily on automatic flight control systems (AFCSs), because its flight dynamical configuration often lacks intrinsic stability over a wide speed range of operation. Failures of AFCSs cause a number of problems in maintaining a desired flight condition. Above all, the problem of inertial coupling must be addressed in detail before the vehicle is put into operation, inasmuch as it is difficult to recover the original flight condition once it develops this problem. Inertial coupling is one of the important nonlinear problems that challenge aircraft stability and control engineers to find recovery control techniques.

Inertial coupling has been known since around 1948 [1]. It is essentially a gyroscopic effect, occurring in high roll-rate maneuvers of modern high-speed airplanes including spinning missiles designed in such a way that most of their masses are concentrated in the fuselage. For such an airplane, a slight deviation of its control surface angle from the steady-state angle may lead to a drastic change in roll-rate, causing damage on its empennage; known as the jump phenomenon. Nonlinear analyses to elucidate this problem have been reported in [2, 3], for example. However, the airplanes treated in these works are stable around the equilibrium point of level flight. On the other hand, an intrinsically unstable reentry vehicle, if combined with a malfunction of its AFCSs, may result in a catastrophe once it falls into a high roll-rate motion. Therefore, the primary objective of this chapter is

0-415-30852-6/04/$0.00+$1.50

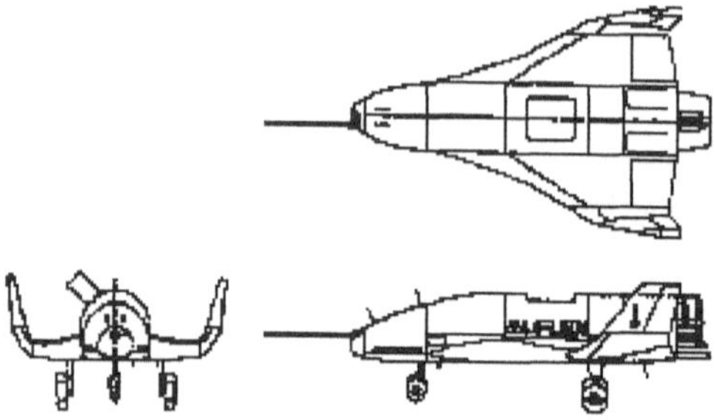

Figure 1 Three views of the ALFLEX plane.

to seek a control technique for an unstable vehicle to recover a zero roll-rate flight from a high roll-rate motion into which it has accidentally fallen.

For this primary objective, a model reentry vehicle is subjected to a bifurcation analysis. The model plane is the Automatic Landing Flight Experiment (ALFLEX) plane [4] as shown in Fig. 1, which is a reduced scale model of an unmanned reusable orbiting spacecraft. The ALFLEX plane is a glider having no means of propulsion aimed at evaluating the flight characteristics of the spacecraft during the final approach and landing phases. Its stable flight is made possible with the help of elaborate AFCSs, without which the bare configuration is unstable both longitudinally and lateral directionally over a wide speed range. Such characteristics of the ALFLEX plane make it important to look into the trimmability of the bare vehicle in case a malfunction of AFCSs should occur.

We first describe the nonlinear equations of motion for the inertial coupling problem of the ALFLEX plane. A bifurcation analysis is then made by the continuation method to determine equilibrium paths (paths of trim points) in the plane of motion variables vs. control surface angles, followed by stability analysis of trim points with the use of Lyapunov's first method. Discussion is also made of the convergence nature of the motion variables' temporal responses for the case where more than one stable trim point exists for a combination of control surface angles. Drawing on these analyses, the chapter finally addresses possible control techniques to recover a zero roll-rate flight from a high roll-rate motion of inertial coupling.

2 EQUATIONS OF MOTION

Because high roll-rate steady states are of primary concern here, the original six-degrees-of-freedom equations of motion of a rigid airplane with respect to an xyz body-axis system [5], where xz is the plane of symmetry, are reduced to five-degrees-of-freedom equations under the assumptions:

1. forward velocity V, weight W, and air density ρ are constant, and
2. angle of attack α and sideslip angle β are small.

The resulting equations of motion are

$$\dot{\beta} = p \sin\alpha - r\cos\alpha + \hat{y} + \frac{g}{V}\sin\phi\cos\theta, \tag{1}$$

$$\dot{\alpha} = -p\beta + q + \hat{z} + \frac{g}{V}\cos\phi\cos\theta, \tag{2}$$

$$\dot{p} - \frac{I_{xz}}{I_x}\dot{r} = i_1\left(-qr + i_a pq + \hat{l}\right), \tag{3}$$

$$\dot{q} = i_2\left(rp + i_b\left(-p^2 + r^2\right) + \hat{m}\right), \tag{4}$$

$$\dot{r} - \frac{I_{xz}}{I_z}\dot{p} = i_3\left(-pq - i_c qr + \hat{n}\right), \tag{5}$$

where the Eulerian angles, roll angle ϕ, and pitch angle θ, are determined by the kinematic relations as

$$\dot{\phi} = p + q\sin\phi\tan\theta + r\cos\phi\tan\theta, \tag{6}$$

$$\dot{\theta} = q\cos\phi - r\sin\phi. \tag{7}$$

In Eqs. (1)–(7)

$$\begin{gathered} i_1 = \frac{I_z - I_y}{I_x}, \quad i_2 = \frac{I_z - I_x}{I_y}, \quad i_3 = \frac{I_y - I_x}{I_z} \\ i_a = \frac{1}{i_1}\frac{I_{xz}}{I_x}, \quad i_b = \frac{1}{i_2}\frac{I_{xz}}{I_y}, \quad i_c = \frac{1}{i_3}\frac{I_{xz}}{I_z}, \\ \hat{y} = \frac{g}{WV}Y_a, \quad \hat{z} = \frac{g}{WV}Z_a, \\ \hat{l} = \frac{L}{i_1 I_x}, \quad \hat{m} = \frac{M}{i_2 I_y}, \quad \hat{n} = \frac{N}{i_3 I_z}, \end{gathered} \tag{8}$$

I_x, I_y, I_z: moments of inertia about x-, y-, and z-axis, respectively,
I_{xy}: product of inertia,
Y_a, Z_a: aerodynamic forces,
L, M, N: aerodynamic moments about the center of gravity,
p, q, r: angular velocities about x-, y-, and z-axis, respectively,
g: gravitational acceleration.

Among the inertial parameters in Eq. (8), i_3 is referred to as the inertial coupling parameter. It can be deduced from Eq. (5) that a large value of i_3 produces large inertial coupling moments. In comparison with the data given in [1], the ALFLEX plane, of which the dimensional data are given in Table 1, has the same magnitude of i_3 as that of the F100A fighter plane that suffered from an inertial coupling problem in its early phase of development.

Aerodynamic forces and moments in Eqs. (1)–(5) are assumed here to be linearly related to motion variables and control surface angles mainly because of lack of nonlinear aerodynamic data. Table 1 summarizes the linear relationships and numerical data necessary for the analysis to follow. The reader is referred to [6] for an example of complete treatment of nonlinear aerodynamic data. Note in Table 1 that a level flight with the steady-state pitch angle θ_0 and angle of attack α_0 is a trivial trim point. Those original equations of motion, Eqs. (1)–(7), can be represented in the general form as

$$\dot{\boldsymbol{x}} = \boldsymbol{H}(\boldsymbol{x}, \boldsymbol{\delta}), \tag{9}$$

Table 4.1 Flight configuration, stability derivatives and numerical data.

$$\hat{y} = (g/WV)(1/2\rho V^2 SC_y) = \hat{y}_\beta \beta + \hat{y}_r r + \hat{y}_{\delta_r}\delta_r$$
$$C_y = (C_{y_\beta})\beta + (C_{y_r})r + (C_{y_{\delta r}})\delta_r$$
$$\hat{z} = (g/WV)(1/2\rho V^2 SC_z) = \hat{z}_0 + \hat{z}_\alpha \Delta\alpha + \hat{z}_{\delta_e}\Delta\delta_e$$
$$C_z = C_{z0} + (C_{z_\alpha})\Delta\alpha + (C_{z_{\delta e}})\Delta\delta_e$$
$$C_{z0} = -W\cos\theta_0/(1/2\rho V^2 S)$$
$$C_{z_\alpha} = -C_{L_\alpha}\cos\alpha_0 + C_L \sin\alpha_0 - C_{D_\alpha}\sin\alpha_0 - C_D\cos\alpha_0$$
$$C_{z_{\delta e}} = -C_{L_{\delta e}}\cos\alpha_0 - C_{D_{\delta e}}\sin\alpha_0$$
$$\hat{l} = (1/2\rho V^2 S b C_l)/(i_1 I_x) = \hat{l}_\beta \beta + \hat{l}_p p + \hat{l}_r r + \hat{l}_{\delta_a}\delta_a + \hat{l}_{\delta_r}\delta_r$$
$$C_l = (C_{l_\beta})\beta + (C_{l_p})p + (C_{l_r})r + (C_{l_{\delta a}})\delta_a + (C_{l_{\delta r}})\delta_r$$
$$\hat{m} = (1/2\rho V^2 S c C_m)/(i_2 I_y) = \hat{m}_\alpha \Delta\alpha + \hat{m}_{\dot\alpha}\Delta\dot\alpha + \hat{m}_q q + \hat{m}_{\delta_e}\Delta\delta_e$$
$$C_m = (C_{m_\alpha})\Delta\alpha + (C_{m_{\dot\alpha}})\Delta\dot\alpha + (C_{m_q})q + (C_{m_{\delta e}})\Delta\delta_e$$
$$\hat{n} = (1/2\rho V^2 S b C_n)/(i_3 I_z) = \hat{n}_\beta \beta + \hat{n}_p p + \hat{n}_r r + \hat{n}_{\delta_a}\delta_a + \hat{n}_{\delta_r}\delta_r$$
$$C_n = (C_{n_\beta})\beta + (C_{n_p})p + (C_{n_r})r + (C_{n_{\delta a}})\delta_a + (C_{n_{\delta r}})\delta_r$$

δ_a: aileron angle, δ_r: rudder angle, $\Delta\delta_e$: incremental elevator angle

$W\,(N) = 760g$	$V(m/s) = 73.84$
$I_x(kg\,m^2) = 407$	$\alpha_0\,(\text{deg}) = 8.18$
$I_y(kg\,m^2) = 1366$	$\gamma_0(\text{deg}) = -17.34$
$I_z(kg\,m^2) = 1643$	$\theta_0\,(\text{deg}) = \gamma_0 + \alpha_0 = -9.16$
$I_{xz}(kg\,m^2) = 10.4$	$\delta_{e0}\,(\text{deg}) = 3.0$
$S\,(m^2) = 9.45$	
$c\,(m) = 3.154$	$\rho\,(kg/m^3) = 1.156$
$b\,(m) = 3.295$	

$C_L = 0.2387$	$C_{L_\alpha} = 2.016$	$C_{L_{\delta e}} = 0.6355$		
$C_D = 0.0745$	$C_{D_\alpha} = 0.2714$	$C_{D_{\delta e}} = 0.1019$		
$C_{m_\alpha} = -0.0134$	$C_{m_q} = -0.0474$	$C_{m_{\delta e}} = -0.2152$		
$C_{y_\beta} = -0.6849$	$C_{y_{\delta r}} = 0.1907$			
$C_{l_\beta} = -0.1774$	$C_{l_p} = -0.0070$	$C_{l_r} = 0.0040$	$C_{l_{\delta a}} = 0.1488$	$C_{l_{\delta r}} = 0.0788$
$C_{n_\beta} = -0.0657$	$C_{n_p} = 0.0032$	$C_{n_r} = -0.0060$	$C_{n_{\delta a}} = -0.0266$	$C_{n_{\delta r}} = -0.0990$

All other derivatives are set equal to zero.

where

$$\boldsymbol{x} = [\beta, \alpha, p, q, r, \phi, \theta]^T, \tag{10}$$
$$\boldsymbol{\delta} = [\delta_e, \delta_a, \delta_r]^T, \tag{11}$$

and δ_e: elevator angle, δ_a: aileron angle , δ_r: rudder angle.

General trim points are determined by solving the transcendental algebraic equation

$$\boldsymbol{H}(\boldsymbol{x}, \boldsymbol{\delta}) = 0. \tag{12}$$

The vehicle, however, cannot attain all of the trim points obtained from Eq. (12). If it has unstable dynamics in the neighborhood of a trim point, a slight external disturbance or a slight deviation of control surface angles from their precise values at the trim point cannot allow the vehicle to stay at the trim point. Stability of the vehicle's dynamics in the neighborhood of a trim point must be examined in order to know whether a steady state can actually be attained. To this end, the stability analysis of trim points is made by using Lyapunov's first method [7], in which the nonlinear equations of motion, Eqs. (1)–(5), are linearized about a trim point as

$$\dot{\boldsymbol{x}}_\varepsilon = \boldsymbol{F}\boldsymbol{x}_\varepsilon, \tag{13}$$

where $\boldsymbol{x}_\varepsilon = [\beta_\varepsilon, \alpha_\varepsilon, p_\varepsilon, q_\varepsilon, r_\varepsilon]^T$: perturbed variable vector, and $\boldsymbol{F}$ is the Jacobian matrix explicitly given by

$$\boldsymbol{F} = \begin{bmatrix} \hat{y}_\beta & \bar{p} & \bar{\alpha} + \sin\alpha_0 & 0 & \hat{y}_r - \cos\alpha_0 \\ -\bar{p} & \hat{z}_\alpha & -\bar{\beta} & 1 & 0 \\ i_p\left(\frac{\hat{l}_\beta}{i_a} + i_3\hat{n}_\beta\right) & 0 & \begin{array}{c} i_p\left[(1-i_3)\bar{q}\right. \\ \left.+\left(\frac{\hat{l}_p}{i_a} + i_3\hat{n}_p\right)\right]\end{array} & \begin{array}{c} i_p\left[(1-i_3)\bar{p}\right. \\ \left.-\left(\frac{1}{i_a} + i_3 i_c\right)\bar{r}\right]\end{array} & \begin{array}{c} i_p\left[-\left(\frac{1}{i_a} + i_3 i_c\right)\bar{q}\right. \\ \left.+\left(\frac{\hat{l}_r}{i_c} + i_3\hat{n}_r\right)\right]\end{array} \\ 0 & i_2\hat{m}_\alpha & i_2(\bar{r} - 2i_b\bar{p}) & i_2\hat{m}_q & i_2(\bar{p} + 2i_b\bar{r}) \\ i_r\left(i_1\hat{l}_\beta + \frac{\hat{n}_\beta}{i_c}\right) & 0 & \begin{array}{c} i_r\left[\left(i_1 i_a - \frac{1}{i_c}\right)\bar{q}\right. \\ \left.+\left(i_1\hat{l}_p + \frac{\hat{n}_p}{i_c}\right)\right]\end{array} & \begin{array}{c} i_r\left[\left(i_1 i_a - \frac{1}{i_c}\right)\bar{p}\right. \\ \left.-(i_1+1)\bar{r}\right]\end{array} & \begin{array}{c} i_r\left[-(i_1+1)\bar{q}\right. \\ \left.+\left(i_1\hat{l}_r + \frac{\hat{n}_r}{i_c}\right)\right]\end{array} \end{bmatrix} \tag{14}$$

In Eq. (14) stability derivatives $\hat{y}_\beta$ and so on are defined in Table 1, and $\bar{\alpha}, \bar{\beta}, \bar{p}, \bar{q}$, and $\bar{r}$ denote the steady-state values, while

$$i_p = \frac{I_{xz}I_z}{I_x I_z - I_{xz}^2}, \quad i_r = \frac{I_{xz}I_x}{I_x I_z - I_{xz}^2}. \tag{15}$$

If all of the eigenvalues of $\boldsymbol{F}$ have negative real parts, the trim point is asymptotically stable, whereas it is unstable if some eigenvalues have positive real parts. It sometimes happens that the real parts of some eigenvalues are zero or close to zero, generating so-called bifurcation points. In such a case, it is necessary to consider the effects of nonlinear terms on the behavior of the solution in the neighborhood of the trim point. The center manifold theory can be applied to such a case [7].

3 CONTINUATION METHOD

An equilibrium path analysis is made using the continuation method [3], which allows one to solve the transcendental algebraic equation, Eq. (12), without any approximation. In the method, Eq. (12) is transformed to an initial value problem such that

$$\frac{d\boldsymbol{c}(s)}{ds} = \dot{\boldsymbol{c}} = \boldsymbol{t}\left(\boldsymbol{H}'(\boldsymbol{c}(s))\right), \tag{16}$$

$$\boldsymbol{c}(0) = (\boldsymbol{x}^T(0), \boldsymbol{\delta}^T(0)), \tag{17}$$

where $\boldsymbol{c}(s)$ is the solution trajectory, s is the arc length, and $\boldsymbol{t}(\boldsymbol{H}')$ denotes the tangent vector to $\boldsymbol{H}'$, which is the derivative of H with respect to s. Because s is the arclength, the initial value problem is accompanied with the constraint condition:

$$\|\dot{\boldsymbol{c}}(s)\| = 1. \tag{18}$$

This initial value problem is solved numerically by the predictor–corrector method with the discretized $\boldsymbol{c}(s)$ as $\boldsymbol{c}(i)$ and the step size as h:

1. Euler method as the predictor:

$$\boldsymbol{c}_p(i+1) = \boldsymbol{c}_c(i) + h\boldsymbol{t}(\boldsymbol{H}'(\boldsymbol{c}_c(i))), \tag{19}$$

where the subscripts p and c denote the value by the predictor and that by the corrector, respectively.

2. Newton's method as the corrector:

$$\boldsymbol{c}_c(i+1) = \boldsymbol{c}_p(i+1) - \boldsymbol{H}'(\boldsymbol{c}_p(i+1))^+\boldsymbol{H}(\boldsymbol{c}_p(i+1)), \tag{20}$$

where $\boldsymbol{H}'(\boldsymbol{u})^+$ is the Moore–Penrose inverse matrix; i.e.,

$$\boldsymbol{H}'(\boldsymbol{u})^+ = \boldsymbol{H}'(\boldsymbol{u})^T(\boldsymbol{H}'(\boldsymbol{u})\boldsymbol{H}'(\boldsymbol{u})^T)^{-1}. \tag{21}$$

Note that the corrector value by Eq. (20) minimizes the norm

$$\|\boldsymbol{c}_c(i+1) - \boldsymbol{c}_p(i+1)\|_{\boldsymbol{H}(\boldsymbol{c}_c(i+1))}. \tag{22}$$

In practice the tangent vectors and the Moore–Penrose inverses that take into account the constraint, Eq. (18), can be obtained by the use of the QR method [8], in which an arbitrary matrix A is factorized so as to be the product of a proper orthogonal matrix Q and an upper triangular matrix R. A good initial guess may lead to a quick convergence of the algorithm. One such initial guess can be given by the pseudosteady-state analysis [2], in which the gravity terms in Eqs. (1) and (2) are ignored to arrive at a polynomial equation with respect to the roll rate p. The greatest merit of the continuation method, in particular that of using the QR method, is that the computation can proceed even at a bifurcation point without difficulty.

4 NUMERICAL RESULTS

Using the flight configuration of Table 1, equilibrium path analyses are made here followed by the stability analyses of trim points. Flight control of the ALFLEX plane actually uses elevator δ_e, aileron δ_a, rudder δ_r, and speed brake δ_s as available control inputs. However, in this analysis the rudder angle is set equal to zero, and the speed brake is fixed at a certain angle. Therefore, the equilibrium paths are determined by varying δ_e and δ_a as parameters. The ranges of variation are -10–10 deg for $\Delta\delta_e$, which denotes an incremental elevator deflection from the initial value of Table 1, and -20–20 deg for δ_a. First of all, the results from the equilibrium path analyses are shown in Figs. 2 and 3, where $\Delta\alpha$ is also an incremental angle of attack. Figures 2(a)–2(e) show the equilibrium paths in the plane of motion variables vs. δ_a for a $\Delta\delta_e$, while Figs. 3(a)–3(b) together with Fig. 2(b) exhibit the variation of p vs. δ_a equilibrium paths for three kinds of $\Delta\delta_e$. It results from the ensuing stability analyses that the trim points on solid equilibrium paths are stable, whereas those on broken paths are unstable. By numerically integrating the original nonlinear equations of motion, these equilibrium paths with stability information have been validated with an additional finding that the origin of $p - \delta_a$ plane for $\Delta\delta_e = -6$ deg actually yields a limit cycle oscillation. The elevator angle should be smaller than -6 deg in order for the origin

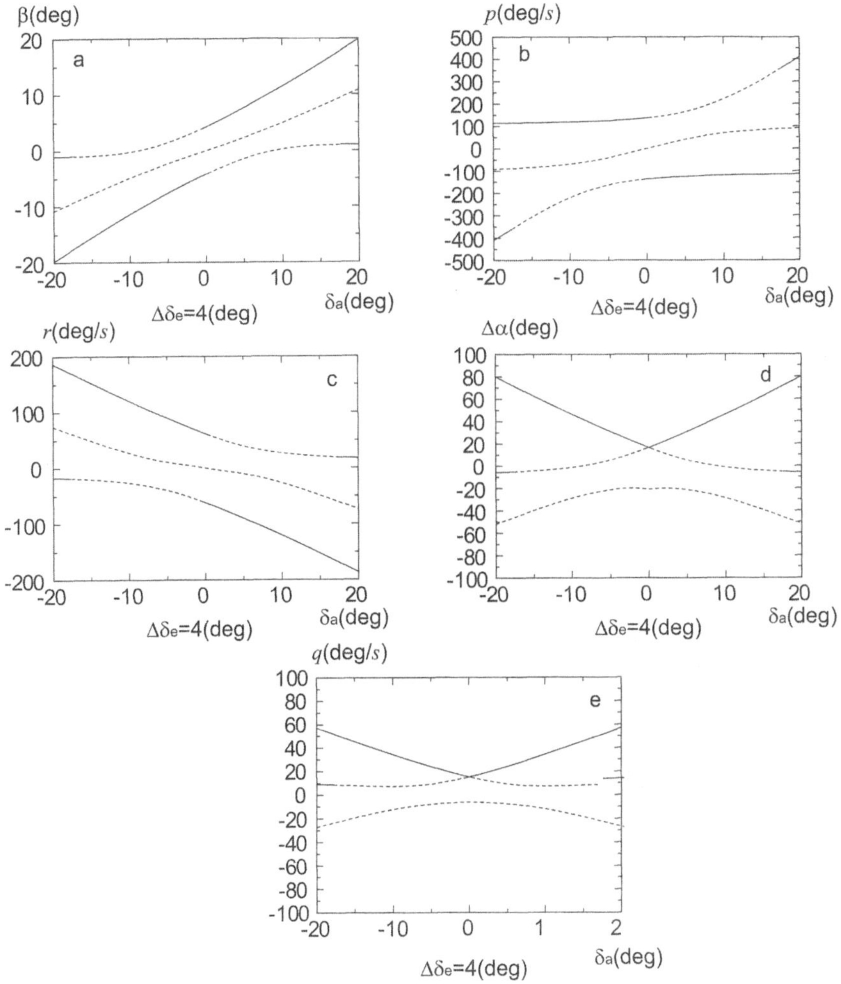

Figure 2 Equilibrium paths of motion variables for $\Delta\delta_e = 4$ deg.

to be a stable trim point. Furthermore, a numerical simulation enables one to visualize the vehicle's dynamic motion. Figure 4 is a snapshot, as seen from a fixed point on the ground, from such a visual three-dimensional flight simulation. The vehicle in Fig. 4 is close to a steady state for a combination of step inputs of $\Delta\delta_e$ and δ_a. Velocity and angular velocity vectors are shown in the figure together with the body-fixed three axes. This kind of simulation helps to understand the flight–dynamic characteristics implicit in the nonlinear equations of motion, while at the same time it may point out the defects of the analysis as well. For instance, it can be learned from Fig. 4 that the angle of attack at this moment is almost 90 deg, violating the assumption that α be small. More realistic analyses should be undertaken in the future, getting rid of the assumption and using nonlinear aerodynamic

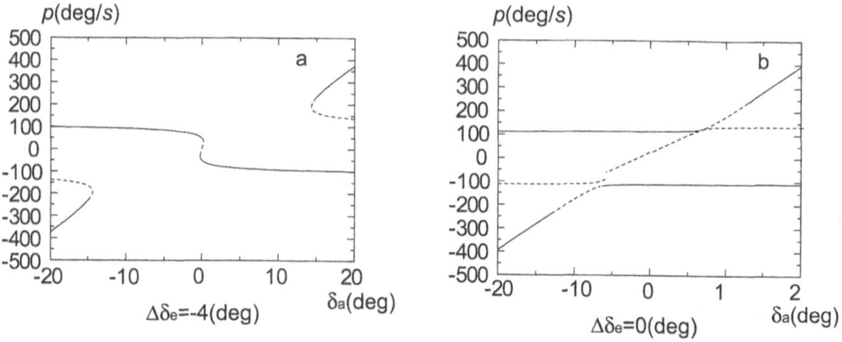

Figure 3 Equilibrium paths of roll rate for $\Delta\delta_e = 4$ and 0 deg.

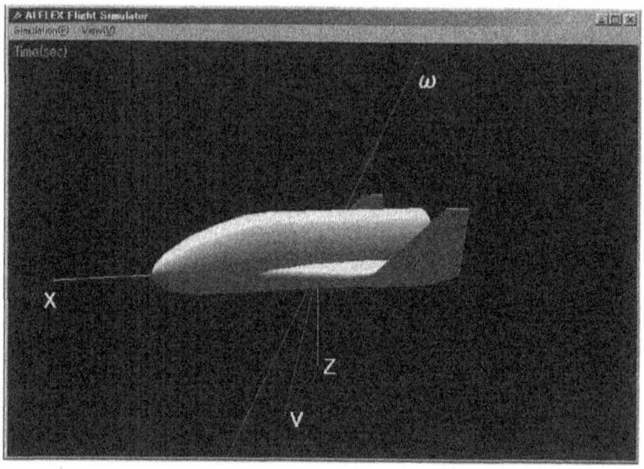

Figure 4 A snapshot of the ALFLEX plane in steady-state motion.

data. Note that this simulation makes use of the quaternion formulation for attitude computation to avoid the singularity problem that turns up for a large θ in Eq. (6).

More importantly, the equilibrium paths with stability information thus obtained may tell the combinations of $(\Delta\delta_e, \delta_a)$ for which jump phenomena are likely to occur. For example, assume that the vehicle is in a steady rolling state of about 100 deg/s for a combination of $(\Delta\delta_e, \delta_a)$= (4 deg, 0 deg) as read from Fig. 2(b). A deviation of the aileron angle to 2 deg brings the roll rate abruptly to about -200 deg/s, because at the new aileron angle the trim point on the same equilibrium path is not stable any more. Looking at Fig. 2(b) again, it can be observed that depending on the combination of $(\Delta\delta_e, \delta_a)$ there exist multiple stable trim points or attractors. An interesting question is to which attractor the motion will converge, given the corresponding combination of control surface angles and an arbitrary set of initial conditions. If the motion starts from within the region of attraction of an attractor, it will settle down on the attractor. For an arbitrary set of initial conditions, however, the general flow pattern of the solution trajectories needs to be known in the phase plane of seven variables. Instead of this general phase plane approach, an attempt is made here at correlating initial conditions with attractors in two-dimensional phase planes by re-

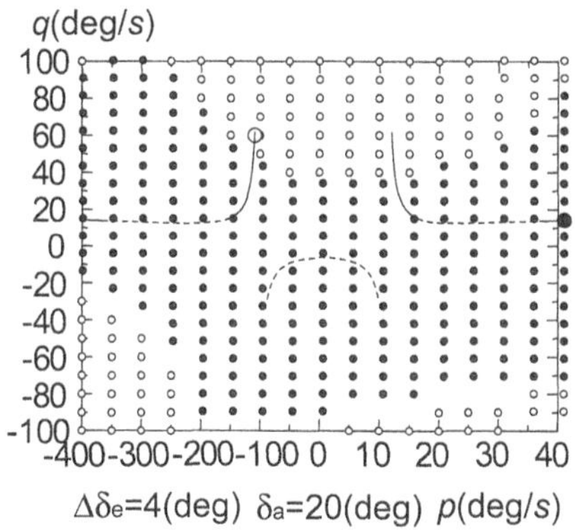

Figure 5 Correlation between initial condition and attractors: ○, •: attractors, °,•: initial conditions.

peating numerical simulations. A result from such an attempt is shown in Fig. 5, which is a $p - q$ phase plane for a combination of $(\Delta\delta_e, \delta_a)$ where two attractors, marked by an open circle and a filled circle, are located on stable equilibrium paths of solid curves, while small circle marks in the plane denote the locations of the initial condition of p and q with the other initial conditions set equal to zero; the initial (p, q) of • takes the motion to the attractor •, while that of ○ eventually goes to ○. A fairly clear pattern can be seen of the distribution of initial conditions. It is a work for the future to find analytical rules that underlie such a pattern.

5 RECOVERY FROM A HIGH ROLL-RATE MOTION

First of all, note that in Fig. 3(a) the equilibrium path that passes through the origin, referred to as the basic solution branch, is partly unstable, while in Fig. 2(b) it is entirely unstable. It turns out that for the flight condition of Table 1 the basic solution branch becomes unstable around the origin for $\Delta\delta_e$ larger than about -6 deg. Therefore, the basic solution branch must be stabilized before any recovery technique is devised. If automatic stabilization is available, certain types of so-called chaotic control may be applicable to such a situation; e.g., the Lyapunov function method [9]. However, this work presupposes that any flight control system is out of order, so that a remote control technique should be worked out through the use of $\delta_e$and δ_a. Looking at Fig. 6(a) where the equilibrium paths for $\Delta\delta_e$= 2 deg are shown together with the stable basic solution branch for $\Delta\delta_e = -8$ deg, such a control technique can be proposed that first moves $\Delta\delta_e$ to -8 deg from 2 deg, at which the vehicle is now experiencing a high roll-rate motion of about -125 deg/s, then shifts δ_a back to 0 deg using the stable basic solution branch. To validate this control technique, the original nonlinear equations of motion are numerically integrated with the initial states (p, δ_a)= (0 deg/s, 5 deg), applying the control sequence shown in Fig. 6(b). Figure 6(c) shows the result that the roll rate subsides to zero, although a large oscillation in p is excited initially, and more noteworthily this new no roll-rate steady state has a nonzero pitch rate. A level flight with no rotational motion at all cannot be attained at the basic flight configuration of Table 1, because it is an unstable trim point. It follows that realizing

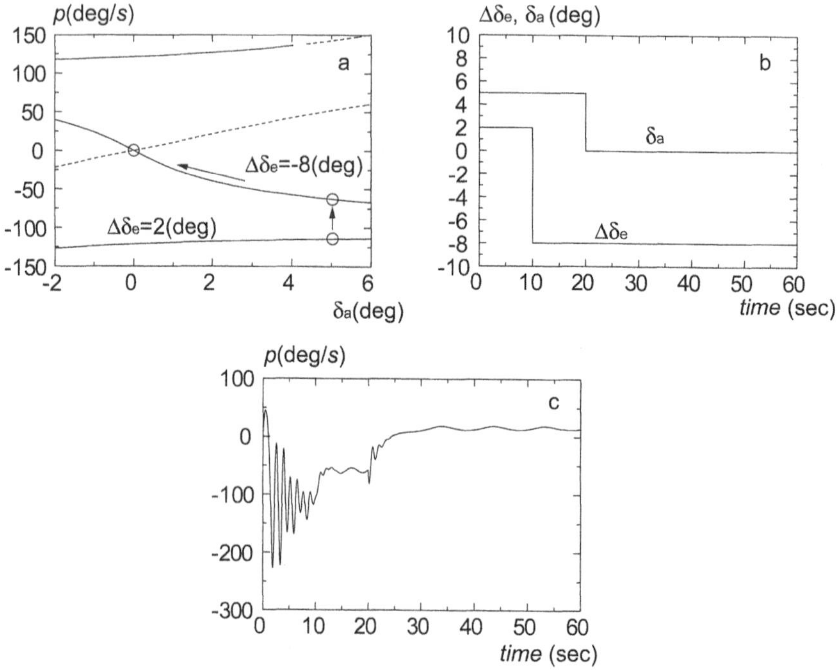

Figure 6 An example of the control technique to recover: (a) equilibrium paths for a recovery control technique; (b) recovery control technique; (c) time history of roll rate with a recovery control technique.

a stable basic solution branch in the hyperplane of motion variables of concern is key to finding a control technique for recovering a true level flight.

6 CONCLUSION

A bifurcation analysis has been made of a model reentry vehicle using the continuation method. The obtained equilibrium paths indicate that jump phenomena of inertial coupling can occur over a wide range of control surface angles. The stability analysis of the equilibrium paths by Lyapunov's first method shows that depending on the combination of control surface angles the basic solution branch is unstable around the zero roll-rate trim point. Discussion is also made on the convergence of the vehicle's motion when multiple attractors exist for a combination of control surface angles. Drawing on the equilibrium paths with stability information, the control sequence of elevator and aileron is finally sought that may bring the vehicle back to a zero roll-rate steady state. It is pointed out that a zero roll-rate steady state could be reached only if it is on a stable basic solution branch. It is a work for the future to devise a remedy control technique for recovery from a general rotational motion for a more realistic case where complete nonlinear equations of motion are utilized together with nonlinear aerodynamic data.

REFERENCES

1. Abzug, M.J. and Larrabee, E.E., *Airplane Stability and Control. A History of the Technologies That Made Aviation Possible*, Cambridge University Press, Cambridge, U.K., Chap. 8, 1997.
2. Schy, A.A. and Hannah, M.E., "Prediction of jump phenomena in rolling coupled maneuvers of airplanes," *Journal of Aircraft*, Vol. 14, pp. 375–382, April 1977.
3. Carrol, J.V. and Mehra, R.D., "Bifurcation analysis of nonlinear aircraft dynamics," *Journal of Guidance, Control, and Dynamics*, Vol. 5, No. 5, pp. 529–536, 1982.
4. Miyazawa, Y., Motoda, T., Izumi, T., and Hata, T., "Longitudinal landing control law for an autonomous reentry vehicle," *Journal of Guidance, Control, and Dynamics*, Vol. 22, No. 6, pp. 791–800, 1999.
5. Etkin, B. and Reid, L.D., *Dynamics of Flight — Stability and Control*, 3rd ed., John Wiley & Sons, New York, Chap. 4, 1996.
6. Sibilski, K., "An agile aircraft non-linear dynamics by continuation methods and bifurcation theory," Proceedings of the 22nd International Congress of Aeronautical Sciences 2000 (CD-ROM), Paper No. ICAS-2000-3.11.2, Aug. 2000.
7. Khalil, H.K., *Nonlinear Systems*, Macmillan, New York, Chaps. 3 and 4, 1992.
8. Mathematical Society of Japan, *Encyclopedic Dictionary of Mathematics*, 2nd ed., Ed. Ito, K., The MIT Press, Cambridge, MA, p. 298F, 1987.
9. Vincent, T.L., "Utilizing chaos in control system design." In *Controlling Chaos and Bifurcation in Engineering Systems*, Ed. Chen, G., CRC Press, New York, pp. 89–106, 1999.

5

Missile Autopilot Design Using Dynamic Fuzzy Gain-Scheduling Technique

Chun-Liang Lin,† Rei-Min Lai,† and Sen-Wei Huang*
†*Department of Electrical Engineering, National Chung Hsing University, Taichung 402, Taiwan*
**Department of Applied Mathematics, National Chung Hsing University, Taichung 402, Taiwan*

A dynamic backpropagation training algorithm for an adaptive fuzzy gain-scheduling feedback control scheme is developed in this chapter. This novel design methodology uses a Takagi–Sugeno fuzzy system to represent the fuzzy relationship between the scheduling variables and controller parameters. Mach number and angle of attack are used as measured, time-varying exogenous scheduling variables injected into the guidance law. By incorporating scheduling parameter variation information, the adaptation law for controller parameters is derived. Results from extensive simulation studies show that the presented approach offers satisfactory controlled system performance.

1 INTRODUCTION

Gain scheduling is an effective way to control systems whose dynamics change with the operating conditions [1, 2]. This technique uses process variables related to dynamics that compensate for the effect of working in different operating regions. It is normally used in the control of nonlinear plants where the relationship between the plant dynamics and operating conditions is known, and for which a single linear time-invariant model is insufficient. This is the reason it was originated in connection with the development of flight control systems.

The missile autopilot design problem is challenging for its wide range of parameter variations and stringent performance requirements. Recently there have been several published studies of gain scheduling in flight control problems. Conventional autopilot gain scheduled designs for tactical missiles can be derived from [3–6]. A robust control method in the framework of μ-synthesis was used in the autopilot design [7]. H_∞ control theory has also been extensively applied to treat the problem [8, 9]. In [10, 11], the concept of linear parameter-varying control techniques was introduced, and a state transformation was proposed to convert nonlinear missile dynamics into a linear parameter-varying form and hence to apply the gain-scheduled control design. Various control laws have been compared in [12] to estimate their potential and applicability in a three-axis missile model.

Typically, a gain-scheduling design involves three issues: partitioning the operating region into several approximately linear regions, local controller design in each linear region, and controller parameter interpolation between the linear regions. The main drawback of most conventional gain scheduling (CGS) is that the parameter change may be

0-415-30852-6/04/$0.00+$1.50

rather abrupt across the region boundaries, which may result in unacceptable or even unstable performance. Another problem is that accurate linear time-invariant models at various operating points may be difficult or impossible to obtain. As a solution to this problem, a fuzzy gain-scheduling (FGS) method has been proposed that utilizes a fuzzy reasoning technique to determine the controller parameters. With this method, human expertise in the linear control design and CGS can be represented by fuzzy rules with a fuzzy inference mechanism used to interpolate the controller parameters in the transition regions [13].

Takagi–Sugeno fuzzy models provide an effective representation of complex nonlinear systems in terms of fuzzy sets and fuzzy reasoning applied to a set of linear input–output submodels. Based on each model, fuzzy gain-scheduling controllers are facilitated using linear matrix inequality methods [14]. The FGS technique has been applied in aircraft flight control design [15, 16]. A robust fuzzy gain scheduler has also been designed for an aircraft flight control system [17].

Neural network gain scheduling (NNGS) could incorporate learning ability into gain-scheduling control [2] where the training example consists of the operating variables and control gains obtained at various operating points and their corresponding desired output. Compared to the FGS technique, the main advantage of NNGS is that it avoids the need to manually design a scheduling program or determine a suitable inferencing system. NNGS techniques have also been applied to the aircraft flight control design in [18].

In summary, the major difficulties in applying FGS consist of (i) difficulty in specifying appropriate fuzzy rules and membership functions, (ii) performance of the gain-scheduled system depends on the accuracy of the missile guidance and control system model under each flight condition. For NNGS, the major difficulties are (i) appropriate network size selection, (ii) difficulty in injecting the designer's expertise into the existing network when necessary.

This paper proposes a new methodology for controlling a class of missile autopilots using an adaptive fuzzy gain-scheduling (AFGS) technique. This design approach uses a Tagaki–Sugeno fuzzy system to represent the fuzzy relationship between the scheduling variables and controller parameters. A dynamic gain-scheduled compensation strategy is used to design the control system. Traditionally the control gains of an autopilot are switched along the missile trajectories according to a function or a table lookup built in the computer. Unlike conventional techniques, the proposed approach offers the advantage of performance improvement for ill-defined flight dynamics through learning using an adaptive fuzzy inferencing mechanism. Rapid adaptivity to environmental changes makes this technique appropriate for flight control systems covering aerodynamic changes during flight.

2 AUTOPILOT MODEL AND PERFORMANCE OBJECTIVE

Consider a representative dynamic nonlinear missile system described by the following state-space equations [8]:

$$\begin{aligned}\dot{\alpha}(t) &= g_1(\alpha, \delta, M) + q \\ \dot{q}(t) &= g_2(\alpha, \delta, M) \\ a_L(t) &= g_3(\alpha, \delta, M)\end{aligned} \tag{1}$$

Table 1 Coefficients of the pitch-axis missile model

K_α	1.18587
K_q	70.586
K_z	0.6661697
a_n	0.000103 deg^{-3}
b_n	−0.00945 deg^{-2}
c_n	−0.1696 deg^{-1}
d_n	−0.034 deg^{-1}
a_m	0.000215 deg^{-3}
b_m	−0.0195 deg^{-2}
c_m	0.051 deg^{-1}
d_m	−0.206 deg^{-1}

where

$$g_1(\alpha,\delta,M) = K_\alpha M C_n[\alpha(t),\delta(t),M(t)]\cos\alpha(t),$$
$$g_2(\alpha,\delta,M) = K_q M^2 C_m[\alpha(t),\delta(t),M(t)],$$
$$g_3(\alpha,\delta,M) = K_z M^2 C_n[\alpha(t),\delta(t),M(t)],$$

α is the angle of attack, q is the pitch angle rate, M is the Mach number, δ is the tail deflection angle, a_L is the actual lateral acceleration; the aerodynamic coefficients C_n and C_m are given by the following polynomial expressions for α and δ given in degrees and M given in Mach:

$$C_n(\alpha,\delta,M) = \text{sgn}(\alpha)\left[a_n|\alpha|^3 + b_n|\alpha|^2 + c_n(2-\frac{M}{3})|\alpha|\right] + d_n\delta$$
$$C_m(\alpha,\delta,M) = \text{sgn}(\alpha)\left[a_m|\alpha|^3 + b_m|\alpha|^2 + c_m(-7+\frac{8M}{3})|\alpha|\right] + d_m\delta.$$

Details of aerodynamic constants for the pitch-axis missile model are given in Table 1. The variables q and a_L are measured variables available for feedback. Let us define the augmented state vector $x(t) = [\alpha(t), q(t)]^T$. Actuator dynamics describing the tail deflection are

$$\frac{d}{dt}\begin{bmatrix}\delta(t)\\ \dot{\delta}(t)\end{bmatrix} = \begin{bmatrix}0 & 1\\ -\omega_a^2 & -2\zeta_a\omega_{na}\end{bmatrix}\begin{bmatrix}\delta(t)\\ \dot{\delta}(t)\end{bmatrix} + \begin{bmatrix}0\\ \omega_{na}^2\end{bmatrix}\delta_c(t), \tag{2}$$

where δ_c and δ are, respectively, the commanded and actual tail deflection angles, and ζ_a and ω_{na} are, respectively, the damping ratio and undamped natural frequency.

For the problem considered here, we wish to design a controller to track commanded acceleration maneuvers and a command with an accuracy no more than 5%. The controller must provide robust performance over a wide range of angles of attack and variations in Mach number. To this aim, an adaptive fuzzy gain-scheduling (AFGS) controller is characterized by

$$\begin{aligned} u(t) &= K_1(\alpha,M)e_a(t)\\ \delta_c(t) &= K_2(\alpha,M)e_v(t), \end{aligned} \tag{3}$$

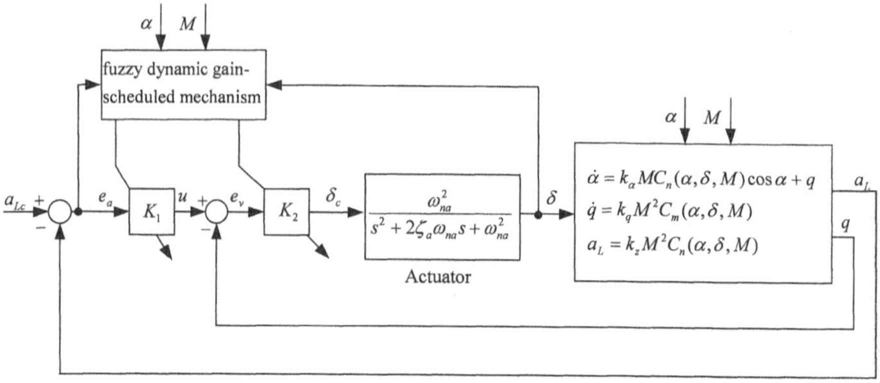

Figure 1 Structure of the fuzzy dynamic gain-scheduled autopilot.

where $e_v = u - q$ and $e_a = a_{Lc} - a_L$ are, respectively, the tracking errors of pitch rate and lateral acceleration command with a_{Lc} denoting the commanded lateral acceleration, K_1 and K_2 are, respectively, the control gains for the acceleration and pitch rate feedback loops. We chose the scheduling variable vector $w = [\alpha, M]^T$ that correlates well with the change in the plant dynamics. The objective of this controller is to minimize a predefined performance measure for the closed-loop system based on the information provided by the scheduling variables. A fuzzy logic-based, adaptive parameter adjustment rule will be proposed in the following paragraphs to train the controller. A schematic diagram for the AFGS control scheme is shown in Fig. 1.

For the flight control scheme, the performance measure is defined by integrating the combination of the tracking error and control command over $[t_0, t]$ as follows:

$$J = \frac{1}{2}\int_{t_0}^{t} \left[e^T(\tau)e(\tau) + \rho\delta^2(\tau)\right] d\tau, \tag{4}$$

where $e(t) = [e_a(t), e_v(t)]^T$, the scalar ρ is included to weight the relative importance of the tracking error and control effort.

To characterize the adaptive controller, a group of adjustable variables σ_{si}^j are introduced. These variables are used as the consequent parameters for the fuzzy rules. The gradient descent algorithm [19] is then used to derive a generalized backpropagation training algorithm:

$$\Delta\sigma_{si}^j = -\eta\nabla_{\sigma_{si}^j} J, \quad \forall\, s,\, i, j, \tag{5}$$

where $\eta > 0$ is the learning rate of the parameter update algorithm, $\Delta\sigma_{si}^j = \sigma_{si}^j(t+\tau) - \sigma_{si}^j(t)$ with τ being the sampling period, the gradient $\nabla_{\sigma_{si}^j} J = \partial J/\partial\sigma_{si}^j$ is calculated using the chain rule:

$$\begin{aligned}\frac{\partial J}{\partial \sigma^j_{si}} &= \left(\frac{\partial J}{\partial e}\right)^T \frac{\partial e}{\partial \sigma^j_{si}} + \frac{\partial J}{\partial \delta}\frac{\partial \delta}{\partial \sigma^j_{si}} \\ &= \int_{t_0}^{t} \left[e^T \frac{\partial e}{\partial \sigma^j_{si}} + \rho\delta \frac{\partial \delta}{\partial \sigma^j_{si}} \right] d\tau \\ &= \int_{t_0}^{t} \left[-e_a \frac{\partial a_L}{\partial \sigma^j_{si}} + e_v \left(\frac{\partial u}{\partial \sigma^j_{si}} - \frac{\partial q}{\partial \sigma^j_{si}} \right) + \rho\delta \frac{\partial \delta}{\partial \sigma^j_{si}} \right] d\tau, \end{aligned} \tag{6}$$

where

$$\frac{\partial u}{\partial \sigma^j_{si}} = \frac{\partial K_1}{\partial \sigma^j_{si}} e_a - K_1 \frac{\partial a_L}{\partial \sigma^j_{si}}.$$

To complete the derivations, the derivatives $\partial a_L/\partial \sigma^j_{si}$, $\partial q/\partial \sigma^j_{si}$ and $\partial \delta/\partial \sigma^j_{si}$ must be found. Determination for these terms is given in the subsequent section.

With minor modifications, the compensation approach developed above can be applied for linear or nonlinear systems with time-varying parameters.

3 FUZZY DYNAMIC GAIN-SCHEDULED AUTOPILOT

The development of an appropriate gain-scheduling law is crucial to the performance of a scheduled system. It is quite often either difficult to find or too complicated to design CGS law. However, this problem can be resolved using fuzzy inference techniques.

In the following application, two term sets describing the scheduling variables α and M are defined respectively as

$$T_\alpha = \{A_1, A_2, A_3, A_4, A_5\}, T_M = \{B_1, B_2, B_3, B_4, B_5\}. \tag{7}$$

where A_i and B_i are the fuzzy sets characterized by certain membership functions. For the purpose of theoretical studies, the physical domain over which the scheduling variable α takes its crisp value is supposed to be $[-25, 25]$ degrees, and the domain for M is $[2, 4]$ Mach.

A Takagi–Sugeno fuzzy model is used next to represent the fuzzy conditional statement between the scheduling variables and controller parameters. This model provides an effective representation of the complex nonlinear relations in terms of fuzzy sets and fuzzy reasoning applied to a set of input–output data.

The complex relationship between the scheduling variables and control gains is constructed through 25 fuzzy rules. For a first-order Sugeno model [13], the ith fuzzy rule is described by

$$\text{If } \alpha \text{ is } A^l_i \text{ and } M \text{ is } B^l_j, \text{ then } K^l_1 = p^l_1\alpha + q^l_1 M + r^l_1,\ K^l_2 = p^l_2\alpha_m + q^l_2\beta + r^l_2, \quad l = 1, \cdots, 25 \tag{8}$$

where (α, M) is the scheduling vector that best fits the description in the premise part of the rule, p^l_1, q^l_1, r^l_1, p^l_2, q^l_2 and r^l_2 are the consequent function parameters. The outputs of this rule are the inferred control gains.

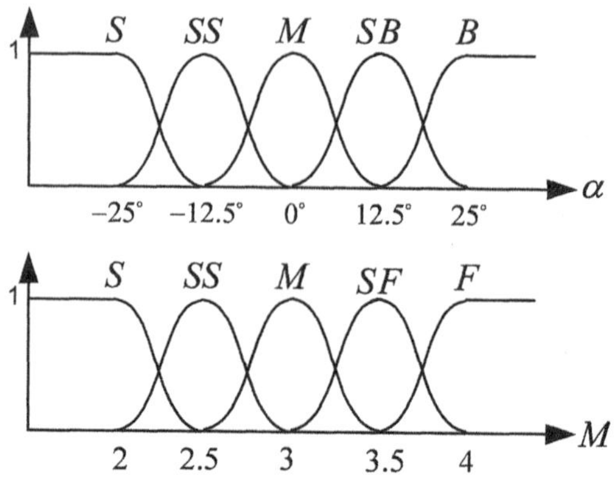

Figure 2 Membership functions of the scheduling variables $\alpha(t)$ and $M(t)$.

To facilitate adaptation to the gain-scheduling control model, it is convenient to put the fuzzy model into a framework of adaptive networks that can compute the gradient vectors systematically. Using the Sugeno inference mechanism [13] the fuzzified control gains K_1^l and K_2^l can be expressed by the weighted sum of all strengths of the fired rules:

$$\begin{aligned} K_1 &= \sum_l \bar{w}^l (p_1^l \alpha + q_1^l M + r_1^l), \\ K_2 &= \sum_l \bar{w}^l (p_2^l \alpha + q_2^l M + r_2^l) \end{aligned} \tag{9}$$

for $l = 1, \cdots, 25$ where the normalized firing strength

$$\bar{w}^l = \frac{w^l}{\sum_l w^l}, \quad w^l = \prod_{i,j} \mu_{A_i}(\alpha)\mu_{B_j}(M),$$

with the membership functions $\mu_{A_i}(\cdot)$ and $\mu_{B_j}(\cdot)$ chosen as follows:

$$\mu_{A_i}(\alpha) = \frac{1}{1 + [(\frac{\alpha - c_i}{a_i})^2]^{b_i}}, \quad \mu_{B_j}(M) = \frac{1}{1 + [(\frac{M - c_j}{a_j})^2]^{b_j}}, \quad i, j = 1, \cdots, 5.$$

Figure 2 shows the bell-shaped membership functions considered in this research. We have constructed an adaptive network that is functionally equivalent to a Sugeno fuzzy model. The consequent function parameters are regulated by the learning algorithm presented in Section 2 when the input information is provided.

Adaptive Parameter Update Law

To minimize the performance index defined in (4) over the flight time, we complete the derivations by introducing (9) and rewriting (6) into the following form:

$$\frac{\partial J}{\partial \sigma_{si}^{j}} = \int_{t_0}^{t} \left[-(e_a + K_1 e_v) \frac{\partial a_L}{\partial \sigma_{si}^{j}} - e_v \frac{\partial q}{\partial \sigma_{si}^{j}} + \rho\delta \frac{\partial \delta}{\partial \sigma_{si}^{j}} + \beta_i e_v \sum_l \bar{w}^l \right] d\tau \tag{10}$$

for $s = 1, 2$, $i = 1, 2, 3$, $j = 1, \ldots, 25$, where

$$\sigma_{1i}^{j} = \begin{cases} p_1^j, & i = 1 \\ q_1^j, & i = 2 \\ r_1^j, & i = 3 \end{cases}, \quad \sigma_{2i}^{j} = \begin{cases} p_2^j, & i = 1 \\ q_2^j, & i = 2 \\ r_2^j, & i = 3 \end{cases}, \quad \beta_i = \begin{cases} \alpha, & i = 1 \\ M, & i = 2 \\ 1, & i = 3 \end{cases}.$$

For the nonlinear dynamic system (1), the derivative $\partial\delta/\partial\sigma_{1i}^{j}$ in (10) is determined via the following derivations. Since $\delta = \delta(e, w, \sigma_{si}^{j})$, thus

$$\begin{aligned} \frac{\partial \delta}{\partial \sigma_{1i}^{j}} &= \frac{\partial \delta}{\partial \delta_c} \left[\frac{\partial \delta_c}{\partial e_v} \left(\frac{\partial u}{\partial \sigma_{1i}^{j}} - \frac{\partial q}{\partial \sigma_{1i}^{j}} \right) + \left(\frac{\partial \delta_c}{\partial w} \right)^T \frac{\partial w}{\partial x} \frac{\partial x}{\partial \sigma_{1i}^{j}} \right] \\ &= \frac{\partial \delta}{\partial \delta_c} \left\{ K_2 \left(\frac{\partial K_1}{\partial \sigma_{1i}^{j}} e_a - K_1 \frac{\partial a_L}{\partial \sigma_{1i}^{j}} - \frac{\partial q}{\partial \sigma_{1i}^{j}} \right) + \left[\begin{array}{cc} \frac{\partial \delta_c}{\partial \alpha} & 2\frac{\partial \delta_c}{\partial M} \end{array} \right] \left[\begin{array}{c} \frac{\partial \alpha}{\partial \sigma_{1i}^{j}} \\ \frac{\partial q}{\partial \sigma_{1i}^{j}} \end{array} \right] \right\} \\ &= \frac{\partial \delta}{\partial \delta_c} \left\{ \beta_i K_2 \sum_l \bar{w}^l e_a - K_1 K_2 \frac{\partial a_L}{\partial \sigma_{1i}^{j}} + \frac{\partial \delta_c}{\partial \alpha} \frac{\partial \alpha}{\partial \sigma_{1i}^{j}} + \left(2\frac{\partial \delta_c}{\partial M} - K_2 \right) \frac{\partial q}{\partial \sigma_{1i}^{j}} \right\}. \end{aligned} \tag{11}$$

Next, we evaluate $\partial a_L/\partial\sigma_{1i}^{j}$ and then proceed with solving the differential equations to obtain $\partial x/\partial\sigma_{si}^{j}$. Introducing (11), we have

$$\begin{aligned} \frac{\partial a_L}{\partial \sigma_{1i}^{j}} &= \frac{\partial g_3}{\partial \alpha} \frac{\partial \alpha}{\partial \sigma_{1i}^{j}} + \frac{\partial g_3}{\partial \delta} \frac{\partial \delta}{\partial \sigma_{1i}^{j}} \\ &= \Omega \left(\frac{\partial g_3}{\partial \alpha} + \frac{\partial g_3}{\partial \delta} \frac{\partial \delta}{\partial \delta_c} \frac{\partial \delta_c}{\partial \alpha} \right) \frac{\partial \alpha}{\partial \sigma_{1i}^{j}} + \Omega \left(2\frac{\partial g_3}{\partial M} - K_2 \frac{\partial g_3}{\partial \delta} \frac{\partial \delta}{\partial \delta_c} \right) \frac{\partial q}{\partial \sigma_{1i}^{j}} \\ &\quad + \beta\Omega K_{2i} \frac{\partial g_3}{\partial \delta} \frac{\partial \delta}{\partial \delta_c} \sum_l \bar{w}^l e_a \\ &= 2\Omega \frac{\partial g_3}{\partial \alpha} \frac{\partial \alpha}{\partial \sigma_{1i}^{j}} + \Omega \left(2\frac{\partial g_3}{\partial M} - K_2 \frac{\partial g_3}{\partial \delta} \frac{\partial \delta}{\partial \delta_c} \right) \frac{\partial q}{\partial \sigma_{1i}^{j}} + \beta_i \Omega K_2 \frac{\partial g_3}{\partial \delta} \frac{\partial \delta}{\partial \delta_c} \sum_l \bar{w}^l e_a, \end{aligned} \tag{12}$$

where $\Omega = (1 + K_1 K_2 \frac{\partial g_3}{\partial \delta} \frac{\partial \delta}{\partial \delta_c})^{-1}$. If the precise plant model is not known, the gradient $\partial g_3/\partial\alpha$ can be approximately calculated using the perturbation method, i.e., $\partial g_3/\partial\alpha \approx \Delta g(\alpha)/\Delta\alpha$. This also applies for $\partial g_3/\partial\delta$. From (12), the problem is transformed to obtain the derivative $\partial x/\partial\sigma_{1i}^{j}$. Differentiating the term $\partial x/\partial\sigma_{1i}^{j}$ with respect to t and applying $\partial\delta/\partial\sigma_{1i}^{j}$ given in (11) results in the gradient dynamics:

$$\frac{d}{dt}\left(\frac{\partial x}{\partial \sigma_{1i}^j}\right) = \begin{bmatrix} \frac{\partial \dot{\alpha}}{\partial \sigma_{1i}^j} \\ \frac{\partial \dot{q}}{\partial \sigma_{1i}^j} \end{bmatrix} = \begin{bmatrix} \frac{\partial g_1}{\partial \alpha}\frac{\partial \alpha}{\partial \sigma_{1i}^j} + \frac{\partial g_1}{\partial \delta}\frac{\partial \delta}{\partial \sigma_{1i}^j} + \frac{\partial q}{\partial \sigma_{1i}^j} \\ \frac{\partial g_2}{\partial \alpha}\frac{\partial \alpha}{\partial \sigma_{1i}^j} + \frac{\partial g_2}{\partial \delta}\frac{\partial \delta}{\partial \sigma_{1i}^j} \end{bmatrix}$$

$$= \begin{bmatrix} 2\frac{\partial g_1}{\partial \alpha}\frac{\partial \alpha}{\partial \sigma_{1i}^j} + \left(1 - K_2\frac{\partial g_1}{\partial \delta}\frac{\partial \delta}{\partial \delta_c} + 2\frac{\partial g_1}{\partial M}\right)\frac{\partial q}{\partial \sigma_{1i}^j} - K_1K_2\frac{\partial g_1}{\partial \delta}\frac{\partial \delta}{\partial \delta_c}\frac{\partial a_L}{\partial \sigma_{1i}^j} + \beta_i K_2\frac{\partial g_1}{\partial \delta}\frac{\partial \delta}{\partial \delta_c}\sum_l \bar{w}^l e_a \\ 2\frac{\partial g_2}{\partial \alpha}\frac{\partial \alpha}{\partial \sigma_{1i}^j} + \left(-K_2\frac{\partial g_2}{\partial \delta}\frac{\partial \delta}{\partial \delta_c} + 2\frac{\partial g_2}{\partial M}\right)\frac{\partial q}{\partial \sigma_{1i}^j} - K_1K_2\frac{\partial g_2}{\partial \delta}\frac{\partial \delta}{\partial \delta_c}\frac{\partial a_L}{\partial \sigma_{1i}^j} + \beta_i K_2\frac{\partial g_2}{\partial \delta}\frac{\partial \delta}{\partial \delta_c}\sum_l \bar{w}^l e_a \end{bmatrix}.$$

Now introducing $\partial a_L/\partial \sigma_{1i}^j$ obtained in (12) yields

$$\frac{d}{dt}\left(\frac{\partial x}{\partial \sigma_{1i}^j}\right) = \begin{bmatrix} 2\left(\frac{\partial g_1}{\partial \alpha} - K_1K_2\Omega\frac{\partial g_1}{\partial \delta}\frac{\partial \delta}{\partial \delta_c}\frac{\partial g_3}{\partial \alpha}\right) & 1 + \left(-K_2\frac{\partial g_1}{\partial \delta}\frac{\partial \delta}{\partial \delta_c} + 2\frac{\partial g_1}{\partial M}\right)\Phi \\ 2\left(\frac{\partial g_2}{\partial \alpha} - K_1K_2\Omega\frac{\partial g_2}{\partial \delta}\frac{\partial \delta}{\partial \delta_c}\frac{\partial g_3}{\partial \alpha}\right) & \left(-K_2\frac{\partial g_2}{\partial \delta}\frac{\partial \delta}{\partial \delta_c} + 2\frac{\partial g_2}{\partial M}\right)\Phi \end{bmatrix}\frac{\partial x}{\partial \sigma_{1i}^j}$$
$$+ \begin{bmatrix} \beta_i K_2 \Phi \frac{\partial g_1}{\partial \delta}\frac{\partial \delta}{\partial \delta_c}\sum_l \bar{w}^l \\ \beta_i K_2 \Phi \frac{\partial g_2}{\partial \delta}\frac{\partial \delta}{\partial \delta_c}\sum_l \bar{w}^l \end{bmatrix} e_a, \tag{13}$$

where $\Phi = 1 - K_1K_2\Omega\frac{\partial g_3}{\partial \delta}\frac{\partial \delta}{\partial \delta_c}$.

Similarly, the adaptation law for the consequent parameter σ_{2i}^j is determined by

$$\frac{\partial \delta}{\partial \sigma_{2i}^j} = \frac{\partial \delta}{\partial \delta_c}\left(\frac{\partial \delta_c}{\partial \sigma_{2i}^j} + \left(\frac{\partial \delta_c}{\partial w}\right)^T \frac{\partial w}{\partial x}\frac{\partial x}{\partial \sigma_{2i}^j}\right)$$
$$= \frac{\partial \delta}{\partial \delta_c}\left\{\frac{\partial K_2}{\partial \sigma_{2i}^j}e_v + K_2\frac{\partial (u-q)}{\partial \sigma_{2i}^j} + \frac{\partial \delta_c}{\partial \alpha}\frac{\partial \alpha}{\partial \sigma_{2i}^j} + 2\frac{\partial \delta_c}{\partial M}\frac{\partial q}{\partial \sigma_{2i}^j}\right\}$$
$$= \frac{\partial \delta}{\partial \delta_c}\left\{\beta_i \sum_l \bar{w}^l e_v - K_1K_2\frac{\partial a_L}{\partial \sigma_{2i}^j} + \frac{\partial \delta_c}{\partial \alpha}\frac{\partial \alpha}{\partial \sigma_{2i}^j} + \left(2\frac{\partial \delta_c}{\partial M} - K_2\right)\frac{\partial q}{\partial \sigma_{21i}^j}\right\}. \tag{14}$$

The term $\partial a_L/\partial \sigma_{2i}^j$ is derived as

$$\frac{\partial a_L}{\partial \sigma_{2i}^j} = \frac{\partial g_3}{\partial \alpha}\frac{\partial \alpha}{\partial \sigma_{2i}^j} + \frac{\partial g_3}{\partial \delta}\frac{\partial \delta}{\partial \sigma_{2i}^j}$$
$$= 2\Omega\frac{\partial g_3}{\partial \alpha}\frac{\partial \alpha}{\partial \sigma_{2i}^j} + \Omega\left(2\frac{\partial g_3}{\partial M} - K_2\frac{\partial g_3}{\partial \delta}\frac{\partial \delta}{\partial \delta_c}\right)\frac{\partial q}{\partial \sigma_{2i}^j} + \beta_i\Omega\frac{\partial g_3}{\partial \delta}\frac{\partial \delta}{\partial \delta_c}\sum_l \bar{w}^l e_v. \tag{15}$$

Differentiating $\partial x/\partial \sigma_{2i}^j$ with respect to t and applying $\partial \delta/\partial \sigma_{2i}^j$ and $\partial a_L/\partial \sigma_{2i}^j$ obtained in (11) and (12) produces the following gradient dynamics:

$$\frac{d}{dt}\left(\frac{\partial x}{\partial \sigma_{2i}^j}\right) = \begin{bmatrix} 2\frac{\partial g_1}{\partial \alpha}\frac{\partial \alpha}{\partial \sigma_{2i}^j} + \left(1 - K_2\frac{\partial g_1}{\partial \delta}\frac{\partial \delta}{\partial \delta_c} + 2\frac{\partial g_1}{\partial M}\right)\frac{\partial q}{\partial \sigma_{2i}^j} - K_1K_2\frac{\partial g_1}{\partial \delta}\frac{\partial \delta}{\partial \delta_c}\frac{\partial a_L}{\partial \sigma_{2i}^j} + \beta_i\frac{\partial g_1}{\partial \delta}\frac{\partial \delta}{\partial \delta_c}\sum_l \bar{w}^l e_v \\ 2\frac{\partial g_2}{\partial \alpha}\frac{\partial \alpha}{\partial \sigma_{2i}^j} + \left(-K_2\frac{\partial g_2}{\partial \delta}\frac{\partial \delta}{\partial \delta_c} + 2\frac{\partial g_2}{\partial M}\right)\frac{\partial q}{\partial \sigma_{2i}^j} - K_1K_2\frac{\partial g_2}{\partial \delta}\frac{\partial \delta}{\partial \delta_c}\frac{\partial a_L}{\partial \sigma_{2i}^j} + \beta_i\frac{\partial g_2}{\partial \delta}\frac{\partial \delta}{\partial \delta_c}\sum_l \bar{w}^l e_v \end{bmatrix}.$$

Introducing (15) gives

$$\frac{d}{dt}\left(\frac{\partial x}{\partial \sigma_{2i}^{j}}\right) = \begin{bmatrix} 2\left(\frac{\partial g_1}{\partial \alpha} - K_1 K_2 \Omega \frac{\partial g_1}{\partial \delta}\frac{\partial \delta}{\partial \delta_c}\frac{\partial g_3}{\partial \alpha}\right) & 1 + \left(-K_2 \frac{\partial g_1}{\partial \delta}\frac{\partial \delta}{\partial \delta_c} + 2\frac{\partial g_1}{\partial M}\right)\Phi \\ 2\left(\frac{\partial g_2}{\partial \alpha} - K_1 K_2 \Omega \frac{\partial g_2}{\partial \delta}\frac{\partial \delta}{\partial \delta_c}\frac{\partial g_3}{\partial \alpha}\right) & \left(-K_2 \frac{\partial g_2}{\partial \delta}\frac{\partial \delta}{\partial \delta_c} + 2\frac{\partial g_2}{\partial M}\right)\Phi \end{bmatrix} \frac{\partial x}{\partial \sigma_{2i}^{j}}$$
$$+ \begin{bmatrix} \beta_i \left(1 - K_1 K_2 \Omega \frac{\partial g_3}{\partial \delta}\frac{\partial \delta}{\partial \delta_c}\right) \frac{\partial g_1}{\partial \delta}\frac{\partial \delta}{\partial \delta_c} \sum_l \bar{w}^l \\ \beta_i \left(1 - K_1 K_2 \Omega \frac{\partial g_3}{\partial \delta}\frac{\partial \delta}{\partial \delta_c}\right) \frac{\partial g_2}{\partial \delta}\frac{\partial \delta}{\partial \delta_c} \sum_l \bar{w}^l \end{bmatrix} e_v. \tag{16}$$

For the previous derivations, $\partial\alpha/\partial\sigma_{si}^{j}$ and $\partial q/\partial\sigma_{si}^{j}$ are viewed as the state variables of the dynamic equation.

PD Type Controller

The previous approach can be directly extended to the proportional-derivative type controller described by

$$u(t) = K_1(\alpha, M)e_a(t) + K_{1d}(\alpha, M)\dot{e}_a(t)$$
$$\delta_c(t) = K_2(\alpha, M)e_v(t) + K_{2d}(\alpha, M)\dot{e}_v(t).$$

For the PD controller the corresponding fuzzy control rules are given by:

If α is A_i^l and M is B_j^l, then $K_1^l = p_1^l\alpha + q_1^l M + r_1^l$, $K_{1d}^l = p_{1d}^l\alpha + q_{1d}^l M + r_{1d}^l$, $K_2^l = p_2^l\alpha + q_2^l M + r_2^l$, $K_{2d}^l = p_{2d}^l\alpha + q_{2d}^l M + r_{2d}^l$, $l = 1, \cdots, 25$.

The inferenced control gains are determined by

$$K_1 = \sum_l \bar{w}^l(p_1^l\alpha + q_1^l M + r_1^l), \qquad K_{1d} = \sum_l \bar{w}^l(p_{1d}^l\alpha + q_{1d}^l M + r_{1d}^l),$$
$$K_2 = \sum_l \bar{w}^l(p_2^l\alpha + q_2^l M + r_2^l), \qquad K_{2d} = \sum_l \bar{w}^l(p_{2d}^l\alpha + q_{2d}^l M + r_{2d}^l). \tag{17}$$

Now considering the situation of $t > 0$, the derivatives $\partial u/\partial\sigma_{1i}^{j}$ and $\partial u/\partial\sigma_{2i}^{j}$ in (11) and (14) should be changed to

$$\frac{\partial u}{\partial \sigma_{1i}^{j}} = \frac{\partial u_1}{\partial \sigma_{1i}^{j}} + \frac{\partial \dot{u}_2}{\partial u_2}\frac{\partial u_2}{\partial \sigma_{1i}^{j}} \approx \frac{\partial K_1}{\partial \sigma_{1i}^{j}} e_a + \left(K_1 + \frac{1}{t}K_{1d}\right)\frac{\partial e_a}{\partial \sigma_{1i}^{j}}$$
$$\frac{\partial u}{\partial \sigma_{1di}^{j}} = \frac{\partial u_1}{\partial \sigma_{1di}^{j}} + \frac{\partial \dot{u}_2}{\partial u_2}\frac{\partial u_2}{\partial \sigma_{1di}^{j}} \approx \frac{1}{t}\frac{\partial K_{1d}}{\partial \sigma_{1di}^{j}} e_a + \left(K_1 + \frac{1}{t}K_{1d}\right)\frac{\partial e_a}{\partial \sigma_{1di}^{j}},$$

where $u_1 = K_1(\alpha, M)e_a$, $u_2 = K_{1d}(\alpha, M)e_a$, and

$$\sigma_{1di}^{j} = \begin{cases} p_{1d}^{j}, & i = 1 \\ q_{1d}^{j}, & i = 2 \\ r_{1d}^{j}, & i = 3 \end{cases}.$$

Similarly, the derivative $\partial\delta_c/\partial\sigma_{2i}^{j}$ in (14) should be changed to

$$\frac{\partial \delta_c}{\partial \sigma_{2i}^{j}} \approx \frac{\partial K_2}{\partial \sigma_{2i}^{j}} e_v + \left(K_2 + \frac{1}{t}K_{2d}\right)\frac{\partial e_v}{\partial \sigma_{2i}^{j}}, \qquad \frac{\partial \delta_c}{\partial \sigma_{2di}^{j}} \approx \frac{1}{t}\frac{\partial K_{2d}}{\partial \sigma_{2i}^{j}} e_v + \left(K_2 + \frac{1}{t}K_{2d}\right)\frac{\partial e_v}{\partial \sigma_{2i}^{j}},$$

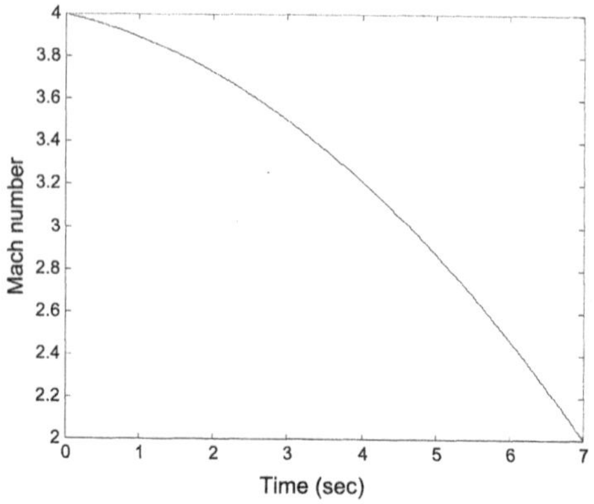

Figure 3 Mach number profile.

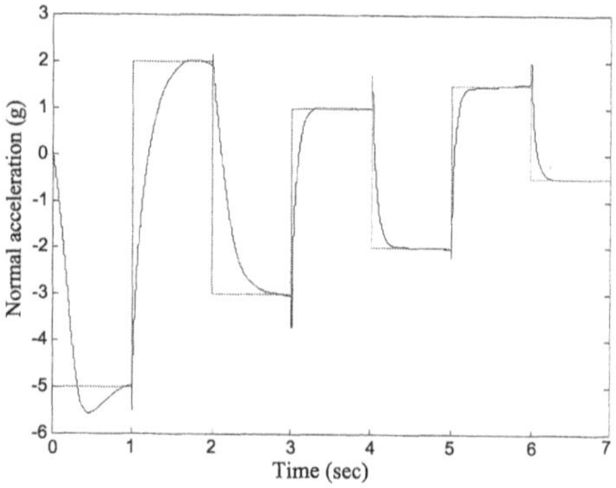

Figure 4 A sequence of step acceleration command $a_{Lc}(t)$ (dashed line) and actual missile acceleration $a_L(t)$ (solid line).

where

$$\sigma_{2di}^j = \begin{cases} p_{2d}^j, & i = 1 \\ q_{2d}^j, & i = 2 \\ r_{2d}^j, & i = 3 \end{cases}.$$

Based on these modifications, the terms used to determine the parameter update algorithm can be derived like those given in (11)–(16).

The adaptive nature of AFGS renders it fundamentally different from the traditional gain-scheduling schemes. The major advantage is that it avoids the need to manually design a scheduling program or to determine a suitable fuzzy inferencing system. Moreover, since the parameters are adaptable, it is more flexible than traditional gain scheduling in the sense that it is not tied to a particular system dynamics and therefore does not require significant modification if the plant is altered. It can thus be expected that the controller becomes more robust and more insensitive to aerodynamic variations.

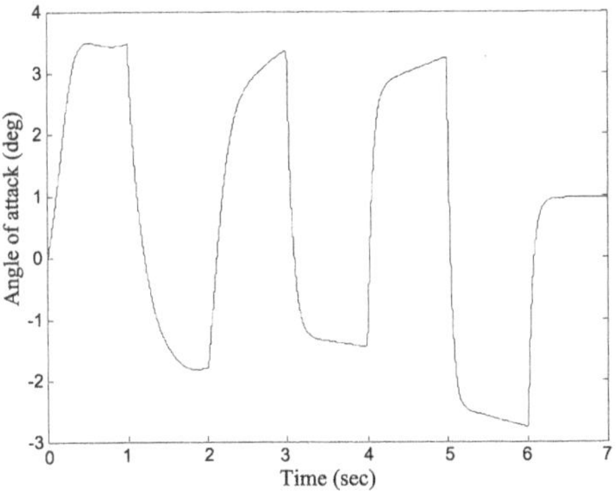

Figure 5 Response of angle-of-attack $\alpha(t)$.

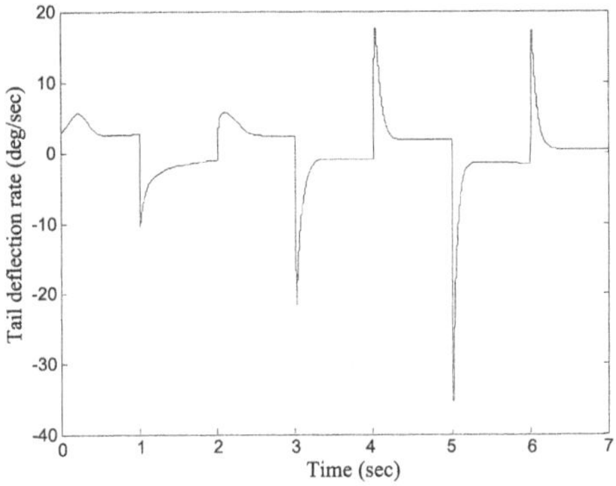

Figure 6 Tail deflection rate $q(t)$.

4 RESULTS AND ANALYSIS

Here, the controller is required to provide robust performance over a wide range of angle of attack and variations in Mach number. The controller is dynamically varying with the evolution of the state of the system. The desired operating range for this example is for Mach numbers from 2 to 4 and angle of attack from -20 deg to 20 deg. The numerical values of various constants used for simulations are listed in Table 1 [7]. These coefficients are valid for the missile travelling between Mach number 2 and 4 at an altitude of 20,000 ft. For simulation purposes, Mach number plotted in Fig. 3 was generated to provide a reasonably realistic missile speed profile. It is assumed to be measurable in real time.

Response $a_L(t)$ of the closed-loop system with our controller to track a series of step commanded acceleration $a_{Lc}(t)$ is illustrated in Fig. 4. The corresponding angle of attack $\alpha(t)$ and tail-deflection rate $q(t)$ are illustrated in Figs. 5 and 6, respectively. Tracking errors of the acceleration and angular velocity loops are illustrated in Figs. 7 and 8, respectively. The step response shows that the time constant is less than 0.047 s, and

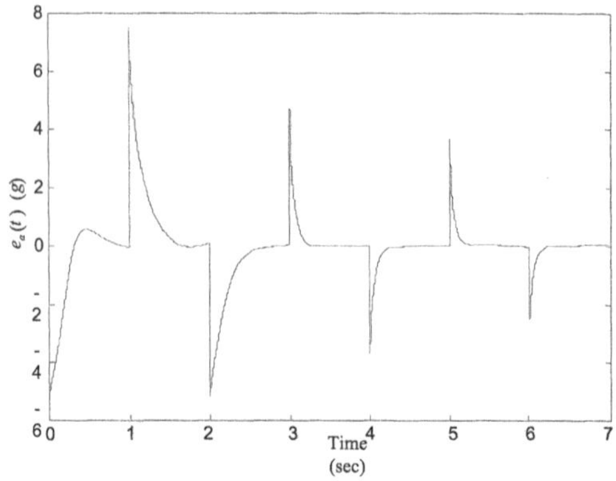

Figure 7 Tracking error $e_a(t)$ of the acceleration loop.

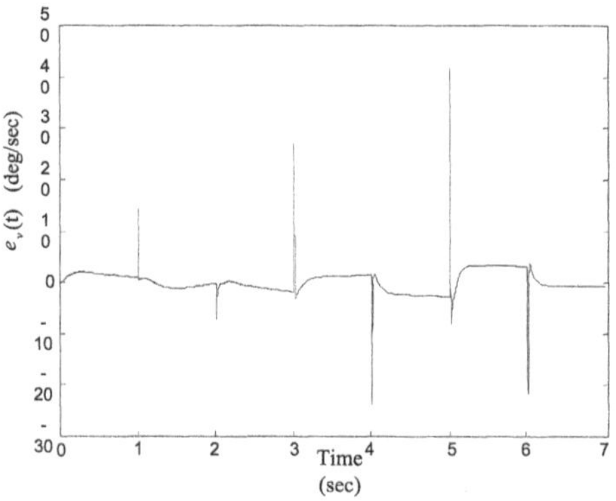

Figure 8 Tracking error $e_v(t)$ of the pitch angular velocity loop.

steady-state error is less than 2%. As one can see from these figures, the tracking errors converge to acceptable levels after a few periods of the step response.

Figures 9 and 10 show, respectively, the transient behavior of the control gains $K_1(\alpha, M)$ and $K_2(\alpha, M)$. The performance measure considered in the simulation study is the minimization of the tracking errors and control effort. Figure 11 shows the resulting performance cost measured over the flight time.

The performance robustness of our proposed controller is verified by perturbing aerodynamic coefficients C_m and C_n. Eleven cases were obtained by multiplying the parameters (a_m, b_m, c_m, d_m) and (a_n, b_n, c_n, d_n) with a factor from 0.5 to 1.5. The resulting acceleration responses compared to the nominal acceleration response are plotted in Fig. 12. We observe reasonable performance variation of the missile under perturbation.

From the extensive simulation studies, we can conclude that the AFGS control law possesses wider design freedom and higher flexibility in the controller architecture. A finely tuned AFGS control law uses less control effort during flight, and therefore, pos-

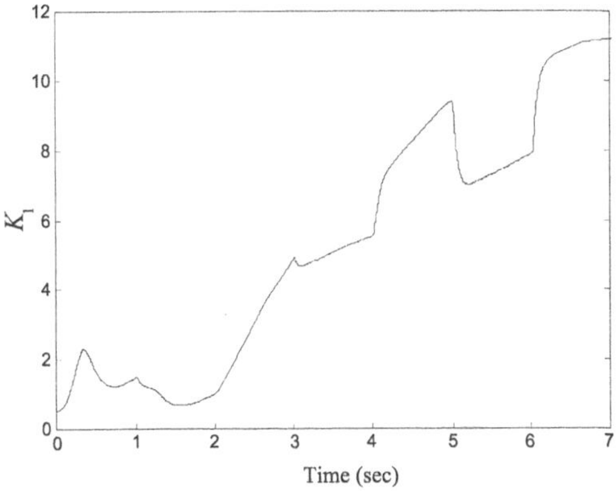

Figure 9 Transient response of gain K_1.

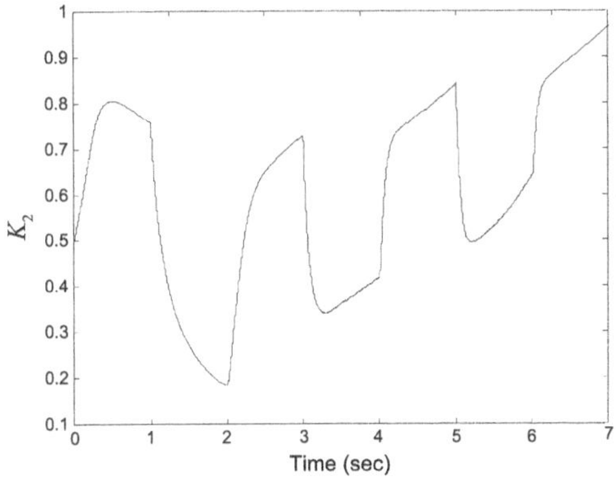

Figure 10 Transient response of gain K_2.

sesses the potential to increase the missile flight speed. Since robustness properties of the design indeed are inherited by our gain-scheduled controller, it can be expected that the flight control system will also give satisfactory performance robustness.

5 CONCLUSIONS

This study developed a novel adaptive fuzzy gain-scheduling control design method for a nonlinear missile system. The design methodology uses a Takagi–Sugeno fuzzy system to represent the fuzzy relationship between the scheduling variables and controller parameters. It extends the underlying state feedback structure to include nonlinear, time-varying dynamics in combination with an adaptive gain-scheduled feedback controller. This technique offers the advantage of performance improvement for ill-defined flight dynamics

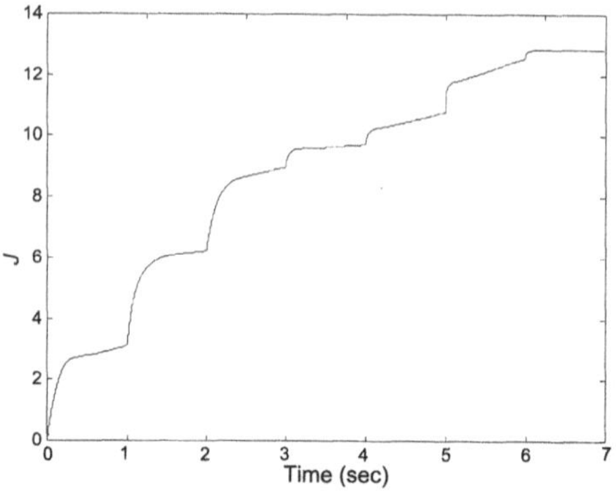

Figure 11 Value of the performance measure.

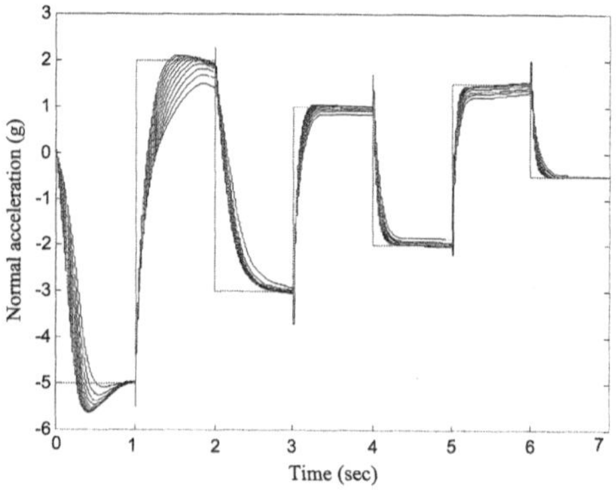

Figure 12 Acceleration $a_L(t)$ with perturbed C_m and C_n.

through learning to use an adaptive fuzzy inferencing mechanism. Simulation results presented successfully demonstrate for the missile autopilot design.

Acknowledgment

This research was sponsored by the National Science Council, Taiwan, R.O.C., under grant NSC 88-2213-E-035-031.

REFERENCES

1. Rugh, W.J., "Analytical framework for gain scheduling," *IEEE Control Systems Magazine*, vol. 11, no. 1, 1991, pp. 79–84.
2. Tan, S., Hang, C. C., and Chai, J.S., "Gain scheduling: from conventional to neuro-fuzzy," *Automatica*, vol. 33, no. 3, 1997, pp. 411–419.

3. Balas, G.J. and Packard, A.K., "Design of robust time-varying controllers for missile autopilot," *Proceedings of the 31st IEEE Conference on Control Applications*, Dayton, OH, 1992, pp. 104–110.
4. Shamma, J.S. and Cloutier, J.R., "Trajectory scheduled missile autopilot design," *Proceedings of the 31st IEEE Conference on Control Applications*, Dayton, OH, 1992, pp. 237–242.
5. White, D.P., Wozniak, J.G., and Lawrence, D.A., "Missile autopilot design using a gain scheduling technique," *Proceedings of the 26th Southeastern Symposium on System Theory*, Athens, OH, 1994, pp. 606–610.
6. Piou, J.E. and Sobel, K.M., "Application of gain scheduling eigenstructure assignment to flight control design," *Proceedings of 1996 IEEE International Conference on Control Applications*, Dearborn, MI, 1996, pp. 101–106.
7. Reichert, R.T., "Robust autopilot design using μ-synthesis," *Proceedings of American Control Conference*, San Diego, CA, 1990, pp. 2368–2373.
8. Nichols, R.A., Reichert, R.T., and Rugh, W.J., "Gain scheduling for H_∞ controllers: a flight control example," *IEEE Transactions on Control Systems Technology*, vol. 1, no. 2, 1993, pp. 69–79.
9. Schumacher, C. and Khargonekar, P.P., "Missile autopilot designs using H_∞ control with gain scheduling and dynamic inversion," *AIAA Journal of Guidance, Control, and Dynamics*, vol. 21, no. 2, 1998, pp. 234–243.
10. Shamma, J.S. and Cloutier, J.R., "Gain-scheduled missile autopilot design using linear parameter varying transformations," *AIAA Journal of Guidance, Control, and Dynamics*, vol. 16, no. 2, 1993, pp. 256–263.
11. Apkarian, P., Gahinet, P., and Becker, G., "Self-scheduled H_∞ control of linear parameter-varying systems: a design example," *Automatica*, vol. 31, no. 9, 1995, pp. 1251–1261.
12. Devaud, E., Harcaut, J.P., and Siguerdidjane, H., "Three-axes missile autopilot design: from linear to nonlinear strategies," *AIAA Journal of Guidance, Control, and Dynamics*, vol. 24, no. 1, 2001, pp. 64–71.
13. Takagi, T. and Sugeno, M., "Fuzzy identification of systems and its applications to modeling and control," *IEEE Transactions on Systems, Man, and Cybernetics*, vol. 15, no. 1, 1985, pp. 116–132.
14. Driankov, D., Hellendoorn, H., and Reinfrank, M., *An Introduction to Fuzzy Control*, Springer, Berlin, 1993, pp. 186–195.
15. Gonsalves, P.G. and Zacharias, G.L., "Fuzzy logic gain scheduling for flight control," *Proceedings of the 3rd IEEE Conference on Fuzzy Systems*, Orlando, FL, 1994, pp. 952–957.
16. Adams, R.J., Sparks, A.G., and Banda, S.S., "A gain scheduled multivariable design for a manual flight control system," *Proceedings of the 31st IEEE Conference on Control Applications*, Dayton, OH, 1992, pp. 584–589.
17. Tanaka, T. and Aizawa, Y., "A robust gain scheduler interpolated into multiple models by membership functions," *AIAA Paper 92-4553*, Aug. 1992.
18. Jonckheere, E.A., Yu, G.R., and Chen, C.C., "Gain scheduling for lateral motion of propulsion controlled aircraft using neural networks," *Proceedings of American Control Conference*, Albuquerque, NM, 1997, pp. 3321–3325.
19. Luenberger, D.G., *Optimization by Vector Space Methods*, John Wiley & Sons, Inc., New York, 1969, pp. 297–299.

6

Model Predictive Control of Nonlinear Rotorcraft Dynamics with Application to the XV-15 Tilt Rotor

Raman K. Mehra and Ravi K. Prasanth
Scientific Systems Company Inc., Woburn, MA

Rotorcraft development presents unique control challenges related to noise, vibration, and flight. Model Predictive Control (MPC) is proposed as a holistic approach to rotorcraft flight control capable of explicitly accounting for operational constraints. The first part of this chapter describes MPC design using linear parameter-varying (LPV) approximation of nonlinear dynamics. This procedure has several advantages including known quality of approximation and polynomial-time computability. Details of MPC problem formulation and its implementation are presented. As a first application, we consider active control of aeromechanical instability, which is a major hurdle in the development of soft-in-plane rotorcraft. A simple nonlinear parameter-varying model that contains a Hopf bifurcation is used to capture the essential features of this instability. The effectiveness of MPC in suppressing nonlinear vibrations subject to control constraints is illustrated. As a second and more involved application, we present the design of an MPC-based flight control system (FCS) for the XV-15 tilt rotor. The FCS was implemented on a 6-DOF high-fidelity XV-15 real-time simulator. Test pilots flew several missions on the simulator with MPC in the loop including a highly nonlinear conversion maneuver. They observed significant reductions in work load, fast response to commands, and rated MPC performance from good to excellent during these simulated maneuvers.

1 INTRODUCTION

NASA, DoD, and the aerospace industry have a very strong interest in the development of advanced rotorcraft. Some unique challenges related to noise, vibration, and flight control are present in their development. A good example comes from tilt rotors, which can hover like helicopter and cruise like a turboprop. When the pilot presses the pedals while in helicopter mode, the tip path plane of one rotor tilts forward and that of the other tilts aft producing a yaw moment. The same pedal input produces a rudder deflection in airplane mode. The flight control system must accommodate these requirements and provide a smooth transition during the conversion process.

0-415-30852-6/04/$0.00+$1.50

Rotorcraft flight control system (FCS) is currently designed in a point-by-point manner using linear principles. Gain scheduling is used to phase-in and phase-out control effects during flight. The flight control computer is also responsible for controlling structural loads on the aircraft. This is achieved, in the V-22 for example, by limiting control surface commands (displacements and rates) produced by pilot stick inputs. Apart from being a complex task, control rate and position limiting have serious effects on aircraft-handling qualities. Pilots evaluating FCS with rate limiting usually mark them down. In fact, control rate limiting is a known trigger mechanism for aircraft-pilot coupling (APC) or pilot-induced oscillation (PIO), which can cause fatalities.

Model Predictive Control (MPC) is a multiinput multioutput digital-control design methodology [6, 7] that uses a highly intuitive principle. At any time step, the control inputs to be applied depend on the current state, control objectives to be achieved, and the current operational constraints. So, a rational procedure is to formulate a constrained optimization problem at each time step and obtain control inputs by solving it. This has the advantage of explicitly accounting for time domain control constraints such as rate limits and state-dependent constraints such as limits on normal acceleration. Frequency domain specifications can also be brought in through MPC design parameters discussed later. Since the control objectives and operational constraints are always updated and accounted for, MPC can provide a holistic approach to rotorcraft flight control. Our aim is to demonstrate this using realistic control problems.

This chapter is organized as follows. The next section describes the MPC problem and implementation in detail. For nonlinear systems, the online optimization problem formulated by MPC is generally nonconvex and cannot be *solved* in real time, especially at the rates required for flight control. We say that a problem is *solvable* if there is an anytime algorithm [13] that terminates in polynomial time for a given accuracy. A procedure for designing real-time implementable MPC using linear-parameter varying (LPV) approximations of rotorcraft dynamics is presented in the next section. Section 3 gives a brief description of active control of aeromechanical instability using MPC. Our results show the effectiveness of MPC in suppressing vibrations in the presence of actuator saturation limits. Section 4 presents the design of an MPC-based flight control system (FCS) for the XV-15 tilt rotor. FCS design specifications include frequency-domain MIL-F-83300 and MIL-F-8785C requirements and time-domain safety and control limits. MPC-based FCS was implemented at 50 Hz on a real-time 6-DOF XV-15 simulator at Bell Helicopter. Results of piloted simulations and evaluations by test pilots are presented. The chapter concludes with some recommendations for future work. Due to space limitations, some details have been omitted; a complete version of this chapter can be obtained from the authors.

2 MODEL PREDICTIVE CONTROL OF NONLINEAR SYSTEMS

This section begins with a description of the control problem and the MPC approach. We then indicate the computational difficulties associated with MPC and present an algorithm that can be readily implemented. The algorithm uses a linear parameter varying (LPV) approximation of the nonlinear system to reduce the computational requirements significantly from those for nonconvex nonlinear programming to that of convex quadratic programming (QP). Thus, our approach trades off optimality for polynomial-time computability.

The MPC implementation is described in detail. Although the presentation is geared toward rotorcraft applications, the algorithm can be applied to problems in other areas as well.

2.1 Control Problem

Consider the nonlinear system:

$$x_{k+1} = f(x_k, u_k) \quad \text{(1a)}$$
$$y_k = g(x_k) \quad \text{(1b)}$$
$$z_k = h(x_k) \quad \text{(1c)}$$
$$u_{\text{lbd}} \le u_k \le u_{\text{ubd}}, \quad \text{for all } k \quad \text{(1d)}$$
$$z_{\text{lbd}} \le z_k \le z_{\text{ubd}}, \quad \text{for all } k \quad \text{(1e)}$$

where $x_k \in \mathbb{R}^{n_x}$ is the state vector, $u_k \in \mathbb{R}^{n_u}$ is the control input, $y_k \in \mathbb{R}^{n_y}$ is the signal that must track a specified command, and $z_k \in \mathbb{R}^{n_z}$ is the vector of signals that are constrained (the control inputs are not included in this vector). It is possible to make the constraint bounds u_{lbd}, z_{lbd}, u_{ubd} and z_{ubd} depend on time, but for simplicity they are taken as constants. As part of the control problem, we are also given a desired trajectory or commanded trajectory y_{cmd}^k.

Control Problem: *Generate a feedback control law so that y_k tracks the desired trajectory y_{cmd}^k subject to the system dynamics and constraints given in (1).*

MPC solution to this problem is the following: *At time k, define the performance index J as follows:*

$$J(x_k, \delta u_k, \delta u_{k+1}, \cdots, \delta u_{k+N-1}) = J_i + J_p + J_u \quad \text{(2a)}$$

$$J_i = \sum_{l=0}^{N-1} \left(\sum_{j=0}^{l} \left(y_{\text{cmd}}^{k+j+1} - y_{k+j+1} \right)^T Q_i \left(y_{\text{cmd}}^{k+j+1} - y_{k+j+1} \right) \right) \quad \text{(2b)}$$

$$J_p = \sum_{i=0}^{N-1} \left(y_{\text{cmd}}^{k+i+1} - y_{k+i+1} \right)^T Q \left(y_{\text{cmd}}^{k+i+1} - y_{k+i+1} \right) \quad \text{(2c)}$$

$$J_u = \delta u_k^T R \delta u_k + \sum_{i=0}^{N-2} \left(\delta u_{k+i+1} - \delta u_{k+i} \right)^T R \left(\delta u_{k+i+1} - \delta u_{k+i} \right) \quad \text{(2d)}$$

and minimize it over control moves subject to the system dynamics and constraints in (1). Apply the control input u_k so obtained, increment k, and repeat the procedure.

The terms J_i and J_p in (2b) and (2c) can be thought of as discrete versions of integral error and tracking error penalties over the control horizon. These terms are expected to yield a proportional-integral (PI) effect in the closed-loop system. Like in classical control, the weights Q_i and Q appearing in (2b) and (2c) can be tuned to obtain satisfactory steady state and transient tracking performances. The third term J_u in (2d) is a penalty for excessive control rate which smoothes out the control inputs. The control inputs are also constrained to satisfy the saturation limits (1d).

Note that the performance index requires knowledge of the desired or commanded trajectory y_{cmd}^k over the prediction horizon. When the application is rotorcraft stability and control augmentation, the commands are pilot stick inputs passed through a command shaping filter. So, their future values $y_{\text{cmd}}^{k+1}, \cdots, y_{\text{cmd}}^{k+N}$ needed in the performance index will not be known at time k. Therefore, we have to make an assumption, and the simplest (in terms of computations) assumption is that

$$y_{\text{cmd}}^{k} = y_{\text{cmd}}^{k+1} = \cdots = y_{\text{cmd}}^{k+N}.$$

This may also be reasonable if the horizon length is chosen to be relatively small. For example, when aircraft dynamics is discretized at 0.02 seconds (50 Hz), a horizon length $N = 10$ corresponds to 0.2 seconds (5 Hz), which is comparable to the pilot's bandwidth so that changes in the command inputs from its current value of y_{cmd}^{k} during the horizon of interest can be neglected. We may be able to extrapolate past pilot commands by fitting polynomials to arrive at better estimates, but this will not be used here. In other applications such as autopilot design, the trajectory is defined by the controller itself and, hence, will be known over the prediction horizon. Both problems fit into the MPC framework and the only difference is in the data used by MPC.

2.2 Linear Parameter-Varying Approximation

When the tracked and constrained variables are written in terms of the control inputs (which are the unknowns), the optimization problem to be solved online becomes nonconvex and corresponds to a general nonlinear programming problem. This difficulty is usually circumvented by linearizing about the current state. The linearized system leads to a convex quadratic programming (QP) that can be readily solved. However, this procedure is flawed for a number of reasons: (a) linearization is performed about a nonequilibrium state, but the linearized system is treated as time-invariant, (b) the quality of approximation provided by linearization is not known because it depends on the states *and* their derivatives, (c) plant model and state estimates have uncertainties, and (d) linearization requires the solution of nonlinear algebraic equations, which can be difficult.

Here, we use a linear parameter varying (LPV) approximation of the nonlinear system:

$$\delta x_{k+1} = A(\theta_k)\delta x_k + B(\theta_k)\delta u_k \tag{3a}$$
$$\delta y_k = C_y(\theta_k)\delta x_k \tag{3b}$$
$$\delta z_k = C_z(\theta_k)\delta x_k \tag{3c}$$
$$x_k = x_{\text{trim}}(\theta_k) + \delta x_k, \quad u_k = u_{\text{trim}}(\theta_k) + \delta u_k \tag{3d}$$
$$y_k = y_{\text{trim}}(\theta_k) + \delta y_k, \quad z_k = z_{\text{trim}}(\theta_k) + \delta z_k \tag{3e}$$
$$u_{\text{lbd}} \leq u_{\text{trim}}(\theta_k) + \delta u_k \leq u_{\text{ubd}}, \quad k = 0, 1, \cdots, N-1 \tag{3f}$$
$$z_{\text{lbd}} \leq z_{\text{trim}}(\theta_k) + \delta z_k \leq z_{\text{ubd}}, \quad k = 1, 2, \cdots, N, \tag{3g}$$

where $\theta_k \in \mathbb{R}^{n_\theta}$ is the scheduling variable (time-varying parameter). This parameter defines the flight condition, and the set of all its possible values defines the flight envelope.

There are several ways to obtain an LPV approximation, some of which depend on the application. See [9, 10, 11, 12] for details on LPV systems. A procedure that confirms to the current practice in aerospace is the following:

1. Subdivide the flight envelope into L polytopic regions. Each polytope is made of m vertices. Let the total number of vertices be M. Let us order the vertices from 1 to M and denote by θ_k the kth vertex.

2. At each of the M vertices, trim the aircraft equations of motion (or the nonlinear system's dynamical equations) to obtain trim states, control settings, tracked variables and constrained variables. Thus at θ_k we obtain:

$$x_{\text{trim}}{}^k, u_{\text{trim}}{}^k, y_{\text{trim}}{}^k \text{ and } z_{\text{trim}}{}^k$$

by solving the algebraic equation obtained from the dynamical equations in (1).

3. At the trim point $(\theta_k, x_{\text{trim}}{}^k, u_{\text{trim}}{}^k, y_{\text{trim}}{}^k, z_{\text{trim}}{}^k)$, linearize the system in (1). This gives a linear system of the form:

$$\delta x_{k+1} = A_k \delta x_k + B_k \delta u_k \tag{4a}$$
$$\delta y_k = C_{yk} \delta x_k \tag{4b}$$
$$\delta z_k = C_{zk} \delta x_k. \tag{4c}$$

4. The collection

$$\left\{\theta_k, x_{\text{trim}}{}^k, u_{\text{trim}}{}^k, y_{\text{trim}}{}^k, z_{\text{trim}}{}^k, A_k, B_k, C_{yk}, C_{zk}\right\}_{k=1}^{M}$$

forms the LPV data.

This is done off-line and the quality of approximation can be checked over admissible trajectories of scheduling variables.

2.3 MPC Implementation

With the LPV approximation in hand, we are now ready to describe the MPC implementation. There are three sequential steps:

1. *Interpolation:* Interpolate system data to obtain a linear system model valid at the current value of scheduling variable.

2. *QP formulation:* Perform computations using interpolated system data and MPC design parameters to set up quadratic programming (QP) problem.

3. *QP solution:* Solve QP problem and obtain optimal control input.

The objective of the first step is to obtain a linear system model that is valid locally. This model depends on the scheduling variable and corresponds to the linear approximation about the trim point associated with the scheduling variable. Thus, we avoid linearization about a nonequilibrium point. In general the interpolation is difficult to perform because the flight envelope need not be a regular region. We shall describe an *efficient algorithm to approximately interpolate*. The second step as stated uses the same state-space model over the prediction horizon. It is possible to combine steps 1 and 2 so that we use an LPV model over the prediction horizon. This will result in better performance. We have separated these steps to keep coding simple. The last step of QP solution is the most computationally intensive task. It takes over 99% of the total computing required in each step of MPC.

2.3.1 Interpolation

The purpose of this task is to determine a linear model that is valid at the current value of scheduling variable θ_k. We use the following approximate interpolator:

(a) Define an $M \times M$ diagonal matrix of weights W:

$$W_{ii} = e^{-\sigma(V_i-\theta_k)^T(V_i-\theta_k)}.$$

This weight gets larger for vertices close to the current point θ_k and smaller for those further away.

(b) Define $M \times 1$ vector α:

$$\alpha = W \begin{bmatrix} V_1^T \\ V_2^T \\ \cdot \\ \cdot \\ V_M^T \end{bmatrix} \left([V_1,\ V_2,\ \cdots,\ V_M]\, W \begin{bmatrix} V_1^T \\ V_2^T \\ \cdot \\ \cdot \\ V_M^T \end{bmatrix} \right)^{-1} \theta_k.$$

This α satisfies $[V_1,\ V_2,\ \cdots,\ V_M]\,\alpha = \theta_k$.

(c) Set data at current point as: $Data^{\text{curr}} = [Data_1,\ Data_2,\ \cdots,\ Data_M]\,\alpha$.

2.3.2 QP Problem Formulation

At the end of interpolation, we have a linear system of the form:

$$\delta x_{k+1} = A^{\text{curr}}\delta x_k + B^{\text{curr}}\delta u_k \tag{5a}$$
$$\delta y_k = {C_y}^{\text{curr}}\delta x_k \tag{5b}$$
$$\delta z_k = {C_z}^{\text{curr}}\delta x_k \tag{5c}$$
$$u_{\text{lbd}} \le {u_{\text{trim}}}^{\text{curr}} + \delta u_k \le u_{\text{ubd}}, \quad k = 0, 1, \cdots, N-1 \tag{5d}$$
$$z_{\text{lbd}} \le {z_{\text{trim}}}^{\text{curr}} + \delta z_k \le z_{\text{ubd}}, \quad k = 1, 2, \cdots, N, \tag{5e}$$

which is used in this task along with the performance index (2) to formulate a QP problem. The computations involved are algebraic and easy. The QP problem has the form:

$$\min_{v}\ v^T\hat{H}v + v^T\hat{g} \tag{6a}$$
$$\text{subject to } U_l \le \hat{A}v \le U_u, \tag{6b}$$

where $\hat{H}$, $\hat{g}$, U_l, $\hat{A}$ and U_u are computed from the LPV state-space model (5).

2.3.3 QP Solver

The optimization problem in (6) is always convex because of the positivity of the matrix $\hat{H}$, which comes from the fact that the control rates are penalized. As a result, this problem can

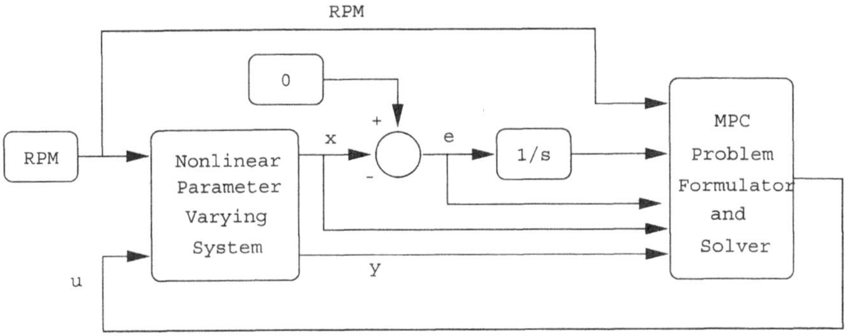

Figure 1 Block diagram for active control of aeromechanical instability using MPC.

be solved in polynomial time. That is, there are algorithms (e.g., ellipsoid method, primal-dual interior point) whose computational cost is a polynomial function of the number of variables, number of constraints, and the (inverse of) the accuracy required. Moreover, there are algorithms that are *anytime*, i.e., they can be stopped at any time to get an answer, the accuracy increases with computational effort, and can be restarted with little overhead. We use a primal-dual interior point method for real-time implementation [8].

3 APPLICATION TO NONLINEAR VIBRATION CONTROL

Consider the nonlinear system:

$$\begin{bmatrix} \dot{x} \\ \dot{y} \end{bmatrix} = \begin{bmatrix} f(\text{RPM}) & -\omega \\ \omega & f(\text{RPM}) \end{bmatrix} + \begin{bmatrix} x \\ y \end{bmatrix} (x^2 + y^2) + \begin{bmatrix} 1 \\ 0 \end{bmatrix} u,$$

where x and y are the states,

$$f(\text{RPM}) = -0.15 + 0.45 e^{-5\left(\frac{RPM}{900} - 1\right)^2}$$

and RPM is the scheduling variable. The scalar control input u is bounded above and below:

$$u_{min} \leq u \leq u_{max}$$

With $u = 0$, the system is in *normal form* and, by a Theorem of Hopf, has a Hopf bifurcation at RPM $= 900$ leading to a stable limit cycle. We are interested in the system behavior as RPM changes with time. As RPM increases from 0 to 900, due to the nature of f the (linear) damping of the system decreases and the system amplifies displacements. Beyond RPM $= 900$, the system damping increases and oscillations become smaller. This behavior imitates a particularly severe vibration problem in rotorcraft known as *aeromechanical instability*.

Aeromechanical instabilities, ground and air resonances, are major design issues and limiting factors in the development of soft-in-plane rotorcraft. Ground resonance, observed in multibladed (more than 2 blades) soft-in-plane rotorcraft, occurs as the rotor RPM is

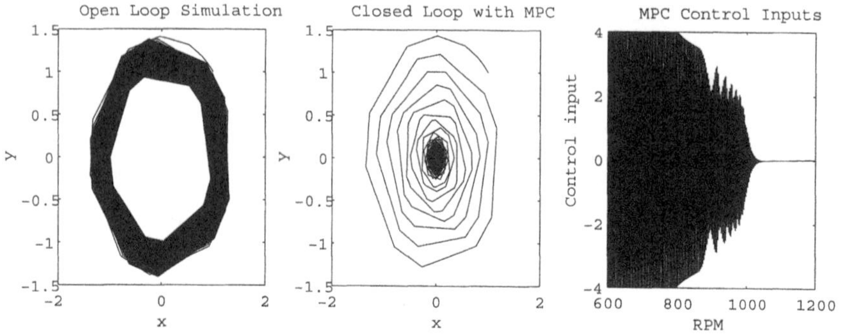

Figure 2 Open- and closed-loop behaviors of aeromechanical system as RPM changes linearly from 600 to 1200. Limit cycling in open loop (left), vibration suppression with MPC (middle), and control inputs generated by MPC (right).

changed from rest to its operational value and is manifested as large uncontrolled rotor tip displacements. Tilt rotor aircraft in airplane mode can, in addition, encounter coupled wing-rotor-pylon instabilities known as air resonance. These instabilities are parameter dependent in that, as the operating conditions change, the rotorcraft changes from stable to unstable behavior. The primary design tool currently in use is frequency placement of vibrational modes by increasing stiffness. This passive approach invariably increases structural weight and reduces aerodynamic performance. The resulting reduction in operational envelope has prompted the rotorcraft industry to vigorously pursue active control of aeromechanical instabilities. See [3, 4, 5] for more details. Typical control objectives are to maintain rotor tip displacements within safety limits (or completely cancel when there is no noise or uncertainty) subject to saturation limits on swashplate motion along any smooth scheduling variable trajectory.

To design MPC, we formulate the control problem as shown in Fig. 1. Since the objective is to make x go to zero asymptotically along smooth RPM trajectories, we introduce a desired trajectory, namely zero, and an integrator to generate tracking error and its integral. The open-loop system augmented with these elements forms the plant for MPC. The next task is to obtain an LPV approximation of the plant, which becomes the internal model used by MPC. This approximation and weights in the performance index (2) form the data read in by the MPC solver prior to closed-loop simulation. Figure 1 shows the complete closed-loop system. Finally, MPC design parameters are tuned using the closed-loop simulation.

Figure 2 shows open- and closed-loop responses. On the left is the open loop response that is typical of aeromechanical instability. Closed-loop response is shown in the middle. The effectiveness of MPC in canceling the instability is clear. The control input on the right clearly indicates how MPC is working. These simulations are for a linearly varying RPM trajectory typical of rotorcraft startup.

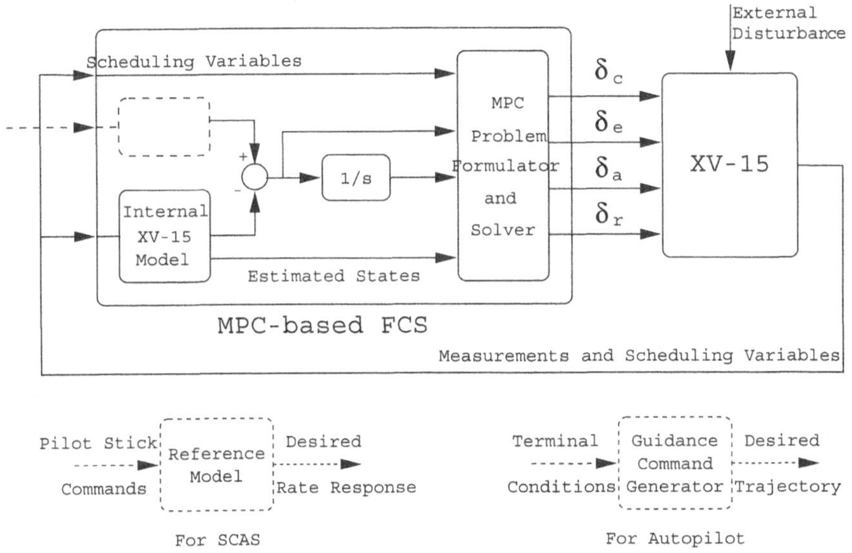

Figure 3 Block diagram representation of MPC-based FCS. Shaded block is toggled by pilot for SCAS or autopilot.

4 APPLICATION TO THE XV-15 TILT ROTOR FLIGHT CONTROL

This section presents the design, real-time implementation, and piloted simulations of an MPC-based flight control system (FCS) for the XV-15 tilt rotor. Current FCS design practices are based on linear time-invariant single-input single-output ideas and do not account for nonlinearities or time-domain constraints. We shall show that the MPC approach can accommodate frequency-domain MIL specifications as well as time-domain requirements. A high-fidelity XV-15 simulator developed at Bell Helicopter using the Generic Tilt Rotor (GTR) model is used for evaluating MPC-based FCS. XV-15/V-22 test pilots flew several missions evaluating various aspects of the design. This section concludes with their evaluation.

4.1 MPC Architectures for FCS

A block diagram of MPC-based FCS is shown in Fig. 3. It can be adapted for two cases that we shall consider by placing a reference model or a guidance command generator in the shaded block. In the first case, MPC interprets pilot stick inputs as desired rate commands and computes the optimal control surface deflections needed to track these commands. MPC can be seen as a Stability and Control Augmentation System (SCAS) in this case. The second case corresponds to an autopilot wherein the pilot can enter desired terminal conditions and MPC generates the guidance commands as well as the stick movements to follow the guidance track.

4.2 Design Procedure

The MPC design parameters for the structure in Fig. 3 are: reference model dynamics, penalties on tracking errors, integrated tracking errors and actuator rates, and horizon length. These parameters were selected as follows:

1. Selection of reference models: This involves the selection of simple low-order dynamic systems that interpret pilot stick inputs and generate a desired rate response as shown in Fig. 3. The frequency and time response of these reference models must satisfy all relevant MIL specifications. Since MPC is a time-domain technique, time-domain specifications such as g-limits and control-surface deflection limits are stated as they are and accounted for in the online optimization. Ideally (in the absence of disturbances, model mismatch, and unbounded controls), MPC design will match the desired rate responses exactly so that the pilot will in effect be flying the reference model. Our experience suggests that decoupled reference models — one for each body angle rate — is better than coupled models from the pilot's perspective. This makes the selection somewhat easier.

2. Formulation of augmented plant and LPV data generation: After selecting the reference models, we generate the tracking error and the integral of tracking error as shown in Fig. 3 by adding integrators. The augmented plant for MPC is given by the system from pilot stick inputs and control inputs to the measurements, tracking error, and its integral. The augmented plant is approximated by an LPV model that forms the system data used by MPC.

3. Selection of weighting matrices and horizon length: For the LPV system, assume that there are no state and control constraints. Design LQG controllers that give good performance by adjusting the weights in the performance index in (2). The LQG weights form the initial set of weights for MPC design. The weights and horizon lengths can be further tuned to balance closed-loop performance and computational requirements.

4.3 Maneuvers for Evaluation

Two maneuvers will be used to evaluate MPC-based FCS in terms of its ability to track pilot commands and reduce pilot workload.

Coordinated Turns: The maneuver begins in airplane mode trimmed at 150 knots. The pilot initiates turn by moving the lateral stick. MPC-based FCS interprets pilot stick inputs as rate commands and provides fast tracking with zero steady-state error and no overshoot for step inputs (we let the pilot fly in actual tests). Level 1 flying qualities requirements for this maneuver can be found in MIL-F-8785C [1].

Conversion-Decel: The maneuver involves converting XV-15 from airplane mode at 150 knots to helicopter mode at zero knots. The maneuver starts at, say, 2500 ft, and during the maneuver the aircraft descends at an acceptable rate to about 1800 ft. At the end of the maneuver, the aircraft hovers at a constant altitude. This maneuver is highly nonlinear as the aircraft dynamics undergo large changes. Controller design is also complicated by the fact that there are three modes of operation — airplane mode, conversion, and hover — that must be considered. Our objective is to design a controller that would allow the pilot to conduct this maneuver with hands off the stick. In addition, we would like to provide the pilot with the ability to regain control of the aircraft at any time. Level 1 flying qualities

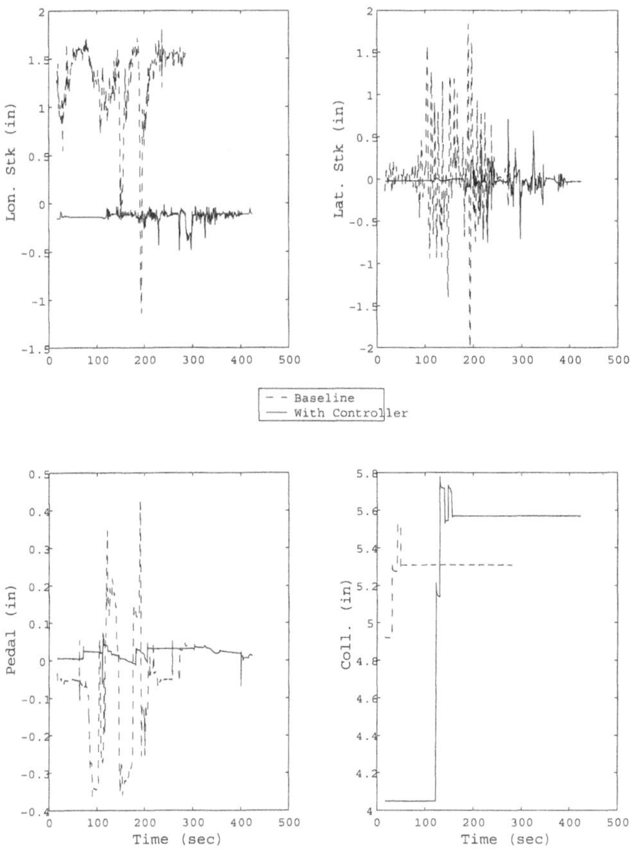

Figure 4 Piloted coordinated turn simulation: pilot stick inputs and angular rates. Mean-square value of baseline (dashed line) stick inputs are much higher than that for MPC-based FCS (solid line) indicating higher workload.

requirements for this maneuver were developed from MIL-F-8785C [1] and MIL-F-83300 [2].

4.4 Piloted Simulations

MPC-based FCS was evaluated by V-22/XV-15 test pilots at Bell Helicopter. The pilots flew simulations with and without MPC. A few initial runs with a digital SCAS were made to familiarize the pilots with the XV-15 simulator cab. During these runs, the pilots noted some differences between the GTR simulator model and actual aircraft. These include higher pitch-power coupling and more delays in the simulation than in the aircraft. For the evaluation maneuvers, the collective-power lever was released to the pilots on their request; MPC computed all four control inputs, but the computed collective stick was not fed back.

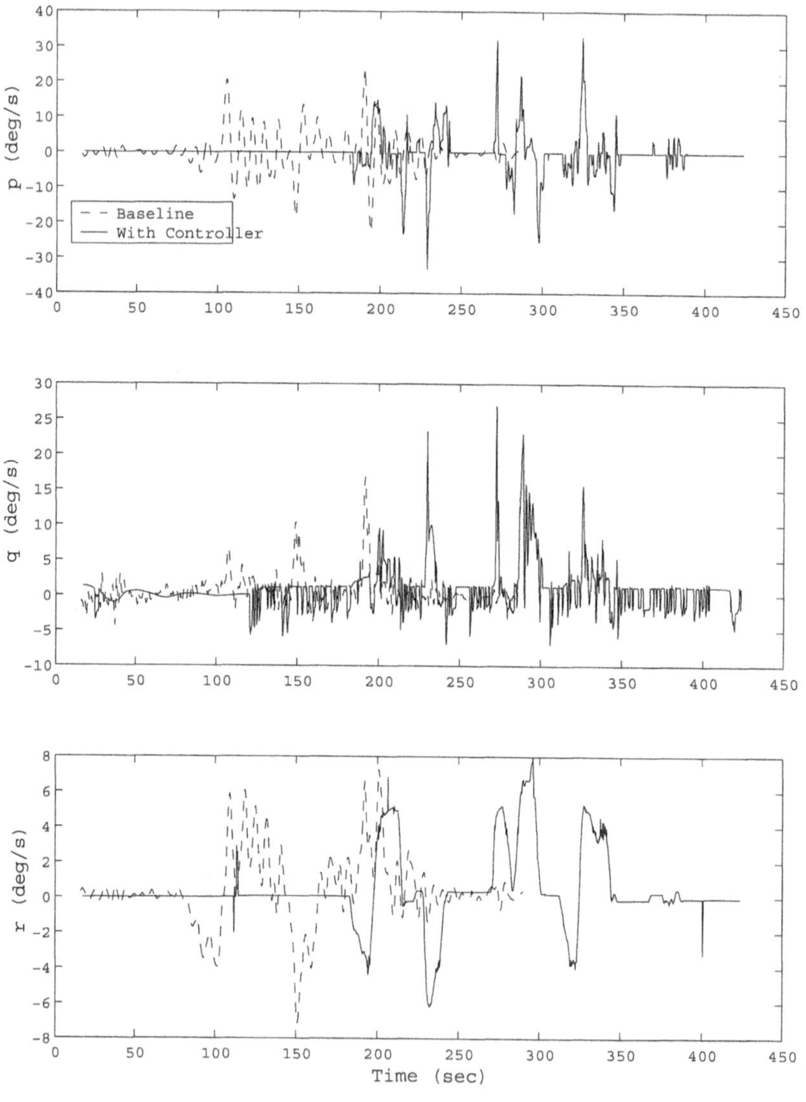

Figure 4 (Continued).

Figure 4 shows sample results from left and right turn simulations. The effectiveness of MPC in reducing pilot work load is very clear. The pilot was happy with the way MPC performed, especially in coordinating the pedal. He noted that very small/no-pedal inputs were required, the aircraft responded well to commands, and rated the controller performance as excellent.

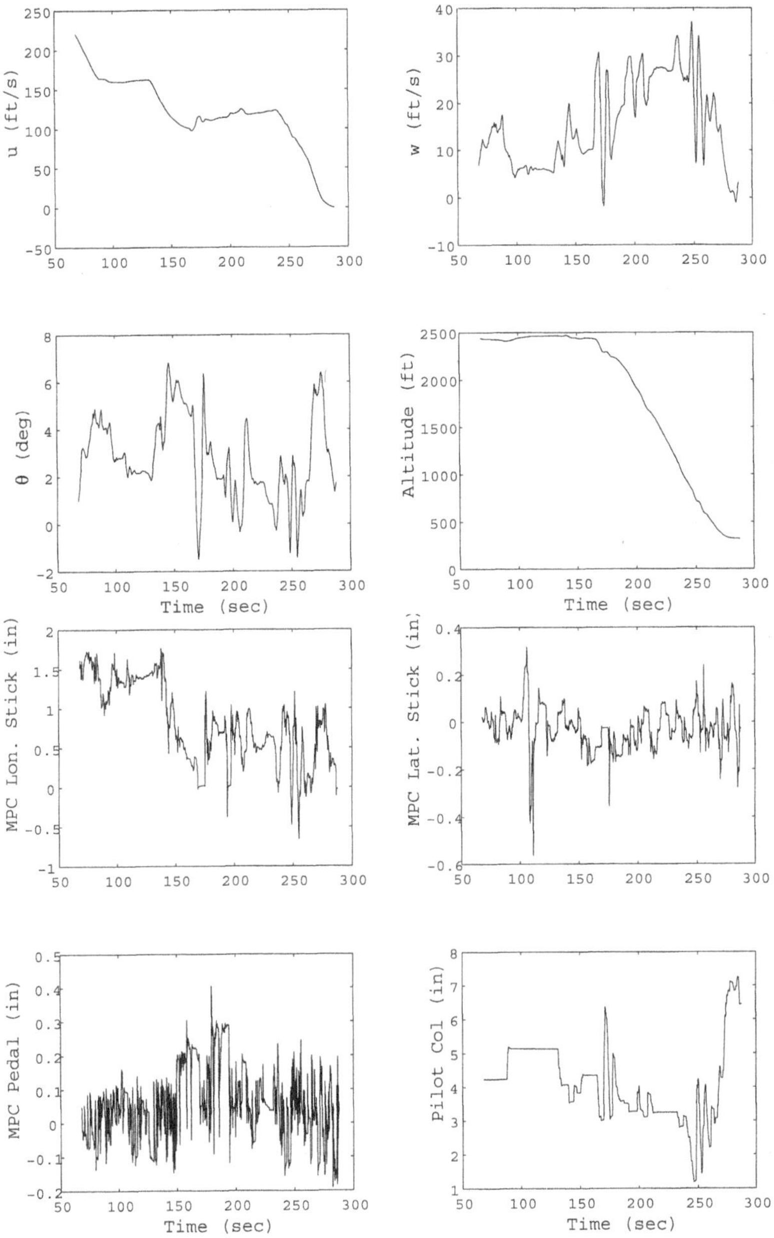

Figure 5 Piloted conversion/decel maneuver with MPC-based FCS in the loop. Note the large variations in airspeed from airplane mode to helicopter mode.

The conversion-decel maneuver began at approximately 145 knots, 2500 feet altitude and 50% power. After bringing the aircraft closer to the landing area, the pilot began converting at which time the MPC-controller was turned on. Sample results are shown in Fig. 5. The pilot noted that the pitch axis control was tight and reasonably good, although the pitch attitude excursions during the maneuver were at times about 4 degrees more than what he would have liked. Over several runs, he rated MPC performance from good to excellent.

5 CONCLUSIONS

We have described a model predictive control paradigm that uses a linear parameter-varying approximation of nonlinear systems instead of on-line linearization. This method has several advantages, the most notable of which are its quality of approximation and polynomial-time computability. These features are essential for successful flight control design. As a demonstration, MPC was implemented on a real-time XV-15 tilt rotor flight simulator. Even though the simulator math model is highly nonlinear, the implementation performed very well in piloted simulations. A coordinated turn maneuver was used to evaluate the multiaxis capability of MPC-based SCAS; while a highly nonlinear conversion maneuver was used to evaluate the autopilot capabilities of MPC. Test pilots rated MPC performance from good to excellent in both cases.

MPC-based flight control system design used several MIL-F-83300 and MIL-F-8785C requirements as well as time-domain saturation type constraints. However, a number of issues such as verification of these requirements were not considered in this study. We have identified the following areas for future research and development:

1. Further Piloted Simulations and Hardware-in-the-Loop (HIL) Testing of MPC-based FCS — During the piloted simulations, the test pilots suggested that further simulations are needed for a complete evaluation of MPC-based FCS. This is also the natural step before HIL testing. Future work should include (a) piloted simulations, (b) implementation of MPC-based FCS on a flight computer, and (c) piloted simulations with a flight computer in the loop.

2. Applications of Virtual Tilt Rotor to Predictive Control[1] — This is a nonlinear dynamical model of a tilt rotor that resides in the flight computer and is executed at rates faster than real time. A high-fidelity virtual tilt rotor model has been developed by Bell Helicopter. This concept has many applications including predictive control. Future work should consider (a) MPC implementations with the virtual tilt rotor as the internal model, and (b) computational requirements for real-time implementation of MPC-based FCS.

3. Multivariable Control of Quad Tilt Rotor (QTR) — QTR is the next generation of tilt rotors being developed by the rotorcraft industry. The QTR has more control inputs than the number of equations/states and represents significant control challenges and opportunities for reconfiguration under control failures.

[1]Proposed by Dr. Richard Bennett of Bell Helicopter.

Acknowledgments

This work was funded by NASA Ames Research Center under contract NAS2-98022. We are grateful to Mr. Stephen Jacklin of NASA Ames for support and encouragement during the project. The real-time simulations were flown at Bell Helicopter Textron Inc. by V-22/XV-15 test pilots Thomas Warren and Roy Hopkins. Their comments and insights were of great help to us in evaluating the controller as well as in identifying future research issues. Dr. Richard Bennett, David Neckels, and Dr. Mark Wasikowski of Bell Helicopter also contributed to this project. We thank them for their efforts.

REFERENCES

1. Military specification flying qualities of piloted airplanes, MIL-F-8785C, November 1980.
2. Military specification flying qualities of piloted V/STOL airplanes, MIL-F-83300, December 1970.
3. W.Y. Chan and I. Chopra, "Aeromechanical stability of hingeless helicopter rotors in forward flight," *AHS specialists meeting*, March 1991.
4. F. Gandhi and W.H. Weller, "Active aeromechanical stability augmentation using fuselage state feedback," *AHS specialists meeting*, 1997.
5. D. Hodges, "An aeromechanical stability analysis for bearingless rotor helicopters," *AHS specialists meeting*, 1978.
6. M. Kothare, V. Balakrishnan and M. Morari, "Robust constrained model predictive control using linear matrix inequalities," *Automatica*, Vol. 32, No. 10, 1996.
7. D.Q. Mayne, "Constrained model predictive control: stability and optimality," *Automatica*, Vol. 36, No. 6, 2000.
8. Yu. Nesterov and A. Nemirovskii, "Interior-point polynomial methods in convex programming," *SIAM*, Philadelphia, 1994.
9. A. Packard, "Gain scheduling via linear fractional transformations," *Systems and Control Letter*, Vol. 22, 1994.
10. W. Rugh and J. Shamma, "Research on gain scheduling," *Automatica*, Vol. 36, No. 9, 2000.
11. J. Shamma and M. Athans, "Guaranteed properties of gain scheduled control for linear parameter varying plants," *Automatica*, Vol. 27, 1991.
12. J. Shamma and J.R. Cloutier, "A linear parameter varying approach to gain scheduled missile autopilot design," *Proceedings of ACC*, 1992.
13. S. Zilberstein, "Using anytime algorithms in intelligent systems," *AI Magazine*, Fall 1996.

7

New Development of Vector Lyapunov Functions and Airplane Control Synthesis

Lyubomir T. Gruyitch
University of Technology Belfort–Montbeliard, Belfort Cedex, France

The chapter presents a conceptual development of vector Lyapunov functions by introducing (semi-) definite vector functions and related various kinds of vector Lyapunov functions. They enable us to extend the Lyapunov method without using a scalar Lyapunov function. Besides, they permit synthesis of robust control of airplanes and other nonlinear systems incorporating all mechanical systems with a variable mass and/or loads (e.g., robots, ships, spacecrafts) to guarantee a high-quality output tracking together with an appropriate Lyapunov stability of a desired motion.

1 INTRODUCTION

The concept of vector Lyapunov functions was coincidentally introduced by Bellman [1] and Matrosov [2]. It appeared very useful for stability studies of complex systems (in particular of interconnected and large-scale systems), [3]–[9]. It was used for a system aggregation, for an effective construction of a scalar Lyapunov function of the overall system and for reduction of a stability study to that of subsystems and to analysis of their interconnections. The use of a scalar Lyapunov function (with or without the comparison principle) characterized such applications of the vector Lyapunov function concept.

The concept of vector Lyapunov functions will be further developed in order to solve a basic problem of control synthesis to simultaneously satisfy stability and output tracking requirements in two different spaces. It enables synthesis of new robust stabilizing and/or output tracking control without a need to use either a scalar Lyapunov function or the comparison principle. This will be achieved in what follows for airplaines and a large class of other nonlinear dynamical systems incorporating mechanical systems with variable mass and/or subjected to actions of variable external perturbations, which are both allowed to be unpredictable and unknown. All system nonlinearities, except those resulting from the inertia matrix, can be also unknown. Various kinds of vector Lyapunov functions will be introduced and defined, for which we will introduce and also define different types of (semi-) definite vector functions. Then, a particular form of a vector Lyapunov function will be introduced. It will enable us to synthesize control that will guarantee global robust

0-415-30852-6/04/$0.00+$1.50

exponential stability of a desired motion or both its global robust asymptotic stability and global robust exponential output tracking.

2 VECTOR LYAPUNOV FUNCTION CONCEPT

2.1 Semidefinite and Definite Time-Independent Vector Functions

Let S be a subset of R^{2m} and ∂S, ClS, InS, $N(S)$ be, respectively, its boundary, closure, interior and neighborhood, $S \subset N(S)$. $N(S) - S$ designates the set difference: $N(S) - InS = \{e^1 : e^1 \in N(S), e^1 \notin InS\}$. Let t denote time, $e = [e_1 \ e_2 \ \dots \ e_m]^T$, $e^{(1)} = \frac{de}{dt}$, $e^1 = [e^T \ \ e^{(1)^T}]^T \in R^{2m}$, $e_i^1 = [e_i \ \ e_i^{(1)}]^T \in R^2$, and $e^2 = [e^{1^T} \ \ e^{(2)^T}]^T \in R^{3m}$. All vector relationships and operations hold elementwise. 0 denotes either scalar zero, or vector or matrix zero of an appropriate dimension.

Definition 1 *A vector function* $w(.) : R^{2m} \rightarrow R^p$, $p \in \{1, 2, ..., 2m\}$, $w(e^1) = [\omega_1(e^1) \ \omega_2(e^1) \ \dots \ \omega_p(e^1)]^T$, *is:*

a) ***positive {negative} semidefinite relative to a set*** S, $S \subset R^{2m}$, *if and only if (i) through (iii) hold, respectively, over some neighborhood* $N(S)$ *of the set* S*:*

(i) *w(.) is defined and continuous on* $N(S) : w(e^1) \in C[N(S)]$,

(ii) $w(e^1) \geq 0 \ \{w(e^1) \leq 0\}$, $\forall e^1 \in [N(S) - InS]$,

(iii) $w(e^1) \leq 0 \ \{w(e^1) \geq 0\}$, $\forall e^1 \in ClS$.

b) ***pairwise positive {negative} semidefinite relative to a set*** S, $S \subset R^{2m}$, *if and only if a) holds, respectively, and both (i) and (ii) are valid:*

(i) $p = m$,

(ii) $\omega_l(e^1) \equiv \omega_l(e_l^1)$, $\forall l = 1, 2, ..., m$.

c) ***elementwise positive {negative} semi-definite relative to a set*** S, $S \subset R^{2m}$, *if and only if a) holds, respectively, and both (i) and (ii) are valid:*

(i) $p = 2m$,

(ii) $\omega_l(e^1) \equiv \omega_l(e_l)$ *and* $\omega_{m+l}(e^1) \equiv \omega_{m+l}(e_l^{(1)})$, $\forall l = 1, 2, ..., m$.

The preceding properties are:

d) ***global (in the whole)*** *if and only if they hold for* $N(S) = R^{2m}$.

e) ***on a set*** A, $A \subset R^{2m}$, *if and only if* A *is a neighborhood of* S *and* $A \subseteq N_L(S)$, *where* $N_L(S)$ *is the largest neighborhood* $N(S)$ *of* S, *which obeys the corresponding conditions under a), or b), or c).*

*The expression "**relative to a set*** S, $S \subset R^{2m}$*," should be omitted if and only if the set* S *is singleton* $O : S = O = \{e^1 : e^1 = 0\}$▲

Let: $i, j, k, l, m, n, p, q, r \in \{1, 2, ...\}$, $\min\{i_l, k_l, j_l, n_l\} \geq 1$, $\max\{i_l, k_l, j_l, n_l\} \leq m$, $m_{ll} = 1$, $\forall l \in \{1, 2, ..., m\}$, $i_1 = j_1 = 1$, $k_p = n_q = m$ and $\sum_{l=1}^{l=p} m_{i_l k_l} = m$ if $p \leq m$. Otherwise, the sum ends with $l = m$. Let $e_{ik} = (e_1 \dots e_{i-1} \ \widetilde{e}_i \ \widetilde{e}_{i+1} \dots \widetilde{e}_k \ e_{k+1} \dots e_m)^T$, $S_{ik} = \{\widetilde{e}_{ik} : \widetilde{e}_{ik} = (\widetilde{e}_i \ \widetilde{e}_{i+1} \dots \widetilde{e}_{k-1} \ \widetilde{e}_k)^T \in R^{m_{ik}} \Longrightarrow \exists e^1 = (e_{ik}^T \ \ e^{(1)^T})^T \in S\}$, $S_{ik}^{(1)} = \{\widetilde{e}_{ik}^{(1)} : \widetilde{e}_{ik}^{(1)} = (\widetilde{e}_i^{(1)} \ \widetilde{e}_{i+1}^{(1)} \dots \widetilde{e}_{k-1}^{(1)} \ \widetilde{e}_k^{(1)})^T \in R^{m_{ik}} \Longrightarrow \exists e^1 = (e^T \ \ e_{ik}^{(1)^T})^T \in S\}$ and $S_{ik}^1 = \{\widetilde{e}_{ik}^1 : \widetilde{e}_{ik}^1 = (\widetilde{e}_{ik}^T \ \ \widetilde{e}_{ik}^{(1)^T})^T \in R^{2m_{ik}} \Longrightarrow \exists e^1 = (e_{ik}^T \ \ e_{ik}^{(1)^T})^T \in S\}$. Let $S_{(.)} \in \{S, S_b, S_c, S_d\}$.

Definition 2 *A vector function* $w(.) : R^{2m} \rightarrow R^p$, $p \in \{1, 2, ..., 2m\}$, $w(e^1) = [\omega_1(e^1)\ \omega_2(e^1)\ ...\ \omega_p(e^1)]^T$, *is:*

a) ***positive {negative} definite relative to a set*** S, $S \subset R^{2m}$, *if and only if (i) through (iv) hold, respectively, over some neighborhood* $N(S)$ *of* S:

(i) *w(.) is defined and continuous on* $N(S) : w(e^1) \in C[N(S)]$,

(ii) $w(e^1) \geq 0\ \{w(e^1) \leq 0\}$, $\forall e^1 \in [N(S) - InS]$,

(iii) $w(e^1) = 0$ *for* $e^1 \in [N(S) - InS]$ *if and only if* $e^1 \in \partial S$,

(iv) $w(e^1) \leq 0\ \{w(e^1) \geq 0\}$, $\forall e^1 \in ClS$.

b) ***positive {negative} definite relative to a set*** S_b, $S_b = S^1_{i_1k_1} \text{x} S^1_{i_2k_2} \text{x} ... \text{x} S^1_{i_pk_p} \subset R^{2m}$, *if and only if (i), (ii) and (iv) of a) hold for* $S = S_b$ *and* $\omega_l(e^1) = 0$ *for* $e^1 = (e^T_{i_lk_l}\ e^{(1)^T}_{i_lk_l})^T \in [N(S_b) - InS_b]$ *if and only if* $\widetilde{e}^1_{i_lk_l} \in \partial S^1_{i_lk_l}$, $\forall l = 1, 2, ..., p \leq m$.

c) ***pairwise positive {negative} definite relative to a set*** S_c, $S_c = S^1_{11} \text{x} S^1_{22} \text{x} ... \text{x} S^1_{mm} \subset R^{2m}$, *if and only if b) holds for* $S_b = S_c$,

(i) $p = m$ *and* $i_l = k_l = l, \forall l = 1, 2, ..., m$, *and*

(ii) $\omega_l(e^1) \equiv \omega_l(e^1_l)$, $\forall l = 1, 2, ..., m$.

d) ***elementwise positive {negative} definite relative to a set*** S_d, $S_d = S_{11} \text{x} S_{22} \text{x} ... \text{x} S_{mm} \text{x} S^{(1)}_{11} \text{x} S^{(1)}_{22} \text{x} ... \text{x} S^{(1)}_{mm} \subset R^{2m}$, *if and only if (i), (ii) and (iv) of a) hold for* $S = S_d$,

(i) $p = 2m$,

(ii) $\omega_l(e^1) \equiv \omega_l(e_l)$ *and* $\omega_{m+l}(e^1) \equiv \omega_{m+l}(e^{(1)}_l)$, $\forall l = 1, 2, ..., m$,

(iii) $\omega_l(\widetilde{e}_l) = 0$ *and* $\omega_{m+l}(\widetilde{e}^{(1)}_l) = 0$ *for* $e^1 = (e^T_{ll}\ e^{(1)^T}_{ll})^T \in [N(S_d) - InS_d]$ *if and only if, respectively,* $\widetilde{e}_l \in \partial S_{ll}$ *and* $\widetilde{e}^{(1)}_l \in \partial S^{(1)}_{ll}$, $\forall l = 1, 2, ..., m$.

The preceding properties are:

e) ***global (in the whole)*** *if and only if they hold for* $N(S_{(.)}) = R^{2m}$.

f) ***on a set*** A, $A \subset R^{2m}$, *if and only if the corresponding conditions hold for* $N(S_{(.)}) = A$.

*The expression "**relative to a set** $S_{(.)}$, $S_{(.)} \subset R^{2m}$," should be omitted if and only if the set* $S_{(.)}$ *is singleton* $O : S_{(.)} = O = \{e^1 : e^1 = 0\}$▲

The preceding definitions are consistent with Lyapunov's original concept of scalar (semi-) definite functions [10] and with the concept of definite matrix functions introduced in [11]. Notice that a) of Definition 2 enables synthesis of control to simultaneously stabilize/track control and optimal control relative to multiple (p) different optimality criteria.

2.2 Semidefinite and Definite Time-Dependent Vector Functions

Definition 3 *A vector function* $v(.) : R\text{x}R^{2m} \rightarrow R^p$, $p \in \{1, 2, ..., 2m\}$, $v(t, e^1) = [v_1(t, e^1)\ v_2(t, e^1)\ ...\ v_p(t, e^1)]^T$, *is:*

a) ***positive {negative} semidefinite relative to a set*** S, $S \subset R^{2m}$, *if and only if (i) through (iii) hold, respectively, over some neighborhood* $N(S)$ *of the set* S:

(i) $v(.)$ *is defined and continuous on* $R\text{x}N(S) : v(t, e^1) \in C[R\text{x}N(S)]$,

(ii) $v(t, e^1) \geq 0\ \{v(t, e^1) \leq 0\}$, $\forall (t, e^1) \in R\text{x}[N(S) - InS]$,

(iii) $v(t, e^1) \leq 0\ \{v(t, e^1) \geq 0\}$, $\forall (t, e^1) \in R\text{x}ClS$.

b) ***pairwise positive {negative} semidefinite relative to a set*** S, $S \subset R^{2m}$, *if and only if a) holds and both (i) and (ii) are valid:*

(i) $p = m$,

(ii) $v_l(t, e^1) \equiv v_l(t, e_l^1), \forall l = 1, 2, ..., m.$

c) ***elementwise positive*** $\{$***negative***$\}$ ***semidefinite relative to a set*** S, $S \subset R^{2m}$, *if and only if a) holds and both (i) and (ii) are valid:*

(i) $p = 2m,$

(ii) $v_l(t, e^1) \equiv v_l(t, e_l)$ *and* $v_{m+l}(t, e^1) \equiv v_{m+l}(t, e_l^{(1)}), \forall l = 1, 2, ..., m.$

The preceding properties are:

d) ***global (in the whole)*** *if and only if they hold for* $N(S) = R^{2m}$.

e) ***on a set*** A, $A \subset R^{2m}$, *if and only if the corresponding conditions hold for* $N(S) = A$.

*The expression "**relative to a set** S, $S \subset R^{2m}$," should be omitted if and only if the set* S *is singleton* $O : S = O = \{e^1 : e^1 = 0\}$▲

Definition 4 *A vector function* $v(.) : RxR^{2m} \to R^p$, $p \in \{1, 2, ..., 2m\}$, $v(t, e^1) = [v_1(t, e^1)\ v_2(t, e^1)\ ...\ v_p(t, e^1)]^T$, *is:*

a) ***positive*** $\{$***negative***$\}$ ***definite relative to a set*** S, $S \subset R^{2m}$, *if and only if (i) and (ii) hold over some neighborhood* $N(S)$ *of the set* S*:*

(i) *v(.) is defined and continuous on* $RxN(S) : v(t, e^1) \in C[RxN(S)]$,

(ii) *there is a time-independent positive definite vector function* $w_1(.)$ *relative to the set* S *on* $N(S)$ *such that 1) and 2) hold, respectively:*

1) $v(t, e^1) \geq w_1(e^1)\ \{w(t, e^1) \leq -w_1(e^1)\}, \forall (t, e^1) \in Rx[N(S) - InS]$,

2) $v(t, e^1) \leq -w_1(e^1)\ \{v(t, e^1) \geq w_1(e^1)\}, \forall (t, e^1) \in RxClS$.

b) ***positive*** $\{$***negative***$\}$ ***definite relative to a set*** S_b, $S_b = S^1_{i_1k_1}xS^1_{i_2k_2}x...x\, S^1_{i_pk_p} \subset R^{2m}$, *if and only if a) is valid, respectively, for* $S = S_b$.

c) ***pairwise positive*** $\{$***negative***$\}$ ***definite relative to a set*** S_c, $S_c = S^1_{11}xS^1_{22}x...xS^1_{mm} \subset R^{2m}$, *if and only if b) holds, respectively, for* $S_b = S_c$, $w_1(.)$ *is pairwise positive* $\{$*negative*$\}$ *definite relative to the set* S_c *on* $N(S_c)$,

(i) $p = m$ *and* $i_l = k_l = l, \forall l = 1, 2, ..., m$, *and*

(ii) $v_l(t, e^1) \equiv v_l(t, e_l^1), \forall l = 1, 2, ..., m.$

d) ***elementwise positive*** $\{$***negative***$\}$ ***definite relative to a set*** S_d, $S_d = S_{11}xS_{22}x...xS_{mm}xS^{(1)}_{11}xS^{(1)}_{22}x...xS^{(1)}_{mm} \subset R^{2m}$, *if and only if a) holds, respectively, for* $S = S_d$, $w_1(.)$ *is elementwise positive* $\{$*negative*$\}$ *definite relative to the set* S_d *on* $N(S_d)$, *and both (i) and (ii) are valid:*

(i) $p = 2m,$

(ii) $v_l(t, e^1) \equiv v_l(t, e_i)$ *and* $v_{m+l}(t, e^1) \equiv v_{m+l}(t, e_l^{(1)}), \forall l = 1, 2, ..., m.$

The preceding properties are:

e) ***global (in the whole)*** *if and only if they hold for* $N(S_{(.)}) = R^{2m}$.

f) ***on a set*** A, $A \subset R^{2m}$, *if and only if the corresponding conditions hold for* $N(S_{(.)}) = A$.

*The expression "**relative to a set** $S_{(.)}$, $S_{(.)} \subset R^{2m}$," should be omitted if and only if the set* $S_{(.)}$ *is singleton* $O : S_{(.)} = O = \{e^1 : e^1 = 0\}$▲

The preceding definitions are consistent with Lyapunov's original concept of scalar (semi-) definite functions [10]. Definition 4 is compatible with the definition of definite matrix functions introduced in [11].

Definition 5 *A vector function* $v(.) : RxR^{2m} \to R^p$, $p \in \{1, 2, ..., 2m\}$, $v(t, e^1) = [v_1(t, e^1)\ v_2(t, e^1)\ ...\ v_p(t, e^1)]^T$, *is:*

*a) **decrescent relative to a set** S, $S \subset R^{2m}$, if and only if (i) and (ii) hold over some neighborhood $N(S)$ of the set S:*

(i) v(.) is defined and continuous on $RxN(S) : v(t, e^1) \in C[RxN(S)]$,

(ii) there is a time-independent positive definite vector function $w_2(.)$ with respect to the set S on $N(S)$ such that:

$v(t, e^1) \leq w_2(e^1), \forall (t, e^1) \in Rx[N(S) - InS]$.

*b) **decrescent relative to a set** S_b, $S_b = S^1_{i_1k_1} x S^1_{i_2k_2} x ... x S^1_{i_pk_p} \subset R^{2m}$, if and only if a) is valid for $S = S_b$.*

*c) **pairwise decrescent relative to a set** S_c, $S_c = S^1_{11} x S^1_{22} x ... x S^1_{mm} \subset R^{2m}$, if and only if b) holds for $S_b = S_c$, $w_2(.)$ is a pairwise positive definite vector function with respect to the set S_c on $N(S_c)$,*

(i) $p = m$ and $i_l = k_l = l, \forall l = 1, 2, ..., m$, and

(ii) $v_l(t, e^1) \equiv v_l(t, e^1_l), \forall l = 1, 2, ..., m$.

*d) **elementwise decrescent relative to a set** S_d, $S_d = S_{11} x S_{22} x ... x S_{mm} x\, S^{(1)}_{11} x S^{(1)}_{22} x ... x S^{(1)}_{mm} \subset R^{2m}$, if and only if a) holds for $S = S_d$, $w_2(.)$ is an elementwise positive definite vector function with respect to the set S_d on $N(S_d)$,*

(i) $p = 2m$,

(ii) $v_l(t, e^1) \equiv v_l(t, e_l)$ and $v_{m+l}(t, e^1) \equiv v_{m+l}(t, e^{(1)}_l), \forall l = 1, 2, ..., m$.

The preceding properties are:

*e) **global (in the whole)** if and only if they hold for $N(S_{(.)}) = R^{2m}$.*

*f) **on a set** A, $A \subset R^{2m}$, if and only if the corresponding conditions hold for $N(S_{(.)}) = A$.*

*The expression "**relative to a set** $S_{(.)}$, $S_{(.)} \subset R^{2m}$," should be omitted if and only if the set $S_{(.)}$ is singleton $O : S_{(.)} = O = \{e^1 : e^1 = 0\}$▲*

This definition is consistent with the definition of a scalar decrescent function [12].

2.3 Kinds of Vector Lyapunov Functions

The initial moment $t_0 = 0$ is accepted and fixed. It will be omitted whenever its omission cannot make a confusion: $q^1(.; 0, q^1_0; d) \equiv q^1(.; q^1_0; d) \equiv q^1(.)$ is a system motion, $e^1(.; 0, e^1_0; d) \equiv e^1(.; e^1_0; d) \equiv e^1(.)$ is a motion error from a desired motion $q^1_d(.)$: $e^1(.) = q^1_d(.) - q^1(.)$, $y^1(.) \equiv y^1(.; y^1_0; d)$ is a system output response together with its first derivative, the error of which relative to $y^1_d(.)$ (a desired output response with its first derivative) is $\varepsilon^1(.; \varepsilon^1_0; d) \equiv \varepsilon^1(.) \equiv y^1_d(.) - y^1(.)$. We will use the right-hand upper Dini derivative $D^+v(.)$ of a vector function $v(.)$ along a system motion error $e^1(.; t_0, e^1_0; d)$,

$$D^+v(t, e^1) = \limsup \left\{ \frac{v[t + \theta, e^1(t + \theta; t, e^1; d)] - v(t, e^1)}{\theta} : \theta \to 0^+ \right\},$$

or along an output error response $\varepsilon^1(.; t_0, \varepsilon^1_0; d)$. If the function $v(.)$ is differentiable then $D^+v(.)$ becomes the Eulerian derivative $v^{(1)}(.)$ of $v(.)$ along $e^1(.; e^1_0; d)$ or along $\varepsilon^1(.; \varepsilon^1_0; d)$, respectively.

Definition 6 *A vector function $v(.) : RxR^{2m} \to R^p$, $p \in \{1, 2, ..., 2m\}$, $v(t, e^1) = [v_1(t, e^1)\ v_2(t, e^1)\ ...\ v_p(t, e^1)]^T$, is:*

a) ***a vector Lyapunov function of a system relative to a set*** S, $S \subset R^{2m}$, *if and only if (i) and (ii) hold:*

(i) it is positive definite relative to the set S,

(ii) its Dini derivative $D^{+}v(t, e^{1})$ *is negative semidefinite with respect to the set* S.

b) ***a vector Lyapunov function of the system relative to a set*** S_b, $S_b = S^1_{i_1k_1} \mathrm{x} S^1_{i_2k_2} \mathrm{x} ... \mathrm{x} S^1_{i_pk_p} \subset R^{2m}$, *if and only if both (i) and (ii) hold:*

(i) it is positive definite relative to the set S_b,

(ii) its Dini derivative $D^{+}v(t, e^{1})$ *is negative semidefinite with respect to the set* S_b.

c) ***a pairwise vector Lyapunov function of the system relative to a set*** S_c, $S_c = S^1_{11} \mathrm{x} S^1_{22} \mathrm{x} ... \mathrm{x} S^1_{mm} \subset R^{2m}$, *if and only if both (i) and (ii) hold:*

(i) it is pairwise positive definite relative to the set S_c,

(ii) its Dini derivative $D^{+}v(t, e^{1})$ *is pairwise negative semidefinite with respect to the set* S_c.

d) ***elementwise vector Lyapunov function of the system relative to a set*** S_d, $S_d = S_{11} \mathrm{x} S_{22} \mathrm{x} ... \mathrm{x} S_{mm} \mathrm{x} S^{(1)}_{11} \mathrm{x} S^{(1)}_{22} \mathrm{x} ... \mathrm{x} S^{(1)}_{mm} \subset R^{2m}$, *if and only if both (i) and (ii) hold:*

(i) it is elementwise positive definite relative to the set S_d,

(ii) its Dini derivative $D^{+}v(t, e^{1})$ *is elementwise negative semi-definite with respect to the set* S_d.

*The expression "**relative to a set** $S_{(.)}$, $S_{(.)} \subset R^{2m}$," should be omitted if and only if the set* $S_{(.)}$ *is singleton* O : $S_{(.)} = O = \{e^1 : e^1 = 0\}$▲

This definition is consistent with the notion of a scalar Lyapunov function and with the notions of vector Lyapunov functions by Bellman [1] and Matrosov [2]. However, it differs essentially from the definitions of vector Lyapunov functions in [5, pp. 34 and 35] and [9, p. 296]. The reasons are explained in [11].

3 SYSTEM DESCRIPTION

Airplanes and other mechanical systems (autonomous vehicles, robots, ships, spacecrafts) with a constant or variable mass and/or subjected to time-varying loads can be described by a 2D mathematical model (1), (2), [13]–[19]:

$$A(t, q)q^{(2)}(t) + h[t, d(t), q^1(t)] = B[t, q(t)]b[u(t)], \tag{1}$$

$$y(t) = g[q(t)]. \tag{2}$$

The vector $q \in R^m$, the inertia matrix function $A(.) : R \mathrm{x} R^m \to R^{m \mathrm{x} m}$ and the nonlinear vector function $h(.) : R \mathrm{x} R^d \mathrm{x} R^{2m} \to R^m$ describe system internal dynamics. The function $h(.)$ incorporates the terms describing all system nonlinearites, except those of the inertia matrix, as well as the terms expressing an influence of variations of both external perturbation vector function $d(.) : R \to R^d$ and the system mass. The perturbation vector function $d(.)$ belongs to a functional family $\mathbf{S}_d$ of all permitted $d(.)$. The matrix function $B(.) : R \mathrm{x} R^m \to R^{m \mathrm{x} r}$, the nonlinear vector function $b(.) : R^r \to R^r$ and the control vector $u \in R^r$. The system output vector function $g(.) : R^m \to R^N$ and the system output vector $y \in R^N$. The function $g(.)$ is well defined, known, and global differentiable. Its Jacobian is denoted by $J_g(.) = [\partial g_i(.)/\partial q_j] : R^m \to R^{N \mathrm{x} m}$. Let $R_+ = [0, \infty[$.

Condition 7 *Systems (1) and (2) obey the following:*

a) The inertia matrix $A(t,q)$ *is well defined, known, and* $\det A(t,q) \neq 0$ *for all* $(t,q) \in R\text{x}R^m$.

b) The elementwise absolute vector value of the nonlinear vector function $h(.)$ *satisfies the next inequality:*

$$\left|h[t,d(t),q^1]\right| \leq h_M(t) + H(t,q^1)\left|q^1\right|, \forall[t,d(.),q^1] \in R\text{x}\mathbf{S}_d\text{x}R^{2m},$$

for some elementwise nonnegative well defined and known both the vector function $h_M(.) : R \rightarrow R_+^m$ *and the matrix function* $H(.) : R\text{x}R^{2m} \rightarrow R_+^{m\text{x}2m}$.

c) The matrix $B(t,q)$ *is well defined, known and rank* $B(t,q) = m$ *for all* $(t,q) \in R\text{x}R^m$.

d) The vector nonlinearity $b(u)$ *is well defined and known for all* $u \in R^r$. *It has the same sign elementwise as the control vector u:* $sign[b(u)] \equiv sign(u)$. *Its inverse function* $b^I(.) : R^r \rightarrow R^r$ *is well defined and known for all* $b \in R^r$ *so that* $b^I\left[b(u)\right] \equiv u$.

e) For a given desired output response $y_d(.)$ *the corresponding desired motion* $q_d^1(.) = q_d^1[.;y_d^1(.);d(.);u_N(.)]$, $q_d^1(.) = [q_d^T(.) \quad q_d^{(1)^T}(.)]^T$, *is well defined and known, where for some* $u(.) = u_N(.)$ *called nominal control:*

$$g[q_d(t)] \equiv y_d(t),\ A[t,q_d(t)]q_d^{(2)}(t) + h[t,d(t),q_d^1(t)] \equiv B[t,q_d(t)]b[u_N(t)].$$

4 CONTROL SYNTHESIS: EXPONENTIAL STABILITY

4.1 Sliding Set

Sliding control attracted significant interest and achieved effective applications due to its advantage to guarantee a desired quality of system motions in the sliding regime independently of the system internal dynamics [20]–[23]. Let S_{ss} be *a sliding set* in the e^1-space R^{2m}:

$$S_{ss} = \left\{e^1 : e^{(1)} = -\Lambda e\right\},\ \Lambda = diag\{\lambda_1\ \lambda_2\ ...\ \lambda_m\} \in R_+^{m\text{x}m}, \tag{3}$$

where Λ is a positive diagonal design matrix. Notice that $S_{ss} = S_{ss11}^1\text{x}S_{ss22}^1\text{x}\ ...\ \text{x}S_{ssmm}^1$, $S_{ssii}^1 = \{e_i^1 : e_i^{(1)} = -\lambda_i e_i\}, \forall i = 1,2,...,m$. The sliding set S_{ss} determines exponential convergence of $e^1(t)$ to $e^1 = 0$, which means that the real motion $q^1(.)$ of system (1), (2) converges exponentially to its desired motion $q_d^1(.)$, both independently of the system internal dynamics. If an initial error vector e_0^1 is not in the sliding set, then the instantaneous error vector $e^1(t;e_0^1;d)$ should reach pairwise the sliding set at the latest at the vector moment $\tau_{RSS}(e_0^1) = [\tau_{RSS1}(e_{10}^1)\ \ \tau_{RSS2}(e_{20}^1)\ \ ...\ \ \tau_{RSSm}(e_{m0}^1)]^T$: $e_i^1(t;e_{i0}^1;d) \in S_{ssii}^1$, $\forall t \in [\tau_{RSSi}(e_{i0}^1),\infty[, \forall i = 1,2,...,m$. Let $\mathbf{1} = (1\ 1\ ...\ 1)^T \in R^m$.

4.2 Stabilizing Control Problem Statement

Problem 8 *Synthesize control* $u(.)$ *so as to ensure that both (i) and (ii) hold for every* $d(.) \in \mathbf{S}_d$*:*

a) the sliding set S_{ss}, *(3), is positive invariant with respect to* $e^1(t)$:

$$if\ e^1(\tau;e_0^1;d) \in S_{ss}\ at\ \tau \in R_+,\ then\ e^1(t;e_0^1;d) \in S_{ss},\ \forall t \in [\tau,\infty[, \tag{4}$$

b) the sliding set S_{ss}, (3), is globally pairwise exponentially stable with a finite vector reachability time $\tau_{RSS}(e_0^1)$, which means that for every $e_0^1 \in R^{2m}$, $e^1(t;e_0^1;d)$ obeys (5):

$$\left|e^{(1)}(t;e_0^1;d)+\Lambda e(t;e_0^1;d)\right| \leq Ke^{-Lt}\left|e_0^{(1)}+\Lambda e_0\right|,\ \forall t\in R_+, \tag{5}$$

for some positive diagonal design mxm matrices K and L, and

$$e^1(t;e_0^1;d)\in S_{ss}, \forall t\mathbf{1}\in[\tau_{RSS}(e_0^1),\infty\mathbf{1}[. \tag{6}$$

4.3 Stabilizing Control Problem Solution

We will use the following notation in order to ensure a requested finite vector reachability time of the sliding set S_{ss}, (3), and its exponential stability: $S(z) = diag\{sign(z_1)\ sign(z_2)\ ...\ sign(z_m)\}$ and $e^{-Lt} = diag\{e^{-\alpha_1 t}\ e^{-\alpha_2 t}\ ...\ e^{-\alpha_m t}\}$ for $L = diag\{\alpha_1\ \alpha_2\ ...\ \alpha_m\}$.

Theorem 9 *If Condition 7 is valid, if T is a positive diagonal matrix such that*

$$\tau_{RSS}(e_0^1) = T\left|e_0^{(1)}+\Lambda e_0\right|,\ \forall e_0^1\in R^{2m},\ T\in R_+^{mxm}, \tag{7}$$

and if the control vector function $u(.)$ is in the form (8),

$$u(t,e^1,q^1,q_d^{(2)}) = b^I\left\{\begin{array}{c} B^T(t,q)\left[B(t,q)B^T(t,q)\right]^{-1}A(t,q)\bullet \\ \left\{\begin{array}{c} S[z(e^1)]\left|A^{-1}(t,q)\right|\left[h_M(t)+H(t,q^1)\left|q^1\right|\right] \\ +q_d^{(2)}(t)+\Lambda e+Lz(e^1)+T^{-1}S[z(e^1)]\mathbf{1} \end{array}\right\}\end{array}\right\}, \tag{8}$$

then:

a) the sliding set S_{ss}, (3), is positive invariant with respect to $e^1(t;e_0^1;d)$ for all $d(.)\in \mathbf{S}_d$,

b) the sliding set S_{ss} is globally pairwise exponentially stable with the finite vector reachability time $\tau_{RSS}(e_0^1)$, (7), for every $d(.)\in\mathbf{S}_d$,

c) the motion error $e^1(.;e_0^1;d)$ obeys (9) for every $[d(.),e_0^1]\in \mathbf{S}_d xR^{2m}$:

$$\left|e^{(1)}(t)+\Lambda e(t)\right| \leq \min\left\{\begin{array}{c} e^{-Lt}\left|e_0^{(1)}+\Lambda e_0\right|,\ t\in[0,\infty[, \\ \left\{\begin{array}{c} \left|e_0^{(1)}+\Lambda e_0\right| - T^{-1}t\mathbf{1},\ \forall t\mathbf{1}\in\left[0,\tau_{RSS}(e_0^1)\right], \\ 0,\ \forall t\mathbf{1}\in\left[\tau_{RSS}(e_0^1),\infty\mathbf{1}\right[\end{array}\right\}\end{array}\right\}. \tag{9}$$

Proof. Let $[d(.),e^1]\in \mathbf{S}_d xR^{2m}$ be arbitrary. Let Condition 7, the conditions of the theorem statement, (10) and (11) hold:

$$z(e^1) = e^{(1)}+\Lambda e = [z_1(e_1^1)\ z_2(e_2^1)\ ...\ z_m(e_m^1)]^T,$$

$$z_i(e_i^1) = e_i^{(1)}+\lambda_i e_i,\ \forall i=1,2,...,m, \tag{10}$$

$$Z(e^1) = diag\left\{z_1(e_1^1)\ z_2(e_2^1)\ ...\ z_m(e_m^1)\right\}. \tag{11}$$

We defined $z(.)$ and $Z(.)$ in order to introduce a vector function $v(.)$, (12),

$$v(z) = (1/2)Zz. \tag{12}$$

This function is global differentiable and elementwise global positive definite (and decrescent) on the z-space R^m. If it is considered on the e^1-space R^{2m} in view of (10) and (11):

$$v[z(e^1)] = (1/2)Z(e^1)z(e^1), \tag{13}$$

then it is also global differentiable, but pairwise global positive definite (and decrescent) with respect to the sliding set S_{ss}, $S_{ss} = S^1_{ss11} \text{x} S^1_{ss22} \text{x} \ldots \text{x} S^1_{ssmm}$, (3). Its Dini derivative $D^+v(.)$ becomes its Eulerian derivative $v^{(1)}(.)$:

$$v^{(1)}[z^1(e^2)] = Z(e^1)z^{(1)}(e^1) = Z(e^1)\left[e^{(2)} + \Lambda e^{(1)}\right], \forall e^2 \in R^{3m}. \tag{14}$$

Thus, (10) and (11) show that $v^{(1)}[z^1(e^2)] = 0$ on the sliding set S_{ss}. At first we eliminate $e^{(2)}$ from (14) in view of $e^{(2)} = q_d^{(2)} - q^{(2)}$. Then, nonsingularity of $A(.)$, a) through d) of Condition 7, (1) and (8) imply:

$$v^{(1)}[z^1(e^2)] \leq -LZ(e^1)z(e^1)] - T^{-1}\left|z(e^1)\right| = -2Lv[z(e^1)] - T^{-1}\sqrt{2v[z(e^1)]}.$$

Hence,

$$v^{(1)}[z^1(e^2)] \leq -\min\left\langle 2Lv[z(e^1)], T^{-1}\sqrt{2v[z(e^1)]}\right\rangle, \forall e^2 \in R^{3m}. \tag{15}$$

This shows that $v^{(1)}[z^1(e^2)]$ is majorized by a global negative definite vector function on R^{2m}, which, together with $v^{(1)}[z^1(e^2)] = 0$ on the sliding set S_{ss} and global positive definiteness of $v(.)$ with respect to the sliding set S_{ss}, proves its global asymptotic stability for every $d(.) \in \mathbf{S}_d$. Hence, the sliding set is also positive invariant relative to the motion error of system (1), (2) for every $d(.) \in \mathbf{S}_d$. This proves the statement under a). We integrate (15) and after that we use (10) and (13). The result is the following:

$$\left|e^{(1)}(t) + \Lambda e(t)\right|$$
$$\leq \min\left\{\begin{array}{c} e^{-Lt}\left|e_0^{(1)} + \Lambda e_0\right|, \ \forall t \in R_+, \\ \left\{\begin{array}{c} \left|e_0^{(1)} + \Lambda e_0\right| - T^{-1}t\mathbf{1}, \ \forall t\mathbf{1} \in \left[0, T\left|e_0^{(1)} + \Lambda e_0\right|\right], \\ 0, \ \forall t\mathbf{1} \in [T\left|e_0^{(1)} + \Lambda e_0\right|, \infty\mathbf{1}[\end{array}\right\} \end{array}\right\}.$$

This and the definition of the sliding set S_{ss}, (3), prove the statements under b) and c) ■

Global finite time reachability and positive invariance of the sliding set S_{ss}, and (3) guarantee that the behavior of controlled system (1), (2) does not depend on its own internal dynamics over the vector time interval $[\tau_{RSS}(e_0^1), \infty\mathbf{1}[$.

5 CONTROL SYNTHESIS: OUTPUT TRACKING AND ASYMPTOTIC STABILITY

5.1 Tracking and Stabilizing Control Problem Statement

Let $S_{ss\epsilon}$ be *a sliding set* in the output space R^N, which is defined by (16),

$$S_{ss\varepsilon} = \left\{\varepsilon^1 : \varepsilon^{(1)} = -\Gamma\varepsilon\right\}, \ \Gamma = \{\gamma_1 \ \gamma_2 \ \ldots \ \gamma_N\} \in R_+^{N\text{x}N}, \tag{16}$$

where Γ is a positive diagonal design matrix. Notice that $S_{ss\varepsilon} = S^1_{ss\varepsilon 11} \mathrm{x} S^1_{ss\varepsilon 22} \mathrm{x} \ldots \mathrm{x} S^1_{ss\varepsilon NN}$, $S^1_{ssii} = \{\varepsilon_i^1 : \varepsilon_i^{(1)} = -\gamma_i \varepsilon_i\}$, $\forall i = 1, 2, ..., N$. The sliding set $S_{ss\varepsilon}$ determines exponential convergence of $\varepsilon^1(t)$ to $\varepsilon^1 = 0$, which means that the real output response $y(.)$ of system (1), (2) converges exponentially to its desired output response $y_d(.)$ over the sliding set $S_{ss\varepsilon}$, (16).

Problem 10 *Synthesize control $u(.)$ so to ensure that (a) through (d) hold:*

a) the sliding set $S_{ss\varepsilon}$, (16), is positive invariant with respect to $\varepsilon^1(t)$:

$$if\ \varepsilon^1(\tau; \varepsilon_0^1; d) \in S_{ss\varepsilon}\ at\ \tau \in R_+,\ then\ \varepsilon^1(t; \varepsilon_0^1; d) \in S_{ss\varepsilon},\ \forall t \in [\tau, \infty[, \tag{17}$$

b) the sliding set $S_{ss\varepsilon}$ is globally pairwise exponentially stable with a finite vector reachability time $\tau_{rss\varepsilon}(\varepsilon_0^1) \in R_+^N$, which means that (18) holds:

$$\left|\varepsilon^{(1)}(t; \varepsilon_0^1; d) + \Gamma\varepsilon(t; \varepsilon_0^1; d)\right| \le K_\varepsilon e^{-L_\varepsilon t} \left|\varepsilon_0^{(1)} + \Gamma\varepsilon_0\right|,\ \forall (t, \varepsilon_0^1) \in R_+ \mathrm{x} R^{2N}, \tag{18}$$

for some positive diagonal design $N\mathrm{x}N$ matrices K_ε and L_ε, and

$$\varepsilon^1(t; \varepsilon_0^1; d) \in S_{ss\varepsilon}, \forall (t\mathbf{1}, \varepsilon_0^1) \in [\tau_{rss\varepsilon}(\varepsilon_0^1), \infty\mathbf{1}[\mathrm{x} R^{2N}, \tag{19}$$

c) the sliding set S_{ss}, (3), is positive invariant with respect to $\mathrm{e}^1(t)$, (4),

d) the sliding set S_{ss} is globally asymptotically stable with a finite scalar reachability time $\tau_{rsss}(\mathrm{e}_0^1)$, which means that $\mathrm{e}^1(t; \mathrm{e}_0^1; d)$ obeys (20),

$$\mathrm{e}^1(t; \mathrm{e}_0^1; d) \in S_{ss}, \forall (t, \mathrm{e}_0^1) \in [\tau_{rsss}(\mathrm{e}_0^1), \infty[\mathrm{x} R^{2m}. \tag{20}$$

5.2 Output Tracking and Stabilizing Control Problem Solution

Condition 11 *The dimensions m of the vector q, N of the output vector y, and r of the control vector u of system (1), (2) obey: $r \ge m > N$.*

Let an auxiliary variable z_ε and the induced matrix Z_ε be defined by (21):

$$z_\varepsilon = (z_{\varepsilon 1}\ \ z_{\varepsilon 2}\ \ldots\ z_{\varepsilon N})^T = \varepsilon^{(1)} + \Gamma\varepsilon,\ Z_\varepsilon = diag\{z_{\varepsilon 1}\ \ z_{\varepsilon 2}\ \ldots\ z_{\varepsilon N}\}. \tag{21}$$

Condition 12 *A subsidiary function $v_s(.) : R^m \to R^p$, $p \in \{1, 2, ..., m - N\}$, (22),*

$$v_s(z) = [v_{s1}(z)\ \ v_{s2}(z)\ \ \ldots\ \ v_{sp}(z)]^T,\ p \in \{1, 2, ..., m - N\}, \tag{22}$$

is a global differentiable positive definite vector function on the z-space R^m, i.e., it is a global differentiable positive definite vector function with respect to the set S_{ss}, (3), if it is considered on the e^1-space R^{2m} in view of (21) and (22).

The function $v_s(.)$ will be used to solve the subproblems c) and d) of Problem 10. Let its Jacobian be denoted by $J_s(.) : R^m \to R^{p\mathrm{x}m}$, $p \in \{1, 2, ..., m - N\}$, $J_s(z) = [\partial v_{si}(z)/\partial z_j]$.

Condition 13 *The Jacobians $J_g(q)$ and $J_s(z)$ of the vector functions $g(.)$, (2), and $v_s(.)$, (22), and $Z_\varepsilon(.)$, (21), obey: $rank\ [W(q, z, z_\varepsilon)] = N + p$, $\forall (q, z, z_\varepsilon) \in R^m \mathrm{x} R^m \mathrm{x} R^N$ and $J_s(0) = 0$, where $W(q, z, z_\varepsilon) = [J_g^T(q) Z_\varepsilon^T(z_\varepsilon)\quad J_s^T(z)]^T \in R^{(N+p)\mathrm{x}m}$.*

Let L_ε and T_s be positive diagonal design $p\text{x}p$ matrices, $T_s = diag\{\tau_{s1}\ \ \tau_{s2}\ \dots\ \tau_{sp}\}$ and $s(v_s) = [sign(v_{s1})\ \ sign(v_{s2})\ \dots\ sign(v_{sp})]^T$.

Theorem 14 *If Conditions 7, 11 through 13 are satisfied, if T_ε is a positive diagonal $N\text{x}N$ matrix such that*

$$\tau_{RSS\varepsilon}(\varepsilon_0^1) = T_\varepsilon \left|\varepsilon_0^{(1)} + \Lambda\varepsilon_0\right|, \ \forall\varepsilon_0^1 \in R^{2N}, \tag{23}$$

and if the control vector function $u(.)$ is in the form (24),

$$\begin{array}{l} u(t,e^1,\varepsilon^1,q^1,q_d^1) = b^I\left\{B^T(t,q)\left[B(t,q)B^T(t,q)\right]^{-1}A(t,q)u_e(t)\right\},\ u_e\in R^m, \\ u_e(t,e^1,\varepsilon^1,q^1,q_d^1) = W^T(q,z,z_\varepsilon)[W(q,z,z_\varepsilon)W^T(q,z,z_\varepsilon)]^{-1}\bullet \\ \left\{\begin{array}{c} Z_\varepsilon(z_\varepsilon)\left\{y_d^{(2)} - J_g^{(1)}(q)q^{(1)} + \Gamma y_d^{(1)} - \Gamma y^{(1)}\right\} \\ +L_\varepsilon Z_\varepsilon(z_\varepsilon)z_\epsilon(\varepsilon^1) + T_\varepsilon^{-1}\left|Z_\epsilon(\varepsilon^1)\right|\mathbf{1} \\ +\left|Z_\epsilon(\varepsilon^1)\right|\left|J_g(q)A^{-1}(t,q)\right|\left[h_M(t) + H(t,q^1)\left|q^1\right|\right] \\ \\ J_s(z)\left[q_d^{(2)} + \Lambda q_d^{(1)} - \Lambda q^{(1)}\right] + L_s v(z) + T_s^{-1}s\left[v_s(z)\right] \\ +\left|J_s(z)A^{-1}(t,q)\right|\left[h_M(t) + H(t,q^1)\left|q^1\right|\right] \end{array}\right\}, \end{array} \tag{24}$$

then for every $d(.) \in \mathbf{S}_d$:

a) the sliding set $S_{ss\varepsilon}$, (16), is positive invariant with respect to $\varepsilon^1(t)$, (17),

b) the sliding set $S_{ss\varepsilon}$ is globally pairwise exponentially stable with the finite vector reachability time $\tau_{rss\varepsilon}(\varepsilon_0^1) \in R_+^N$, (18), (19), (23),

c) the sliding set S_{ss}, (3), is positive invariant with respect to $e^1(t)$, (4),

d) the sliding set S_{ss} is globally asymptotically stable with the finite scalar reachability time $\tau_{rsss}(e_0^1) = \max\{\tau_{si}(e_0^1) : i = 1, 2, \dots, p\}$, (20),

e) the vector function $v_s(.)$ and the motion error $e^1(.; e_0^1; d)$ obey (25):

$$\begin{array}{l} \left[\ v_s\left\{z[e^1(t;e_0^1;d)]\right\}\ \right] \\ \leq\ \min\left\{\begin{array}{c} e^{-L_s t}v_s[z(e_0^1)],\ \forall t\in R_+, \\ v_s[z_s(e_0^1)] - T_s^{-1}t\mathbf{1},\ \forall t\mathbf{1}\in\left[0,T_s v_s[z_s(e_0^1)]\right] \\ 0,\ \forall t\mathbf{1}\in\left[T_s v_s[z_s(e_0^1)],\infty\mathbf{1}\right[. \end{array}\right\}. \end{array} \tag{25}$$

Proof. Let $[d(.), e^1, \varepsilon^1] \in \mathbf{S}_d\text{x}R^{2m}\text{x}R^{2N}$ be arbitrary. Let all the conditions of the theorem statement hold. Let an extended vector function $v_e(.)\ :\ R^{m+N}\ \rightarrow\ R^{N+p}$ be composed of the vector functions $v_\varepsilon(.) = (1/2)Z_\varepsilon(.)z_\varepsilon(.)$ and $v_s(.)$:

$$v_e(z_e) = \left[v_\varepsilon^T(z_\varepsilon)\ \ v_s^T(z)\right]^T,\ z_e = \left(z_\varepsilon^T\ \ z^T\right)^T \in R^{m+N}. \tag{26}$$

It is global differentiable positive definite vector function on the z_e-space R^{m+N}. Its Jacobian $J_e(.) : R^{m+N} \rightarrow R^{(N+p)\text{x}(m+N)}$ has the next form:

$$J_e(z_e) = blockdiag\left\{Z_\varepsilon(z_\varepsilon)\ \ J_s(z)\right\}. \tag{27}$$

Hence, along an output error response and a motion error of system (1), (2):

$$v_e^{(1)}(z_e^1) = \begin{bmatrix} Z_\varepsilon(z_\varepsilon)z_\varepsilon^{(1)} \\ \\ J_s(z)z^{(1)} \end{bmatrix} = \begin{bmatrix} Z_\varepsilon(z_\varepsilon)\left[\varepsilon^{(2)} + \Gamma\varepsilon^{(1)}\right] \\ \\ J_s(z)\left[e^{(2)} + \Lambda e^{(1)}\right] \end{bmatrix} \Longrightarrow v_e^{(1)}(0) = 0,$$

or, by using: $\varepsilon^{(2)} = y_d^{(2)} - y^{(2)}$, $e^{(2)} = q_d^{(2)} - q^{(2)}$, $y^{(2)} = [g(q)]^{(2)} = J_g^{(1)}(q)q^{(1)} + J_g(q)q^{(2)} = J_g^{(1)}(q)q^{(1)} + J_g(q)A^{-1}(t,q)[B(t,q)b(u) - h(t,d,q^1)]$ [due to (1), (2) and Condition 7], the definition of $W(q,z,z_\varepsilon)$, Condition 13 and (24), we find:

$$v_e^{(1)}(z_e^1) \leq \begin{bmatrix} -L_\varepsilon Z_\epsilon(\varepsilon^1) z_\epsilon(\varepsilon^1) - T_\varepsilon^{-1} \left| Z_\epsilon(\varepsilon^1) \right| \mathbf{1} \\ -L_s v(z) - T_s^{-1} s\left[v_s(z)\right] \end{bmatrix}.$$

Hence,

$$v_e^{(1)}(z_e^1) = \begin{bmatrix} v_\varepsilon^{(1)}[z_\epsilon^1(\varepsilon^2)] \\ v_s^{(1)}[z^1(e^2)] \end{bmatrix} \leq - \begin{bmatrix} \min\left\{L_\varepsilon Z_\epsilon(\varepsilon^1) z_\epsilon(\varepsilon^1),\ T_\varepsilon^{-1}\left|Z_\epsilon(\varepsilon^1)\right| \mathbf{1}\right\} \\ \min\left\{L_s v_s[z(e^1)],\ T_s^{-1} s\left\langle v_s[z(e^1)]\right\rangle\right\} \end{bmatrix}$$

$$= - \begin{bmatrix} \min\left\{2L_\varepsilon v_\varepsilon[z_\epsilon(\varepsilon^1)], T_\varepsilon^{-1}\sqrt{2v_\varepsilon^{(1)}[z_\epsilon(\varepsilon^1)]}\right\} \\ \min\left\{L_s v_s[z(e^1)], T_s^{-1} s\left\langle v_s[z(e^1)]\right\rangle\right\} \end{bmatrix}, \forall (e^2, \varepsilon^2) \in R^{3m} \mathrm{x} R^{3N}.$$

This shows that $v_e^{(1)}(z_e^1)$ is majorized by a global negative definite vector function on $R^m \mathrm{x} R^N$, which, together with (21), $v_e^{(1)}[z_e^1(e^2,\varepsilon^2)] = 0$ on the Cartesian product sliding set $S_{ss} \mathrm{x} S_{ss\varepsilon}$ (in $R^{2m} \mathrm{x} R^{2N}$ due to $Z_\varepsilon(0) = 0$ and $J_s(0) = 0$) and global positive definiteness of $v_e(.)$ with respect to the product sliding set $S_{ss} \mathrm{x} S_{ss\varepsilon}$, proves global asymptotic stability of the product sliding set $S_{ss} \mathrm{x} S_{ss\varepsilon}$ for every $d(.) \in \mathbf{S}_d$. Hence, the product sliding set $S_{ss} \mathrm{x} S_{ss\varepsilon}$ is also positive invariant relative to the error motions and error output responses of system (1), (2) for every $d(.) \in \mathbf{S}_d$. This proves the statements under a) and c). Integration of the preceding inequality, (21), $v_\varepsilon(.) = (1/2) Z_\varepsilon(.) z_\varepsilon(.)$ and (26) yield:

$$\begin{bmatrix} \left|\varepsilon^{(1)}(t) + \Gamma\varepsilon(t)\right| \\ v_s\left\{z[e^1(t; e_0^1; d)]\right\} \end{bmatrix}$$

$$\leq \begin{bmatrix} \min\left\{ \begin{array}{c} e^{-L_\varepsilon t}\left|\varepsilon_0^{(1)} + \Gamma\varepsilon_0\right|,\ \forall t \in R_+, \\ \left\{ \begin{array}{c} \left[\left|\varepsilon_0^{(1)} + \Gamma\varepsilon_0\right| - T_\varepsilon^{-1} t\mathbf{1}\right],\ \forall t\mathbf{1} \in \left[0, T_\varepsilon\left|\varepsilon_0^{(1)} + \Gamma\varepsilon_0\right|\right] \\ 0,\ \forall t\mathbf{1} \in [T_\varepsilon\left|\varepsilon_0^{(1)} + \Gamma\varepsilon_0\right|, \infty\mathbf{1}[. \end{array} \right\} \end{array} \right\} \\ \min\left\{ \begin{array}{c} e^{-L_s t} v_s[z(e_0^1)],\ \forall t \in R_+, \\ \left\{ \begin{array}{c} v_s[z_s(e_0^1)] - T_s^{-1} t\mathbf{1},\ \forall t\mathbf{1} \in [0, T_s v_s[z_s(e_0^1)]] \\ 0,\ \forall t\mathbf{1} \in [T_s v_s[z_s(e_0^1)], \infty\mathbf{1}[\end{array} \right\} \end{array} \right\} \end{bmatrix}.$$

This, (3), (16), and (23) complete the proof ■

6 CONCLUSION

The new vector development of the concepts of (semi-) definite scalar functions and decrescent scalar functions, and the development of the concept of vector Lyapunov functions enable an effective resolution of a basic complex problem of control synthesis in two different spaces. Such a control ensures simultaneous global asymptotic stability of a system desired motion and a good quality of output tracking that is accepted to be exponential. Various kinds of vector Lyapunov functions, which are introduced herein, additionally permit an effective sliding control synthesis. Control guarantees that the motion error and the output error response of the controlled system are independent of the system internal dynamics over the sliding sets in the e^1-error space and in the output ε^1-error space. Asympotic (or exponential) stability of the e^1-sliding set S_{ss} and of the ε^1-sliding set $S_{ss\varepsilon}$ with pre-specified finite reachability times provide an engineering sense to the results and enable their effective applications. Synthesized controls are directly applicable to airplanes and to other nonlinear dynamical systems (autonomous vehicles, robots, ships, and spacecrafts). All nonlinearities and parameters of their mathematical models can be unknown, except those of their inertia matrices and the ouput function $g(.)$. Besides, information about the form and real values of external disturbances is not requested. The synthesized controls, stability, and tracking properties are highly robust relative to such uncertainties. The paper opens new directions in studies of stability, tracking, and control of the nonlinear dynamical systems.

REFERENCES

1. R. Bellman, "Vector Lyapunov functions", *J. S. I. A. M. Control*, Ser. A, **1**, No. 1, pp. 32–34, 1962.
2. V.M. Matrosov, "To the theory of stability of motion" (in Russian), *Prikl. Math. Mekh.*, **26**, No. 5, pp. 885–895, 1962.
3. V.M. Matrosov, "Vector Lyapunov functions in the analysis of nonlinear interconnected systems," *Proc. Symp. Mathematica*, Bologna, **6**, pp. 209–242, 1971.
4. V. Lakshmikantham, "Vector Lyapunov functions and conditional stability," *J. Mathematical Analysis and Applications*, pp. 368–377, 1975.
5. J.P. La Salle, *The Stability of Dynamical Systems*, SIAM, Philadelphia, 1976.
6. A.N. Michel and R. Miller, *Qualitative Analysis of Large-Scale Dynamical Systems*, Academic Press, New York, 1977.
7. D.D. Šiljak, *Large-Scale Dynamic Systems: Stability and Structure*, North-Holland, New York, 1978.
8. Lj.T. Grujić, A.A. Martynyuk, and M. Ribbens-Pavella, *Large-Scale Systems under Structural and Singular Perturbation* (in Russian: Naukova Dumka, Kiev, 1984), Springer Verlag, Berlin, 1987.
9. V.M. Matrosov, *Vector Lyapunov Function Method: Analysis of Dynamical Properties of Nonlinear Systems* (in Russian), FIZMATLIT, Moscow, 2001.
10. A.M. Lyapunov, *General Problem of Stability of Motion*, Taylor and Francis, London, 1992.
11. Lj.T. Grujić, "On large-scale systems stability." In: *Computing and Computers for Control Systems*, Eds. P. Borne et al., J.C. Baltzer AG, Scientific Publishing Co., IMACS, pp. 201–206, 1989.
12. W. Hahn, *Stability of Motion*, Springer Verlag, Berlin, 1967.
13. J. Wertz, Ed., *Spacecraft Attitude Determination and Control*, R. Reidel Publishing Company, Boston, 1978.

14. B. Etkin, *Dynamics of Flight — Stability and Control*, John Wiley & Sons, New York, 1982.
15. P.C. Hughes, *Spacecraft Attitude Dynamics*, John Wiley & Sons, New York, 1986.
16. M.W. Spong and M. Vidyasagar, *Robot Dynamics and Control*, John Wiley and Sons, New York, 1989.
17. T.I. Fossen, *Guidance and Control of Ocean Vehicles*, John Wiley & Sons, Chichester, 1994.
18. L. Sciavicco and B. Siciliano, *Modeling and Control of Robot Manipulators*, The McGraw Hill Companies, Inc., New York, 1996.
19. L. Cvetičanin, *Dynamics of Machines with Variable Mass*, Gordon and Breach Science Publishers, Amsterdam, 1998.
20. V.I. Utkin, "Variable structure systems with sliding modes," *IEEE Transactions on Automatic Control*, Vol. AC-23, No. 2, pp. 212–222, 1977.
21. J.-J. E. Slotine, "The robust control of robot manipulators," *The International Journal of Robotics Research*, Vol. 4, No. 2, pp. 49–64, 1985.
22. V.I. Matyukhin and E.S. Pyatnitskii, "Controlling the motion of robot manipulators by decomposition taking into account the actuator dynamics," *Avtomatika i Telemekhanika* (in Russian), No. 9, pp. 67–81, 1989.
23. C.-Y. Su, T.-P. Leung and Y. Stepanenko, "Real-time implementation of regressor-based sliding mode control algorithm for robotic manipulators," *IEEE Transactions on Industrial Electronics*, Vol. 40, No. 1, pp. 71–79, 1993.

8

Stabilization of Unstable Aircraft Dynamics under Control Constraints

M.G. Goman† and M.N. Demenkov*

†*De Montfort University, Faculty of Computing Sciences and Engineering, Leicester U.K.*

**Department of Computing Sciences and Control, Bauman Moscow State Technical University, Moscow, Russia*

Stabilization of the unstable dynamic linear system with control constraints is considered in terms of maximizing the size of the closed-loop system stability region. A controllability region for the open-loop system is a natural limit for the stability region of the closed-loop system with any designed controller. A relation between the controllability and stability regions is considered as a performance metric for controller assessment. The linear control laws maximizing the stability region for the constrained linear system are derived for two aircraft stabilization problems.

1 INTRODUCTION

Stability augmentation of an aircraft's dynamics by control system allows one not only to improve its handling quality characteristics, but also to expand the flight envelope and increase its performance characteristics. For example, an aircraft with aerodynamically unstable configuration stabilized by control system can provide higher lift-to-drag ratio or lower signature [1–3]. Critical flight regimes such as high-incidence departures or aeroelastic instabilities can be significantly relaxed or even eliminated by an active control approach [4].

The central problem in control law design for regimes with aircraft dynamic instability is in taking into account the realistic constraints for control effectors such as deflection limits and rate saturation. Unfortunately, in control design works the stabilization problem of unstable aircraft dynamics is usually considered as a linear one without the direct effect of nonlinear control constraints [5].

The control saturation can lead to significant degradation of dynamic characteristics or even to the loss of stability in the closed-loop system. The unstable constrained linear system has a bounded controllability region, where the stabilization problem can be solved. This fundamental fact is normally ignored or assumed insignificant. However, in many practical situations associated with a high level of instability and low control authority, the controllability region decreases significantly and classical linear control design methods lose their efficiency.

0-415-30852-6/04/$0.00+$1.50

The closed-loop system stability region (or region of attraction) presents only part of the open-loop system controllability region and therefore is even more bounded. The "good" controller in terms of linear system criteria can provide the closed-loop system with a very small and unsatisfactory stability region.

The computation of the controllability region for the open-loop system and region of attraction for the closed-loop system will extend understanding of the stabilization problem.

All controllers designed using linear methods provide different stability regions for the closed-loop system and the relationship between the stability region and controllability region can be considered as an additional metric for assessment of linear controller robustness to external disturbances.

The allowable level of external disturbances is directly connected with the size of the stability region. The bigger the region of attraction, the higher the level of external disturbances that can be rejected by the system. If external disturbances move the system beyond the controllability region, any excellent controller will not be able to keep the system stable.

The number of works devoted to control design methods maximizing stability region of constrained linear system is not high; however, the interest in this area is growing rapidly [7–13].

The best controller in terms of rejection of external disturbances has to keep the system stable in the whole controllability region. Such a controller is called a "maximum stabilizer" [12] or as "stability optimal" [7, 8]. The controller of some particular structure designed with additional criteria for maximizing stability region [10, 11, 13] can be called the "stability suboptimal" one.

In this chapter the problem of linear controller design maximizing the closed-loop system stability region is considered for the linearized system with one and two unstable eigenvalues and only deflection constraints. The results presented in [8] are extended to the case of multiple control inputs considering two design examples for an aircraft's dynamics. The basic concepts and methods of controllability region computation are also discussed.

Hereafter the following notations are used. The capital letters denote matrices and sets, lower case letters denote column vectors or scalars, R^n denotes the Euclidean n-space and $R^{n\times m}$ the set of all real $n \times m$ matrices. Vectors are supposed to be column, subscripts denote elements of a sequence (x_i), superscripts in parentheses denote component of a vector ($x^{(i)}$), and x^T the transposed vector x (in the same way as transposition of the matrix). If $x \in R^p$ and $y \in R^q$, are row vectors, then (x, y) denotes the row vector, which is the union of the vectors x and y: $(x, y) \in R^{p+q}$. The following vector 2-norm is used for vector $x \in R^n$: $||x||_2 = \sqrt{x^T x}$.

2 SOME BASIC PROPERTIES OF UNSTABLE LINEAR SYSTEMS WITH CONTROL CONSTRAINTS

It is supposed that the following continuous time-invariant linear system

$$\dot{x}(t) = Ax(t) + Bu(t), \tag{1}$$

with $A \in R^{n\times n}$, $B \in R^{n\times m}$, $x \in R^n$, $u \in R^m$, $t \in [0, \infty)$ has $q \leq n$ unstable eigenvalues λ_i, $Re\lambda_i > 0, i = 1, \ldots, q$ and the control vector is bounded:

$$u \in U, \tag{2}$$

where $U \in R^m$ is a compact and convex set, which is defined in the form of amplitude constraints:

$$|u^{(i)}| \leq u^{(i)}_{max}, i = \overline{1, m}. \tag{3}$$

It is additionally supposed that matrix A has no neutral eigenvalues and the state vector is fully observable.

All eigenvalues of matrix A can be divided into stable ($Re\lambda_i < 0, i = 1, \ldots, (n - q)$) and unstable ($Re\lambda_i > 0, i = 1, \ldots, q$) ones. Considering the problem of system stabilization and maximization of the closed-loop system stability region, it is natural to take into account only the subspace corresponding to unstable eigenvalues.

There are two well-known transformations of the state variables suitable for system decomposition. The first one is based on the Schur transformation [14] and the second one is based on the block Jordan transformation of matrix A. In this chapter we will use the Schur decomposition as the most robust and reliable one.

Suppose that n eigenvalues of matrix A are ordered into two groups according to their real part magnitudes. The first one consists of $n - q$ stable eigenvalues and the second one consists of q unstable eigenvalues.

Consider the following transformation of the basis in (1):

$$x = [Q_1|Q_2] \begin{bmatrix} s \\ z \end{bmatrix},$$

where matrix $Q = [Q_1|Q_2]$ is the orthogonal one ($Q^T Q = I$) and the columns of $Q_1 \in R^{n\times(n-q)}$ span the subspace V_1 associated with $n-q$ stable eigenvalues, while the columns of $Q_2 \in R^{n\times q}$ span the subspace V_2, which is a complementary subspace of V_1. This matrix Q can be obtained by the Schur decomposition of matrix A with reordering, if necessary, the eigenvalues of its diagonal blocks [14] in accordance with the formed eigenvalue order. For numerical computation one may use the function *blkrsch* from MATLAB Robust Control Toolbox [15].

In the new basis the open-loop system is represented as

$$\begin{bmatrix} \dot{s}(t) \\ \dot{z}(t) \end{bmatrix} = Q^T A Q \begin{bmatrix} s(t) \\ z(t) \end{bmatrix} + Q^T B u(t) = \begin{bmatrix} A_{11} & A_{12} \\ 0 & A_z \end{bmatrix} \begin{bmatrix} s(t) \\ z(t) \end{bmatrix} + \begin{bmatrix} B_s \\ B_z \end{bmatrix} u(t),$$

where the eigenvalues of matrices A_{11} and A_z contain the stable and unstable eigenvalues of original system (1), respectively.

Note that $z(t)$ dynamics associated with the unstable eigenvalues is decoupled from $s(t)$ and described by the following open-loop, reduced-order system:

$$\dot{z}(t) = A_z z(t) + B_z u(t), \tag{4}$$

where $A_z \in R^{q\times q}$, $B_z \in R^{q\times m}$, $u(t) \in U$ and $z(t) = Q_2^T x(t)$.

If system (4) is stabilized and its states approach the equilibrium point with $z(t) = 0$ and $u(t) = 0$, it is trivial to show that subsystem $s(t)$ will also asymptotically approach the origin and thus to stabilize system (1) it is enough to stabilize only subsystem $z(t)$ (4).

The following definition and two basic properties of the *controllability* region, formulated in [6], are given below to assist the main content of the paper.

Definition 1 *The controllability region C of system (1) with constraint (2) is the set of all points $x_0 \in R^n$, from which the system can be moved to the origin by applying the bounded control $u(t) \in U$:*

$$C = \{x_0 \in R^n : \exists u(t) \in U \;\; \textit{so that} \; \lim_{t \to \infty} x(t, x_0) = 0\}.$$

Theorem 1 *The controllability region C of system (1) with constraint (2) is convex.*

Theorem 2 *Let linear system (1) have only unstable eigenvalues and control vector be constrained as (3). Then the system controllability region C is a bounded set:*

$$\max_{x \in C} ||x||_2 < \infty.$$

If system (1) has stable and unstable eigenvalues, it contains unstable subsystem (4) with the bounded controllability region C in a subspace of R^n and as a result system (1) cannot be stabilized in the full state-space R^n. Bounded controllability region C of the unstable subsystem produces a cylinder in the full state-space of original system, which is unbounded in some directions.

Suppose that the following linear controller is applied to the system (1):

$$u(t) = Kx(t), \tag{5}$$

where $K \in R^{mxn}$ stabilizes the closed-loop system (1), (5) in the absence of control saturation.

To take into account control constraints, consider the saturation function defined for scalar $u^{(i)} \in R^1$ as:

$$\text{sat}(u^{(i)}) = \begin{cases} u_{max}^{(i)}; & u^{(i)} > u_{max}^{(i)} \\ u^{(i)}; & |u^{(i)}| \leq u_{max}^{(i)} \\ -u_{max}^{(i)}; & u^{(i)} < -u_{max}^{(i)} \end{cases}$$

and for vector $u = (u^{(1)}, ..., u^{(m)})^T$ as:

$$\text{sat}(u) = (\text{sat}(u^{(1)}), ..., \text{sat}(u^{(m)}))^T.$$

The constrained linear closed-loop system is defined by (1) and the following nonlinear control:

$$u(t) = \text{sat}(Kx(t)). \tag{6}$$

We will operate with the following definition of *stability* region.

Definition 2 *The stability region S of the closed-loop system (1), (6) is the set of all points $x_0 \in R^n$ from which the closed-loop system asymptotically approaches the origin:*

$$S = \{x_0 \in R^n : \lim_{t \to \infty} x(t, x_0) = 0\}.$$

The performance of the closed-loop system is changed when the control input saturates, so that the phase trajectories of the closed-loop system starting inside the controllability region C can leave it. As a result the closed-loop system stability region S can be less than C. The optimal stability region S_{opt} is equal to controllability region C.

In this chapter the computation of controllability region C and the linear controller design maximizing the stability region S of the constrained closed-loop system is considered for cases when matrix A has one and two unstable eigenvalues and system (1) has multiple control inputs $m \geq 1$.

3 MAXIMUM STABILIZER DESIGN FOR ONE UNSTABLE EIGENVALUE

The unstable subsystem is described by the first-order differential equation:

$$\dot{z}(t) = a_z z(t) + b_z u(t),$$

where $a_z > 0$, $b_z \in R^{1\times m}$, $u(t) \in U$ and $z(t) = Q_2^T x(t)$.

The closed-loop system is stable if the following algebraic stability conditions hold:

$$\begin{aligned} z > 0 &\Rightarrow \dot{z} < 0; \\ z = 0 &\Rightarrow \dot{z} = 0; \\ z < 0 &\Rightarrow \dot{z} > 0. \end{aligned}$$

Consider the upper limit on z, while $\dot{z} < 0$:

$$z < \frac{1}{a_z}(-\sum_{i=1}^{m} b_z^{(i)} u^{(i)}) \leq \max_{u\in U} \frac{1}{a_z}(-\sum_{i=1}^{m} b_z^{(i)} u^{(i)}) = \frac{1}{a_z}\sum_{i=1}^{m} b_z^{(i)} u_{max}^{(i)} \text{sign}(b_z^{(i)}).$$

Applying the same reasoning to obtain the lower limit on z, while $\dot{z} > 0$, one can see that the controllability region of the unstable subsystem is an interval: $-z_{max} < z < z_{max}$, where

$$z_{max} = \frac{1}{a_z}(\sum_{i=1}^{m} |b_z^{(i)}| u_{max}^{(i)}). \tag{7}$$

In the state-space of the original system the controllability region looks like a "strip" that is defined by the following linear inequality:

$$-z_{max} < Q_2^T x < z_{max}. \tag{8}$$

To derive a saturated linear control law $u^{(i)}(t) = \text{sat}(k_i z(t))$ maximizing the closed-loop system stability region, it is necessary to satisfy the following two requirements:

1. In the absence of saturation the closed-loop system must be stable:

$$a_z + \sum_{i=1}^{m} b_z^{(i)} k_i < 0. \tag{9}$$

2. On the boundary of controllability region the saturated control must satisfy the boundary conditions, which prevent leaving the controllability region. They result in the following inequalities for each k_i:

$$\begin{aligned} b_z^{(i)} < 0 &\Rightarrow k_i z_{max} \geq u_{max}^{(i)} \\ b_z^{(i)} > 0 &\Rightarrow k_i z_{max} \leq -u_{max}^{(i)} \\ b_z^{(i)} = 0 &\Rightarrow k_i = 0. \end{aligned} \tag{10}$$

Any linear control law that satisfies the system of linear inequalities (9) and (10) is the maximum stabilizer, i.e., it provides the system with the maximal stability region, which is equal to controllability region C. This control law can be computed, for example, via linear programming.

Example 1. To illustrate the method let us consider the following linearized lateral-directional dynamics of an aircraft without a vertical tail [2]. The stability and controllability matrices correspond to flight at Mach number $M = 0.4$, altitude $H = 15000$ ft, and angle of attack $\alpha = 8.8$ deg:

$$[A|B] =$$
$$\left[\begin{array}{ccc|ccccc} -0.004 & 0.154 & -0.988 & -0.009 & 0.013 & 0.006 & 0.011 & -0.004 \\ -8.210 & -0.785 & 0.117 & 7.570 & -4.970 & 3.650 & 0.079 & -0.614 \\ -0.889 & -0.030 & -0.016 & 0.091 & -0.181 & -0.416 & -0.804 & 0.188 \end{array}\right].$$

The state vector $x = (\beta, p, r)^T$, where β is the sideslip angle in degrees, p, r are the body-axis roll and yaw rates in degrees per second. The control vector $u = (\delta_e, \delta_s, \delta_t, \delta_{ytv}, \delta_f)^T$, where δ_e are the differential elevons, δ_s are the differential spoiler/slot-deflectors, δ_t are the differential all moving tips, δ_{ytv} is the yaw thrust vectoring, and δ_f is the differential outboard leading-edge flaps, all control deflections are expressed in degrees.

The following unstable subsystem can be obtained via the Schur decomposition of matrix A:

$$[a_z|b_z] = \left[\begin{array}{c|ccccc} 0.434 & -0.485 & 0.206 & -0.654 & -0.737 & 0.218 \end{array}\right].$$

The variable z of unstable subspace is connected with the system state vector x as

$$z(t) = Q_2^T x(t), \quad Q_2^T = [\ -0.421 \quad -0.075 \quad 0.904\ ].$$

All control variables are constrained by the same value: $|u^{(i)}| \leq 5, i = \overline{1,5}$. In accordance with (7) and (8), the controllability region of unstable subsystem is defined by $z_{max} = 26.472$.

To compute the linear maximum stabilizer the following system of linear inequalities must be solved:

$$0.434 - 0.485k_1 + 0.206k_2 - 0.654k_3 - 0.737k_4 + 0.218k_5 \leq \gamma,$$

$$26.472k_1 \geq 5, \quad 26.472k_2 \leq -5, \quad 26.472k_3 \geq 5, \quad 26.472k_4 \geq 5, \quad 26.472k_5 \leq -5,$$

where $\gamma < 0$.

One of the possible solutions for $\gamma = -1$ is

$$k_1 = 0.5365, k_2 = -0.2276, k_3 = 0.7236, k_4 = 0.816, k_5 = -0.2412.$$

The final step is to convert the control law to a linear function of the original system state vector:

$$K_{max} = \begin{bmatrix} k_1 Q_2^T \\ \dots \\ k_5 Q_2^T \end{bmatrix} = \begin{bmatrix} -0.2259 & -0.0405 & 0.4850 \\ 0.0958 & 0.0172 & -0.2058 \\ -0.3046 & -0.0546 & 0.6541 \\ -0.3435 & -0.0615 & 0.7377 \\ 0.1015 & 0.0182 & -0.2180 \end{bmatrix}.$$

Figure 1 shows the cross-sections of the controllability region (solid lines) in the planes of two state variables. The controllability region and the stability region for the maximum stabilizer $u(t) = \text{sat}(K_{max}x(t))$ coincide.

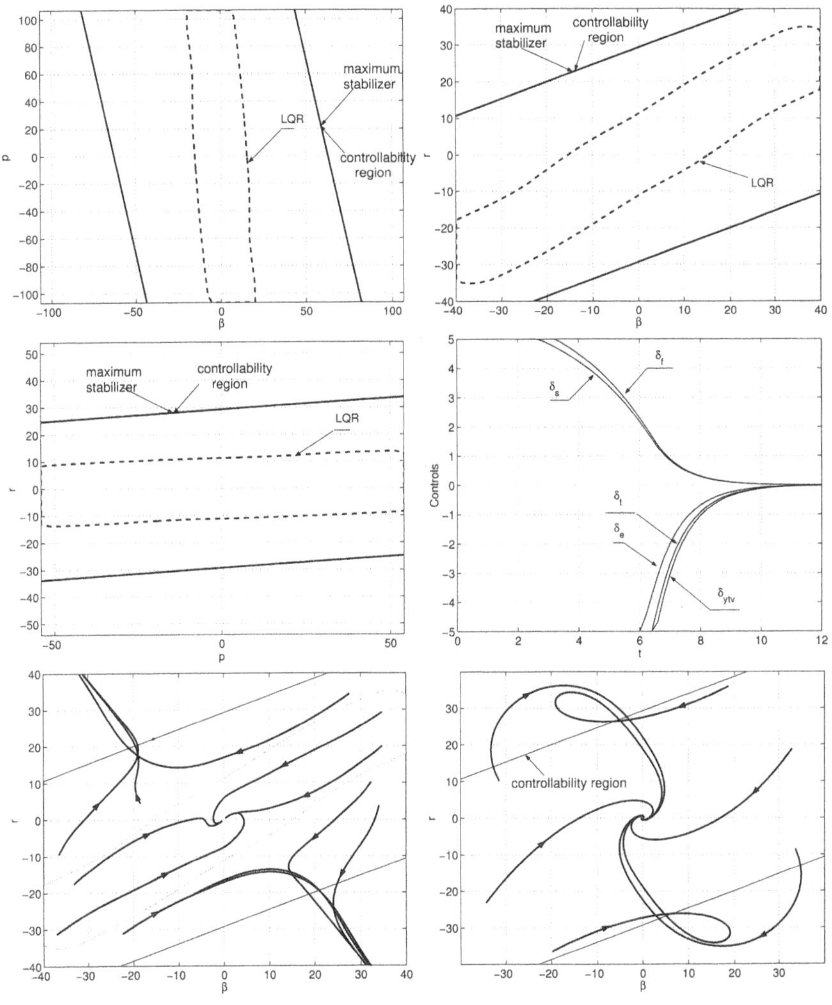

Figure 1 Cross-sections of stability regions for maximum stabilizer and linear-quadratic regulator (LQR), example of control functions produced by maximum stabilizer (middle right plot), closed-loop system trajectories for LQR (bottom left plot) and maximum stabilizer (bottom right plot) in the (β, r) plane.

The linear-quadratic regulator (LQR) minimizing the performance index,

$$J = \int_0^\infty (x^T(t)x(t) + u^T(t)u(t))dt, \tag{11}$$

has been synthesized for comparison purposes. The LQR stability region boundaries (dashed lines) have been computed by direct closed-loop system simulation and they are shown in Fig. 1 (see the top and left middle plots). One can see that the maximum stabilizer rejects the level of external disturbances approximately three times bigger than LQR. In the bottom plots in Fig. 1 several trajectories of the closed-loop system with LQR and maximum stabilizer controllers are shown to illustrate the size of stability regions.

4 STABILITY SUBOPTIMAL CONTROLLER FOR TWO UNSTABLE EIGENVALUES

In this case the unstable subsystem is described by a second-order linear system

$$\dot{z}(t) = A_z z(t) + B_z u(t),$$

where $A_z \in R^2$, $B_z \in R^{2\times m}$, $u(t) \in U$ and $z(t) = Q_2^T x(t)$.

4.1 Controllability Region Computation

A variational method for controllability region computation for an unstable system with only unstable eigenvalues has been developed by Formalsky [6].

The idea of this method is in maximization of a scalar linear function $f(z) = l^T z$ considering points in the controllability region. The maximization process is performed toward the direction l, so the final solution coincides with the point z_l on the boundary of controllability region C having vector l normal to the boundary at this point (see the top left plot in Fig. 2):

$$z_l = \arg\max_{z\in C}(l^T z).$$

A solution of this task always exists, because the controllability region from Theorem 2 is totally bounded. The optimization task may be solved for different directions l_k and obtained points z_{l_k} can approximate the controllability region boundary as well, as necessary, because according to Theorem 1 the controllability region is convex.

To compute the controllability region boundary, let us consider a *null-controllable* set C_t of all points in the state-space, from which the system can be moved to the origin in the presence of control constraints in a time less than or equal to t. Note that the boundaries of null-controllable sets approach the controllability region boundary as $t \to \infty$.

The points in a *null-controllable* set C_t can be expressed explicitly as initial points $z(0)$ for the Cauchy problem through the applied control function $u(\tau)$ on the time interval $[0, t]$. The general solution of system (1),

$$z(t) = e^{A_z t} z(0) + \int_0^t e^{A_z(t-\tau)} B_z u(\tau) d\tau, z(t) = 0,$$

can be used to describe null-controllable sets as follows:

$$C_t = \{z|z = -\int_0^t e^{-A_z\tau} B_z u(\tau) d\tau, u(\tau) \in U\}.$$

The optimization task for maximizing a linear function $l^T z$ on C_t has only one solution z_l:

$$z_l = \arg\max_{z \in C_t} l^T z = -\sum_{i=1}^{m} \int_0^t e^{-A_z\tau} b_{zi} u_{max}^{(i)} \text{sign}(-l^T e^{-A_z\tau} b_{zi}) d\tau, \tag{12}$$

where b_{zi} is a column of matrix B_z (see the top left plot in Fig. 2).

Computing boundary points z_{l_k} of a *null-controllable* set C_t for a number of different directions l_k,

$$l_k = (\cos\varphi_k, \sin\varphi_k)^T; \ \varphi_k = \frac{2\pi k}{N}, \ k = \overline{1, N},$$

we can approximate C_t rather accurately.

Considering a sequence of $t_i = t_{i-1} + \Delta$ approaching infinity $t_i \to \infty$ we will have $C_{t_i} \to C$. This convergence process can be stopped if

$$\max_k ||z_{l_k}(t) - z_{l_k}(t + \Delta)||_2 \leq \epsilon,$$

where Δ is a small time increment and ϵ is required accuracy.

4.2 Stability Suboptimal Linear Controller

The design objectives in this case are similar to the previous design with one real eigenvalue. A saturated linear controller $u = \text{sat}(Kz)$ has to stabilize the closed-loop system in the absence of control saturation and prevent leaving the controllability region when system operates close to its boundary. We will consider a *null-controllable* set C_t with large enough t as an approximation of controllability region C.

In every boundary point of C_t the control function following (12) is expressed in the form of relay law:

$$u^{(i)}(0) = u_{max}^{(i)} \text{sign}(-l^T b_{zi}), i = \overline{1, m}, \tag{13}$$

where l is a vector normal to the boundary of C_t at this point.

Each component of control vector takes on the boundary of C_t at its positive or negative extreme value and there are some points where it changes sign. Because any *null-controllable* set is symmetrical, the points on the boundary of C_t, where control changes its sign, are also symmetrical.

These points $\pm z_{0i}$ for every component of control vector $u^{(i)}$ are defined by the condition that the vectors $\pm l_i$, which are normal to the boundary of C_t at these points, are also normal to controllability vector b_{zi}, i.e., $l_i^T b_{zi} = 0$ (see the top right plot in Fig. 2).

A relay control law

$$u^{(i)}(z) = u_{max}^{(i)} \text{sign}(k_i^T z), i = \overline{1, m}, \tag{14}$$

where vectors k_i are normal to switch lines passing through $\pm z_{0i}$, can produce at every boundary point of C_t the same direction of control vector as in (13), and therefore prevent the system from moving outside of C_t. The direction of k_i is defined from inequality $k_i^T b_{zi} < 0$. In this case produced control vector $b_{zi}u^{(i)}(z)$ will "push" the system state toward the controllability region interior.

We will consider the linear control law with k_i vectors from relay control law (14):

$$u^{(i)} = \gamma k_{i*}^T z, \qquad \text{where} \qquad k_{i*} = \frac{k_i u_{max}^{(i)}}{||k_i||_2}. \tag{15}$$

At the presence of control deflection constraints the control functions become nonlinear and can be expressed as

$$u^{(i)}(z) = u_{max}^{(i)} \text{sat}(\gamma k_{i*}^T z), \; i = \overline{1, m}.$$

The switching lines of relay control law (14) in the phase plane, defined by $k_i^T z = 0$, are transformed into "strips," where the closed-loop system is free from control saturation and therefore linear.

Parameter γ, which is the same for all control components, allows one to vary the width of these "strips" and thus to find the compromise between the local stability characteristics expressed in terms of eigenvalues location and the size of the closed-loop system stability region. Of course, the stability region of this saturated linear control law is less than the controllability region, but in practice it can approach it very closely.

Example 2. As a design example, consider the longitudinal dynamics of an aircraft trimmed at altitude $H = 25000$ ft and Mach number $M = 0.9$ [1, 15]. The linearized system of motion equations has the following stability and controllability matrices:

$$[A|B] = \left[\begin{array}{cccccc|cc} -0.0226 & -36.6170 & -18.8970 & -32.0900 & 3.2509 & -0.7626 & 0 & 0 \\ 0.0001 & -1.8997 & 0.9831 & -0.0007 & -0.1708 & -0.0050 & 0 & 0 \\ 0.0123 & 11.7200 & -2.6316 & 0.0009 & -31.6040 & 22.3960 & 0 & 0 \\ 0 & 0 & 1 & 0 & 0 & 0 & 0 & 0 \\ 0 & 0 & 0 & 0 & -30 & 0 & 30 & 0 \\ 0 & 0 & 0 & 0 & 0 & -30 & 0 & 30 \end{array}\right].$$

The first part of the state vector consists of the aircraft's basic rigid body variables: perturbation in the forward velocity δV, the angle of attack α, the pitch rate q, and the pitch attitude angle θ. Two first-order lags $x^{(5)}$ and $x^{(6)}$ are appended to the state vector to represent actuators dynamics. The control variables δ_e and δ_c are signals sent to elevon and canard actuators. In the control law design the following constraints are considered: $|\delta_e| \leq 0.035, |\delta_c| \leq 0.035$. Note that the angles and angular rates are expressed in radians and radians per second, respectively.

The system due to aircraft aerodynamic static instability has an oscillatory unstable mode in the phugoid frequency range, which can be separated off via the Schur decomposition:

$$[A_z|B_z] = \left[\begin{array}{cc|cc} 0.7067 & 2.6227 & 5.1766 & -3.6583 \\ -0.0237 & 0.6729 & 6.7815 & -4.6968 \end{array}\right],$$

$$z(t) = Q_2^T x(t), Q_2^T$$
$$= \begin{bmatrix} 0.0098 & -0.0377 & -0.1852 & -0.9589 & 0.1726 & -0.1219 \\ -0.0036 & -0.9269 & -0.2146 & 0.1385 & 0.2260 & -0.1566 \end{bmatrix}.$$

Using the Formalsky method, the *null-controllable* sets C_t have been computed until convergence to the boundary of controllability region C. In this example, the convergence process has been stopped at $t = 10$ sec. In the middle left plot in Fig. 2 the *null-controllable* sets are plotted with time increment $\Delta = 0.2$ sec in the interval $t = [0, 10]$ sec. The null-controllable sets were approximated by 120 points.

To design linear control law (15), the vectors l_i, which are normal to the boundary of C_{10} at the points where control functions change the sign ($l_i^T b_{zi} = 0$, $i = 1, 2$), have been defined:

$$l_1 = (b_{z1}^{(2)}, -b_{z1}^{(1)})^T = (6.7815, -5.1766)^T,$$
$$l_2 = (b_{z2}^{(2)}, -b_{z2}^{(1)})^T = (-4.6968, 3.6583)^T.$$

Now the control switching points on the boundary of controllability region, approximated by C_{10}, can be computed:

$$z_{01} = \arg \max_{z \in C_{10}} l_1^T z$$
$$= -0.035 \sum_{i=1}^{2} \int_0^{10} e^{-A_z \tau} b_{zi} \text{sign}(-l_1^T e^{-A_z \tau} b_{zi}) d\tau = (1.5703, -0.5421)^T,$$
$$z_{02} = \arg \max_{z \in C_{10}} l_2^T z$$
$$= -0.035 \sum_{i=1}^{2} \int_0^{10} e^{-A_z \tau} b_{zi} \text{sign}(-l_2^T e^{-A_z \tau} b_{zi}) d\tau = (-1.5703, 0.5421)^T.$$

In our case there are two switching lines passing through points $\pm z_{01}$ and $\pm z_{02}$ on the controllability region boundary, and these lines are numerically the same for each of two controls. These switching lines are described by the following equations:

$$k_1^T z = 0, \quad k_2^T z = 0,$$

where k_i are defined as

$$k_1 = (z_{01}^{(2)}, -z_{01}^{(1)})^T = (-0.5421, -1.5703)^T,$$
$$k_2 = (z_{02}^{(2)}, -z_{02}^{(1)})^T = (0.5421, 1.5703)^T.$$

Since $k_1^T b_{z1} < 0$, $k_2^T b_{z2} < 0$, these k_i already have correct signs.

In the middle right plot in Fig. 2 the root-locus for the closed-loop system with control law (15) is shown for $\gamma = 0 \div 10$. Parameter $\gamma = 6$ is taken as a candidate for the "stability suboptimal" control law.

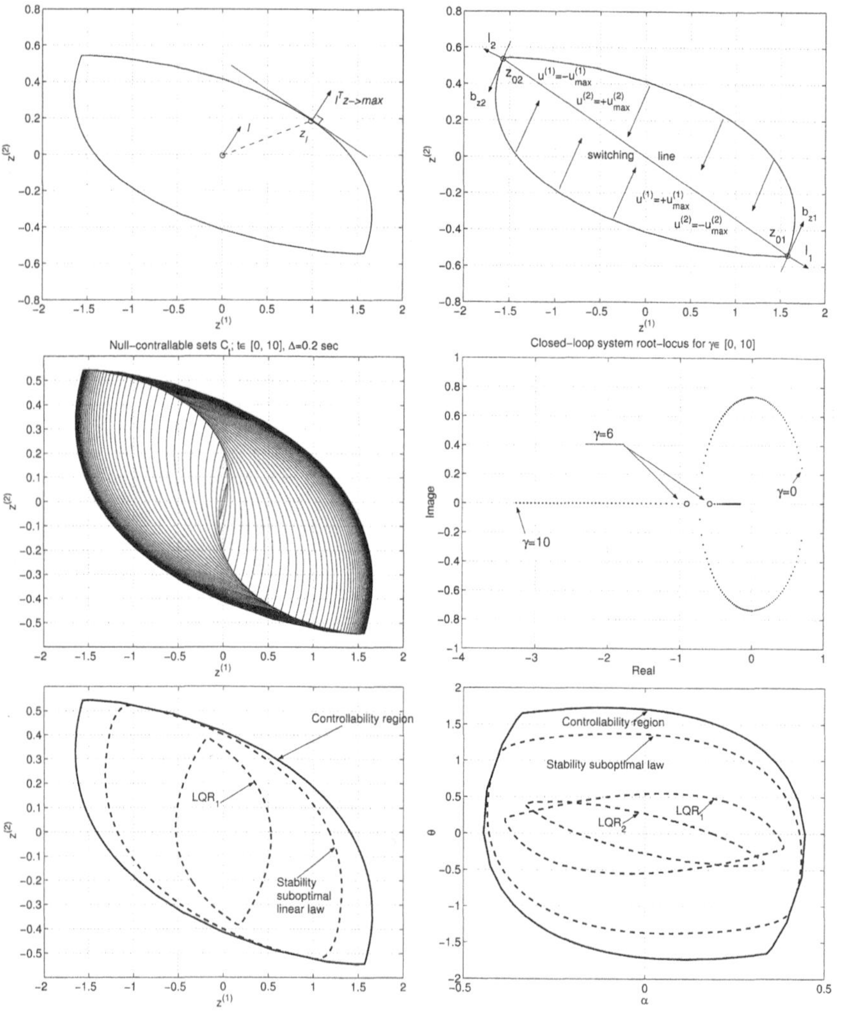

Figure 2 The Formalsky method (top left plot), relay law computation (top right plot), null-controllable sets (middle left plot), closed-loop system root locus (middle right plot), stability and controllability regions for unstable subsystem (bottom left plot), cross-sections of stability regions in the plane (α, θ) (bottom right plot).

In the state-space of the original system this suboptimal control law has the form $u(x) = \text{sat}(K_{sub}x)$, where

$$K_{sub} = \begin{bmatrix} 6\frac{0.035}{||k_1||_2} k_1^T Q_2^T \\ \\ 6\frac{0.035}{||k_2||_2} k_2^T Q_2^T \end{bmatrix} = \begin{bmatrix} 0 & 0.186 & 0.0552 & 0.0382 & -0.0566 & 0.0394 \\ 0 & -0.186 & -0.0552 & -0.0382 & 0.0566 & -0.0394 \end{bmatrix}.$$

The comparison of the stability region for the designed "stability suboptimal" control law and the open-loop system controllability region is presented in the bottom left and right plots in Fig. 2 in the plane of unstable variables and in physical plane (α, θ). It is clearly seen that the closed-loop system stability region with designed linear control is rather close to the controllability region.

In the two bottom plots in Fig. 2 one can see also the cross-sections of stability regions for two linear quadratic regulators designed for comparison purposes. Linear quadratic regulators have been computed by minimizing the performance index (11): one considering only unstable subspace LQR_1 and the other for the original system including stable and unstable subspaces LQR_2.

One can see that the stability region for LQR_1 is bigger than the region for the LQR_2, because LQR_1 stabilizes only the unstable subsystem, while LQR_2 allocates some control resourses for the stable subspace and leaves less for the unstable one. Meanwhile the designed stability suboptimal control law provides a much larger stability region than both linear quadratic regulators and its size is very close to that of the controllability region.

CONCLUDING REMARKS

1. Stabilization of an aircraft's linearized dynamics with one unstable eigenvalue and a redundant number of control effectors can be performed in the whole controllability region by the saturated linear control law. The maximum stabilizer controller can be designed by application of the linear programming method.

2. In the case of two unstable eigenvalues the saturated linear controller can be only the "stability suboptimal" one. The maximum stabilizer providing the necessary stability characteristics can be designed only in nonlinear form.

3. The problem of constrained stabilization of unstable dynamics requires further development of control design methods including more unstable modes and control rate saturation effects.

REFERENCES

1. G.L. Hartmann, M.F. Barrett, and C.S. Greene, "Control design for an unstable vehicle," NASA Dryden Flight Research Center, Contract Rep. NAS 4-2578, Dec. 1979.
2. A.D. Ngo, W.C. Reigelsperger, S.S. Banda, and J.S. Bessolo, "Multivariable control law design of a tailless airplane," AIAA Paper 96-3866, July 1996.

3. D. Cameron and N. Princen, "Control allocation challenges and reguirements for the Blended Wing Body," *AIAA Guidance, Navigation and Control Conference*, August 2000, AIAA Paper 2000-4539.
4. P.P. Friedmann, "Renaissance of aeroelasticity and its future," *Journal of Aircraft*, Vol. 36, No. 1, Jan.–Feb. 1999.
5. V. Mukhopadhyay, "Transonic flutter suppression control law design and wind tunnel test results," *Journal of Guidance, Control, and Dynamics*, Vol. 23, No. 5, Sept.–Oct. 2000.
6. A.M. Formalsky, *Controllability and Stability of Systems with Control Constraints*, Nauka, Moscow, 1974 (in Russian).
7. G.A. Stepanjantz, *Theory of Dynamical Systems*, Mashinostroenie, Moscow, 1985 (in Russian).
8. M.G. Goman, E.V. Fedulova, and A.V. Khramtsovsky, "Maximum stability region design for unstable aircraft with control constraints," *AIAA Guidance, Navigation and Control Conference*, July 1996, AIAA Paper 96-3910.
9. F. Blanchini, and S. Miani, "Constrained stabilization of continuous-time linear systems," *Systems and Control Letters*, Vol. 28, 1996, pp. 95–102.
10. V.A. Kamenetskiy, "Construction of a constrained stabilizing feedback by Lyapunov functions," *Proceedings of 13th IFAC World Congress*, San Francisco, 1996 (2b-04 6), pp. 139–144.
11. T.E. Pare, H. Hindi, J.P. How, and S.P. Boyd, "Synthesizing stability regions for systems with saturating actuators," *Proceedings of the 37th IEEE Conference on Decision & Control*, 1998, pp. 1981–1982.
12. L. Scibile and B. Kouvaritakis, "Stability region for a class of open-loop unstable linear systems: theory and application," *Automatica*, Vol. 36, No. 1, 2000, pp. 37–44.
13. D. Henrion, S. Tarbouriech, and V. Kučera, "Control of linear systems subject to input constraints: a polynomial approach," *Automatica*, Vol. 37, 2001, pp. 597–604.
14. G.H. Golub and C.F. Van Loan, *Matrix Computations*, North Oxford Academic, London, 1986.
15. R.Y. Chiang and M.G. Safonov, *MATLAB Robust Control Toolbox User's Guide*, The MathWorks, Natick, MA, 1998.

9

QUEST Algorithms Case Study: GPS-Based Attitude Determination of Gyrostat Satellite

Jinlu Kuang and Soonhie Tan

Satellite Engineering Center, The School of Electrical and Electronic Engineering, Nanyang Technological University, Singapore

The theoretical methodologies and numerical results of quaternion estimation (QUEST) algorithms for the GPS-based attitude determination of a low-earth-orbit gyrostat satellite are presented. The attitude tumbling of the gyrostat satellite is suppressed with the proposed linear regulators. The numerical techniques for the generation of angular velocities of the gyrostat satellite and the determination of the attitude are investigated. The simulation results indicate that QUEST algorithms can be employed to accurately estimate the attitude of the gyrostat satellite point by point, when combined with the GPS phase-difference signals. Under the appropriate assumption of measurement noises at the GPS phase-difference signals the precision of the attitude estimates is assessed to be at the level of 0.5 degrees for the attitude motions of the gyrostat satellite. The results from the theoretic simulation in this chapter can be used to validate the algorithms of the integer offset time biases and the assumption of line-of-sight measurement noises and baseline calibration.

1 INTRODUCTION

The utilization of phase-difference measurements from the GPS receivers to determine the attitude of satellites, aircraft, and ships has recently attracted the attention of many scientists, because of the GPS sensors' potential to provide effective cost in terms of dollars, weight, and power. In particular, the application of the GPS-based attitude determination for satellites has been focused upon. Crassidis, Lightsey, and Markley (1999) developed an approach for the efficient and optimal attitude determination by using recursive GPS signal operations. They derived their new algorithms from the generalized predictive filter for nonlinear systems proposed by Lu (1994). The distinct advantages of their algorithms are that they can provide optimal attitude solutions even for coplanar baseline configurations. Crassidis and Markley (1997) created another novel technique for finding a point-by-point (deterministic) attitude solution of vehicle by using GPS phase-difference measurements, based upon the algorithms for attitude determination by using GPS phase measurements from the research by Bar-Itzhack, Montgomery, and Garrick (1997). In their paper the general GPS cost function was transformed into a Wahba cost function. They formulated

0-415-30852-6/04/\$0.00+\$1.50

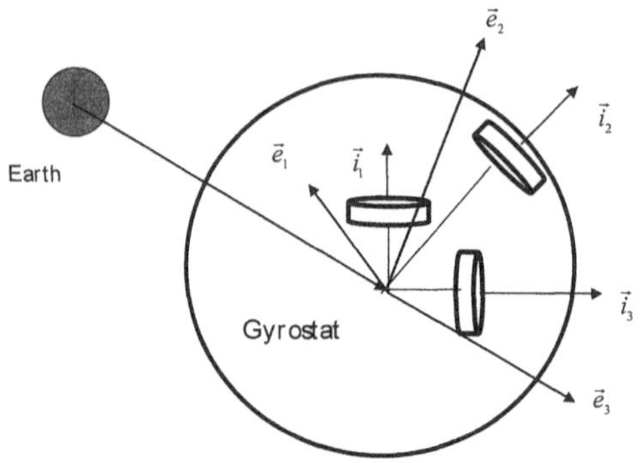

Figure 1 The configuration of the gyrostat rotating about the center of mass O.

the covariance equations for their new method. Crassidis, Markley, and Lightsey (1999) investigated the GPS integer ambiguity resolution without attitude knowledge. This work was extended from the one by Hill and Euler (1996). The algorithm proposed by Crassidis, Markley, and Lightsey (1999) was sequential and independent of attitude, so that it may be implemented in real time. Bar-Itzhack (1996) constructed a recursive QUEST algorithm for sequential attitude determination. Axelrad and Ward (1996) presented the methodology and results for the RADCAL satellite's attitude estimation using the GPS signals. Their research was based upon the method of Kalman filtering. The precision of the final attitude estimation is tested to be at the level of 0.4 deg for pitch and roll and 0.7 deg for yaw. Fujikawa and Zimbelman (1995) also studied the spacecraft attitude determination by Kalman filtering of the GPS signals based upon the observation of the phase-difference measurement. They estimated the vehicle attitude, body rate, disturbance torque, and antenna phase center shift. The factors affecting the system accuracy are also examined, which include the GPS satellite visibility, the number of receiving antennas, and the attitude error observability. Cohen, Lightsey, and Parkinson (1994) investigated the preliminary space flight results (RADCAL) in a low earth orbit. Conway et al. (1996) invented a new motion-based algorithm for the GPS attitude integer resolution, which was suitable for conditions of satellite visibility.

After reviewing the representative references, we discovered that the approaches used to estimate the attitude angles or quaternions could be classified into two types. One is the point-by-point solution for the attitude, e.g., QUEST; the other is an iterative method for the attitude, e.g., discrete Kalman filtering algorithms. The theory of the Kalman filtering algorithms can be found in the book written by Balakrishnan (1987). The Kalman filtering method can be used to deal with the estimation of the attitude, angular velocities, cable biases, and baseline calibration at the same time. The well-known method of point-by-point solutions (e.g., QUEST) is a welcome algorithm for least-squares fitting of the attitude quaternions of a satellite to vector measurements. In this chapter, assuming that the integer offset line biases, line-of-sight measurement noises and the baseline-calibration are known and compensated in advance, we discover that QUEST will provide good solutions of attitude for the gyrostat satellite by using the GPS signals. The motivation for this chapter is based upon the work written by Crassidis and Markley (1997). We formulated the complete skeleton of how to fulfill the QUEST algorithms by using the GPS signals.

The chapter is organized as follows. First, the basic concepts of the GPS phase-difference measurements and the idea of the GPS-based attitude determination of the gyrostat satellites by using the QUEST algorithms are introduced. Next, the attitude propagation equations of the gyrostat satellite are described. In the case of no external disturbance torques being applied to the gyrostat satellite, the exactly analytical solutions from Volterra (1898) are utilized. In the case of external disturbance torques being applied to the gyrostat, the adaptive fourth-order Runge–Kutta integration algorithm is employed to generate the propagation solutions of its angular velocities and quaternions. Then, in order to show the dynamics encountered in the control design of the attitude of a gyrostat satellite whose dynamical equations have multiple equilibrium positions, the equations of multiple equilibrium positions are derived and the stability of the equilibrium positions is also analyzed. Finally, the control law (linear regulator) for the suppression of the attitude tumbling of the gyrostat satellite is proposed. The simulation for the motion of the low Earth orbital gyrostat satellite is given to validate the QUEST method and the proposed linear regulators. During the run there are at least 5 GPS satellites available. The lines of sight and baselines are used to form simulated phase-difference measurements with Gaussian measurement errors. The GPS orbit data are generated from the GPSoft software.

2 QUEST ALGORITHMS FOR GPS-BASED ATTITUDE DETERMINATION

Now consider a gyrostat satellite in a circular, low earth orbit possessing a body-fixed reference with basis vectors $\{\vec{i}_1, \vec{i}_2, \vec{i}_3\}$. The local vertical and local horizontal (LVLH) reference frame with its origin at the center of mass of an orbiting user satellite has a set of vectors $\{\vec{e}_1, \vec{e}_2, \vec{e}_3\}$, with $\vec{e}_1$ perpendicular to the orbit plane, $\vec{e}_2$ along the orbit direction, and $\vec{e}_3$ outward from the Earth center (see Fig. 1). The attitude of the user satellite is assumed to be tumbling in a corresponding circular orbit. This situation represents a realistic scenario when the attitude of the user satellite is out of control. The GPS constellation consists of 24 satellites (21 active plus 3 spares) oriented in six evenly spaced orbit planes inclined 55 degrees relative to the equator [Cohen, 1992].

The measurement utilized for the attitude determination is supposed to be the phase difference of the GPS signal received from two antennae separated by a baseline. Cohen (1992) formulated the attitude determination problem in terms of the GPS observable. The theory of the wavefront angle and wavelength are used to develop the phase difference. The phase measurement is described by

$$b_l \cos(\theta) = \lambda\,(k + \Delta\varphi_a/2\pi), \tag{1}$$

where b_l is the baseline length; θ is the angle between the baseline and the line of sight to the same GPS satellite; k is the number of integer wavelengths of the phase difference between two antennae of the receivers; $\Delta\varphi_a$ is the actual phase-difference measurement; and λ is the wavelength of the GPS signal (see Fig. 2). All GPS satellites generate a carrier signal at the L1 frequency. This frequency is centered on 1575.42 MHz, which corresponds to a wavelength of 19.03 cm. For the details of the GPS phase-difference signals' application to the attitude determination of the satellites, the readers are referred

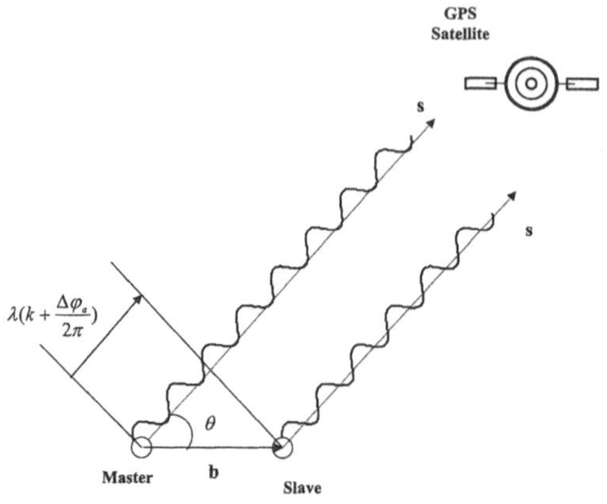

Figure 2 GPS phase difference geometry.

to Crassidis, Lightsey, and Markley (1999). Assuming that the integer offset is known and compensated, the measure normalized fractional phase difference $\Delta\varphi$ can be defined by

$$\Delta\varphi = \boldsymbol{b}^T \boldsymbol{C}\, \boldsymbol{s} + v; \tag{2}$$

here $\boldsymbol{s} \in R^3$ is the normalized line-of-sight vector to the GPS satellite in a local vertical and local horizontal (LVLH) frame; $\boldsymbol{b} \in R^3$ is the normalized baseline vector, which is the relative position vector from one antenna (master) to another antenna (slave) installed in the user gyrostat satellite, and $\boldsymbol{C}$ is the orthogonal attitude matrix, which is the transformation matrix from the LVLH reference frame to the body-fixed reference frame; and v is the measurement error, which is assumed to be a zero-mean random Gaussian process with the standard deviation given by σ, which is $0.5\ \text{cm}/\lambda = 0.026$ wavelengths for the typical phase noise. Equation (2) is used to form equations of observation that are nonlinear and time-variant.

In order to determine the attitude of the gyrostat satellite using Eq. (2), Crassidis and Markley (1997), Bar-Itzhack et al. (1997) converted the phase measurements into vector measurements. The method extended by them involves finding the proper orthogonal matrix $\boldsymbol{C}$, which minimizes the following generalized loss function:

$$J(\boldsymbol{C}) = 0.5 \sum_{i=1}^{m} \sum_{j=1}^{n} \sigma_{ij}^{-2} (\Delta\varphi - \boldsymbol{b}_i^T \boldsymbol{C}\, \boldsymbol{s}_j)^2, \tag{3}$$

where m stands for the number of baselines and n stands for the number of observed GPS satellites, and σ_{ij} stands for the standard deviation of the ijth measurement error. The vectorized measurement problem involves determining the sight line vector in the body frame represented by

$$\boldsymbol{S}_j = \boldsymbol{C}\, \boldsymbol{s}_j, \tag{4}$$

or the baseline in a LVLH frame represented by

$$\boldsymbol{B}_i = \boldsymbol{C}^T \boldsymbol{b}_i. \tag{5}$$

For the sight line case, the following cost function is minimized:

$$J_j(\boldsymbol{S}_j) = 0.5\sum_{i=1}^{m} \sigma_{ij}^{-2}(\Delta\varphi_{ij} - \boldsymbol{b}_i^T\,\boldsymbol{S}_j)^2; \quad \text{for} \quad j = 1,\ 2,\ 3,\ \cdots,\ n. \tag{6}$$

The minimization of Eq. (6) leads to

$$\boldsymbol{S}_j = \boldsymbol{M}_j^{-1}\,\boldsymbol{y}_j, \tag{7}$$

where

$$\boldsymbol{M}_j = \sum_{i=1}^{m} \boldsymbol{b}_i\,\boldsymbol{b}_i^T/\sigma_{ij}^2; \qquad \boldsymbol{y}_j = \sum_{i=1}^{m} \Delta\varphi_{ij}\,\boldsymbol{b}_i/\sigma_{ij}^2 \qquad \text{for } j = 1,\ 2,\ 3,\ \cdots,\ n. \tag{8}$$

According to Eq. (6) it is required that at least three baselines should not be coplanar for the sight lines to be determined in the body frame. Readers are referred to Crassidis and Markley (1997) for the extensive treatment of the case of baselines and the relevant special cases. Wahba (1965) posed the problem of attitude matrix determination as a least-squares problem as follows:

$$J_w(\boldsymbol{C}) = 0.5\sum_{j=1}^{n} a_j\,|\boldsymbol{S}_j - \boldsymbol{C}\,\boldsymbol{s}_j|^2. \tag{9}$$

We can weigh each measurement respectively according to the accuracy of the particular vector measurement. The weight coefficients a_j are positive numbers assigned to each measurement. In order to find the solution of attitude quaternions, rather than attitude matrix (the representation of the attitude), Eq. (9) can be replaced by

$$J_q(\boldsymbol{q}) = 0.5\sum_{j=1}^{n} a_j\,|\boldsymbol{S}_j - \boldsymbol{C}(\boldsymbol{q})\,\boldsymbol{s}_j|^2, \tag{10}$$

where $\boldsymbol{q}$ represents the column vector of the attitude quaternions $(q_1, q_2, q_3, q_4)^T$, which describe the orientation of the body-fixed frame to the LVLH reference frame; and $\boldsymbol{C}$ is defined as follows (Wie, 1995):

$$\boldsymbol{C} = \begin{bmatrix} C_{11} & C_{12} & C_{13} \\ C_{21} & C_{22} & C_{23} \\ C_{31} & C_{32} & C_{33} \end{bmatrix} = \begin{bmatrix} q_4^2 - q_3^2 - q_2^2 + q_1^2 & 2(q_1q_2 + q_3q_4) & 2(q_1q_3 - q_2q_4) \\ 2(q_1q_2 - q_3q_4) & q_4^2 - q_3^2 + q_2^2 - q_1^2 & 2(q_2q_3 + q_1q_4) \\ 2(q_2q_4 + q_1q_3) & 2(q_2q_3 - q_1q_4) & q_4^2 + q_3^2 - q_2^2 - q_1^2 \end{bmatrix}. \tag{11}$$

Instead of minimizing the function in Eq. (9) or (10), we can maximize the following function (Wertz, 1978):

$$g(\boldsymbol{q}) = \boldsymbol{q}^T\boldsymbol{K}\boldsymbol{q}, \tag{12}$$

where $\boldsymbol{K}$ is constructed as follows. Define

$$\boldsymbol{B} = \sum_{j=1}^{n} a_j\boldsymbol{S}_j\,\boldsymbol{s}_j^T/w; \quad \boldsymbol{z} = \sum_{j=1}^{n} a_j\boldsymbol{S}_j \times \boldsymbol{s}_j/w = (B_{23} - B_{32}, B_{31} - B_{13}, B_{12} - B_{21})^T, \tag{13}$$

$$w = \sum_{j=1}^{n} a_j; \quad \boldsymbol{F} = \boldsymbol{B} + \boldsymbol{B}^T; \quad \sigma_a = \sum_{j=1}^{n} a_j \boldsymbol{S}_j^T \boldsymbol{s}_j / w = \text{trace}(\boldsymbol{B}) = B_{11} + B_{22} + B_{33}. \tag{14}$$

Then

$$\boldsymbol{K} = \begin{bmatrix} \boldsymbol{F} - \sigma_a \boldsymbol{I} & \boldsymbol{z} \\ \boldsymbol{z}^T & \sigma_a \end{bmatrix}, \tag{15}$$

where $\boldsymbol{I}$ is the third-order identity matrix, and $B_{ij}(i, j = 1, 2, 3)$ are elements of the matrix $\boldsymbol{B}$. The extrema of $g(\boldsymbol{q})$, subject to the normalization constraint

$$q_1^2 + q_2^2 + q_3^2 + q_4^2 = 1, \tag{16}$$

can be found by the method of Lagrange multipliers. It is shown that the optimal quaternion $\boldsymbol{q}^*$ satisfies

$$\boldsymbol{K}\,\boldsymbol{q}^* = \chi\,\boldsymbol{q}^*, \tag{17}$$

where χ is a yet undetermined Lagrange multiplier.

By choosing the largest eigenvalue $\chi_{\max}$ as the desired optimal eigenvalue, the function $g(\boldsymbol{q})$ can be maximized and the optimal quaternion $\boldsymbol{q}^*$ corresponding to the largest eigenvalue $\chi_{\max}$ can be found using the known relation

$$\boldsymbol{q}^* = \begin{bmatrix} \boldsymbol{y}^* / \sqrt{1 + \boldsymbol{y}^{*2}} \\ 1 / \sqrt{1 + \boldsymbol{y}^{*2}} \end{bmatrix}, \tag{18}$$

where

$$\boldsymbol{y}^* = \left[(\chi_{\max} + \sigma_a)\boldsymbol{I} - \boldsymbol{F} \right]^{-1} \boldsymbol{z}. \tag{19}$$

The algorithm for obtaining $\chi_{\max}$ and $\boldsymbol{q}^*$ from the vector observations discussed is known as the quaternion estimation (QUEST) algorithm. Shuster (1990) showed how to deal with the singularity matrix case in Eq. (19). The weight coefficients $a_j(j = 1, 2, \cdots, n)$ can be selected from the assumed values of the standard deviation $\sigma_{ij}$of the measured phase-difference signals (Bar-Itzhack, 1996). QUEST is a famous algorithm for least-squares fitting of the attitude quaternion of a satellite to vector measurements. QUEST is also a single-point attitude determination algorithm, which yields a closed-form solution of the quaternion and therefore experiences no divergence problems. The divergence problems are sometimes encountered in the use of extended Kalman filter approach. Bar-Itzhack (1996) developed an algorithm of a recursive QUEST (REQUEST) algorithms for sequential attitude determination, which considered all past measurements based upon the fact that the so-called $\boldsymbol{K}$ matrix, one of whose eigenvectors is the sought quaternion, is linearly related to the measured vector pairs and on the ability to propagate $\boldsymbol{K}$. REQUEST yields the quaternion estimates that are functions of gyro bias in practical situations. In this chapter we combined the QUEST algorithms and GPS resources to determine the attitude of the gyrostat satellite.

3 ATTITUDE PROPAGATION EQUATIONS OF GYROSTAT SATELLITES

A model of the attitude dynamics of the gyrostat satellite is needed to smooth the state estimate and to provide an attitude reference between GPS updates and during periods of poor observing geometry. Here, between measurement epochs, the quaternion and angular velocity states are advanced to the current measurement time by numerically integrating the nonlinear equations of motion for a gyrostat satellite. These ordinary differential equations are described as follows from the principles of analytical mechanics:

$$\dot{q}_1 = 0.5(\omega_1 q_4 - \omega_2 q_3 + \omega_3 q_2 - \Omega q_4), \tag{20}$$
$$\dot{q}_2 = 0.5(\omega_1 q_3 + \omega_2 q_4 - \omega_3 q_1 + \Omega q_3), \tag{21}$$
$$\dot{q}_3 = 0.5(-\omega_1 q_2 + \omega_2 q_1 + \omega_3 q_4 - \Omega q_2), \tag{22}$$
$$\dot{q}_4 = 0.5(-\omega_1 q_1 - \omega_2 q_2 - \omega_3 q_3 + \Omega q_1), \tag{23}$$

$$I_1 \dot{\omega}_1 - (I_2 - I_3)\omega_2 \omega_3 + \omega_2 h_3 - \omega_3 h_2 = u_1 + d_1, \tag{24}$$
$$I_2 \dot{\omega}_2 - (I_3 - I_1)\omega_3 \omega_1 + \omega_3 h_1 - \omega_1 h_3 = u_2 + d_2, \tag{25}$$
$$I_3 \dot{\omega}_3 - (I_1 - I_2)\omega_1 \omega_2 + \omega_1 h_2 - \omega_2 h_1 = u_3 + d_3, \tag{26}$$

where I_1, I_2, and I_3 are the principal moments of inertia of the gyrostat satellite; ω_1, ω_2, and ω_3 are the components of the angular velocities along the body-fixed reference frame; h_1, h_2, and h_3 are assumed to be constant components of the angular momenta of the wheels in the body-fixed coordinate system; Ω is the orbit rate; u_1, u_2, and u_3 are components of the control torques; d_1, d_2, and d_3 are the components of external disturbance torques, which may be made up of the gravity-gradient torques, the aerodynamic torques, torques due to solar radiation pressure, and residual magnetic dipole moment.

The three Euler angles of $C_1(\theta_1) \leftarrow C_3(\theta_3) \leftarrow C_2(\theta_2)$ sequence from the LVLH reference frame to a body-fixed reference of a gyrostat satellite in a circular orbit are related to quaternions (Wie, 1998):

$$q_1 = \cos(0.5\theta_1)\sin(0.5\theta_2)\sin(0.5\theta_3) + \sin(0.5\theta_1)\cos(0.5\theta_2)\cos(0.5\theta_3), \tag{27}$$
$$q_2 = \cos(0.5\theta_1)\sin(0.5\theta_2)\cos(0.5\theta_3) + \sin(0.5\theta_1)\cos(0.5\theta_2)\sin(0.5\theta_3), \tag{28}$$
$$q_3 = \cos(0.5\theta_1)\cos(0.5\theta_2)\sin(0.5\theta_3) - \sin(0.5\theta_1)\sin(0.5\theta_2)\cos(0.5\theta_3), \tag{29}$$
$$q_4 = \cos(0.5\theta_1)\cos(0.5\theta_2)\cos(0.5\theta_3) - \sin(0.5\theta_1)\sin(0.5\theta_2)\sin(0.5\theta_3). \tag{30}$$

The inverse relationships can be derived as follows:

$$\theta_1 = \tan^{-1}[2(q_1 q_4 - q_2 q_3)/(1 - 2q_1^2 - 2q_3^2)], \tag{31}$$
$$\theta_2 = \tan^{-1}[2(q_2 q_4 - q_1 q_3)/(1 - 2q_2^2 - 2q_3^2)], \tag{32}$$
$$\theta_3 = \sin^{-1}[2(q_1 q_2 + q_3 q_4)]. \tag{33}$$

During the long-term mission, a fourth-order Runge–Kutta integration algorithm was used to give the sequential solutions of the angular velocities and attitude angles or quaternions. If we neglect the external disturbance/control torques including the terms related to Ω in Eqs. (20)–(26), the angular velocities can be solved analytically (Volterra, 1989; Wittenburg, 1975). Assume that the kinetic energy is T(constant) and the moment of momentum is $2DT$ (D is also a known constant). Also assume that the real numbers $\lambda_1 < \lambda_2 < \lambda_3 < \lambda_4$ are the eigenvalues of the following generalized characteristic equations:

$$Z(\lambda) = \det(\boldsymbol{A} - \lambda \boldsymbol{B}) = 0, \tag{34}$$

where

$$A = \frac{1}{\sqrt{(I_1 I_2 I_3)}} \begin{bmatrix} I_1^2 & 0 & 0 & I_1 h_1 \\ 0 & I_2^2 & 0 & I_2 h_2 \\ 0 & 0 & I_3^2 & I_3 h_3 \\ I_1 h_1 & I_2 h_2 & I_3 h_3 & h^2 - 2DT \end{bmatrix}, \tag{35}$$

$$B = \frac{1}{\sqrt{(I_1 I_2 I_3)}} \begin{bmatrix} I_1 & 0 & 0 & 0 \\ 0 & I_2 & 0 & 0 \\ 0 & 0 & I_3 & 0 \\ 0 & 0 & 0 & -2T \end{bmatrix}; \quad h^2 = h_1^2 + h_2^2 + h_3^2, \tag{36}$$

then

$$\omega_i(t) = \frac{A_{i1} sn(\tau,k) + A_{i2} cn(\tau,k) + A_{i3} dn(\tau,k) + A_{i4}}{B_1 sn(\tau,k) + B_2 cn(\tau,k) + B_3 dn(\tau,k) + B_4}, \tag{37}$$

where sn, cn, and dn are the Jacobi elliptic functions with modulus k.

$$\tau = t\sqrt{\frac{2T(\lambda_1-\lambda_3)(\lambda_2-\lambda_4)}{I_1\, I_2\, I_3}} + \tau_0, \qquad k^2 = \frac{(\lambda_1-\lambda_2)(\lambda_3-\lambda_4)}{(\lambda_1-\lambda_3)(\lambda_2-\lambda_4)}, \tag{38}$$

$$B_1 = (\lambda_2-\lambda_3)(\lambda_3-\lambda_4)\sqrt{(\lambda_1-I_1)(\lambda_1-I_2)(\lambda_1-I_3)(\lambda_4-\lambda_2)}, \tag{39}$$

$$B_2 = (\lambda_1-\lambda_3)(\lambda_3-\lambda_4)\sqrt{(\lambda_2-I_1)(\lambda_2-I_2)(\lambda_2-I_3)(\lambda_1-\lambda_4)}, \tag{40}$$

$$B_3 = (\lambda_2-\lambda_4)\sqrt{(\lambda_3-I_1)(\lambda_3-I_2)(\lambda_3-I_3)(\lambda_1-\lambda_2)(\lambda_1-\lambda_3)(\lambda_1-\lambda_4)}, \tag{41}$$

$$B_4 = (\lambda_1-\lambda_3)(\lambda_2-\lambda_3)\sqrt{(\lambda_4-I_1)(\lambda_4-I_2)(\lambda_4-I_3)(\lambda_1-\lambda_2)}, \tag{42}$$

$$A_{ij} = h_i B_j/(\lambda_j - I_i)\,; \qquad \text{for} \quad i = 1,2,3; \quad j = 1,2,3,4, \tag{43}$$

and τ_0 depending upon the initial conditions. If we set the initial angular velocities as ω_{10}, ω_{20}, and ω_{30}, then τ_0 satisfies the following equation:

$$sn(\tau_0,k) = \sqrt{\frac{(\lambda_4-\lambda_2)(\lambda_1-I_1)(\lambda_1-I_2)(\lambda_1-I_3)}{(\lambda_1-\lambda_2)(\lambda_4-I_1)(\lambda_4-I_2)(\lambda_4-I_3)}\,\frac{p_{11}\frac{\omega_{10}}{h_1}+p_{12}\frac{\omega_{20}}{h_2}+p_{13}\frac{\omega_{30}}{h_3}+p_{14}}{p_{41}\frac{\omega_{10}}{h_1}+p_{42}\frac{\omega_{20}}{h_2}+p_{43}\frac{\omega_{30}}{h_3}+p_{44}}}, \tag{44}$$

where

$$p_{ij} = \begin{cases} (I_k - I_l)(\lambda_\alpha - I_j)(\lambda_\beta - I_j)(\lambda_\gamma - I_j), \\ \text{for } i = 1,4;\ (j,k,l) = (1,2,3) \text{ cyclic};\ (\alpha,\beta,\gamma) = (1,2,3,4) - (i) \text{ cyclic}, \\ (I_1 - I_2)(I_2 - I_3)(I_3 - I_1), \text{ for } \ \mathrm{i} = 1{,}4; \ \ \mathrm{j} = 4. \end{cases} \tag{45}$$

From the characteristic Eq. (34), assuming that the real numbers

$$\lambda_1 < \lambda_2 < \lambda_3 < \lambda_4 \tag{46}$$

are known, we could derive the formulae of the wheel's moments of momentum as follows:

$$h_1^2 = 2T(\lambda_1-I_1)(\lambda_2-I_1)(\lambda_3-I_1)(\lambda_4-I_1)/[I_1(I_1-I_2)(I_1-I_3)], \tag{47}$$

$$h_2^2 = 2T(\lambda_1-I_2)(\lambda_2-I_2)(\lambda_3-I_2)(\lambda_4-I_2)/[I_2(I_2-I_1)(I_2-I_3)], \tag{48}$$

$$h_3^2 = 2T(\lambda_1-I_3)(\lambda_2-I_3)(\lambda_3-I_3)(\lambda_4-I_3)/[I_3(I_3-I_1)(I_3-I_2)], \tag{49}$$

$$D = \lambda_1\left(2T + h_1^2/(\lambda_1-I_1) + h_2^2/(\lambda_1-I_2) + h_3^2/(\lambda_1-I_3)\right)/(2T). \tag{50}$$

For the practical application, the sampling interval should be well below Nyquist's limit. The determined quaternion can be found by integrating the ordinary differential Eqs. (20)–(26) from time t_k to time t_{k+1}. If the angular velocity vector ω_k is assumed to be constant over this interval, a discrete propagation for Eqs. (20)–(23) can be used, given by Wertz (1978):

$$\boldsymbol{q}_{k+1} = [\beta_k \boldsymbol{I}_{4\times4} + \gamma_k \Lambda(\rho_k)]\boldsymbol{q}_k, \tag{51}$$

where

$$\beta_k = \cos(0.5\,\|\omega_k\|\,\Delta t); \quad \gamma_k = \sin(0.5\,\|\omega_k\|\,\Delta t); \quad \rho_k = \omega_k/\|\omega_k\|; \quad \Delta t = t_{k+1} - t_k, \tag{52}$$

$$\Lambda(\rho_k) = \begin{bmatrix} 0 & \rho_{k3} & -\rho_{k2} & \rho_{k1} \\ -\rho_{k3} & 0 & \rho_{k1} & \rho_{k2} \\ \rho_{k2} & -\rho_{k1} & 0 & \rho_{k3} \\ -\rho_{k1} & -\rho_{k2} & -\rho_{k3} & 0 \end{bmatrix}. \tag{53}$$

4 STABILITY ANALYSIS OF A KELVIN GYROSTAT SATELLITE

Now consider the attitude dynamics model of a gyrostat satellite without the external torques. In this case the gyrostat is called a Kevin gyrostat whose Euler equations are described as follows:

$$\dot{\omega}_1 = [(I_2 - I_3)\omega_2\omega_3 - \omega_2 h_3 + \omega_3 h_2]/I_1 \equiv f_1(\omega_1, \omega_2, \omega_3), \tag{54}$$
$$\dot{\omega}_2 = [(I_3 - I_1)\omega_3\omega_1 - \omega_3 h_1 + \omega_1 h_3]/I_2 \equiv f_2(\omega_1, \omega_2, \omega_3), \tag{55}$$
$$\dot{\omega}_3 = [(I_1 - I_2)\omega_1\omega_2 - \omega_1 h_2 + \omega_2 h_1]/I_3 \equiv f_3(\omega_1, \omega_2, \omega_3). \tag{56}$$

From the theory of the dynamical system, the equilibrium positions should satisfy the following equations:

$$f_1(\omega_{1e}, \omega_{2e}, \omega_{3e}) = 0, \tag{57}$$
$$f_2(\omega_{1e}, \omega_{2e}, \omega_{3e}) = 0, \tag{58}$$
$$f_3(\omega_{1e}, \omega_{2e}, \omega_{3e}) = 0. \tag{59}$$

According to the law of conservation of kinetic energy of the gyrostat satellite and Eqs. (57) and (59), the following equilibrium positions are derived:

$$\omega_{3e} = \omega_{2e}\, h_3/[h_2 + (I_2 - I_3)\,\omega_{2e}], \tag{60}$$
$$\omega_{1e} = \omega_{2e}\, h_1/[h_2 + (I_2 - I_1)\,\omega_{2e}], \tag{61}$$

where ω_{2e} are the roots of the following polynomial equation of the 6th power:

$$\beta_6\,\omega_{2e}^6 + \beta_5\,\omega_{2e}^5 + \beta_4\,\omega_{2e}^4 + \beta_3\,\omega_{2e}^3 + \beta_2\,\omega_{2e}^2 + \beta_1\,\omega_{2e} + \beta_0 = 0, \tag{62}$$

where

$$\beta_6 = I_2(I_2 - I_1)^2(I_2 - I_3)^2; \quad \beta_5 = 2I_2h_2(2I_2 - I_1 - I_3)(I_2 - I_1)(I_2 - I_3), \tag{63}$$

$$\beta_4 = I_1h_1^2(I_2 - I_3)^2 + I_2h_2^2\left[(2I_2 - I_1 - I_3)^2 + 2(I_2 - I_1)(I_2 - I_3)\right] + I_3h_3^2(I_2 - I_1)^2 - 2T\,(I_2 - I_1)^2(I_2 - I_3)^2, \tag{64}$$

$$\beta_3 = 2h_2\left[I_1h_1^2(I_2 - I_3) + I_2h_2^2(2I_2 - I_1 - I_3) + I_3h_3^2(I_2 - I_1)\right] - 4T\,h_2(2I_2 - I_1 - I_3)(I_2 - I_1)(I_2 - I_3), \tag{65}$$

$$\beta_2 = h_2^2(I_1h_1^2 + I_2h_2^2 + I_3h_3^2) - 2T\,h_2^2\left[(2I_2 - I_1 - I_3)^2 + 2(I_2 - I_1)(I_2 - I_3)\right], \tag{66}$$

$$\beta_1 = -4T\,h_2^3(2I_2 - I_1 - I_3); \quad \beta_0 = -2T\,h_2^4. \tag{67}$$

The Jacobi matrix of the vector functions appearing on the right side of Eqs. (54)–(56) at the equilibrium positions can be defined as follows:

$$\left.\frac{\partial(f_1, f_2, f_3)}{\partial(\omega_1, \omega_2, \omega_3)}\right|_{(\omega_{1e}, \omega_{2e}, \omega_{3e})} = \begin{bmatrix} f_{11} & f_{12} & f_{13} \\ f_{21} & f_{22} & f_{23} \\ f_{31} & f_{32} & f_{33} \end{bmatrix}, \tag{68}$$

where

$$f_{11} = 0; \quad f_{12} = [(I_2 - I_3)\omega_{3e} - h_3]/I_1; \quad f_{13} = [(I_2 - I_3)\omega_{2e} + h_2]/I_1, \tag{69}$$

$$f_{21} = [(I_3 - I_1)\omega_{3e} + h_3]/I_2; \quad f_{22} = 0; \quad f_{23} = [(I_3 - I_1)\omega_{1e} - h_1]/I_2, \tag{70}$$

$$f_{31} = [(I_1 - I_2)\omega_{2e} - h_2]/I_3; \quad f_{32} = [(I_1 - I_2)\omega_{1e} + h_1]/I_3; \quad f_{33} = 0. \tag{71}$$

The characteristic equation corresponding to the Jacobi matrix can be given as follows:

$$\begin{vmatrix} f_{11} - \Gamma & f_{12} & f_{13} \\ f_{21} & f_{22} - \Gamma & f_{23} \\ f_{31} & f_{32} & f_{33} - \Gamma \end{vmatrix} = 0, \tag{72}$$

where Γ is the eigenvalue. The instability or stability of the equilibrium positions will depend on the positive or negative signs of the real parts of the eigenvalue Γ. The full understanding of the multiple equilibrium states of the gyrostat satellite dynamic system in Eqs. (60)–(62) will be conducive to the design of the control law. The nonlinear attitude motions of the gyrostat satellite subject to the appropriate perturbations will exhibit chaotic features (Kuang et al., 2000, 2001).

5 SUPPRESSION OF ATTITUDE TUMBLING OF THE GYROSTAT SATELLITE

Assume that no disturbance or external control torque is applied to the gyrostat satellite, then the attitude of the gyrostat satellite will be tumbling. How to suppress the tumbling attitude of the gyrostat satellite is one of our concerns from the point of view of dynamics and control. Here, we propose the following feedback controls that can be used to suppress the tumbling attitude of the gyrostat satellite:

$$u_1 = -\xi_1\,\omega_1 - \psi\,q_1, \tag{73}$$

$$u_2 = -\xi_2\,\omega_2 - \psi\,q_2, \tag{74}$$

$$u_3 = -\xi_3\,\omega_3 - \psi\,q_3, \tag{75}$$

where ξ_i, ψ $(i = 1, 2, 3)$ are positive numbers to be selected (Kuang and Leung, 2001) .

The equations of states for the closed-loop control system consist of Eqs. (20)–(26) (in the case of $d_x = d_y = d_z = 0$) with the additional constraint (Eq. (16)). Mortensen (1968) used the linear regulator in Eqs. (73)–(75) to control the attitude error of the pure rigid model.

To analyze the stability of the equilibrium point,

$$\boldsymbol{x}^* = (\omega_1, \omega_2, \omega_3, q_1, q_2, q_3, q_4) = (0,\ 0,\ 0,\ 0,\ 0,\ 0, 1), \tag{76}$$

we construct the function,

$$E = 0.5(I_1\omega_1^2 + I_2\omega_2^2 + I_3\omega_3^2) + \psi(q_1^2 + q_2^2 + q_3^2 + (q_4 - 1)^2). \tag{77}$$

It is easy to show that E is a positive definite quadratic form. The total time derivative can be computed by the chain rule. Considering the state equations of the closed-loop system the function $\dot{E} = \frac{dE}{dt}$ can be reduced to the following form:

$$\dot{E} = -\xi_1\omega_1^2 - \xi_2\omega_2^2 - \xi_3\omega_3^2. \tag{78}$$

Consequently, we know the derivative with respect to time $\dot{E} \leqslant 0$ for all t. According to the Lyapunov stability theory, the perturbed motions with respect to the equilibrium point are globally asymptotically stable for any positive values of ξ_i, ψ $(i = 1,\ 2,\ 3)$. This can be validated using the numerical simulation experiment given in the next section.

6 SIMULATION PROCEDURES AND RESULTS

The following section contains a compilation of attitude solutions and numerical simulation results of the attitude determination using the GPS phase-difference signals from the gyrostat satellite that is assumed to be a small satellite model in a low-Earth, circular, equatorial orbit at an altitude of 870 km, and the weight is 50 kg. The simulation procedures are outlined as follows. First using the GPSoft software to produce the GPS satellite orbits and the user satellite orbit when given the angle of mask, then using the principles of GPS phase-measurement geometry to construct the line of sight, we can simulate the GPS phase-difference measurement. It is very important to obtain the GPS phase-difference signals as accurately as possible so as to identify the right attitude of the gyrostat satellite. The simulation results show that the attainable precision of the Euler angles' estimates using QUEST is assessed to be at the level of 10^{-15} for amplitudes if there are no measurement noises at the GPS phase-difference signals. The physical parameters are given as follows: $I_1 = 9$ kg m^2, $I_2 = 7$ kg m^2, $I_3 = 3$ kg m^2, $h_1 = 10\sqrt{21}$ N m s, $h_2 = 9\sqrt{5}$ N m s, $h_3 = 15\sqrt{7}$ N m s. The normalized baseline vectors are assumed to be $\mathrm{b}_1 = [1/\sqrt{2},\ 1/\sqrt{2}, 0]$, $\mathrm{b}_2 = [0,\ 1, 0], \mathrm{b}_3 = [0, 0, 1]$ (Kuang and Tan, 2002).

CASE 1. Attitude Determination of the Gyrostat Satellite under Torque-Free Conditions

Assume that the kinetic energy constant is $2T = 3780$ N m, the moment of momentum (constant) is $2DT = 1920$ (N m s)^2. The attitude of the gyrostat satellite will be tumbling

Table 1 Equilibrium positions of kelvin gyrostat satellite

λ^* (kg m^2)	D^* (N m s^2)	ω_1^* (rad/s)	ω_2^* (rad/s)	ω_3^* (rad/s)	
5.5999	5.1538	−13.478	−14.373	15.265	Unstable
4.3799	4.9967	−9.9187	−7.6807	28.761	Stable
1.8052	0.9990	−6.3693	−3.8740	−33.216	Stable
11.305	14.878	19.884	4.6751	4.7788	Stable

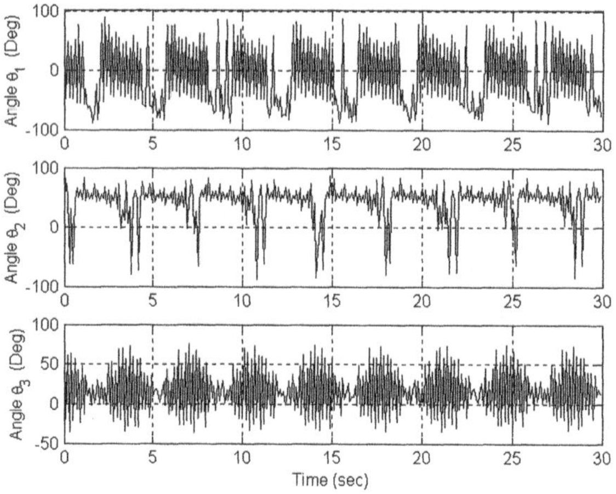

Figure 3 The evolution of the Euler angles (*theoretical solutions A*).

in the orbit. According to the characteristic Eq. (34), the corresponding eigenvalues are computed as follows: $\lambda_1 = 8, \quad \lambda_2 = 6, \quad \lambda_3 = 5, \quad \lambda_4 = 4.$

The equilibrium positions can be calculated according to the Eqs. (60)–(62) and are given in Table 1.

In Table 1, the following results apply according to Wittenburg (1975):

$$I_1\,\omega_1^* + h_1 = \lambda^*\omega_1^* \,; \quad I_2\,\omega_2^* + h_2 = \lambda^*\omega_2^* \,; \quad I_3\,\omega_3^* + h_3 = \lambda^*\omega_3^*; \tag{79}$$

$$I_1\,(\omega_1^*)^2 + I_2\,(\omega_2^*)^2 + I_3\,(\omega_3^*)^2 = 2T; \tag{80}$$

$$D^* = \left[(I_1\,\omega_1^* + h_1)^2 + (I_2\,\omega_2^* + h_2)^2 + (I_3\,\omega_3^* + h_3)^2\right] \,/(2T). \tag{81}$$

The nominal parameters of moments of inertia and constant moments of momenta of the wheels are calculated from Eqs. (47)–(50).

In the case of no external disturbance torques, Fig. 3 shows the evolution history of the Euler angles in Eqs. (31)–(33) (called *Theoretical Solutions A*), directly from the analytical solutions in Eqs. (37) corresponding to the dynamic attitude Eqs. (20)–(26). Figure 4

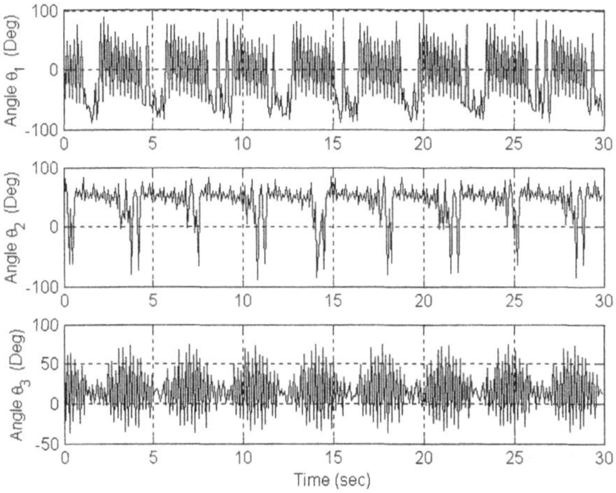

Figure 4 The evolution of the Euler angles (*QUEST solutions A*).

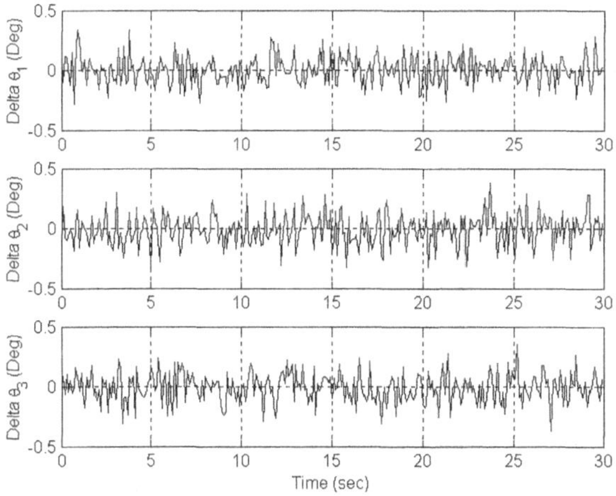

Figure 5 The errors' evolution of the Euler angles.

shows the evolution history of the Euler angles in Eqs. (31)–(33) (called *QUEST Solutions A*), directly using GPS signals from the approach of QUEST. Figure 5 shows the errors' evolution of the Euler angles between Theoretical Solutions A and QUEST Solutions A. By comparison, we found out that the precision of the attitude estimates was tested to be within 0.5 degrees for the attitude motions.

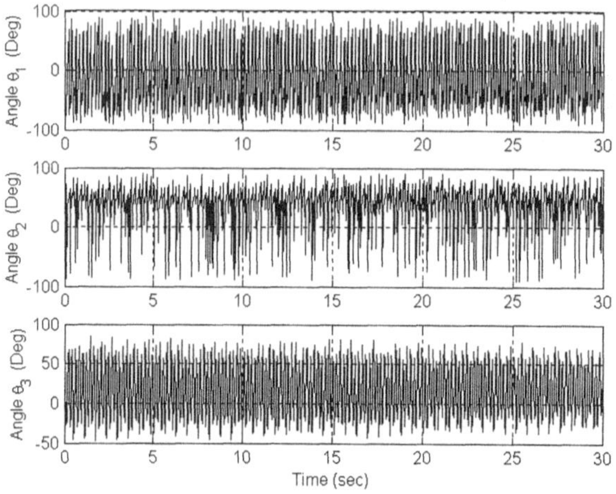

Figure 6 The evolution of the Euler angles (*theoretical solutions B*).

CASE 2. Attitude Determination of Gyrostat Satellites under Gravity-Gradient Torques

For simplicity here, the aerodynamic torques, torques due to solar radiation pressure, and residual magnetic dipole moment applied to the user satellite are neglected. Assume that the gravity-gradient torques and control torques are as follows:

$$d_1 = -3\Omega^2(I_2 - I_3)C_{23}C_{33}\,, \qquad u_1 = 0, \tag{82}$$

$$d_2 = -3\Omega^2(I_3 - I_1)C_{13}C_{33}\,, \qquad u_2 = 0, \tag{83}$$

$$d_3 = -3\Omega^2(I_1 - I_2)C_{13}C_{23}\,, \qquad u_3 = 0, \tag{84}$$

where $\Omega = \sqrt{\mu/R^3}$, and R is the distance of the gyrostat satellite mass center from the Earth's center, and μ is the gravitational parameter of the Earth. Figure 6 shows the evolution history of the Euler angles (called *Theoretical solutions B*) directly numerically integrated from dynamical equations excited by the gravity-gradient torques using the Runge–Kutta fourth-order method. Figure 7 shows the evolution history of the Euler angles from QUEST method (called *QUEST solutions B*) with consideration of the gravity-gradient torques. Figure 8 shows the errors' evolution history of the Euler angles between Theoretical solutions B and QUEST solutions B. By comparison, we also found out that the precision of the attitude estimates was tested to also be within 0.5 degrees for the attitude motions.

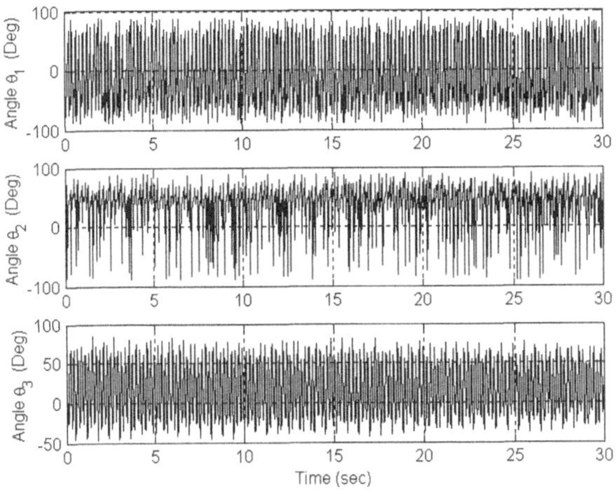

Figure 7 The evolution of the Euler angles (*QUEST solutions B*).

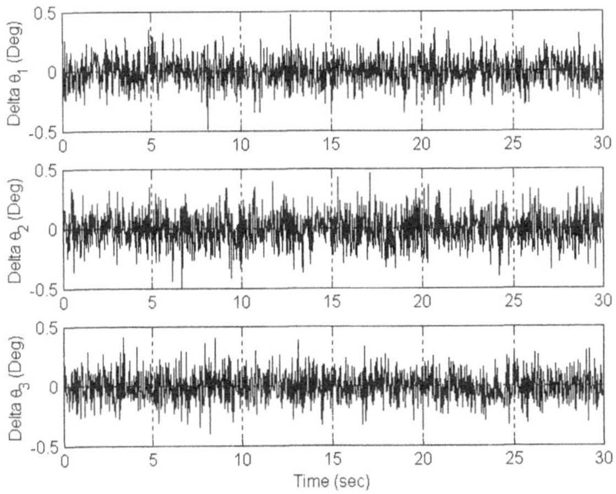

Figure 8 The error's evolution of the Euler angles.

CASE 3. Attitude Determination of Gyrostat Satellites with Linear Feedback Control Torques

In the case $\xi_i = 24$ $(i = 1, 2, 3)$ N m s , $\psi = 18$ N m, the linear regulator are defined in Eqs. (73)–(75), and the simulated start time of control is from $t = 10$ s. Figure 9

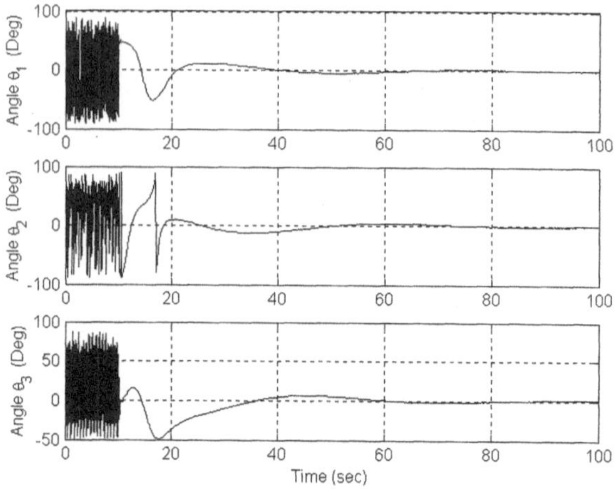

Figure 9 The evolution of the Euler angles (*theoretical solutions C*).

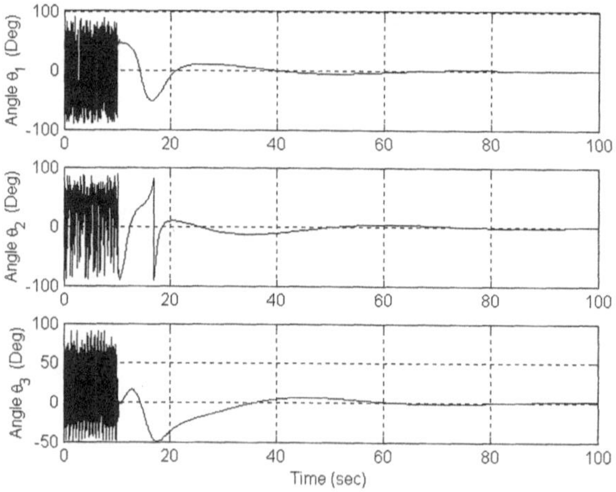

Figure 10 The evolution of the Euler angles (*QUEST solutions C*).

shows the evolution history of the Euler angles (called *Theoretical solutions C)* directly numerically integrated from dynamical equations excited by the linear feedback controls using the Runge–Kutta fourth-order method. Figure 10 shows the evolution history of the Euler angles (called *QUEST solutions C*) from the QUEST method with consideration of the linear feedback controls. Figure 11 shows the errors' evolution history of the Euler

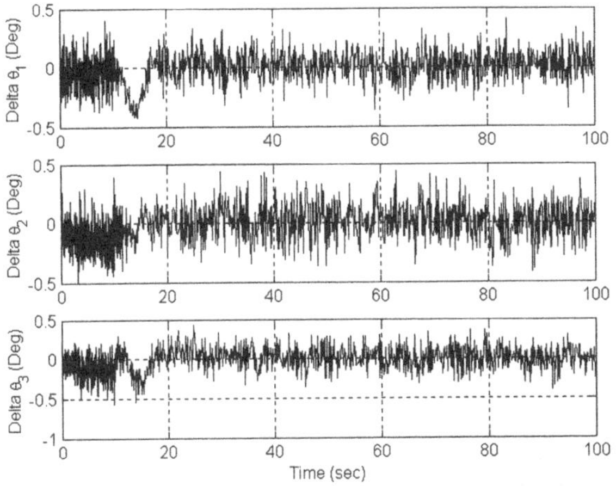

Figure 11 The errors' evolution of the Euler angles.

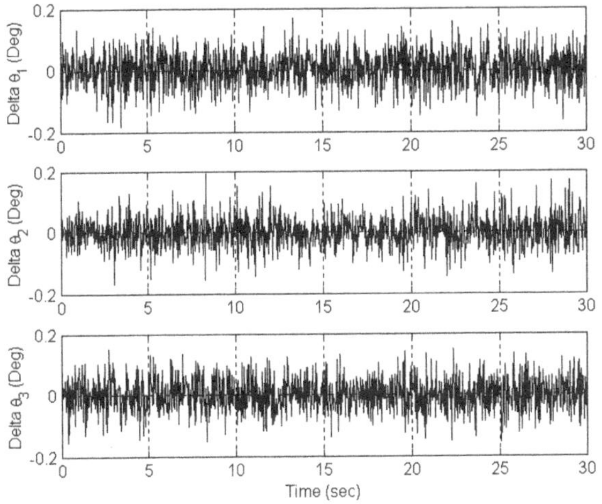

Figure 12 Error evolution of the Euler angles when the noise level is halved; the other parameters are the same as in Case 2.

angles between Theoretical solutions C and QUEST solutions C. By comparison, we also found out that the precision of the attitude estimates was tested to be within 0.5 degrees for the attitude motions.

CASE 4. Effect of Measurement Noises

The amplitudes of the Euler angles' error are sensitive to the standard square-root deviation of the measurement noises. Figure 12 shows that when the σ is reduced to be half of the typical phase noise, i.e., $0.25\ \text{cm}/\lambda = 0.013$ wavelengths, the maximum amplitudes of the Euler angles' error are greatly reduced by 70%. The other parameters are the same as in CASE 2.

7 CONCLUSIONS

In this paper we presented the theoretical methodologies of QUEST algorithms and numerical results for the GPS-based attitude determination of a low-earth-orbit gyrostat satellite, which has multiple equilibrium positions in the Kelvin gyrostat case. The feedback control law of the attitude tumbling is proposed. By applying the linear feedback control along the body-fixed frame of the gyrostat satellite, the suppression effect of the attitude tumbling of the gyrostat satellite is good enough. For the case of no external disturbance/control torques applied to the gyrostat satellite, by using the exactly theoretical solutions of the angular velocities from V. Volterra, the theoretical attitude solutions are determined from the discrete propagation equations for quaternions. In this way, the GPS theoretical phase-difference signals are simulated accurately through using GPSoft software, and the QUEST solutions for the Euler angles are computationally easier than the solutions by using the Kalman filtering method. For the case of external disturbance/control torques applied to the gyrostat satellite, the fourth-order Runge–Kutta algorithms are employed to generate the theoretical, point-by-point solutions of the angular velocities and quaternions based upon the GPS-simulated phase-difference signals numerically simulated. Using the famous QUEST method, the precision of the Euler angles can be found within about 0.5 degrees for the attitude motions under the appropriate assumption of measurement noises at the GPS phase-differences. All of the simulation results show that the QUEST method is effective for the determination of the Euler angles of the gyrostat satellite using the GPS signals if the practical GPS phase-difference signals are measured with sufficient accuracy. The nonlinear dynamics of the gyrostat satellite encountered in this chapter is also suitable for the nonlinear dynamics of the liquid-filled solid body model of the satellite. The work done here will be implemented for the future mission of the Singapore X-satellite that is being developed.

REFERENCES

1. Axelrad, P. and Ward, L.M. (1996) "Spacecraft attitude estimation using the Global Positioning System: methodology and results for RADACL," *Journal of Guidance, Control, and Dynamics,* Vol. 19, pp. 1201–1209.
2. Balakrishnan, A.V. (1987) *Kalman Filtering Theory*, Optimization Software, Inc., Publications Division, New York.
3. Bar-Itzhack, I.Y. (1996) "REQUEST: a recursive QUEST algorithm for sequential attitude determination," *Journal of Guidance, Control, and Dynamics*, Vol. 19, pp. 1034–1038.

4. Bar-Itzhack, I.Y., Montgomery, P.Y., and Garrick, J.C. (1997) "Algorithms for attitude determination using GPS," *Proceedings of the AIAA Guidance, Navigation, and Control Conference*, New Orleans, LA.
5. Cohen, C.L. (1992) *Attitude Determination Using GPS*, Ph.D. dissertation, Dept. of Aeronautics and Astronautics, Stanford University, Stanford, CA.
6. Cohen, C.L., Lightsey, E.G., and Parkinson, B.W. (1994) "Space flight tests of attitude determination using GPS," *International Journal of Satellite Communications*, Vol. 12, pp. 427–433.
7. Conway, A., Montgomery, P., Rock, S., Cannon, R., and Parkinson, B.W. (1996) "A new motion-based algorithm for GPS attitude integer resolution," *Navigation, Journal of the Institute of Navigation*, Vol. 43, pp. 179–190.
8. Crassidis, J.L., Lightsey, E.G., and Markley, F.L (1999) "Efficient and optimal attitude determination using recursive Global Positioning System signal operations," *Journal of Guidance, Control, and Dynamics*, Vol. 22, pp. 193–201.
9. Crassidis, J.L. and Markley, F.L. (1997) "New algorithm for attitude determination using Global Positioning System signals," *Journal of Guidance, Control, and Dynamics*, Vol. 20, pp. 891–896.
10. Crassidis, J.L., Markley, F.L., and Lightsey, E.G. (1999) "Global Positioning System integer ambiguity resolution without attitude knowledge," *Journal of Guidance, Control, and Dynamics*, Vol. 22, pp. 212–218.
11. Fujikawa, S.J. and Zimbelman, D.F. (1995) "Spacecraft attitude determination by Kalman filtering of Global Positioning System signals," *Journal of Guidance, Control, and Dynamics*, Vol. 18, pp. 1365–1371.
12. *GPSoft Software Manual Book*, Princeton Satellite Systems, Inc., Princeton, NJ.
13. Hill, C.D. and Euler, H.J. (1996) "An optimal ambiguity resolution technique for attitude determination," *Proceedings of the IEEE, Position, Location, and Navigation Symposium*, Atlanta, GA, pp. 62–269.
14. Hoots, F.R. and Roehrich, R.L. (1988) *Models for Propagation of NORAD Element Sets*, SpaceTrack Report No. 3, Compiled by T.S. Kelso.
15. Kuang, J.L. and Tan, S.H. (2002) "The GPS-based attitude determination of gyrostat satellites by quaternion estimation algorithms," *Acta Astronautica*, Vol. 51, pp. 743–759.
16. Kuang, J.L., Tan, S.H., and Leung, A.Y.T. (2000) "Chaotic attitude motion of satellites under small perturbation torques," *Journal of Sound and Vibration*, Vol. 235, pp. 175–200.
17. Kuang, J.L., Tan, S.H., and Arichandran, K. (2000) "Chaotic attitude motion of symmetric gyrostat via Melnikov integral," *Proceedings of the Third Non-linear Problems in Aviation and Aerospace*, Daytona Beach, FL, Edited by S. Sivasundaram (in press).
18. Kuang, J.L. and Leung, A.Y.T. (2001) "H_∞ feedback for attitude control of liquid-filled spacecraft," *Journal of Guidance, Control, and Dynamics*, Vol. 24, pp. 46–53.
19. Kuang, J.L., Tan, S.H., Arichandran, K., and Leung, A.Y.T. (2001) "Chaotic dynamics of an asymmetrical gyrostat," *International Journal of Non-linear Mechanics*, Vol. 36, pp. 35–55.
20. Lu, P. (1994) "Non-linear prediction controllers for continuous systems," *Journal of Guidance, Control, and Dynamics*, Vol. 17, pp. 553–560.
21. Mortensen, R.E. (1968) "A globally stable linear attitude regulator," *International Journal of Control*, Vol. 8, pp. 297–302.

22. Shuster, M.D. (1990) "Kalman filter of spacecraft attitude and QUEST model," *Journal of the Astronautical Sciences*, Vol. 38, pp. 377–393.
23. Volterra, V. (1898) "Sur la theories des variations des latitudes," *Acta Math.*, Vol. 22, pp. 201–357.
24. Wahba, G. (1965) "A least-squares estimates of spacecraft attitude," *SIAM Review*, Vol. 7, p. 409.
25. Wertz, J.R. (1978) *Spacecraft Attitude Determination and Control*, D. Reidel, Dordrecht, The Nethelands, pp. 414–416.
26. Wie, B. (1998) *Space Vehicle Dynamics and Control*, AIAA Education Series, AIAA, Reston, VA, pp. 307–459.
27. Wittenburg, J. (1975) Beitrage zur Dynamik von Gyrostaten, *Acc. Naz. Dei Lincei*, Vol. 217, pp. 1–187.

10

Asymptotic Controllability in Terms of Two Measures of Hybrid Dynamic Systems

V. Lakshmikantham† and S. Sivasundaram*

†*Florida Institute of Technology, Department of Mathematical Sciences, Melbourne, FL*

**Embry-Riddle Aeronautical University, Department of Mathematics, Daytona Beach, FL*

1 INTRODUCTION

The relationship between asymptotic controllability to the origin in R^n of a nonlinear system,

$$x' = f(x, u), \tag{1.1}$$

exhibited by an open-loop control $u : [0, \infty) \to U$, and the existence of a feedback control $k : R^n \to U$ that stabilizes trajectories of the system

$$x' = f(x, k(x)), \tag{1.2}$$

with respect to the origin, have been recently studied by many authors in this field. It is well known that continuous feedback laws may not exist even for simple asymptotically controllable systems. General results regarding the nonexistence of continuous feedback were presented in [1], which led to the search for feedback laws that are not necessarily of the form $u = k(x)$, where k is a continuous function. It is natural to ask about the existence of discontinuous feedback laws $u = k(x)$ (optimal control problems), which led to the search for general theorems ensuring their existence. Difficulty arises in defining the meaning of solution $x(.)$ of (1.2) with a discontinuous right-hand side. The Filippov solution, namely the solution of a certain differential inclusion with multivalued right-hand side that is built from $f(x, k(x))$, is one possibility. But there is no hope of obtaining general results if one insists on the use of the Filippov solution. In [2] this was generalized for arbitrary feedback $k(x)$. The main objective is to study the relationship between the existence of stabilizing (discontinuous) feedback and asymptotic controllability of the open-loop system (1.1), considering the feedback law, which, for fast enough sampling, drives all states asymptotically to the origin with small overshoot. Feedback law constructed can be robust with respect to the actual errors as well as to perturbations of the system dynamics, and also may be highly sensitive to errors in the measurement of the state vector. In [3] these drawbacks were avoided by designing a dynamic hybrid stabilizing controller which, while preserving robustness to external perturbations and actuator error, is also robust with respect to measurement error. Recently two measures have been used to unify various stability concepts

0-415-30852-6/04/$0.00+$1.50

and to offer a more general framework [4]. Partial stability, for example, can be discussed by using two measures. In this chapter, we attempt to utilize the concept of two measures to extend the results of [2].

2 PRELIMINARIES

We need the following notation for our discussions:

(1) $\mathcal{K} = \{a \in C[R_+, R_+] : a(u)$ is strictly increasing in u and $a(u) = 0\}$.

(2) $\mathcal{L} = \{a \in C[R_+, R_+] : a(u)$ is strictly decreasing in u and $\lim_{u\to\infty} a(u) = 0\}$.

(3) $\mathcal{KL} = \{a \in C[R_+^2, R_+] : a(t,s) \in \mathcal{K}$ for each s and $a(t,s) \in \mathcal{L}$ for each $t\}$.

(4) $C\mathcal{K} = \{a \in C[R_+^2, R_+] : a(t,s) \in \mathcal{K}$ for each $t\}$.

(5) $\Gamma = \{h \in C[R_+ \times R^n, R_+] : inf h(t,x) = 0\}$.

(6) $S(h,\rho) = \{(t,x) \in R_+ \times R^n,\ h(t,x) < \rho,\ \rho > 0\}$.

(7) $S^c(h_0,\rho) = \{(t,x) \in R_+ \times R^n,\ h_0(t,x) \geq \rho,\ \rho > 0\}$.

We shall now define stability concepts in terms of two measures $h_0, h \in \Gamma$.

Definition 2.1. The differential system (1.1) is said to be (h_0, h)-equistable if for each $\epsilon > 0$ and $t_0 \in R_+$ there exists a function $\delta = \delta(t_0, \epsilon) > 0$ that is continuous in t_0 for each ϵ such that $h_0(t_0, x_0) < \delta$ implies $h(t, x(t)) < \epsilon$, $t \geq t_0$ where $x(t) = x(t, t_0, x_0)$ is any solution of (1.1).

Definition 2.2. (h_0, h) – asymptotically stable if Definition 2.1 holds and given $t_0 \in R_+$ there exists a $\delta_0 = \delta_0(t_0) > 0$ such that $h_0(t_0, x_0) < \delta_0$ implies $\lim_{t\to\infty} h(t, x(t)) = 0$.

It is easy to see that the above definitions reduce to

(1) the well-known stability of the trivial solution $x(t) \equiv 0$ of (1.1) or equivalently, of the invariant set $\{0\}$, if $h(t,x) = h_0(t,x) = \|x\|$;

(2) the stability of prescribed motion $x_0(t)$ of (1.1) if $h(t,x) = h_0(t,x) = \|x - x_0(t)\|$;

(3) the partial stability of the trivial solution of (1.1) if $h_0(t,x) = \|x\|$ and $h(t,x) = \|x\|_s$, $1 \leq s \leq n$;

(4) the stability of the asymptotically invariant set $\{0\}$, if $h(t,x) = h_0(t,x) = \|x\| + \sigma(t)$, where $\sigma \in \mathcal{L}$.

Definition 2.3. The system (1.1) is said to be (h_0, h)-equibounded if for each $\alpha > 0, t_0 \in R_+$, there exists a positive function $\beta = \beta(t_0, \alpha)$ that is continuous in t_0 for each α such that $h_0(t_0, x_0) \leq \alpha$ implies $h(t, x(t)) < \beta$, $t \geq t_0$, where $x(t) = x(t, t_0, x_0)$ is any solution of (1.1).

Definition 2.4. (h_0, h) is uniformly bounded if β in the above definition is independent of t_0.

Definition 2.5. Let $h_0, h \in \Gamma$. Then, we say that

(i) h_0 is finer than h if there exists a $\rho > 0$ and a function $\phi \in C\mathcal{K}$ such that $h_0(t, x) < \rho$ implies $h(t, x) \leq \phi(t, h_0(t, x))$.

(ii) h_0 is uniformly finer than h if in (i) ϕ is independent of t.

(iii) h_0 is asymptotically finer than h if there exists a $\rho > 0$ and a function $\phi \in \mathcal{KL}$ such that $h_0(t, x) < \rho$ implies $h(t, x) \leq \phi(t, h_0(t, x))$.

Note 2.1. h_0, h and other functions defined above are independent of t in further discussions.

We assume that U is a locally compact metric space, and the mapping $f : R^n \times U \to R^n$ such that f is continuous, locally Lipschitz on x, uniformly on compact subsets of $R^n \times U$. $|x|$ denote the Euclidean norm of $x \in R^n$. $\langle x, y\rangle$ denotes the inner product of two vectors.

A locally bounded function $b : R^n \to U$ will be called a feedback.

Any finite sequence $p = \{t_i\}_{i\geq 0}$ consisting of numbers $0 = t_0 < t_1 < t_2 ...$ with $\lim_{i\to\infty} t_i = \infty$ is called a partition of $[0, \infty)$, $d(p) = \sup_{i\geq 0}(t_{i+1} - t_i)$ is its diameter.

Consider the nonlinear system

$$x' = f(x, u), \tag{2.1}$$

where $u : [0, \infty) \to U$ and the feedback control system

$$x' = f(x, b(x)), \tag{2.2}$$

where $b : R^n \to U$.

Definition 2.6. The p-trajectory of (2.1) starting from x_0 is the function $x(.)$ obtained by recursively solving

$$x' = f(x(t), b(x(t_i))), \quad t \in [t_i, t_{i+1}],\ i = 0, 1, 2..., \tag{2.3}$$

using $x(t_i)$ as the initial value [the end point of the solution on the preceding interval, starting with $x(t_0) = x_0$] for given feedback b, a partition p and $x_0 \in R^n$.

Definition 2.7. The feedback $b : R^n \to U$ is said to be (h_0, h)-sample-stabilizing the system (2.1) for each pair $0 < r < R$, if there exist $M = M(R) > 0$, $\lambda = \lambda(r, R) > 0$ and $T = T(r, R) > 0$ such that, for every partition p with $d(p) < \lambda$ and for any initial state $x_0$such that $h_0(x_0) \leq R$ the p-trajectory $x(.)$ of (2.2) starting from x_0 is well defined and

$$h(x(t)) \leq r \text{ for } t \geq T; \tag{2.3a}$$

$$h(x(t)) \leq M(R) \text{ for all } t \geq 0; \tag{2.3b}$$

$$\lim_{R\to 0} M(R) = 0. \tag{2.3c}$$

Definition 2.8. The system (2.1) is (h_0, h)-asymptotically controllable if

(a) for each $x_0 \in R^n$, there exists some control u such that the trajectory $x(t) = x(t, x_0, u)$ is defined for all $t \geq 0$ and $h(x(t)) \to 0$ as $t \to \infty$;

(b) for each $\epsilon > 0$ there exist $\delta > 0$ such that for each $x_0 \in R^n$ with $h_0(x_0) < \delta$ there is a control u in (a) such that $h(x(t)) < \epsilon$ for all $t \geq 0$;

(c) there are compact subsets X_0 and U_0 of R^n and U, respectively, such that if the initial state $x_0 \in X_0$, then the control in (b) can be chosen with $u(t) \in U_0$ for almost all t.

Given a function $V : R^n \to R$ and a vector $v \in R^n$ the lower directional derivative of V in the direction of v' is

$$DV(x,v) = \lim_{t\to 0} \inf_{v'\to v} \frac{1}{t}[V(x+tv') - V(x)].$$

The function $v \mapsto DV(x,v)$ is lower semicontinuous. For a set $F \subseteq R^n$, CoF denotes its convex hull.

Definition 2.9. A control Lyapunov pair for the system (2.1) consists of two continuous functions $V, W : R^n \to R$ such that the following properties hold:

(a) $V(x) > 0$ and $W(x) > 0$ for all $x \neq 0$, and $V(0) = 0$.

(b) The set $\{x : V(x) \leq \beta\}$ is bounded for each β.

(c) For each bounded subset $G \subseteq R^n$, there is some compact subset $U_0 \subseteq U$ such that

$$\min_{v \in Cof(x,U_0)} DV(x;v) \leq -W(x) \text{ for every } x \in G. \tag{2.4}$$

If V is part of the control Lyapunov pair (V,W), then V is called the control Lyapunov function.

Theorem 2.1. System (2.1) is asymptotically controllable if it admits a control Lyapunov function.

Remark 2.1. If the function V is smooth (2.4) can be written as

$$\min_{u \in U_0} \langle \nabla V(x); f(x,u)\rangle \leq -W(x). \tag{2.5}$$

Definition 2.10. A vector $\xi \in R^n$ is a proximal subgradient (supergradient) of the function $V : R^n \to (-\infty,\infty)$ at x if there exists some $\sigma > 0$ such that for all y in some neighborhood of x:

$$V(y) \geq V(x) + \langle \xi, y-x\rangle - \sigma\, h_0(x-y) \tag{2.6}$$

$$(V(y) \leq V(x) + \langle \xi, y-x\rangle + \sigma\, h_0(x-y)).$$

Definition 2.11. The set of proximal subgradient of V at x (may be empty) is denoted by $\partial_p V(x)$ and is called the proximal subdifferential of V at x.

Remark 2.2. (2.5) can be replaced by

$$\min_{u \in U_0} \langle \xi; f(x,u)\rangle \leq -W(x) \tag{2.7}$$

for every $\xi \in \partial_p V(x)$ where ξ and $\partial_p V(x)$ are the proximal subgradient and subdifferential, respectively, of the function V at the point x.

Remark 2.3. For any $\xi \in \partial_p V(x)$ and any $v \in R^n$,

$$\langle \xi; v \rangle \leq DV(x, v). \tag{2.8}$$

Definition 2.12. Let h_0 satisfy the following conditions:

$$h_0(u+v) \leq h_0(u) + h_0(v) + h_{0x}(v)u, \tag{2.9}$$
$$h_0(v) \leq h_0(u) + h_0(u-v), \tag{2.10}$$
$$k_1|x|^2 \leq h_0(x) \leq k_2|x|^2, \tag{2.11}$$
$$|h_{0x}(x - y_\alpha)| \leq L(R^0), \text{ where } x, y_\alpha \in S(h, R^0), \tag{2.12}$$

where we assume that $h_{0x}(x)$ exists and is continuous.

3 MAIN RESULTS

Define the function

$$V_\alpha(x) = \inf_{y \in R^n} \left[V(y) + \frac{1}{\alpha^2} h_0(x-y) \right], \ \alpha \in (0, 1], \tag{3.1}$$

which is called the inf-convolution of the function V. Then the function V_α is locally Lipschitz and

$$\lim_{\alpha \to 0} V_\alpha(x) = V(x).$$

Let $y_\alpha(x) \neq \phi$ be an element of the set of minimizing points of y in (3.1). Then

$$V_\alpha(x) = V(y_\alpha(x)) + \frac{1}{\alpha^2} h_0(x - y_\alpha(x)) \leq V(y) + \frac{1}{\alpha^2} h_0(x-y) \tag{3.2}$$

for all $y \in R^n$. Let $\xi_\alpha(x) = \frac{1}{\alpha^2} h_{0x}(x - y_\alpha)$, then we have the following lemmas.

Lemma 3.1. For any $x \in R^n$, $\xi_\alpha(x) \in \partial_p V(y_\alpha(x))$.

Proof. By (3.2) we have

$$V(y) \geq V(y_\alpha(x)) + \frac{1}{\alpha^2}[h_0(x - y_\alpha(x)) - h_0(x-y)].$$

By definition 2.8,

$$h_0(x-y) \leq h_0(y_\alpha - y) + h_0(x - y_\alpha) + h_{0x}(x - y_\alpha)(y_\alpha - y).$$

Hence

$$V(y) \geq V(y_\alpha(x)) + \langle \xi_\alpha, y - y_\alpha \rangle - \frac{1}{\alpha^2}(y_\alpha - y),$$

and by definition 2.8,

$$\xi_\alpha(x) \in \partial_p V(y_\alpha(x)).$$

□

Lemma 3.2. For any $x \in R^n$, $\xi_\alpha(x)$ is a proximal supergradient of V_α at x.

Proof. We shall prove this by showing that

$$V_\alpha(y) \le V_\alpha(x) + \langle \xi_\alpha(x), y - x\rangle + \sigma\, h_0(y - x) \text{ for every } x, y \in R^n. \tag{3.3}$$

By definition,

$$V_\alpha(y) - V_\alpha(x) \le \frac{1}{\alpha^2}[\, h_0(y - y_\alpha(x)) - h_0(x - y_\alpha(x))].$$

By definition 2.8 we have

$$h_0(y - y_\alpha(x)) \le h_0(y - x) + h_0(x - y_\alpha(x)) + h_{0x}(x - y_\alpha)(y - x).$$

Hence

$$V_\alpha(y) \le V_\alpha(x) + \langle \xi_\alpha(x), y - x\rangle + \frac{1}{\alpha^2} h_0(y - x),$$

which completes the proof with $\sigma = \frac{1}{\alpha^2}$. □

Let us define the following notations for convenience.

$$\begin{aligned} G_R &= \left\{ x : V(x) \le \frac{1}{2} b(R) \right\}, \\ G_R^\alpha &= \left\{ x : V_\alpha(x) \le \frac{1}{2} b(R) \right\}, \\ \rho(R) &= \max[\rho : S(h_0, \rho) \subseteq G_R], \\ R_0 &= \max(M(R), \phi(R)), \\ S(h, R_0) &= [x : V(x) \le b(R_0)], \\ S_\alpha(h, R_0) &= [x : V_\alpha(x) \le b(R_0)]. \end{aligned}$$

Also,

$$\lim_{R\to 0} a(R) = \lim_{R\to 0} \rho(R) = 0, \tag{3.4}$$

$$\lim_{R\to\infty} b(R) = \lim_{R\to\infty} \rho(R) = \infty. \tag{3.5}$$

Let

$$w_R(\lambda) = \max\{V(x) - V(\tilde{x}) : h_0(x - \tilde{x}) \le \lambda;\ x, \tilde{x} \in S(h_0, R^0), \tag{3.6}$$

where $R^0 = R + a(R)$,

$$b(h(x)) \le V(x) \le a(h_0(x)). \tag{3.7}$$

Lemma 3.3.

$$h_0(x) \le R \text{ implies } h_0(x - y_\alpha) \le a(R)\alpha^2. \tag{3.8}$$

Proof. Since $V_\alpha(x) \le V(x)$, we have

$$V_\alpha(x) = V(y_\alpha(x)) + \frac{1}{\alpha^2} h_0(x - y_\alpha(x)),$$

By (3.7), this implies $h_0(x - y_\alpha(x)) \le a(R)\alpha^2$. Hence by definition 2.8 $h_0(y_\alpha(x)) \le a(R)\alpha^2 + R$. Recall that $\alpha \in (0, 1]$, which implies that

$$h_0(y_\alpha(x)) \le [a(R) + R] = R^0.$$

That is $x \in S(h_0, R^0)$ implies $h_0(y_\alpha(x)) \le [a(R) + R] = R^0$, which yields $y_\alpha \in S(h_0, R^0)$. □

Lemma 3.4. For any $x \in S(h_0, R)$,

$$V_\alpha(x) \le V(x) \le V_\alpha(x) + w_R(a(R)\alpha^2). \tag{3.9}$$

Proof. We know $V_\alpha(x) \le V(x)$, and hence $V_\alpha(x) \ge V(y_\alpha(x))$. By (3.6),

$$w_R(a(R)\alpha^2) = \max[V(x) - V(y_\alpha(x)) : h_0(x - y_\alpha(x)) \le a(R)\alpha^2,\ x, y_\alpha \in S(h_0, R^0)].$$

Hence $V(y_\alpha(x)) \ge V(x) - V(y_\alpha(x))$, which implies that

$$V_\alpha(x) \le V(x) \le V_\alpha(x) + w_R(a(R)\alpha^2).$$

Thus $S(h_0, \rho(R)) \subseteq G_R \subseteq G_R^\alpha$ true for all $R > 0$, $\alpha > 0$. □

Lemma 3.5. For any $R > 0$ and α satisfying

$$w_R(a(R)\alpha^2) < \frac{1}{2}b(R_0) \tag{3.10}$$

we have

$$G_{R_0}^\alpha \subseteq int\, S(h, R_0). \tag{3.11}$$

Proof. Let $x \in G_{R_0}^\alpha$. Then by definition,

$$V_\alpha(x) \le \frac{1}{2}b(R_0) < b(R_0).$$

But

$$V_\alpha(x) = V(y_\alpha(x)) + \frac{1}{\alpha^2}h_0(x - y_\alpha(x)),$$

which implies that

$$V(y_\alpha(x)) \le \frac{1}{2}b(R_0),$$

and

$$\frac{1}{\alpha^2}h_0(x - y_\alpha(x)) \le \frac{1}{2}b(R_0).$$

By definition 2.7 and lemma 3.3 it follows that $y_\alpha(x) \in S(h_0, R^0)$. Then

$$\begin{aligned} V(x) &\le V_\alpha(y_\alpha(x)) + w_R(a(R)\alpha^2) \\ &\le \frac{1}{2}b(R_0) + \frac{1}{2}b(R_0), \end{aligned}$$

which yields by definition 3.7 $h(x) < R_0$. □

Theorem 3.1. Let V be a control Lyapunov function. Then for any $0 < r < R$ there are $\alpha_0 = \alpha_0(r, R)$ and $T = T(r, R)$ such that for any $\alpha \in (0, \alpha_0)$ there exists $\lambda > 0$ such that for any $x_0 \in G_R^\alpha$ and any partition p with $d(p) < \lambda$ the trajectory $x(.)$ of $x' = f(x, k_\nu(x))$ starting at x_0 must satisfy $x(t) \in G_R^\alpha, \forall t \ge 0$ and $x(t) \in S(h, r), \forall t \ge T$.

Proof. By continuity of $f(x,u)$ and the local Lipschitz property assumed, we know that there are some constants l, m such that

$$|f(x,u) - f(\tilde{x},u)| \le l|x-\tilde{x}|,\ |f(x,u)| \le m \tag{3.12}$$

for all $x, \tilde{x} \in S(h, R_0)$ and all $u \in U_0$.

Let

$$\Delta = \frac{1}{3}\min\left\{w(y) : \frac{1}{2}\rho(r) \le h_0(y) \text{ and } h(y) \le R^0\right\}. \tag{3.13}$$

Note that $\Delta > 0$.

Claim 1. Let $\alpha \in (0,1]$ satisfy

$$a(R)\alpha^2 < \frac{1}{2}\rho(r), \quad \frac{L(R^0)}{k_1} l w_R(a(R)\alpha^2) < \Delta. \tag{3.14}$$

Then, for any $x \in S(h, R^0)\backslash S(h_0, \rho(r))$,

$$\langle \xi_\alpha(x); f(x, k_\nu(x))\rangle \le -2\Delta. \tag{3.15}$$

Proof. We have

$$\begin{aligned}\langle \xi_\alpha(x); f(x, k_\nu(x))\rangle &= \min_{u\in U_0}\langle \xi_\alpha(x); f(x,u)\rangle \\ &= \min_{u\in U_0}\langle \xi_\alpha(x); f(y_\alpha,u) + f(x,u) - f(y_\alpha,u)\rangle \\ &\le \min_{u\in U_0}\langle \xi_\alpha(x); f(y_\alpha(x),u)\rangle + |\xi_\alpha(x)|\, l\, |x - y_\alpha(x)|.\end{aligned}$$

But by definition 2.7, it follows that

$$\begin{aligned}|\xi_\alpha(x)|\,|x - y_\alpha(x)| &\le \frac{L(R^0)}{k_1}[V(x) - V(y_\alpha(x))] \\ &\le \frac{L(R^0)}{k_1} w_R(a(R)\alpha^2),\end{aligned}$$

which implies that

$$\langle \xi_\alpha(x); f(x, k_\nu(x))\rangle \le -W(y_\alpha(x)) + l\frac{L(R^0)}{k_1} w_R(a(R)\alpha^2).$$

Since $x \in S(h, R^0)\backslash S(h_0, \rho(r))$, we obtain

$$\frac{1}{2}\rho(r) \le h_0(y) \text{ and } h(y) \le R^0,$$

which implies, by the definition of Δ, $W(y_\alpha) \ge 3\Delta$, and

$$\begin{aligned}\langle \xi_\alpha(x); f(x, k_\nu(x))\rangle &\le -3\Delta + \Delta \\ &\le -2\Delta.\end{aligned}$$

□

Now consider any p-trajectory of $x' = f(x, k_\nu(x))$ corresponding to a partition $p = \{t_i\}_{i\geq 0}$ with $d(p) \leq \lambda$ satisfying the inequality

$$\left(\frac{1}{\alpha^2} L(R^0) lm + k_2 m^2\right) \lambda \leq \Delta. \tag{3.16}$$

Claim 2. Let a, λ satisfy (3.10), (3.14) and (3.16) and assume that for some index i, if $x(t_i) \in G_R^\alpha \backslash S(h_0, \rho(r))$, then

$$V_\alpha(x(t)) - V_\alpha(x(t_i)) \leq -\Delta(t - t_i) \tag{3.17}$$

for all $t \in [t_i, t_{i+1}]$. In particular $x(t) \in G_R^\alpha$ for all such t.

Proof. Let $x_i \equiv x(t_i)$ and consider the largest $\bar{t} \in (t_i, t_{i+1})$ such that $x(\tau) \in S(h, R_0)$ for all $\tau \in [t_i, \bar{t}]$. By (3.11), we have $x(\tau) \in int\, S(h, R_0)$. Choose any $t \in [t_i, \bar{t}]$. By (3.12) we have

$$|x(\tau) - x_i| \leq m(\tau - t_i). \tag{3.18}$$

In general

$$x(t) = x_i + (t - t_i) f_i, \tag{3.19}$$

where

$$f_i = \frac{1}{t - t_i} \int_{t_i}^{t} f(x(\tau), k_\nu(x_i))\, d\tau = f(x_i, k_\nu(x_i)) + \eta_i;$$

(3.12) and (3.18) imply that

$$|\eta_i| \leq lm(\tau - t_i), \; |f_i| \leq m. \tag{3.20}$$

Also, by (3.3) we get

$$V_\alpha(y) \leq V_\alpha(x) + \langle \xi_\alpha(x), y - x\rangle + \frac{1}{\alpha^2} h_0(y - x)$$

and

$$\begin{aligned} V_\alpha(x(t)) - V_\alpha(x(t_i)) &= V_\alpha(x_i + (t - t_i) f_i) - V_\alpha(x(t_i)) \\ &\leq (t - t_i)\langle \xi_\alpha(x_i), f_i\rangle + \frac{1}{\alpha^2} h_0((t - t_i) f_i). \end{aligned}$$

But we find that

$$\begin{aligned} \langle \xi_\alpha(x_i), f_i\rangle &\leq \langle \xi_\alpha(x_i), f(x_i, k_\nu(x_i))\rangle + |\xi_\alpha(x_i)|\, |\eta_i| \\ &\leq -2\,\Delta + \left|\frac{1}{\alpha^2} h_{0x}(x - y_\alpha)\right| lm\lambda, \end{aligned}$$

which implies

$$V_\alpha(x(t)) - V_\alpha(x(t_i)) \leq (t - t_i)\left[-2\Delta + \left|\frac{1}{\alpha^2} h_{0x}(x - y_\alpha)\right| lm\lambda\right] + \frac{1}{\alpha^2} h_0((t - t_i) f_i).$$

Now by definition 2.7 and (3.16), the R.H.S. of the above inequality reduces to

$$\begin{aligned}&\leq (t-t_i)\left[-2\,\Delta + \frac{1}{\alpha^2}L(R^0)lm\lambda + \frac{1}{\alpha^2}k_2m^2\lambda\right]\\&\leq -\Delta(t-t_i),\end{aligned}$$

which implies that $x(\bar{t}) \in S(h, R_0)$, which contradicts the maximality of $\bar{t}$ unless $\bar{t} = t_{i+1}$.
Therefore it holds for all $t \in [t_i, t_{i+1}]$. □

Claim 3. Assume that α satisfies (3.10), (3.14), and choose any λ such that (3.16) is valid. Then for any p-trajectory $x(.)$ with $d(p) \leq \lambda$ and every $x(0) \in G_R^\alpha$,

$$t_N \leq T = \frac{b(R)}{2\Delta}. \tag{3.21}$$

Proof. From claim 1 and the minimality of N, we have

$$x(t_i) \in G_R^\alpha \setminus G_r.$$

Using the fact that $S(h_0, \rho(r)) \subseteq G_r$ and

$$\begin{aligned}0 \leq V_\alpha(x(t_N)) &\leq V_\alpha(x(0)) - \Delta t_N\\&\leq \frac{1}{2}b(R) - \Delta t_N\end{aligned}$$

yields

$$t_N \leq T = \frac{b(R)}{2\Delta}.$$

□

Claim 4. Assume that α and λ satisfy all the previous conditions and in addition the following two conditions are satisfied:

$$w_R(a(R)\alpha^2) < \frac{1}{4}b(r) \tag{3.22}$$

$$w_R(k_2m^2\lambda) < \frac{1}{4}b(r). \tag{3.23}$$

Then $x(t) \in S(h, r)$ for all $t \geq t_N$.

Proof. If $x(t_i) \in G_r$, then by (3.12) and (3.23), we get

$$V_\alpha(x(t)) \leq V(x(t)) \leq V(x(t_i)) + w_R(k_2m^2\lambda) < \frac{3}{4}b(r)$$

for all $t \in [t_i, t_{i+1}]$. Then

$$V_\alpha(x(t_{i+1})) < \frac{3}{4}b(r).$$

If $x(t_{i+1}) \notin G_r$ by claim 2, we have

$$V_\alpha(x(t)) < \frac{3}{4}b(r) \text{ for all } t \in [t_i, t_j], \tag{3.24}$$

where j is the least integer such that $t_i > t_j$ and $x(t_j) \in G_r$ (if there is any).

Starting from t_j, we may repeat the argument and show that (3.24) holds for all $t \geq 0$. The relation (3.22) and Lemma 3.4 imply that

$$V(x(t)) \leq V_\alpha(x(t))w_R(a(R)\alpha^2) < b(r),$$

which shows $x(t)) \in S(h, r)$ for all $t \geq t_N$. □

Now finally, let $\bar{\alpha}$ be the supremum of the set of all $\alpha > 0$ that satisfy the conditions (3.14) and (3.22). Then for any $\alpha \in (0, \bar{\alpha})$, one can choose λ satisfying (3.16) and (3.23), so that for each partition with $d(p) \leq \lambda$ and every p-trajectory starting from a state in G_R^α, satisfies $x(t) \in G_R^\alpha$, it follows that for all $t \geq 0$ and $x(t) \in S(h, r)$ for all $t \geq T$.

This completes the proof of the theorem. □

Given an asymptotically controllable system and a control Lyapunov function V for it, we pick an arbitrary $R_0 > 0$. There is then a sequence $\{R_j\}_{j=-\infty}^{\infty}$ satisfying $2R_j \leq \rho(R_{j+1})$, $j = 0, \pm 1, \pm 2, ...$ $\lim_{R\to\infty} b(R) = \lim_{R\to\infty} \rho(R) = \infty$ implies having R_0 fixed one can find R_1, R_2 in a way such that $R_0 < 2R_0 < \rho(R_1) < R_1 < 2R_1 < \rho(R_2) < \ldots.$

Let $r_j = \frac{1}{2}\rho(R_{j-1})$. Pick $R_{-1} < \frac{1}{2}\rho(R_0)$ and $R_{-2} < 2R_{-2} < \rho(R_{-1}) < R_{-1} < 2R_{-1} < \rho(R_0) < R_0 < 2R_0 \ldots$, which implies that $\lim_{R\to-\infty} R_j = 0$, $\lim_{R\to\infty} R_j = \infty$.

Consider any integer j, for the pair (r_j, R_j), then Theorem 3.1 implies that there exist $\alpha_j > 0, \delta_j > 0$ and $T_j > 0$ a map $k_j : S(h, R_j) \to U_j, k_j \equiv k(\alpha_j, r_j, R_j)$ such that $G_{R_j}^{\alpha_j}$ is invariant with respect to all p-trajectories when

$$k_\nu = k_j \text{ and } d(p) \leq \lambda_j \tag{3.25}$$

and for each such trajectory it holds that $h(t, x(t)) \leq r_j$ for all $t \geq T_j$.

Recall that in the construction of k_j, we used the fact that there is some compact subset $U_0 \subseteq U$, U_j satisfying condition (3.25) for Lyapunov fair for all $x \in S(h, R^0)$. Since R_j, an increasing sequence, and if the min in (3.25) also holds if we enlarge U_0, we can assume $U_j = U_0$ for all $j < 0$ and $U_j \subseteq U_{j+1}$ for all $j \geq 0$. For $x \in S(h,\ R_j)$ pick a bound $|f(x, u)| \leq m_j$ for all $u \in U_j$ and note that for all j, $m_j \leq m_{j+1}$, because of the monotonicity of U_j and $S(h, R_j)$.

Finally let $A_j \equiv G_{R_{j+1}}^{\alpha_{j+1}} \backslash G_{R_j}^{\alpha_j}$.

Lemma 3.6. For $i \neq j$, $A_i \cap A_j = \phi$, and $R^n \backslash \{0\} = \cup_{j=-\infty}^{\infty} A_j$.

Proof. We know that $G_{R_j}^{\alpha_j} \subseteq S(h, R_j) \subseteq S(h, 2R_j) \subseteq G_{R_{j+1}}^{\alpha_{j+1}}$ for all j, which implies that $A_j \subseteq G_{R_j}^{\alpha_j}$ and $A_i \cap A_j = \phi$ whenever $i < j$.

For any two integers $M < N$, we get

$$\{x : R_M \leq h(t, x) \leq R_N\} \subseteq \bigcup_{j=M}^{N} A_j = G_{R_{N+1}}^{\alpha_{N+1}} \backslash G_{R_M}^{\alpha_M}.$$

□

Also we find that $S(h, r_j) \subseteq IntG^{\alpha_{j-1}}_{R_{j-1}}$ by the definition of r_j.

Since $A_j \cup \{0\}$ constitute a partition of the state space, define a map $k : R \to U$ by $k(x) = k_j(x)$ for all $x \in A_{j-1}$ for each j, $k(0) = u_0$ where u_0 is any fixed element of U_0.

Now to show the feedback is sample stabilizing, we need the following.

Claim 5. The set $G^{\alpha_j}_{R_j}$ is invariant with respect to p-trajectories of (2.2) when (3.25) is used and $d(p) = \min(\delta_j, \frac{R_{j-1}}{m_j})$.

Proof. Consider any p-trajectory starting at a state in $G^{\alpha_j}_{R_j}$, where p satisfies (3.25).

It is enough to show that $x(t) \in G^{\alpha_j}_{R_j}$ for all $t \in [t_i, t_{i+1}]$ if $x(t) \in G^{\alpha_j}_{R_j}$.

For any such i, we have

(a) $x(t_i) \in A_{j-1}$,

(b) $x(t_i) \in A_{l-1}$ for some $l \leq j-1$,

(c) $x(t_i) = 0$.

(a) Since $k(x) = k_j(x)$ is known to have $G^{\alpha_j}_{R_j}$ invariant, there is nothing to prove.

(b) p – may not be fine enough to guarantee $G^{\alpha_l}_{R_l}$ invariant under the feedback k_l (trajectory cannot go too far because sampling time is small). Since $x(t_i) \in A_{l-1} \subset G^{\alpha_l}_{R_l} \subset S(h, R_l), h(x(t_i)) < R_l$, pick $\bar{t} \in [t_i, t_{i+1}]$ so that $h(x(t)) < R_j$ for all $t \in [t_i, \bar{t}]$. The choice of m_j implies that $k(x(t_i)) \in U_l \subseteq U_j$, $h(x(t)) - h(x(t_i)) \leq m_j(t - t_i)$ for all $t \in [t_i, \bar{t}]$ which implies that

$$\begin{aligned} h(x(t)) &\leq m_j(t - t_i) + R_l \\ &\leq R_{j-1} + R_l \\ &\leq 2R_{j-1} \text{ for all } t \in [t_i, \bar{t}]. \end{aligned} \tag{3.26}$$

Claim: $h(x(t)) < R_j$.

If not, there is some $\bar{t} \in [t_i, t_{i+1}]$ such that $h(x(t)) < R_j$ for $t \in [t_i, \bar{t}]$ and $h(x(\bar{t})) = R_j$. Since $h(x(t)) < R_j$ for all $t \in [t_i, \bar{t}]$, (3.26) yields $R_j \leq 2R_{j-1}$, which is a contradiction. This implies that the argument holds with $t = t_{i+1}$, which shows the trajectory stays in $S(h, 2R_{j-1})$, that is included in $G^{\alpha_j}_{R_j}$.

(c) Since $k(0) \in U_0 \subseteq U_j$, the argument in (b) is repeated with 0 instead of R_l, which proves that the trajectory stays in $S(h, R_{j-1})$, so it is in $G^{\alpha_j}_{R_j}$. □

To show feedback k is sample stabilizing, we consider the following.

Choose an integer $K = K(r)$, and least integer $N = N(R)$, such that $G^{\alpha_K}_{R_K} \subseteq S(h, r)$ and $S(h, R) \subseteq G^{\alpha_N}_{R_N}$.

Define

$$\delta = \min\{\min_{K \leq j \leq N} \delta_j, \frac{R_{K-1}}{m_N}\}, \quad T = \sum_{j=K+1}^{N} T_j, \quad M = R_{N(R)}.$$

Claim 6. For any $0 < r < R$, there exist $\delta = \delta(r, R) > 0$, $T = T(r, R) > 0$, and $M = M(R)$, $d(p) < \lambda$, $h(x(0)) < R$, such that

$$h(x(t)) \leq r, \ \forall t \geq T,$$

and

$$h(x(t)) \leq M, \ \forall t \geq 0.$$

Proof. Consider the p-trajectory starting from $x_0 \in S(h, R)$.

Claim: For some $t_i \in p$ such that $t_i \leq T$, $x(t_i) \in G^{\alpha_K}_{R_K}$. If $x_0 \notin G^{\alpha_K}_{R_K}$, then $x_0 \in A_{N'}$, for $K \leq N' \leq N-1$. Claim 5 implies that $G^{\alpha'_N}_{R_{N'}}$ is invariant, which yields that the trajectory remaining in $H_{N'}$ until it enters $G^{\alpha_{\mu-1}}_{R_{\mu-1}}$ in some $(t_{i-1}, t_i]$ for some $\mu < N'$. Note $t_i \leq T_{N'}$.

Theorem 3.1 implies that $x(t)$ remains in $H_{N'}$ until moment $T_{N'}$ then $x(T_{N'}) \in S(h, r_{N'})$, which is a contradiction to $x(t)$ remaining in $H_{N'}$. If $\mu \leq K+1$ there is no more to be done.

Otherwise repeat the argument to get $x(t_i) \in G^{\alpha_K}_{R_K}$ for $t_i \leq T$.

Claim 5 implies that $x(t)$ stays in this set for all $t \geq t_i$ by (3.25), which proves that $x(t)$ remains in $G^{\alpha_N}_{R_N} \subseteq S(h, R_N)$ for all $t \geq 0$, hence $h(x(t)) \leq M(R)$ for all t. This completes the proof of the claim. □

Definition 2.7 implies that feedback has attractiveness and bounded overshoot.

Finally we have to show that k has Lyapunov stability property. To prove this,

$$M(R) \text{ must satisfy } \lim_{R \to 0} M(R) = 0.$$

Claim 7. For $N = N(R)$, $S(h, R)$ is not contained in $G^{\alpha_{N-1}}_{R_{N-1}}$ and therefore $\rho(R_{N-1}) < R$.

Proof. If not $\rho(R_{N-1}) \geq R$, which implies, $R_{N-1} \geq R$, since $\rho(r) < r$ for all r.

By definition of $\rho(t)$, $S(h, R) \subseteq G_R$, and consequently, $S(h, R) \subseteq G_R \subseteq G_{R-1} \subseteq G^{\alpha_{N-1}}_{R_{N-1}}$, since $R \leq R_{N-1}$, which in contradiction to $S(h, R)$ is not contained in $G^{\alpha_{N-1}}_{R_{N-1}}$, and this yields $\rho(R_{N-1}) < R$. □

Thus it follows that $\lim_{R \to 0} N(R) = -\infty$, and $\lim_{j \to -\infty} R_j = 0$, which yields $\lim_{R \to 0} M(R) = 0$. The proof is complete.

Robust Dynamic Hybrid Controller

The aim is to design a dynamic hybrid stabilizing controller which, while preserving robustness to external perturbations and actual error, is also robust with respect to measurement error.

In the presence of measurement errors $e(t)$ only values $\tilde{x}(t)$ of the measured estimate of state vector $x(t)$, $\tilde{x}(t) \equiv x(t) + e(t)$ can be used for control.

Define a tracking controller $k_t : R^n \times R^n \to U$ as follows:

$$\langle z - \tilde{x}, f(\tilde{x}, k_t(\tilde{x}, z)) \rangle = \max u \in U \langle z - \tilde{x}, f(\tilde{x}, u) \rangle. \tag{3.27}$$

Important features:

$$x' = f(x, k_t(\tilde{x}, z)) + d_1(t) \tag{3.28}$$

can track any trajectory of the control system

$$z' = f(z, u(t)) + d_2(t), \tag{3.29}$$

for any control $u(t)$ if the measurement errors $e(t)$, disturbances $d_1(t), d_2(t)$, and the diameter of the sampling partition p_x of (3.28) are small enough.

Reinitialization: There are monotone sequences of positive numbers $\{R_j\}_{-\infty}^{+\infty}$ and closed sets $\{G_j\}_{-\infty}^{+\infty}$:

$$\lim_{j\to-\infty} R_j = 0, \quad \lim_{j\to\infty} R_j = +\infty, \tag{3.30}$$

$G_j \subset B_{R_j} \subset B_{2R_j} \subset G_{j+1}$, and the set G_{j+1} is invariant with respect to p_z-trajectories of the internal model

$$z(t) \in G_{j+1} \text{ for all } t \geq 0, \tag{3.31}$$

if $d(p_z) \leq \lambda_j$ and $|d_2(t)| \leq \theta_j$ for some positive numbers λ_j, θ_j .

Moreover, there exist moments $T_j > 0$, such that, for any p_z-trajectory as above with $z(0) \in G_{j+1}$, we have the relation

$$z(t') \in G_{j-1}, \ |z(t)| > R_{j-2} \text{ for all } t \in [0, t'] \tag{3.32}$$

at some moment $t' \leq T_j$.

A dynamic hybrid controller consists of:

(i) a tracking controller k_t that drives the system (3.28);

(ii) an internal model of the system (2.1) driven by a sample-stabilizing feedback

$$z' = f(z, k(z)) + d_2(t); \tag{3.33}$$

(iii) a set of reinitialization rules that define the moments t'_k from the sampling partition p_x for (3.33) for reinitialization of internal model (3.33),

$$z(t'_k) = \tilde{x}(t'_k); \tag{3.34}$$

(iv) a sampling rule to choose sampling moments $t_i \in p_x$, and $s_i \in p_z$:

$$t_{i+1} - t_i \leq \lambda_x(\tilde{x}(t_i)), \tag{3.35}$$

$$s_{i+1} - s_i \leq \lambda_z(z(s_i)). \tag{3.36}$$

We have $R^n \backslash \{0\} = \cup_{j=-\infty}^{+\infty} A'_j$, for the sets $A'_j = G_{j+1} \backslash G_j$.

Now define reinitialization rules for determining sequential moments $t'_k \in p_z$, $k = 0, 1, \ldots$ of reinitialization (3.34) of the internal model (3.33).

Let $t'_0 = 0$, and assume t'_k has already been determined.

Choose an index j_k that satisfies $\tilde{x}(t'_k) \in A'_{jk}$. Then the next moment t'_{k+1} of the reinitialization is defined as the first moment $t' > t'_k$ from p_z such that one of the following two events occur:

$$\tilde{x}(t') \in A'_{jk+1} \text{ for some } j_{k+1} \leq j_k - 1, \tag{3.37}$$

$$\tilde{x}(t') \notin G_{jk+2}. \tag{3.38}$$

Lemma 3.7 (Tracking Lemma [3]). Let us assume that function $f : R^n \times U \to R^n$ satisfies the following growth and Lipschitz conditions:

$$|f(x,u)| \le m, \text{ on } R^n \times U, \tag{3.39}$$

$$|f(x_1,u) - f(x_2,u)| \le l|x_1 - x_2| \text{ for all } x_1, x_2 \in R^n,\ u \in U. \tag{3.40}$$

For any positive T, β, any measured estimate $x'(t)$ and disturbances $d_1(t), d_2(t)$ satisfying

$$|x'(t) - x(t)| \le \beta \text{ for any } t \ge 0, \tag{3.41}$$

$$|d_1(t)| \le \beta,\ |d_2(t)| \le \beta \text{ for any } t \ge 0, \tag{3.42}$$

and any trajectory $z(t)$ of (3.33) defined on $[0,T]$, an arbitrary p_x-trajectory of (3.27) with

$$d(p_x) \le \beta^2, \tag{3.43}$$

and initial conditions

$$|x(0) - z(0)| \le \beta \tag{3.44}$$

is defined on $[0,T]$ and satisfies

$$|x(t) - z(t)| \le \alpha(\beta) \text{ for all } t \in [0,T], \tag{3.45}$$

where

$$\alpha(\beta) = e^{2lT}(1 + l(1+m\beta)^2 + 4m)^{\frac{1}{2}}\beta. \tag{3.46}$$

Theorem 3.2. Let the system (2.1) be (h_0,h)-asymptotically controllable, and k be sample - (h_0,h) – stabilizing feedback. Then there exist continuous functions

$$\psi : R^n \to R_+,\ \lambda_x : R^n \to R_+,\ \lambda_z : R^n \to R_+,$$

as well as reinitialization rules (3.37), (3.38), such that for any measured estimate $\tilde{x}$, of state vector x, and any disturbances $d_1(t), d_2(t)$ satisfying

$$h(\tilde{x}(t) - x(t)) \le \psi(x(t)) \text{ for any } t \ge 0, \tag{3.47}$$

$$|d_1(t)| \le \psi(x(t)),\quad |d_2(t)| \le \psi(x(t)), \text{ for any } t \ge 0. \tag{3.48}$$

The p_x trajectories $x(t)$ of (3.28) are (h_0,h)-stable.

Proof. Let m_j and l_j denote, respectively, the upper bound for $|f(x,u)|$ and Lipschitz constant of f on the set $B_{R_{j+3}} \times U$ for every integer j.

Define $\beta_j = \max\{\beta > 0 : \alpha_j(\beta) + 2\beta \le R_{j-2} - R_{j-3}\}$, where the function α_j is given by (3.46) for $T = T_j$, $m = m_j$, $l = l_j$.

Let

$$\widehat{\psi}(x) = \min\{\beta_i : j-3 \le i \le j+3\} \text{ for } R_j < |x| \le R_{j+1},$$

$$\widehat{\lambda}(x) = \min\{\beta_i^2 : j-3 \le i \le j+3\} \text{ for } R_j < |x| \le R_{j+1},$$

$$\psi(x) = \min\left\{\widehat{\psi}(y) + \frac{1}{2}|y-x| : y \in R^n\right\},$$

$$\lambda_x(x) = \min\left\{\widehat{\lambda}(y) + \frac{1}{2}|y-x| : y \in R^n\right\}.$$

It is easy to see that functions ψ and λ_x are positive for $x \neq 0$ with the Lipschitz constant $\frac{1}{2}$, and

$$\psi(x) \leq \beta_i,\ \lambda_x(x) \leq \beta_i^2, \tag{3.49}$$

for any $j-3 \leq i \leq j+3$ and $R_j < |x| \leq R_{j+1}$.

We assume that without loss of generality $\psi \leq \theta$.

By applying the Lipschitz condition for the function ψ, and for any x, that satisfy $|x - x'| \leq \psi(x)$, we get

$$|x - x'| \leq 2\psi(x'). \tag{3.50}$$

Let us consider two sequential moments t'_k, t'_{k+1}, of the reinitialization of the internal model (3.33).

Let $t^* = \min\{t'_{k+1}, t'_k + T_{jk}\}$ and let $\widehat{t}$ the maximal $t' \in [t'_k, t^*]$ such that

$$R_{jk-3} < |x(t)| \leq R_{jk+3}, \tag{3.51}$$

$$R_{jk-3} < |x'(t)| \leq 2R_{jk+2}, \tag{3.52}$$

for all $t \in [t'_k, t']$.

Since $|x(t'_k) - x'(t'_k)| \leq 2\psi(x'(t'_k)) \leq 2\beta_{jk}$, (3.47), (3.49), and (3.50) imply that (3.51) and (3.52) hold for $t = t'_k$ and $\widehat{t} > t'_k$.

Then by (3.49) we get

$$\psi(x(t)) \leq \beta_{jk}, \quad \lambda_x(x'(t'_k)) \leq \beta_{jk}^2 \text{ for all } t \in [t'_k, \widehat{t}].$$

Because of (3.47), by applying the Lemma (tracking lemma), we get

$$|x(t) - z(t)| \leq \alpha_j(\beta_j) \text{ true for all } t \in [t'_k, \widehat{t}]. \tag{3.53}$$

Then by using (3.31), (3.32) and by the definition of β_j, we show that (3.51), (3.52) hold for $t = \widehat{t}$, which implies that both these relations are true in the entire interval $[t'_k, t^*]$.

To show that $t'_{k+1} < t'_k + T_{jk}$, where the moment t'_{k+1} is defined by the relation (3.37), let us assume that this is not true. Then (3.53) holds and $x'(t) \notin G_{jk}$ for all $t \in [t'_k, t'_k + T_{jk}]$.

Recall that in this case, for the trajectory $z(t)$ of the internal model, there exists $t' \in (t'_k, t'_k + T_{jk}]$, such that the first inclusion in (3.32) holds.

This yields that

$$|x'(t'_k)| \leq |z(t')| + \alpha_j(\beta_j) + \beta_j < R_{jk-1} + R_{jk-2} - R_{jk-3} < 2R_{jk-1}.$$

Because of (3.30) we obtain that $x'(t') \in G_{jk}$. This contradiction proves that the relation (3.37) determines the reinitialization moment

$$t'_{k+1} \text{ and } t'_{k+1} - t'_k < T_{jk}, \tag{3.54}$$

Hence we have proved the following lemma.

Lemma 3.8. For any $k = 0, 1, \ldots$ (3.50) holds and

$$x'(t) \in B_{2R_{jk+2}} \subset G_{jk+3} \text{ for all } t \in [t'_k, t'_{k+1}], \tag{3.55}$$

$$x'(t'_{k+1}) \in H_{jk+1} \subset G_{jk}. \tag{3.56}$$

It is obvious that by using Lemma 3.8, we can prove that the dynamic hybrid controller provides the uniform convergence of $x'(t)$ to the origin in R^n. Then it follows from (3.43) and (3.46) that $x(t)$ uniformly converges to the origin as well.

To prove this, let us consider any $0 < r < R$ and initial point $x_0 \in B_R$ for the p_x-trajectory of (3.27) with $x(0) = x_0$.

Then we have from (3.43) and definition β_j that there exists an integer $N = N(R)$ which does not depend upon x_0 such that $x'(0) \in G_N$, and $\lim_{R\to 0} N(R) = -\infty$.

Now define $M(R) = 4R_{N(R)+2}$, and note that it satisfies (2.3c).

In accordance with the Lemma 3.8, we have $|x'(t)| \leq 2R_{N(R)+2}$, which implies (2.3b) because of (3.46).

Define the maximal $K(r)$, such that $4R_{N(R)+2} < r$. By applying Lemma 3.8, we deduce that there is a moment t' such that $t' \leq T = \sum_{i=K(r)}^{N(R)} T_i$, and $x'(t') \in G_{K(r)}$, which yields due to Lemma 3.8 that $|x(t)| \leq 4R_{K(r)+2}$ for all $t \leq T$. Then $x(t)$ satisfies (2.3a). This completes the proof of the theorem. □

REFERENCES

1. R.W. Brockett, "Asymptotic stability and feed back stabilization," in *Differential Geometric Control Theory* (R.W. Brockett, R.S. Millman, and H.J. Sussmann, eds.), Birkhauser, Boston, 1983, pp. 181–191.
2. F. H. Clarke, Yu. S. Ledyaev, E.D. Sontag, and A.I. Subbotin, "Asymptotic controllability implies feedback stabilization," *IEEE Transactions on Automatic Copntrol*, Vol. XX, 1999, pp 1–13.
3. N.N. Krasovskii and A.I. Sabotin, *Positive Differential Games*, Nauka, Moscow, 1974 [in Russian]. Revised English translation: *Game-Theoretical Control Problems*, Springer-Verlag, New York, 1988.
4. V. Lakshmikantham and X. Liu, *Stability Analysis in Terms of Two Measures*, World Scientific, Singapore, 1993.

11

Exact Boundary Controllability of a Hybrid PDE System Arising in Structural Acoustic Modeling

George Avalos[†] and Irena Lasiecka[*]

[†]*Department of Mathematics and Statistics, University of Nebraska-Lincoln, Lincoln, NE*

[*]*Department of Mathematics, University of Virginia, Charlottesville, VA*

In this chapter, we consider the active control of a pair of wave equations, each defined on different geometries: one wave equation holds on the interior of a bounded domain Ω; the other wave equation is satisified on a portion of the boundary Γ_0 of $\partial\Omega$. The respective wave equations are coupled by trace terms on the boundary interface Γ_0. For this coupled system of equations, we present results of exact controllability in the case that the controllers are exerted strictly on the boundary $\partial\Omega$. In particular, we give precise geometric conditions under which control on the "active portion" Γ_0 only gives exact controllability for the dynamics of *both* wave equations, the interior as well as the boundary wave.

1 STATEMENT OF THE PROBLEM AND MAIN RESULTS

1.1 Description of the Problem

We shall consider here a hybrid partial differential equation (PDE) system that models acoustic pressure in a cavity — taken to be an n-dimensional entity; this cavity in turn is interacting with another environment (manifold), taken to be a wall of dimension $n - 1$. The coupling between the structures is accomplished through the back pressure exerted on the wall. These kinds of systems are typical in structural acoustic modeling (see, e.g., [8], [24]). In this chapter, we shall consider a canonical model, in which the hybrid consists of two disparate wave equations that are coupled at the interface. This interface consists of the boundary of the region (cavity). We are thus dealing here with interactions of two hyperbolic PDEs, but other configurations have also been considered in the literature; e.g., a coupling of a parabolic PDE to a hyperbolic PDE, where the hyperbolic component is a wave equation satisfied within the cavity, but with the wall (structure) satisfied by a beam (rather than a wave) equation under the influence of various degrees of structural damping (see [1]).

Our goal here is to study *exact* controllability properties, with a (natural) emphasis being on boundary control with respect to initial and terminal data in physically meaningful *finite energy states*. (However, there is generally no guarantee that during the transit time

0-415-30852-6/04/$0.00+$1.50

from the initial to the terminal data, the corresponding state will remain in the space of finite energy, that is the underlying control operator for our control operator is not "admissible." In part, this is a consequence of the fact that the Lopatinski conditions are not satisfied for the wave equation with Neumann boundary conditions (see [15]).

In order to clearly convey our main ideas and the technical difficulties associated with the exact controllability of PDEs that govern interactive structures, we have chosen to analyze here a relatively simple-looking model; namely, a wave/wave model that displays a variety of the phenonema intrinsic to coupled PDE structures in general. Moreover, this canonical PDE model provides a comfortable setting from which to study the delicate issue of "propagating" controllability properties from one PDE component to the other.

As we shall see below, the geometry of the region plays a critical role in our analysis; and it will turn out that some geometries can be particularly exploited for the sake of controllability. Indeed, our results lead to the conclusion that there are favorable geometric configurations that should be kept in mind when one wishes to design acoustic chambers (such as the cabin of a helicopter) that are amenable to active controllers distributed on a portion of the chamber walls.

Other, higher-order equations (of the fourth order on the interface), can and have been considered by the authors as well. See [1], [2], and [5] for an account of these results.

1.2 The PDE Model

Let Ω be a bounded open subset of $\mathbb{R}^n$, $n \geq 2$, with Lipschitz boundary $\partial\Omega = \Gamma_0 \cup \Gamma_1$, with each Γ_i nonempty, and $\Gamma_0 \cap \Gamma_1 = \emptyset$. There will eventually be additional assumptions imposed on the Γ_i (see (A1) and (A2) below). For this geometry in place, we shall study the solutions $[z(t,x), v(t,\tau)]$ to the following (controlled) PDE model:

$$\begin{cases} z_{tt}(t,x) = \Delta z(t,x) \quad \text{on } (0,T) \times \Omega \\ \begin{cases} \dfrac{\partial}{\partial \nu} z(t,x) = v_t(t,x) \;\; \text{on } (0,T) \times \Gamma_0 \\ \dfrac{\partial}{\partial \nu} z(t,x) = u_1 \quad \text{on } (0,T) \times \Gamma_1 \end{cases} \\ [z(0,x),\; z_t(0,x)] = \vec{z}_0 \;\; \text{on } \Omega. \end{cases}$$

This PDE is coupled to the following on the interface Γ_0: (1)

$$\begin{cases} v_{tt}(t,\tau) = \frac{\partial^2}{\partial \tau^2} v(t,\tau) + u_0(t,\tau) - z_t|_{\Gamma_0} \quad \text{on } (0,T) \times \Gamma_0 \\ \dfrac{\partial v}{\partial n} = 0 \;\; \text{on } (0,T) \times \partial\Gamma_0 \\ [v(0,x), v_t(0,x)] = \vec{v}_0 \;\; \text{on } \partial\Gamma_0. \end{cases}$$

As usual, $\frac{\partial}{\partial \tau}$ denotes the (unit) tangential, and $\frac{\partial}{\partial \nu}$ the outward normal derivative with respect to Γ. $\frac{\partial}{\partial n}$ is here the unit exterior normal derivative with respect to the $n-1$ manifold $\partial\Gamma_0$.

To account for the initial data of the problem, we define the spaces,

$$\begin{aligned} &H_1 \equiv H^1(\Omega) \times L^2(\Omega); \\ &H_0 \equiv H^1(\Gamma_0) \times L^2(\Gamma_0); \\ &\mathcal{H} \equiv \Big\{ [z_0, z_1] \in \mathbf{H}_1 \text{ and } [v_0, v_1] \in \mathbf{H}_0 \text{ such that } \textstyle\int_\Omega z_1 = \int_{\Gamma_0} v_0 \\ &\qquad \text{and } \textstyle\int_\Omega z_0 = \int_{\Gamma_0} v_1 \Big\}. \end{aligned} \tag{2}$$

It can be readily shown that $\mathcal{H}$ is a Hilbert space, with the inner product

$$\left(\begin{bmatrix} z_0 \\ z_1 \\ v_0 \\ v_1 \end{bmatrix}, \begin{bmatrix} \tilde{z}_0 \\ \tilde{z}_1 \\ \tilde{v}_0 \\ \tilde{v}_1 \end{bmatrix} \right)_{\mathcal{H}}$$

$$= \int_\Omega \nabla z_0 \cdot \nabla \tilde{z}_0 d\Omega + \int_\Omega z_1 \tilde{z}_1 d\Omega + \int_{\Gamma_0} \frac{\partial v_0}{\partial \tau} \frac{\partial \tilde{v}_0}{\partial \tau} + \int_{\Gamma_0} v_1 \tilde{v}_1 d\Gamma_0.$$

With this definition of the inner product, one can then proceed to use the Lumer–Phillips Theorem to show the existence of a C_0-*group* $\{e^{\mathcal{A}t}\}_{t\geq 0}$ associated with this coupled system of wave equations. Accordingly, with regard to the PDE (1), a straightforward consequence of these dynamics $\{e^{\mathcal{A}t}\}_{t\geq 0}$ (see, e.g., [1], [25]) yields the continuity of the mapping

$$\{[\vec{z}_0, \vec{v}_0] \in \mathcal{H}, u_1 = 0, u_0 = 0\} \Rightarrow [\vec{z}, \vec{v}] \in C([0, T]; \mathcal{H}).$$

In short, the uncontrolled problem (i.e., the PDE (1) with $u_i = 0$) is well posed in the basic space of energy $\mathcal{H}$. Thus, the well-posedness of the system is not an issue here; the goal of our chapter is rather to determine exact controllability properties[1] of (1) with boundary controls u_1, u_0 taken in prescribed spaces of controls $\mathcal{U}_1, \mathcal{U}_0$, say, under the imposition of specific conditions upon the geometry Ω.

1.3 The Main Results

Our first result in Theorem 1 below states that *all finite energy states are controlled exactly with controls located on Γ_0 alone (with these controls acting only on the v-component).* This result does require, however, that the geometry be "appropriate" to the situation; viz., the domain Ω is convex and the "roof" of the acoustic chamber is not too "deep." (See Assumption (A1) and Fig. 1.) In addition, the control u_0 must be of the appropriate topological strength; i.e., $u_0 \in [H^1(0, T; L^2(\Gamma_0))]'$. So for our first result (Theorem 1(a) below), we assume the following:

Assumption (A1) Assume that Ω is a bounded subset of $\mathbb{R}^n$, with boundary $\Gamma = \overline{\Gamma}_0 \cup \overline{\Gamma}_1$, $\overline{\Gamma}_0 \cap \overline{\Gamma}_1 = \emptyset$, with Γ_0 being flat. Moreover assume the following:

[1] The classical definition of exact controllability is intended here. Namely, the PDE (1) is exactly controllable if there is a $T^* > 0$ such that for terminal time $T > T^*$, one has the following reachability property: for all initial data $[\vec{z}_0, \vec{v}_0] \in \mathcal{H}$ and preassigned target data $[\vec{z}_T, \vec{v}_T] \in \mathcal{H}$, there exist control functions $[u_1, u_0] \in \mathcal{U}_1 \times \mathcal{U}_0$ (to be specified), such that at terminal time T the corresponding solution $[z, v]$ to (1) satisfies $[\vec{z}(T), \vec{v}(T)] = [\vec{z}_T, \vec{v}_T]$.

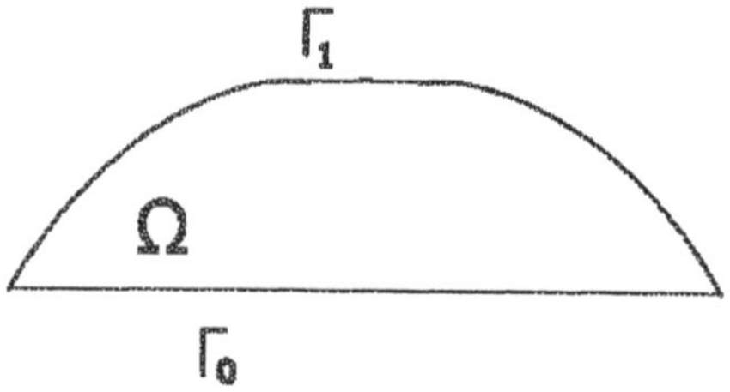

Figure 1 A triple $\{\Omega, \Gamma_0, \Gamma_1\}$ that satisfies Assumption (A1).

(i) Ω is convex; (ii) there exists a point $x_0 \in \mathbb{R}^2$ such that

$$(x - x_0) \cdot \nu \leq 0 \text{ for all } x \in \Gamma_1.$$

The special vector field that is available, in case that Assumption (A1) holds true — constructed in [19] and denoted below as h in (13) — will be used in the derivation of the observability inequality associated with exact controllability (see (3) below). In particular, this special h appears in the wave multipliers classically used to estimate the energy of the z-wave equation (see, e.g., [13], [14], [23], [27], [28]). The behavior of h on the inactive portion of the boundary i.e., $h \cdot \nu|_{\Gamma_1} = 0$, is a key driver in our first result. With control on Γ_0 only, and under Assumption (A1), the PDE (1) is exactly controllable on $\mathcal{H}$. A discussion concerning the possible configuration of those triples $\{\Omega, \Gamma_0, \Gamma_1\}$ that satisfy Assumption (A1) is given in Appendix C in [19]. A canonical example of such a triple is given in Fig. 1.

On the other hand, if the geometry of the acoustic chamber is unrestricted (see Assumption (A2)), then from our previous discussion, we clearly must have additional control on Γ_1. In this case, the second part of Theorem 1 states that *all finite energy states are controlled exactly with controls* $u_0 \in L^2(0, T; H^{-\frac{1}{4}}(\Gamma_0))$ *(located on* Γ_0*) and* $u_1 \in L^2(0, T; L^2(\Gamma_1))$ *(located on the roof of a chamber of arbitrary geometrical configuration).* So in our second result (Theorem 1(b) below), we make the following assumption:

Assumption (A2) Assume that Ω is either a bounded subset of $\mathbb{R}^n$ with smooth boundary Γ, or else Ω is a parallelepiped. Moreover, assume boundary $\Gamma = \Gamma_0 \cup \Gamma_1$, where Γ_0 is flat. No assumptions are made on Γ_1 (see Fig. 2).

If Assumption (A2) holds true, then one has exact controllability of (1) for arbitrary initial data of finite energy, with the control region taken to be $\Gamma_0 \cup \Gamma_1$. The point of making the Assumption (A2) is that in this case, one can take a radial vector field h to assist in the multiplier method to be employed to estimate the (acoustic) wave energy. However, since Assumption (A2) is much less restrictive than (A1) — in particular, no impositions are made on the hard walls Γ_1 — the corresponding h cannot be expected to help with the high-order terms on Γ_1, and hence the need for control on the hard walls. A common feature in both of our main results is the critical use of the "sharp" regularity theory, which has been developed to handle the tangential derivatives (on the boundary) of solutions to wave equations (see [17] and Lemma 4) below).

With these assumptions, we now state our main results concisely.

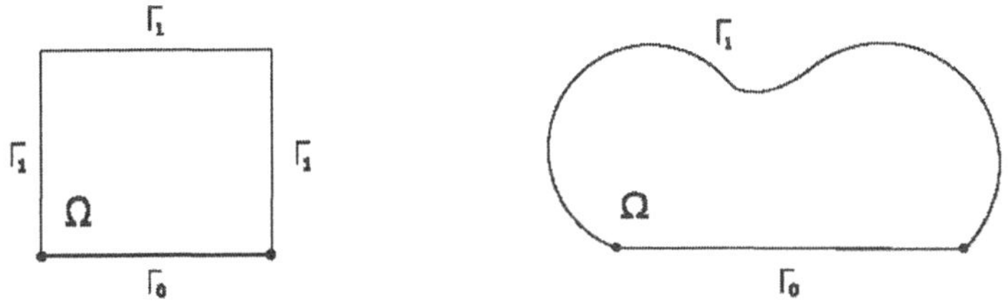

Figure 2 A triple $\{\Omega, \Gamma_0, \Gamma_1\}$ that satisfies Assumption (A2).

Theorem 1 *(a) Let Assumption (A1) stand, and set*

$$\begin{aligned} \mathcal{U}_1 &= \{0\}\,; \\ \mathcal{U}_0 &= \left[H^1\left(0,T;L^2(\Gamma_0)\right)\right]'. \end{aligned}$$

Then for long enough terminal time T, the problem (1) is exactly controllable on $\mathcal{H}$ within the class of $\mathcal{U}_1 \times \mathcal{U}_0$-controls.

(b) Let Assumption (A2) stand, and set

$$\begin{aligned} \mathcal{U}_1 &= L^2(0,T;L^2(\Gamma_1)) \\ \mathcal{U}_0 &= L^2(0,T;H^{-\frac{1}{4}}(\Gamma_0)). \end{aligned}$$

Then likewise for long enough terminal time T, the problem (1) is exactly controllable on $\mathcal{H}$ within the class of $\mathcal{U}_1 \times \mathcal{U}_0$-controls.

Remark 2 *The precise specification of the controllability time is $T > 2diam(\Omega)$. This is on account of the terminal time needed for Holmgren's Uniqueness Theorem to hold true. This theorem is needed at the level of a "compactness/uniqueness" argument so as to eliminate lower-order terms that corrupt the associated observability estimate (see (35) and (37)).*

Remark 3 *The specification here of natural boundary conditions for the v-component in (1) is not critical in the derivation of our observability results. In fact, one could obtain a similar exact controllability result for (1), with instead $v|_{\Gamma_0} = 0$ on $\partial\Gamma_0$.*

1.4 About the Problem and the Literature

The PDE system (1) and other classes of PDE models seen in the literature and that are associated with the mathematical descriptions of stuctural acoustic flow, are principally characterized as a coupling of distinct dynamics (be it of hyperbolic/hyperbolic or hyperbolic/parabolic type), with the coupling being accomplished across boundary interfaces (see [1], [2], [6], [8], [9], [20], [24]). In the present case, we have a chamber Ω, with hard, rigid walls Γ_1 and its flexible wall portion Γ_0. Moreover, the flow within this chamber Ω is assumed to be of the acoustic wave type. Given this assumption of acoustic-type flow, an interior wave equation (in z) consequently appears in the governing of the PDE system, this wave equation being under the influence of Neumann forcing data v_t. In turn, the acoustic wave is coupled to the boundary wave equation (in v) on Γ_0 (this boundary wave is to account for the dynamics involved in the flexible Γ_0). The coupling mechanism between the two dynamics is brought about through the boundary traces of the respective wave velocities.

In accordance with the classical notion of exact controllability for PDEs, we shall concern ourselves here with deriving reachability properties with respect to initial and terminal data in the *finite energy space* $\mathcal{H}$, as defined in (2). Moreover in line with the applied problem addressed in the literature, a principal objective is the attainment of some statement that assures the exact controllability of the PDE system, in the case that control is exerted on the active wall Γ_0 only (that is, Γ_1 is to be inactive). However, this objective, which is really an accommodation to the needs of the given physical problem, is at odds with the geometric conditions necessary for exact controllability, and which are prescribed in [7]. In fact, in the structural acoustic control problem, as stated in [6] and [1], the active control region Γ_0 comprises one side of the chamber wall only. In consequence, for an *arbitrary* triple $\{\Omega, \Gamma_0, \Gamma_1\}$, one will generally not have exact controllability with control implemented solely on Γ_0, as the necessary conditions of geometric optics will not be satisfied. In particular, it is now known that in order to control the z-wave equation from the boundary, it is necessary that the support of the control region be sufficiently large (see [7]). Thus prescribing control only on Γ_0 will generally be insufficient, unless Γ_0 is large relative to Ω.

Another (topological) complication presented here is a direct consequence of the coupling involved between the two dynamics, i.e., the acoustic interaction (1) presents a situation where the component $\vec{z}$ on Ω, as the solution of the wave equation with forcing boundary data v_t, is subject to the "smoothing effects" due to the (L^2) regularity of the Neumann boundary data v_t (see, e.g., [16], [18]). As such, it will have a certain measure of regularity. Therefore, a sole control u_0 — that is $u_1 = 0$ in (1) — acting strictly on the wave component $\vec{v}$, in the absence of geometric assumptions, will generally not be strong enough to drive the acoustic variable $\vec{z}$ to an arbitrary state of finite energy.

To the best of our knowledge, Theorem 1(a) and (b) constitutes the first exact controllability results for structural acoustic interactions in finite energy spaces, and with general spatial domains Ω. All other results (see [22] and references therein) pertain to controllability on specified subspaces of finite energy — such as those described by the asymptotic behavior of Fourier coefficients — and moreover these results are proved for very special geometries only — a 2D rectangle — with very large classes of controls: $H^{-2}(0,T; L^2(\Gamma_0))$.

From the mathematical point of view, the key ingredients in our proofs are the following:

(i) Sharp trace regularity for the wave equation in the absence of Lopatinski conditions (a distinguishing feature of the Neumann case); see Lemma 5.

(ii) Microlocal analytical estimates that allow the absorption of tangential (wave) traces by time derivatives on the boundary; see Lemma 4.

(iii) A recent result in [19] concerning Carleman's estimates for the wave equation with the controlled Neumann part of the boundary. These estimates lead to the aforementioned special vector field h that allows us, in this chapter, to handle the uncontrolled portion of the boundary so as to derive the requisite observability estimates.

2 THE NECESSARY INEQUALITY

With the control spaces $\mathcal{U}_1, \mathcal{U}_0$ as prescribed in Theorem 1(a) or (b), let $\mathcal{L}_T : \mathcal{U}_1 \times \mathcal{U}_0 \supset D(\mathcal{L}_T) \to \mathcal{H}$ denote the *control to terminal state map,*

$$\mathcal{L}_T \begin{bmatrix} u_0 \\ u_1 \end{bmatrix} = \begin{bmatrix} \vec{z}(T) \\ \vec{v}(T) \end{bmatrix}.$$

Then for the asserted controllability result Theorem 1(a) (resp. (b)) is equivalent to showing the surjectivity of $\mathcal{L}_T$ as a mapping between the said spaces. In turn, by the classical functional analysis (see, e.g., Lemma 3.8.18 and Theorem 6.5.10 of [12]), this ontoness is equivalent to establishing the following inequality for all $\left[\vec{\phi}_0, \vec{\psi}_0\right] \in D(\mathcal{L}_T^*) \subset \mathcal{H}$ (where $\mathcal{L}_T^*$ denotes the Hilbert space adjoint of $\mathcal{L}_T$):

$$\left\| \mathcal{L}_T^* \begin{bmatrix} \vec{\phi}_0 \\ \vec{\psi}_0 \end{bmatrix} \right\|_{\mathcal{U}_1 \times [\mathcal{U}_0]'} \geq C_T \left\| \left[\vec{\phi}_0, \vec{\psi}_0\right] \right\|_{\mathcal{H}}. \tag{3}$$

In PDE terms, this inequality assumes the following form:

Let $\left[\vec{\phi}(t,x), \vec{\psi}(t,\tau)\right]$ denote the solution to the backward system (*adjoint* with respect to the PDE (1)):

$$\begin{cases} \phi_{tt}(t,x) = \Delta\phi(t,x) \quad \text{on } (0,T)\times\Omega \\ \begin{cases} \dfrac{\partial}{\partial\nu}\phi(t,x) = \psi_t \text{ on } (0,T)\times\Gamma_0 \\ \dfrac{\partial}{\partial\nu}\phi(t,x) = 0 \text{ on } (0,T)\times\Gamma_1 \end{cases} \\ [\phi(T,x),\, \phi_t(T,x)] = \vec{\phi}_0 \text{ on } \Omega; \end{cases}$$

$$\begin{cases} \psi_{tt}(t,\tau) = \frac{\partial^2}{\partial\tau^2}\psi(t,\tau) - \phi_t|_{\Gamma_0} \quad \text{on } (0,T)\times\Gamma_0 \\ \dfrac{\partial\psi}{\partial n} = 0 \text{ on } (0,T)\times\partial\Gamma_0 \\ [\psi(T,x), \psi_t(T,x)] = \vec{\psi}_0. \end{cases} \tag{4}$$

(By standard semigroup theory, the homogenous system above is well posed for terminal data $\left[\vec{\phi}_0, \vec{\psi}_0\right] \in \mathcal{H}$.)

To prove Theorem 1(a), the necessary abstract inequality (3) takes the form

$$\left\| \left[\vec{\phi}_0, \vec{\psi}_0\right] \right\|_{\mathcal{H}}^2 \leq C_T \left\|\psi\right\|_{H^2(0,T;L^2(\Gamma_0))}^2, \tag{5}$$

for all $\left[\vec{\phi}_0, \vec{\psi}_0\right] \in D(\mathcal{L}_T^*)$.

On the other hand, the reverse inequality (3) needed to obtain the exact controllability statement Theorem 1(b) takes the explicit form

$$\left\| \left[\vec{\phi}_0, \vec{\psi}_0\right] \right\|_{\mathcal{H}}^2 \leq C_T \left\{ \left\|\phi_t\right\|_{L^2(0,T;L^2(\Gamma_1))}^2 + \left\|\psi_t\right\|_{L^2(0,T;H^{\frac{1}{4}}(\Gamma_0))}^2 \right\}, \tag{6}$$

for all $\left[\vec{\phi}_0, \vec{\psi}_0\right] \in D(\mathcal{L}_T^*)$.

This derivation of these inequalities is the objective of the sequel.

3 DERIVATION OF THE MAIN ESTIMATE

3.1 Preliminary Notation

We will use throughout the standard denotations for the "energy" of the system (4):

$$\begin{aligned}
\mathcal{E}_\phi(t) &= \int_\Omega \left[|\nabla\phi(t)|^2 + |\phi_t(t)|^2\right] d\Omega; \\
\mathcal{E}_\psi(t) &= \int_{\Gamma_0} \left[\left|\frac{\partial}{\partial\tau}\psi(t)\right|^2 + |\psi_t(t)|^2\right] d\Gamma_0; \\
\mathcal{E}(t) &= \mathcal{E}_\phi(t) + \mathcal{E}_\psi(t). \qquad (7)
\end{aligned}$$

Under the assumptions made above on the geometry Ω, we can assume throughout that there exists a dense set of data corresponding to smooth (enough) solutions to the PDE (4). Indeed, if Ω has a smooth boundary (the first part of Assumption (A2)), this assertion follows from classical elliptic and semigroup theory. On the other hand, if Ω is either a parallelepiped (the second part of Assumption (A2)) or else satisfies Assumption (A1) (so that in particular Ω is convex)], then one can appeal to [10]). In this way, one can justify the computations to be done below. Also, we will use the standard denotations:

$$\begin{aligned}
Q_{\epsilon_0} &= (\epsilon_0, T-\epsilon_0) \times \Omega; \\
\Sigma_0 &= (0,T) \times \Gamma_0; \Sigma_1 = (0,T)\times\Gamma_1.
\end{aligned}$$

In addition, with $\vec{\phi} = [\phi, \phi_t]$ and $\vec{\psi} = [\psi, \psi_t]$, we will use below the standard denotation for terms that are "below the level of energy"; namely,

$$\text{l.o.t.}(\vec{\phi}, \vec{\psi}) \equiv C \left\| \left[\vec{\phi}, \vec{\psi}\right] \right\|_{C([0,T];H^{1-\epsilon}(\Omega)\times H^{-\epsilon}(\Omega)\times H^{1-\epsilon}(\Gamma_0)\times H^{-\epsilon}(\Gamma_0))},$$

for some constant C, where $\epsilon > 0$.

Finally, we will need throughout the cutoff function $\xi(t) \in C_0^\infty(\mathbb{R})$, which is defined by having for arbitrary $\epsilon_0 > 0$,

$$\xi(t) = \begin{cases} 1, & \text{for } t \in [\epsilon_0, T-\epsilon_0] \\ \text{a } C^\infty \text{ function with range in } (0,1), & \text{for } t \in (0,\epsilon_0) \cup (T-\epsilon_0, T) \\ 0, & \text{for } t \in (-\infty, 0) \cup (T, \infty). \end{cases} \qquad (8)$$

3.2 Step 1 (The Conservation Relation)

Multiplying the first equation of (4) by ϕ_t, the second by ψ_t, and subsequently integrating in time and space, and integrating by parts we obtain

$$\frac{1}{2}\left(\phi_t(\sigma), \phi_t(\sigma)\right)_{L^2(\Omega)}\Big]_{\sigma=s}^{\sigma=t} = -\frac{1}{2}\left(\nabla\phi(\sigma), \nabla\phi(\sigma)\right)_{L^2(\Omega)}\Big]_{\sigma=s}^{\sigma=t} + \int_s^t \left(\frac{\partial\phi}{\partial\nu}, \phi_t\right)_{L^2(\Gamma_0)} dt;$$

$$\frac{1}{2}\left(\psi_t(\sigma), \psi_t(\sigma)\right)_{L^2(\Omega)}\Big]_{\sigma=s}^{\sigma=t} = -\frac{1}{2}\left(\frac{\partial}{\partial\tau}\psi(\sigma), \frac{\partial}{\partial\tau}\psi(\sigma)\right)_{L^2(\Omega)}\Big]_{\sigma=s}^{\sigma=t} - \int_s^t (\phi_t, \psi_t)_{L^2(\Gamma_0)} dt.$$

Applying the Neumann boundary condition in (4) to the first equation above and summing the two yields the expected conservation of the system; i.e.,

$$\mathcal{E}(t) = \mathcal{E}(s) \text{ for all } 0 \leq s, t \leq T.$$

In particular then,

$$\mathcal{E}(s) = \left\| \left[\vec{\phi}_0, \vec{\psi}_0 \right] \right\|_{\mathcal{H}}^2 \quad \text{for all } 0 \leq s \leq T. \tag{9}$$

3.3 Step 2 (The Acoustic Wave Estimates)

Let h be a $[C^2(\overline{\Omega})]^n$-vector field, which will be eventually specified. With this h, we apply the "classical" wave multipliers (see, e.g., [14], [23], [27], [28]). Multiplying the ϕ-wave equation of (4) by $h \cdot \nabla\phi$, integrating in time and space and using the Neumann boundary condition, we have

$$\begin{aligned}\int_{Q_{\epsilon_0}} H\nabla\phi \cdot \nabla\phi dQ_{\epsilon_0} &= \int_{\epsilon_0}^{T-\epsilon_0} \int_{\Gamma_0} \psi_t \left(h \cdot \nabla\phi\right) dt d\Gamma_0 + \frac{1}{2} \int_{\epsilon_0}^{T-\epsilon_0} \int_{\Gamma} \phi_t^2 h \cdot \nu dt d\Gamma \\ &- \frac{1}{2} \int_{\epsilon_0}^{T-\epsilon_0} \int_{\Gamma} |\nabla\phi|^2 h \cdot \nu dt d\Gamma + \frac{1}{2} \int_{Q_{\epsilon_0}} \left(|\nabla\phi|^2 - \phi_t^2 \right) div(h) dQ_{\epsilon_0} \\ &- \left[(\phi_t, h \cdot \nabla\phi)_{L^2(\Omega)} \right]_{\epsilon_0}^{T-\epsilon_0}. \end{aligned} \tag{10}$$

Next, we consider again the first wave equation in (4), this time multiplying by the quantity $\phi div(\tilde{h})$, where $\tilde{h}(x) \in \left[C^2(\overline{\Omega})\right]^n$ is arbitrary. Integrating in time and space, and invoking Green's Theorem and the identity $\nabla\left(\phi div(\tilde{h})\right) = \phi\nabla\left(div(\tilde{h})\right) \cdot \nabla\phi + |\nabla\phi|^2 div(\tilde{h})$, we obtain

$$\begin{aligned} \int_{Q_{\epsilon_0}} \left(\phi_t^2 - |\nabla\phi|^2\right) div(\tilde{h}) dQ_{\epsilon_0} &= \left\langle \phi_t, \phi div(\tilde{h}) \right\rangle_{H^{-\epsilon}(\Omega) \times H^{\epsilon}(\Omega)} \Big]_{\epsilon_0}^{T-\epsilon_0} \\ &+ \int_{Q_{\epsilon_0}} \phi \nabla\left(div(\tilde{h})\right) \nabla\phi dQ_{\epsilon_0} - \int_{\epsilon_0}^{T-\epsilon_0} \int_{\Gamma} \frac{\partial\phi}{\partial\nu} \phi div(\tilde{h}) dt d\Gamma. \end{aligned} \tag{11}$$

Using now the Neumann boundary condition for ϕ in (4), along with Sobolev Trace Theory, we have the following inequality for any C^2-vector field $\tilde{h}$:

$$\begin{aligned} &\left| \int_{Q_{\epsilon_0}} \left(\phi_t^2 - |\nabla\phi|^2\right) div(\tilde{h}) dQ_{\epsilon_0} \right| \\ &\leq C \int_{\Sigma_0} \psi_t^2 d\Sigma_0 + \epsilon \int_{Q_{\epsilon_0}} |\nabla\phi|^2 dQ_{\epsilon_0} \\ &\quad + C_{\tilde{h},\epsilon} \left(\int_0^T \|\phi\|_{H^{\frac{1}{2}+\epsilon}(\Omega)}^2 dt + \|\phi_t\|_{C([0,T]:H^{-\epsilon}(\Omega))}^2 + \|\phi\|_{C([0,T]:H^{\epsilon}(\Omega))}^2 \right) \\ &\leq C_\epsilon \int_{\Sigma_0} \psi_t^2 d\Sigma_0 + \epsilon \int_{Q_{\epsilon_0}} |\nabla\phi|^2 dQ_{\epsilon_0} + \text{l.o.t.}\left(\vec{\phi}, \vec{\psi}\right). \end{aligned} \tag{12}$$

We will now consider the two cases, corresponding to Assumptions (A1) and (A2).

3.3.1 Case I. Assumption (A1) is in Force

Given the assumptions on both Γ_0 and Γ_1, it is shown in [19] that there exists a vector field $h(x) = [h_1(x), h_2(x), ..., h_n(x)] \in [C^2(\overline{\Omega})]^n$ such that

$$\begin{array}{l} \text{(i) } h \cdot \nu = 0 \text{ on } \Gamma_1; \\ \text{(ii) The matrix } H(x) \text{, defined by} \\ \quad [H(x)]_{ij} = \frac{\partial h_i(x)}{\partial x_j}, \quad 1 \leq i, j \leq n, \\ \text{satisfies } H(x) \geq \rho_0 I \text{ on } \Omega \text{, for some positive constant } \rho_0. \end{array} \tag{13}$$

Applying this particular h to (10), we obtain

$$\begin{aligned} \rho_0 \int\limits_{Q_{\epsilon_0}} |\nabla\phi|^2 \, dQ_{\epsilon_0} \leq & \int\limits_{\epsilon_0}^{T-\epsilon_0} \int\limits_{\Gamma_0} \psi_t (h \cdot \nabla\phi) \, dt d\Gamma_0 + \frac{1}{2} \int\limits_{\epsilon_0}^{T-\epsilon_0} \int\limits_{\Gamma_0} \phi_t^2 h \cdot \nu dt d\Gamma_0 \\ & - \frac{1}{2} \int\limits_{\epsilon_0}^{T-\epsilon_0} \int\limits_{\Gamma_0} |\nabla\phi|^2 h \cdot \nu dt d\Gamma_0 + \frac{1}{2} \int\limits_{Q_{\epsilon_0}} \left(|\nabla\phi|^2 - \phi_t^2 \right) div(h) dQ_{\epsilon_0} \\ & - \left[(\phi_t, h \cdot \nabla\phi)_{L^2(\Omega)} \right]_{\epsilon_0}^{T-\epsilon_0}. \end{aligned}$$

Using the relation $|\nabla\phi|^2 = \left(\psi_t^2 + \frac{\partial\phi}{\partial\tau}^2 \right)$ on Γ_0, as well as the conservation relation (9), we obtain from this the majorization,

$$\begin{aligned} \rho_0 \int\limits_{Q_{\epsilon_0}} |\nabla\phi|^2 \, dQ_{\epsilon_0} \leq & \, C_h \left\{ \int\limits_{\Sigma_0} \psi_t^2 d\Sigma_0 + \int\limits_{\epsilon_0}^{T-\epsilon_0} \int\limits_{\Gamma_0} \left(\phi_t^2 + \frac{\partial\phi}{\partial\tau}^2 \right) dt d\Gamma_0 \right\} \\ & + 2 \max_{1 \leq i \leq n} \|h_i\|_{L^\infty(\Omega)} \mathcal{E}(T) + \frac{1}{2} \left| \int\limits_{Q_{\epsilon_0}} \left(\phi_t^2 - |\nabla\phi|^2 \right) div(h) dQ_{\epsilon_0} \right|, \end{aligned} \tag{14}$$

where h is the vector field in (13).

Combining (14) and (12) (with $\tilde{h} = h$ therein) now yields

$$\begin{aligned} (\rho_0 - \epsilon) \int\limits_{Q_{\epsilon_0}} |\nabla\phi|^2 \, dQ_{\epsilon_0} \quad \leq \quad & C_{\epsilon,h} \left\{ \int\limits_{\Sigma_0} \psi_t^2 d\Sigma_0 + \int\limits_{\epsilon_0}^{T-\epsilon_0} \int\limits_{\Gamma_0} \left(\phi_t^2 + \frac{\partial\phi}{\partial\tau}^2 \right) dt d\Gamma_0 \right\} \\ & + 2 \max_{1 \leq i \leq n} \|h_i\|_{L^\infty(\Omega)} \mathcal{E}(T) + \text{l.o.t.}(\vec{\phi}, \vec{\psi}). \end{aligned} \tag{15}$$

In turn, the inequalities (12) (taking therein $\tilde{h}$, which satisfies $div(\tilde{h}) = 1$) and (15) (taking therein $\epsilon > 0$ small enough) give the intermediate estimate,

$$\begin{aligned} \int\limits_{\epsilon_0}^{T-\epsilon_0} \mathcal{E}_\phi(t) dt \quad \leq \quad & C_{\epsilon,h} \left\{ \int\limits_{\Sigma_0} \psi_t^2 d\Sigma_0 + \int\limits_{\epsilon_0}^{T-\epsilon_0} \int\limits_{\Gamma_0} \left(\phi_t^2 + \frac{\partial\phi}{\partial\tau}^2 \right) dt d\Gamma_0 \right\} \\ & + C_h \mathcal{E}(T) + \text{l.o.t.}(\vec{\phi}, \vec{\psi}). \end{aligned} \tag{16}$$

Now in estimating the tangential derivative $\left.\frac{\partial\phi}{\partial\tau}\right|_{\Gamma_0}$ on the right-hand side of (16), there is no appeal to classical Sobolev trace theory. Instead, we recall the following estimate, a product of microlocal machinery.

Lemma 4 *(see [17], Lemma 7.2). Let $\epsilon_0 > 0$ be arbitrarily small. Let w be a solution of the wave equation on $(0,T)\times\Omega$, or more generally, any second-order hyperbolic equation with smooth space-dependent coefficients. Then,with $w_c \equiv \xi w$ (with $\xi(t)$ being the cutoff function defined in (8)), we have the estimate*

$$\int_0^T\int_{\Gamma_*}\left(\frac{\partial w_c}{\partial\tau}\right)^2 d\Sigma_0 \leq C_{T,\epsilon_0}\left(\int_0^T\int_{\Gamma_*}\frac{\partial w_c}{\partial t}^2 d\Sigma_0 + \int_0^T\int_{\Gamma}\frac{\partial w}{\partial\nu}^2 d\Sigma\right) + \text{l.o.t.}(w), \tag{17}$$

where Γ_* is a smooth connected segment of boundary Γ.

Then, invoking Lemma 4 (with $\Gamma_* = \Gamma_0$ therein), so as to handle the tangential derivative in (16), gives the following: *under Assumption (A1), one has the integral estimate for the energy of the component ϕ of (4),*

$$\begin{aligned}&\int_{\epsilon_0}^{T-\epsilon_0}\mathcal{E}_\phi(t)dt\\ &\leq C_{T,\epsilon_0,h}\left\{\int_{\epsilon_0}^{T-\epsilon_0}\int_{\Gamma_0}\phi_t|_{\Gamma_0}^2\,dtd\Gamma_0 + \int_{\Sigma_0}\xi^2\,\phi_t|_{\Gamma_0}^2\,d\Sigma_0 + \int_{\Sigma_0}\psi_t^2 d\Sigma_0\right\}\\ &\quad + C_h\mathcal{E}(T) + \text{l.o.t.}(\vec{\phi},\vec{\psi}),\end{aligned}\tag{18}$$

where the cutoff function ξ is as defined in (8). Here, $C_{T,\epsilon_0,h}$ depends on time T, but C_h does not.

3.3.2 *Case II. Assumption (A2) is in Force*

In this case, since Γ_0 is flat, one can construct a *radial* vector field $h(x)\in[C^2(\overline{\Omega})]^n$ such that

$$h(x)\cdot\nu = 0 \text{ on } \Gamma_0. \tag{19}$$

(Indeed, if $x_0\in\Gamma_0$, then we can take $h(x) = x - x_0$.) Applying this vector field h to the relation (10), we obtain the inequality,

$$\begin{aligned}\int_{Q_{\epsilon_0}}|\nabla\phi|^2\,dQ_{\epsilon_0} \leq& \int_{\epsilon_0}^{T-\epsilon_0}\int_{\Gamma_0}\psi_t(h\cdot\nabla\phi)dtd\Gamma_0 + \frac{1}{2}\int_{\epsilon_0}^{T-\epsilon_0}\int_{\Gamma_1}\phi_t^2 h\cdot\nu dtd\Gamma_1\\ &-\frac{1}{2}\int_{\epsilon_0}^{T-\epsilon_0}\int_{\Gamma_1}|\nabla\phi|^2\,h\cdot\nu dtd\Gamma_1 + \frac{n}{2}\int_{Q_{\epsilon_0}}\left(|\nabla\phi|^2-\phi_t^2\right)dQ_{\epsilon_0}\\ &-\left[(\phi_t, h\cdot\nabla\phi)_{L^2(\Omega)}\right]_{\epsilon_0}^{T-\epsilon_0}.\end{aligned}$$

Using $|\nabla\phi|^2 = \left(\frac{\partial\phi}{\partial\nu}^2 + \frac{\partial\phi}{\partial\tau}^2\right)$ on Γ, the conservation relation (9), and the fact that radial h is parallel to Γ_0, we obtain

$$\begin{aligned}\int_{Q_{\epsilon_0}} |\nabla\phi|^2 \, dQ_{\epsilon_0} \leq C_h &\left\{ \int_{\Sigma_0} \psi_t^2 d\Sigma_0 + \int_{\Sigma_1} \phi_t^2 d\Sigma_1 + \int_{\epsilon_0}^{T-\epsilon_0} \int_{\Gamma_1} \frac{\partial\phi}{\partial\tau}^2 \, dt d\Gamma_1 \right\} \\ &+ C \left| \int_{\epsilon_0}^{T-\epsilon_0} \int_{\Gamma_0} \psi_t \frac{\partial\phi}{\partial\tau} dt d\Gamma_0 \right| + 2 \max_{1\leq i\leq n} \|h_i\|_{L^\infty(\Omega)} \mathcal{E}(T) \\ &+ \frac{n}{2} \left| \int_{Q_{\epsilon_0}} \left(\phi_t^2 - |\nabla\phi|^2\right) dQ_{\epsilon_0} \right|. \end{aligned} \tag{20}$$

Now combining (20) and (12) (with $\tilde{h} = h$ and $\epsilon = 1$ therein) gives

$$\begin{aligned}\frac{1}{2} \int_{Q_{\epsilon_0}} |\nabla\phi|^2 \, dQ_{\epsilon_0} \leq \; C_h &\left\{ \int_{\Sigma_0} \psi_t^2 d\Sigma_0 + \int_{\Sigma_1} \phi_t^2 d\Sigma_1 + \int_{\epsilon_0}^{T-\epsilon_0} \int_{\Gamma_1} \frac{\partial\phi}{\partial\tau}^2 \, dt d\Gamma_1 \right\} \\ &+ C \left| \int_{\epsilon_0}^{T-\epsilon_0} \int_{\Gamma_0} \psi_t \frac{\partial\phi}{\partial\tau} dt d\Gamma_0 \right| + 2 \max_{1\leq i\leq n} \|h_i\|_{L^\infty(\Omega)} \mathcal{E}(T) \\ &+ \text{l.o.t.}(\vec{\phi}, \vec{\psi}). \end{aligned} \tag{21}$$

Now to handle the term $\int_{\epsilon_0}^{T-\epsilon_0} \int_{\Gamma_0} \psi_t \frac{\partial\phi}{\partial\tau} dt d\Gamma$: again, Sobolev trace theory attaches no meaning to the boundary trace $\left.\frac{\partial\phi}{\partial\tau}\right|_{\Gamma_0}$ for initial data of finite energy. Instead, we deal with this term by the following "sharp" trace regularity result:

Lemma 5 *(see [18], p. 113, Corollary 3.4(b) and Theorem 3.3(a) [with $\alpha = \beta = \frac{3}{4}$ therein]). Let Γ_0 be a flat portion of the boundary, and let w solve the following wave equation:*

$$\begin{cases} w_{tt} = \Delta w & \text{on } (0,T) \times \Omega \\ \dfrac{\partial w}{\partial\nu} = g \in L^2(0,T; H^{\frac{1}{4}}(\Gamma_0)) & \text{on } (0,T) \times \Gamma_0 \\ [w(0), w_t(0)] = [w_0, w_1] \in H_1. \end{cases} \tag{22}$$

Then one has the following estimate:

$$\left\| \frac{\partial w}{\partial\tau} \right\|_{L^2(0,T;H^{-\frac{1}{4}}(\Gamma_0))} \leq C_T \left(\|g\|_{L^2(0,T;H^{\frac{1}{4}}(\Gamma))} + \|[w_0, w_1]\|_{H_1} \right).^2$$

[2] As explicitly stated in [18], Theorem (3.3)(a) mandates that Neumann boundary data g be smoother in space *and* time in order to obtain the requisite tangential regularity; namely, $g \in H^{\frac{1}{4}}((0,T) \times \Gamma_0)$ (again Γ_0 is flat). However, in the details of proof, it is evident that indeed continuously, $g \in L^2(0,T; H^{\frac{1}{4}}(\Gamma_0)) \Rightarrow \frac{\partial w}{\partial\tau} \in L^2(0,T; H^{-\frac{1}{4}}(\Gamma_0))$.

Applying Lemma 5 (taking $g \equiv \psi_t$ therein), along with $ab \leq \frac{\delta}{2}a^2 + \frac{1}{2\delta}b^2$ (taking $\delta \equiv C_T$ of (22)), we have

$$\begin{aligned}
\int_{\epsilon_0}^{T-\epsilon_0}\int_{\Gamma_0} \psi_t \frac{\partial \phi}{\partial \tau} dt d\Gamma &= \int_{\epsilon_0}^{T-\epsilon_0} \left\langle \psi_t, \frac{\partial \phi}{\partial \tau} \right\rangle_{H^{\frac{1}{4}}(\Gamma_0)\times H^{-\frac{1}{4}}(\Gamma_0)} dt \\
&\leq \int_{\epsilon_0}^{T-\epsilon_0} \|\psi_t\|_{H^{\frac{1}{4}}(\Gamma_0)} \left\| \frac{\partial \phi}{\partial \tau} \right\|_{H^{-\frac{1}{4}}(\Gamma_0)} dt \\
&\leq \frac{1}{2}\mathcal{E}_\phi(T) + (\frac{C_T}{2} + \frac{1}{2}) \int_{\epsilon_0}^{T-\epsilon_0} \|\psi_t\|^2_{H^{\frac{1}{4}}(\Gamma_0)} dt. \qquad (23)
\end{aligned}$$

Coupling (23) with (21) yields

$$\begin{aligned}
\frac{1}{2}\int_{Q_{\epsilon_0}} |\nabla \phi|^2 dQ_{\epsilon_0} \leq\; & C_{\epsilon,h,T}\left\{ \int_0^T \|\psi_t\|^2_{H^{\frac{1}{4}}(\Gamma_0)} dt + \int_{\Sigma_1} \phi_t^2 d\Sigma_1 + \int_{\epsilon_0}^{T-\epsilon_0}\int_{\Gamma_1} \frac{\partial \phi}{\partial \tau}^2 dt d\Gamma_1 \right\} \\
& + C_h \mathcal{E}(T) + \text{l.o.t.}(\vec{\phi}, \vec{\psi}). \qquad (24)
\end{aligned}$$

In turn, combining this estimate with that in (12) (again with $\tilde{h} = h$), we obtain

$$\begin{aligned}
\int_{\epsilon_0}^{T-\epsilon_0} \mathcal{E}_\phi(t) dt \leq\; & C_{\epsilon,h,T}\left\{ \int_0^T \|\psi_t\|^2_{H^{\frac{1}{4}}(\Gamma_0)} dt + \int_{\Sigma_1} \phi_t^2 d\Sigma_1 + \int_{\epsilon_0}^{T-\epsilon_0}\int_{\Gamma_1} \frac{\partial \phi}{\partial \tau}^2 dt d\Gamma_1 \right\} \\
& C_h \mathcal{E}(T) + \text{l.o.t.}(\vec{\phi}, \vec{\psi}).
\end{aligned}$$

A subsequent application of the tangential estimate (17) (with $\Gamma_* = \Gamma_1$) produces the following: *Under Assumption (A2), the ϕ-component of the adjoint system (4) obeys the integral estimate,*

$$\begin{aligned}
\int_{\epsilon_0}^{T-\epsilon_0} \mathcal{E}_\phi(t) dt \leq\; & C_{T,\epsilon_0,h}\left\{ \int_0^T \|\psi_t\|^2_{H^{\frac{1}{4}}(\Gamma_0)} dt + \int_{\Sigma_1} \phi_t^2 d\Sigma_1 \right\} \\
& + C_h \mathcal{E}(T) + \text{l.o.t.}(\vec{\phi}, \vec{\psi}), \qquad (25)
\end{aligned}$$

where $C_{T,\epsilon_0,h}$ depends on T, but C_h does not.

3.4 Step 3 (An Estimate for ψ)

Applying the multiplier ψ to the second wave equation in (4), and integrating in time and space yields

$$\left[\int_{\Gamma_0} \psi_t(t)\psi(t)d\Gamma_0\right]_{t=0}^{t=T} - \int_{\Sigma_0} \psi_t^2 d\Sigma_0$$

$$= -\int_{\Sigma_0}\left(\frac{\partial\psi}{\partial\tau}\right)^2 d\Sigma_0 - \left[\int_{\Gamma_0} \phi(t)|_{\Gamma_0}\psi(t)d\Gamma_0\right]_{t=0}^{t=T} + \int_{\Sigma_0} \phi|_{\Gamma_0}\psi_t d\Sigma_0.$$

A rearrangement of terms, combined with a use of Sobolev trace theory, yields

$$\begin{aligned}
\int_{\Sigma_0}\left(\frac{\partial\psi}{\partial\tau}\right)^2 d\Sigma_0 &= \int_{\Sigma_0} \psi_t^2 d\Sigma_0 + \int_{\Sigma_0} \phi|_{\Gamma_0}\psi_t d\Sigma_0 \\
&\quad - \left[\left\langle \psi_t(t) + \phi(t)|_{\Gamma_0}, \psi(t)\right\rangle_{H^{-\epsilon}(\Gamma_0)\times H^{\epsilon}(\Gamma_0)}\right]_{t=0}^{t=T} \\
&\leq C\left(\int_{\Sigma_0} \psi_t^2 d\Sigma_0 + \int_{\Sigma_0} \phi|_{\Gamma_0}^2 d\Sigma_0\right. \\
&\quad \left. + \left[\|\psi_t(t)\|_{H^{-\epsilon}(\Gamma_0)}^2 + \|\phi(t)|_{\Gamma_0}\|_{H^{-\epsilon}(\Gamma_0)}^2 + \|\psi(t)\|_{H^{\epsilon}(\Gamma_0)}^2\right]_{t=0}^{t=T}\right) \\
&\leq C\int_{\Sigma_0} \psi_t^2 d\Sigma_0 + \text{l.o.t.}(\vec{\phi},\vec{\psi}). \qquad (26)
\end{aligned}$$

3.5 Step 4 (Final Statement of Derived Estimate)

Combining (18) and (26) (in the case that Assumption (A1) holds true), and (25) and (26) (if Assumption (A2) holds true), we have the preliminary estimate for the energy:

Lemma 6 *(a) Under Assumption (A1), the solution to the PDE (4) satisfies the following estimate for all $\epsilon_0 > 0$:*

$$\begin{aligned}
&\int_{\epsilon_0}^{T-\epsilon_0} \mathcal{E}(t)dt \\
&\leq C_{T,\epsilon_0}\left\{\int_{\epsilon_0}^{T-\epsilon_0}\int_{\Gamma_0} \phi_t|_{\Gamma_0}^2 dt d\Gamma_0 + \int_{\Sigma_0} \xi^2\, \phi_t|_{\Gamma_0}^2 d\Sigma_0 + \int_{\Sigma_0} \psi_t^2 d\Sigma_0\right\} \\
&\quad + C\mathcal{E}(T) + \text{l.o.t.}(\vec{\phi},\vec{\psi}), \qquad (27)
\end{aligned}$$

where constant C is independent of time T.

(b) Under Assumption (A2), the solution to the PDE (4) satisfies the following estimate for all $\epsilon_0 > 0$:

$$\int_{\epsilon_0}^{T-\epsilon_0} \mathcal{E}(t)dt$$

$$\leq \quad C_{T,\epsilon_0,h}\left\{\int_0^T \|\psi_t\|^2_{H^{\frac{1}{4}}(\Gamma_0)} dt + \int_{\Sigma_1} \phi_t^2 d\Sigma_1\right\} + C\mathcal{E}(T) + \text{l.o.t.}(\vec{\phi},\vec{\psi}), \qquad (28)$$

where constant C is independent of time T.

Recall that our aim is to attain the inequalities (5) (under Assumption (A2)), and (6) (under Assumption (A2)). We can handle the terms $\int_{\epsilon_0}^{T-\epsilon_0} \mathcal{E}(t)dt$ and $\mathcal{E}(T)$ in (27) and (28) by the conservation relation (9), and the lower-order terms by a compactness-uniqueness argument. Hence, the "bad terms" appear only in the inequality (27); i.e., $\int_{\epsilon_0}^{T-\epsilon_0}\int_{\Gamma_0} \phi_t|^2_{\Gamma_0} dtd\Gamma_0$ and $\int_{\Sigma_0} \xi^2\phi_t^2 d\Sigma_0$. Accordingly, we concern ourselves next with $\phi_t|_{\Gamma_0}$.

4 ANALYSIS OF $\phi_T|_{\Gamma_0}$

4.1 Supporting Results

In [4], we show outright the derivation of the following sequence of supporting Propositions:

Proposition 7 *The ψ-component of the PDE (4) satisfies the following estimate for arbitary $\epsilon_1, \epsilon_2 > 0$:*

$$\int_{\Sigma_0} \left(\xi\frac{\partial\psi_t}{\partial\tau}\right)^2 d\Sigma_0 \leq C\,\|\psi\|^2_{H^2(0,T;L^2(\Gamma_0))} + \epsilon_1\int_{\Sigma_0} \xi^2\phi_t^2 d\Sigma_0 + \epsilon_2\int_{\Sigma_0} \xi^2\left(\frac{\partial^2\psi}{\partial\tau^2}\right)^2 d\Sigma_0. \qquad (29)$$

Proposition 8 *The ψ-component of the PDE system (4) obeys the estimate:*

$$\int_{\Sigma_0} \xi^2\left(\frac{\partial^2\psi}{\partial\tau^2}\right)^2 d\Sigma_0$$

$$\leq \quad \{\epsilon_1 C_{\epsilon_5} + \epsilon_5 C_T\}\int_{\Sigma_0} \xi^2\; \phi_t|^2_{\Gamma_0} d\Sigma_0$$

$$+(\epsilon_2 C_{\epsilon_5} + \epsilon_3 + \epsilon_4)\int_{\Sigma_0} \xi^2\left(\frac{\partial^2\psi}{\partial\tau^2}\right)^2 d\Sigma_0$$

$$+C\,\|\psi\|^2_{H^2(0,T;L^2(\Gamma_0))} + \text{l.o.t.}(\vec{\phi},\vec{\psi}), \qquad (30)$$

where the ϵ_i, $i = 1,..5$, are arbitrarily small.

Proposition 9 *The ϕ-component of the solution to the system (4) obeys the following estimate:*

$$\int_{\Sigma_0} \xi^2 \, \phi_t|^2_{\Gamma_0} \, d\Sigma_0 \;\leq\; (\epsilon_2 C + \epsilon_3 + \epsilon_4) \int_{\Sigma_0} \xi^2 \left(\frac{\partial^2 \psi}{\partial \tau^2}\right)^2 d\Sigma_0$$
$$+ C \, \|\psi\|^2_{H^2(0,T;L^2(\Gamma_0))} + \text{l.o.t.}(\vec{\phi}, \vec{\psi}), \tag{31}$$

where ϵ_i, $i = 2, 3, 4$ are arbitrarily small.

4.2 The Main Estimate for $\phi_t|_{\Gamma_0}$

The Proposition 9 above can be used to prove the following lemma:

Lemma 10 *Let $\epsilon > 0$ be arbitrary. Then the ψ-component of the system (4) satisfies the inequality*

$$\int_{\Sigma_0} \xi^2 \psi^2_{\tau\tau} d\Sigma_0 dt \leq C \, \|\psi\|^2_{H^2(0,T;L^2(\Gamma_0))} + \text{l.o.t.}(\vec{\phi}, \vec{\psi}),$$

where ξ is the cutoff function defined in (8).

Proof of Lemma 10: Squaring both sides of the ψ_c-wave equation in (4), integrating in time and space, and applying Cauchy–Schwartz, we obtain the inequality

$$\int_{\Sigma_0} \xi^2 \left(\frac{\partial^2 \psi}{\partial \tau^2}\right)^2 d\Sigma_0 \leq 2 \int_{\Sigma_0} \xi^2 \left(\phi_t|_{\Gamma_0}\right)^2 d\Sigma_0 + C \, \|\psi\|^2_{H^2(0,T;L^2(\Gamma_0))} \cdot \tag{32}$$

Now applying the trace estimate (31) gives

$$\int_{\Sigma_0} \xi^2 \left(\frac{\partial^2 \psi}{\partial \tau^2}\right)^2 d\Sigma_0$$
$$\leq \; 2 (\epsilon_2 C + \epsilon_3 + \epsilon_4) \int_{\Sigma_0} \xi^2 \left(\frac{\partial^2 \psi}{\partial \tau^2}\right)^2 d\Sigma_0$$
$$+ C \, \|\psi\|^2_{H^2(0,T;L^2(\Gamma_0))} + \text{l.o.t.}(\vec{\phi}, \vec{\psi}).$$

Taking ϵ_i to be small enough yields the estimate

$$\int_{\Sigma_0} \xi^2 \left(\frac{\partial^2 \psi}{\partial \tau^2}\right)^2 d\Sigma_0 \leq C \, \|\psi\|^2_{H^2(0,T;L^2(\Gamma_0))} + \text{l.o.t.}(\vec{\phi}, \vec{\psi}).$$

This completes the proof of Lemma 10.

Since

$$\phi_t|_{\Gamma_0} = \psi_{\tau\tau} - \psi_{tt} \quad \text{on } (0, T) \times \Gamma_0,$$

Lemma 10 immediately gives the following necessary estimate:

Corollary 11 *Let $\epsilon > 0$ be arbitrary. Then the ϕ-component of the system (4) satisfies the inequality*

$$\int_{\Sigma_0} \xi^2 \ \phi_t|^2_{\Sigma_0} \, d\Sigma_0 \leq C \, \|\psi\|^2_{H^2(0,T;L^2(\Gamma_0))} + \text{l.o.t.}(\vec{\phi}, \vec{\psi}).$$

5 PROOF PROPER OF THEOREM 1

5.1 Completion of the Proof of Theorem 1(a)

Recall that Theorem 1(a), which provides for exact controllability of the PDE system (1) under Assumption (A1), is equivalent to obtaining the inequality (5) for the adjoint system (4).

Combining the estimates from Lemma 6 and Corollary 11, along with the definition of the cutoff function ξ in (8), we obtain

$$\begin{aligned} & \int_{\epsilon_0}^{T-\epsilon_0} \mathcal{E}(t)dt \\ \leq \ & C_{T,\epsilon_0} \, \|\psi\|^2_{H^2(0,T;L^2(\Gamma_0))} + C\mathcal{E}(T) + \text{l.o.t.}(\vec{\phi}, \vec{\psi}), \end{aligned} \tag{33}$$

where again C_{T,ϵ_0} depends on time T, but C does not. Using the conservation relation (9), we have

$$\begin{aligned} & (T - 2\epsilon_0) \, \mathcal{E}(T) \\ \leq \ & C_T \, \|\psi\|^2_{H^2(0,T;L^2(\Gamma_0))} + C\mathcal{E}(T) + \text{l.o.t.}(\vec{\phi}, \vec{\psi}). \end{aligned}$$

From this inequality, we then have

Lemma 12 *Under Assumption (A1), the solution of (4) obeys the following estimate for long enough terminal time T:*

$$\left\| \left[\vec{\phi}_0, \vec{\psi}_0 \right] \right\|^2_{\mathcal{H}} \leq C \, \|\psi\|^2_{H^2(0,T;L^2(\Gamma_0))} + \text{l.o.t.}(\vec{\phi}, \vec{\psi}). \tag{34}$$

A compactness-uniqueness argument can be used to eliminate the lower-order terms in (34). That is, given the existence of the estimate (34), one can ascertain the existence of a positive constant C_T such that

$$\left\| \left[\vec{\phi}, \vec{\psi} \right] \right\|_{C([0,T];H^{1-\epsilon}(\Omega)\times H^{-\epsilon}(\Omega)\times H^{1-\epsilon}(\Gamma_0)\times H^{-\epsilon}(\Gamma_0))} \leq C_T \, \|\psi\|^2_{H^2(0,T;L^2(\Gamma_0))} \tag{35}$$

(see [4] for the full details of the argument). Combining (34) and (35) completes the proof of Theorem 1(a).

5.2 Completion of the Proof of Theorem 1(b)

Applying the conservation relation (9) to the inequality (28), valid under Assumption (A2), we have

$$\begin{aligned}&(T-2\epsilon_0)\mathcal{E}(T)\\ \leq\ & C_{T,\epsilon_0,h}\left\{\int_0^T \|\psi_t\|^2_{H^{\frac{1}{4}}(\Gamma_0)}\,dt+\int_{\Sigma_1}\phi_t^2 d\Sigma_1\right\}+C\mathcal{E}(T);\end{aligned} \tag{36}$$

whence the inequality

$$\mathcal{E}(T)\leq C_{T,\epsilon_0,h}\left\{\int_0^T \|\psi_t\|^2_{H^{\frac{1}{4}}(\Gamma_0)}\,dt+\int_{\Sigma_1}\phi_t^2 d\Sigma_1\right\}+\text{l.o.t.}(\vec{\phi},\vec{\psi}). \tag{37}$$

By an argument virtually identical to that used to prove the estimate (35), we can eliminate the lower-order terms in (37) to thereby obtain the requisite reverse inequality (6). This completes the proof of Theorem 1(b).

Acknowledgment

Research was partially supported by NSF Grant No. DMS-0196359, NSF Grant No. DMS-0104305, and ARO DAAD 19-02-1-0179.

REFERENCES

1. G. Avalos and I. Lasiecka, Differential Riccati equation for the active control of a problem in structural acoustics, *Journal of Optimization Theory and Applications*, Vol. 91, No. 3 (1996), pp. 695–728.
2. G. Avalos, The exponential stability of a coupled hyperbolic/parabolic system arising in structural acoustics, *Applied and Abstract Analysis*, Volume 1, Number 2 (1996), pp. 203–217.
3. G. Avalos and I. Lasiecka, Boundary controllability of thermoelastic plates via the free boundary conditions, *SIAM J. Control Optim.*, Vol. 38, No. 2 (2000), pp. 337–383.
4. G. Avalos and I. Lasiecka, Exact controllability of structural acoustic interactions, *IMA Preprint Series #1771* (June 2001) and submitted to *J. Math. Pures Appl.*
5. G. Avalos, I. Lasiecka, and R. Rebarber, *Boundary Controllability of a Coupled Wave/Kirchoff System*, preprint.
6. H. T. Banks, W. Fang, R. J. Silcox, and R. Smith, *Approximation Methods for Control of Acoustic/Structure Models with Piezoceramic Actuators*, Contract Report 189578, NASA (1991).
7. C. Bardos, G. Lebeau, and J. Rauch, Sharp sufficient conditions for the observation, control and stabilization of waves from the boundary, *SIAM J. Control Optim.*, 30 (1992), pp. 1024–1065.
8. J.T. Beale, Spectral properties of an acoustic boundary condition, *Indiana Univ. Math. J.*, 25 (1976), pp. 895–917.

9. M. Camurdan and R. Triggiani, Sharp regularity of a coupled system of a wave and Kirchoff equation with point control arising in noise reduction, *Differential and Integral Equations*, Vol. 12, No. 1, January (1999), pp. 101–118.
10. P. Grisvard, *Elliptic Problems in Nonsmooth Domains*, Pitman Advanced Publishing, Boston (1985).
11. L Hörmander, *Linear Partial Differential Operators*, Springer-Verlag, New York (1969).
12. V. Hutson and J.S. Pym, *Applications of Functional Analysis and Operator Theory*, Academic Press, New York (1980).
13. V. Komornik, *Exact Controllability and Stabilization. The Multiplier Method*, Masson, Paris (1994).
14. J. Lagnese, Decay of solutions of wave equations in a bounded region with boundary dissipation, *J. Differential Equations*, 50 (1983), pp. 106–113.
15. I. Lasiecka and R. Triggiani, Exact controllability of the wave equation with Neumann boundary control, *Applied Mathematics and Optimization*, Vol. 19 (1989), pp. 243–290.
16. I. Lasiecka and R. Triggiani, Sharp regularity results for mixed second-order hyperbolic equations of Neumann type: The L_2-boundary case, *Annali di Matem. Pura e Appl.*, IV, CLVII (1990), pp. 285–367.
17. I. Lasiecka and R. Triggiani, Uniform stabilization of the wave equation with Dirichlet or Neumann feedback control without geometrical conditions, *Applied Mathematics and Optimization*, Vol. 25 (1992), pp. 189–224.
18. I. Lasiecka and R. Triggiani, Recent advances in regularity of second-order hyperbolic mixed problems, and applications, *Dynamics Reported: Expositions in Dynamical Systems*, Vol. 3 (1994), pp. 104–158.
19. I. Lasiecka, R. Triggiani, and X. Zhang, Nonconservative wave equations with unobserved Neumann B.C.: Global uniqueness and observability on one shot, *Contemporary Mathematics*, Vol. 268 (2000), pp. 227–325.
20. W. Littman and B. Liu, On the spectral properties and stabilization of acoustic flow, *IMA Preprint Series #1436*, University of Minnesota, November (1996).
21. W. Littman and L. Markus, Stabilization of hybrid system of elasticity by feedback boundary damping, *Annali di Mathematica Pure et Applicata*, 152 (1988), pp. 281–330.
22. S. Micu, Boundary controllability of a linear hybrid system arising in the control of noise, *SIAM J. Control Optim.*, Vol. 35 (1997), pp. 531–555.
23. C.S. Morawetz, Energy identities for the wave equation, *Math. Sci. Res. Rep. No. IMM 346*, NYU Courant Institute (1976).
24. P.M. Morse and K.U. Ingard, *Theoretical Acoustics*, McGraw-Hill, New York (1968).
25. A. Pazy, *Semigroups of Linear Operators and Applications to Partial Differential Equations*, Springer-Verlag, New York (1983).
26. J. Simon, Compact sets in the space $L^p(0,T;B)$, *Ann. Mat. Pura Appl.*, (4), 148 (1987), pp. 65–96.
27. W. Strauss, Dispersal of waves vanishing on the boundary of an exterior domain, *Comm. Pure Appl. Math.* 28 (1976), pp. 265–278.
28. R. Triggiani, Wave equation on a bounded domain with boundary dissipation: An operator approach, *Journal of Mathematical Analysis and Applications* 137 (1989), pp. 438–461.

12

Flow Field, Temperature, and Dopant Concentration Evolution in a Bridgman–Stockbarger Crystal Growth System in a Strictly Zero-Gravity and a Low-Gravity Environment

St. Balint[†] and A.M. Balint[*]

[†]*Department of Mathematics, University of West Timisoara, Timisoara, Romania*
[*]*Department of Physics, University of West Timisoara, Timisoara, Romania*

In order to describe the evolution of the flow, temperature, and dopant concentration in the melt in a Bridgman–Stockbarger crystal growth system, a continuum mechanical model is presented. The equations of the model are the classical Navier–Stokes equation in the Boussinesq approximation, the classical incompressibility equation as well as a modified energy equation and a modified dopant dispersion equation. The classical energy and dopant transport equations were modified due to the thermotransport and precrystallization-zone influences. We give a model-based simulation of the flow, heat transport, and Ga dispersion evolution in the melt and prediction of axial and radial Ga distribution in a Ga-doped Ge semiconductor crystal grown in a strictly zero-gravity and a low-gravity environment. We present adequate initial Ga distribution in the melt for obtaining uniform Ga distribution in the crystal.

1 INTRODUCTION

The prototype Bridgman–Stockbarger growth system consists of crystal and melt contained in a cylindrical ampoule of radius R and length L pulled slowly through a vertically aligned furnace with hot and cold isothermal zones, separated by an adiabatic region of length L_g, designed to promote steep axial temperature gradient and to maintain a flat solidification interface, as shown in Fig. 1 [1].

In [2] there is a characterization of the sample obtained in such a system in the shuttle mission 51-G, experiment MRS77F075. The axial dopant distribution in the sample, reported in [2], exhibits an important nonlinear variation. This compositional nonuniformity influences the crystallographic quality of the sample. Therefore it is extremely important to identify the origin of this nonuniformity.

0-415-30852-6/04/$0.00+$1.50

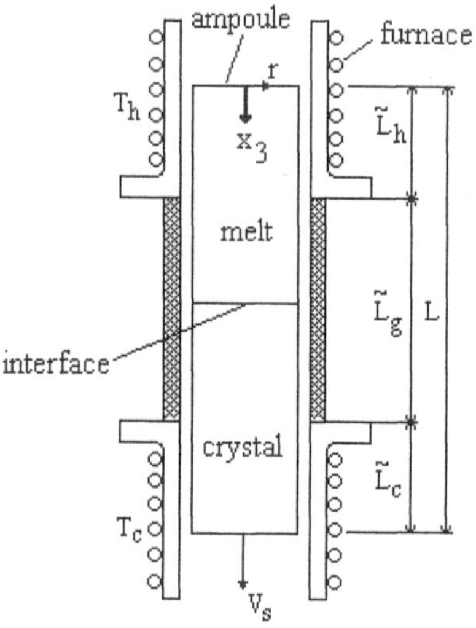

Figure 1 Prototype Bridgman–Stockbarger growth system.

In a general context, the axial and radial dopant distribution in a semiconductor crystal grown in a Bridgman–Stockbarger system is determined by the thermal convection in the melt [1], [3], [4]; the thermotransport [5], [6]; the decrease of the melt in the ampoule [5], [6]; the morphology of the solidified front [7], [8]; the initial dopant distribution in the melt [6], [9].

During the growth on Earth, the temperature gradients in the melt generate buoyancy driving forces and thermal convection in the melt [1], [3], [4], which results in a "nearly completely" mixed melt [4]. Crystals grown from well-mixed melts exhibit a nonlinear variation of the dopant concentration along the growth axis [9]. On the other hand, crystals grown from a quiescent melt, after an initial transient exhibit a uniform axial dopant distribution [10]. Thus, reduction of the magnitude of the buoyancy forces by processing semiconductors in a low-gravity environment has been pursued over the past decades. The effectiveness of space processing for the growth of chemically uniform crystals is supported by experimental and theoretical studies. For example, the InSb crystal reported in [11] exhibits axial segregation profiles that are characteristic of diffusion-controlled mass transfer growth. In [1], [3] it is shown, based on modeling studies, that the low-gravity levels achieved in space are sufficient to inhibit interference of thermal convection with segregation in small-diameter Ge and GeSi melts. The above analysis has been performed using constant values of the gravitational acceleration. In [12], [13] there is a study of the influence of nonsteady gravity on the thermal convection during microgravity solidification of semiconductors. The study corresponds to the use of low-duration low-gravity vehicles, such as sounding rockets and KC-135 aircraft, for low-gravity crystal growth experiments. In these vehicles, low-gravity periods of 20 seconds and 6 minutes, for KC-135 aircraft and sounding rockets, respectively, are achieved at costs that are at orders of magnitude smaller than space experiments.

In [5], the authors consider a strictly zero-gravity environment, in the absence of convective mixing, and identify the thermotransport and the decrease of the melt in the ampoule as the origin of the nonlinear variation of the dopant concentration along the axis of the sample, reported in [2]. In [6], there is also an analysis of the role of the decrease of the melt and of the role of thermotransport in axial dopant segregation of Ga in Ge.

Concerning the morphology of the solidified front, in [7] it is proved that significant radial segregation occurs in the system without convection tangent to the crystal surface when the radius of curvature of this interface is the same or less than the length-scale of the concentration gradient adiacent to the interface. Curvature-induced segregation was shown to be an important contribution to the dopant inhomogeneity in capillary growth systems [14], [15]. All the above-mentioned studies were made in crystal growth models in which the solidification interface is a smooth mathematical surface. But, physically the melt $\rightarrow$ solid phase transition takes place in a thin layer of width 10^{-9} m, masking the crystal, where both phases, melt and solid, coexist. In [8] it has been shown that such a boundary layer reduces the molecular diffusion from D to $D_{\text{eff}} = D\times 0.917$, increases the heat conductivity from K to $K_{\text{eff}} = K\times 1.1$, and does not influence the flow in the region. In [6] it is shown that such a boundary layer has a significant influence on the dopant distribution in the crystal.

The influence of the initial dopant distribution in the melt on the dopant distribution in the crystal was pointed out in [9]. In [6], this influence is used to show how it is possible to reduce the axial segregation (nonuniformity) produced by the decrease of the melt, thermotransport, and precrystallization zone, respectively, in the case of crystals grown in strictly zero gravity.

In this chapter we present a continuum mechanical model that takes into account the thermoconvection in the melt, the thermotransport, the decrease of the melt, the morphology of the solidified front and the initial dopant distribution in the melt. We give a model-based simulation of the flow evolution, heat transport, dopant dispersion in the melt and we predict axial and radial dopant distribution in a Ga-doped Ge semiconductor crystal grown in a strictly zero-gravity and in a low-gravity environment. We present adequate initial Ga distribution in the melt for obtaining uniform Ga distribution in the Ga-doped Ge semiconductor crystal.

2 THE CONTINUUM MECHANICAL MODEL

A continuum mechanical model that takes into account the thermoconvection in the melt, the thermotransport, the decrease of the melt in the ampoule, the influence of the precrystallization-zone and the initial dopant distribution in the melt, is defined in a mobile reference frame, as that shown in Fig. 1, by the following dimensionless equations, boundary conditions, and initial conditions written in cylindrical polar cordinates:

$$\frac{\partial \boldsymbol{u}}{\partial t} + (\boldsymbol{u}\nabla)\boldsymbol{u} = -\nabla p + \mathrm{Pr}\cdot\Delta\boldsymbol{u} + \mathrm{Pr}\cdot Ra\cdot\theta\cdot\boldsymbol{e}_3, \tag{1}$$

$$\nabla\boldsymbol{u} = 0, \tag{2}$$

$$b_\theta \frac{\partial \theta}{\partial t} + \boldsymbol{u} \cdot \nabla \theta = K_* \Delta \theta - 2\frac{K_a}{K}\frac{L}{\delta}\frac{L}{R}(\theta_w - \theta_f), \tag{3}$$

$$b_c \frac{\partial c}{\partial t} + \boldsymbol{u} \cdot \nabla c = D_* \Delta c + \sigma(T_H - T_C) \cdot \nabla(c \cdot \nabla \theta), \tag{4}$$

$$\boldsymbol{u}|_{\partial\Omega_t} = 0, \tag{5}$$

$$\boldsymbol{u}(0, r\varphi, x_3) = \boldsymbol{u}_0(r, \varphi, x_3), \tag{6}$$

$$\left.\frac{\partial \theta}{\partial x_3}\right|_{(\partial\Omega_t)_{top}} = 0, \tag{7}$$

$$\theta|_{(\partial\Omega_t)_w} = \theta_w, \tag{8}$$

$$\theta|_{(\partial\Omega_t)_i} = \theta_i, \tag{9}$$

$$\theta(0, r, \varphi, x_3) = \theta_0(r, \varphi, x_3), \tag{10}$$

$$\left.\frac{\partial c}{\partial x_3}\right|_{(\partial\Omega_t)_{top}} = 0, \tag{11}$$

$$\left.\frac{\partial c}{\partial r}\right|_{(\partial\Omega_t)_w} = 0 \tag{12}$$

$$\left.\frac{\partial c}{\partial x_3}\right|_{(\partial\Omega_t)_i} = \phi_i, \tag{13}$$

$$c(0, r, \varphi, x_3) = c_0(r, \varphi, x_3). \tag{14}$$

Equation (1) is the dimensionless Navier–Stokes equation describing the thermoconvection in the melt due to the buoyancy driving forces $\Pr \cdot Ra \cdot \theta \cdot e_3$, in the Boussinesq approximation. The significance of the quantities $\boldsymbol{u}$, p, $\Pr Ra$ and θ is given in Table 1. The equation is considered in the time-dependent domain D_t defined by

$$\begin{aligned} D_t = &\left\{(t, r, \varphi, x_3) \,\middle|\, 0 \leqslant t \leqslant \frac{1}{u_p}\frac{L_g}{2L};\ 0 \leqslant r \leqslant \Lambda;\ 0 \leqslant \varphi \leqslant 2\pi;\ 0 \leqslant x_3 \leqslant 1\right\} \\ &\cup \left\{(t, r, \varphi, x_3) \,\middle|\, \frac{1}{u_p}\frac{L_g}{2L} \leqslant t \leqslant \frac{1}{u_p}\left(1 + \frac{L_g}{2L}\right);\right. \\ &\left. 0 \leqslant r \leqslant \Lambda;\ 0 \leqslant \varphi \leqslant 2\pi;\ 0 \leqslant x_3 \leqslant 1 + \frac{L_g}{2L} - u_p t\right\}. \end{aligned} \tag{15}$$

This is because the domain Ω_t occupied by the melt is time dependent and is defined by

$$\Omega_t = \begin{cases} \{(r, \varphi, x_3) \,|\, 0 \leqslant r \leqslant \Lambda;\ 0 \leqslant \varphi \leqslant 2\pi;\ 0 \leqslant x_3 \leqslant 1\} \text{ for } 0 \leqslant t \leqslant \frac{1}{u_p}\frac{L_g}{2L} \\ \left\{(r, \varphi, x_3) \,\middle|\, 0 \leqslant r \leqslant \Lambda;\ 0 \leqslant \varphi \leqslant 2\pi;\ 0 \leqslant x_3 \leqslant 1 + \frac{L_g}{2L} - u_p t\right\} \\ \text{for } \frac{1}{u_p}\frac{L_g}{2L} \leqslant t \leqslant \frac{1}{u_p}\left(1 + \frac{L_g}{2L}\right) \end{cases}. \tag{16}$$

The restriction imposed on x_3, $0 \leqslant x_3 \leqslant 1 + L_g/2L - u_p t$ for $1/u_p \cdot L_g/2L \leqslant t \leqslant 1/u_p \cdot (1 + L_g/2L)$, expresses the decrease of the melt in the ampoule with constant ratio u_p.

Equation (2) is a dimensionless incompressibility equation of the melt. This equation is considered in D_t, too.

Table 1 List of Symbols

Group	Definition	Value
$u = vL/\chi_m$	dimensionless velocity field	
v	velocity field	
L	length of the ampoule	2 cm
χ_m	thermal diffusivity in the melt	0.13 cm^2/ s
$p = p_m L^2/(\rho_m \chi_m)$	dimensionless pressure field	
p_m	pressure field	
$\rho_{m,s}$	density of melt and solid, respectively	5.6 g/ cm^3
$t = \tau \chi_m / L^2$	dimensionless time	
τ	time	
$\theta = (T - T_C)/(T_H - T_C)$	dimensionless temperature	
T_H	temperature of the hot zone of the furnace	1097°C
T_C	temperature of the cold zone of the furnace	819°C
T	temperature	
c	dopant concentration	
R	radius of the ampoule	0.5 cm
L_g	length of the adiabatic region	0.25 cm
$\Lambda = R/L$	aspect ratio	0.25
$\Pr = \nu/\chi_m$	Prandtl number	0.01
$Ra = \beta g (T_H - T_C) L^3/\chi_m \nu$	Rayleigh number	0; 10^3
β	thermal expansion coefficient	0.25×10^{-3} K^{-1}
g	gravity acceleration	0; 0.264 cm/ s^2
ν	kinematic viscosity of the melt	1.3×10^{-3} cm^2/ s
$Pe = V_s L/\chi_m$	Péclet number	0.01
$Sc = \nu/D$	Schmidt number	10
Sc_{eff}	effective Schmidt number	10.9
D	diffusivity of Ga in Ge	1.3×10^{-4} cm^2/ s
$K = k_s/k_m$	conductivity ratio	1
$k_{m,s}$	thermal conductivity of melt and solid, respectively	0.17 W/K cm
K_a	heat conductivity of the ampoule	3.26 W/K cm
$c_{m,s}$	heat capacity of melt and solid, respectively	0.39 J/K g
$u_p = V_s L/\chi_m$	ratio of decrease of the melt	0.01
V_s	growth rate	16 μm/s
δ	wall thickness	0.2 cm
σ	Soret coefficient	0.005
k	segregation coefficient	0.1

Equation (3) is a modified dimensionless heat transport equation in the melt (D_t). The significance of the quantities θ, $\boldsymbol{u}$, L, R, δ, K_a, and K is given in Table 1. The quantities b_θ and K_* are related to the precrystallization zone. Physically, this zone is a thin layer (masking the crystal) of width 10^{-9} m in which there are periodically distributed solid inclusions of size 10^{-10} m. The porosity of this layer is $\Pi_Y = 0.9$ and its thermal effect is the increase of the heat conductivity from K to $K_{\text{eff}} = 1.1 \times K$. For more details, see [8]. The quantities b_θ and K_* are defined as follows:

$$b_\theta = \begin{cases} 1 \text{ on } D_t^1 \\ 1 \text{ on } D_t^{2,1} \\ \Pi_Y + (1 - \Pi_Y)\frac{\rho_s \cdot c_s}{\rho_m \cdot c_m} \text{ on } D_t^{2,2} \\ \Pi_Y + (1 - \Pi_Y)\frac{\rho_s \cdot c_s}{\rho_m \cdot c_m} \text{ on } D_t^{2,3} \end{cases} ; \quad K_* = \begin{cases} 1 \text{ on } D_t^1 \\ 1 \text{ on } D_t^{2,1} \\ 1.1 \text{ on } D_t^{2,2} \\ 1.1 \text{ on } D_t^{2,3} \end{cases} , \tag{17}$$

where the sets D_t^1, $D_t^{2,1}$, $D_t^{2,2}$, and $D_t^{2,3}$ are defined by

$$D_t^1 = \left\{ (t, r, \varphi, x_3) \,\middle|\, 0 \leqslant t \leqslant \frac{1}{u_p}\frac{L_g}{2L};\ 0 \leqslant r \leqslant \Lambda;\ 0 \leqslant \varphi \leqslant 2\pi;\ 0 \leqslant x_3 \leqslant 1 \right\}$$

$$D_t^{2,1} = \left\{ \begin{array}{c} (t, r, \varphi, x_3) | \frac{1}{u_p}\frac{L_g}{2L} \leqslant t \leqslant \frac{1}{u_p}\left(1 + \frac{L_g - 2\cdot 10^{-9}}{2L}\right);\ 0 \leqslant r \leqslant \Lambda;\ 0 \leqslant \varphi \leqslant 2\pi; \\ 0 \leqslant x_3 \leqslant 1 + \frac{L_g - 2\cdot 10^{-9}}{2L} - u_p t \end{array} \right\}$$

$$D_t^{2,2} = \left\{ \begin{array}{c} (t, r, \varphi, x_3) | \frac{1}{u_p}\frac{L_g}{2L} \leqslant t \leqslant \frac{1}{u_p}\left(1 + \frac{L_g - 2\cdot 10^{-9}}{2L}\right);\ 0 \leqslant r \leqslant \Lambda;\ 0 \leqslant \varphi \leqslant 2\pi; \\ 1 + \frac{L_g - 2\cdot 10^{-9}}{2L} - u_p t \leqslant x_3 \leqslant 1 + \frac{L_g}{2L} - u_p t \end{array} \right\}$$

$$D_t^{2,3} = \left\{ \begin{array}{c} (t, r, \varphi, x_3) | \frac{1}{u_p}\left(1 + \frac{L_g - 2\cdot 10^{-9}}{2L}\right) \leqslant t \leqslant \frac{1}{u_p}\left(1 + \frac{L_g}{2L}\right); \\ 0 \leqslant x_3 \leqslant 1 + \frac{L_g}{2L} - u_p t;\ 0 \leqslant r \leqslant \Lambda;\ 0 \leqslant \varphi \leqslant 2\pi; \end{array} \right\}. \tag{18}$$

For $\rho_s c_s = \rho_m c_m$, b_θ is constant, equal to one, on D_t.

The significance of the quantities θ_w and θ_f will be discussed later, together with the boundary condition (8).

Equation (4) is a modified dopant dispersion equation in the melt (D_t). The term $\sigma \cdot (T_H - T_C) \cdot (c\nabla\theta)$ expresses the thermotransport. The significance of the quantities c, $\boldsymbol{u}$, σ, T_H, T_C, and θ is given in Table 1. The quantities b_c and D_* are related to the precrystallization zone (which reduces the molecular diffusivity of the dopant from D to $D_{\text{eff}} = D \times 0.917$) and are defined as follows:

$$b_c = \begin{cases} 1 \text{ on } D_t^1 \\ 1 \text{ on } D_t^{2,1} \\ \Pi_Y \text{ on } D_t^{2,2} \\ \Pi_Y \text{ on } D_t^{2,3} \end{cases};\quad D_* = \begin{cases} \Pr/Sc \text{ on } D_t^1 \\ \Pr/Sc \text{ on } D_t^{2,1} \\ \Pr/Sc_{eff} \text{ on } D_t^{2,2} \\ \Pr/Sc_{eff} \text{ on } D_t^{2,3} \end{cases}. \tag{19}$$

The boundary condition (5) is the nonslip condition on the boundary $\partial\Omega_t$. This boundary can be represented as follows:

$$\partial\Omega_t = (\partial\Omega_t)_{top} \cup (\partial\Omega_t)_w \cup (\partial\Omega_t)_i. \tag{20}$$

The set $(\partial\Omega_t)_{top}$ is the boundary of the melt limited by the top of the ampoule and is defined by

$$(\partial\Omega_t)_{top} = \{(r, \varphi, 0) | 0 \leqslant r \leqslant \Lambda;\ 0 \leqslant \varphi \leqslant 2\pi\}. \tag{21}$$

The set $(\partial\Omega_t)_w$ is the melt boundary limited by the walls of the ampoule and is defined by

$$(\partial\Omega_t)_w = \left\{ \begin{array}{l} \{(\Lambda, \varphi, x_3) \,|\, 0 \leqslant \varphi \leqslant 2\pi;\ 0 \leqslant x_3 \leqslant 1\} \text{ for } 0 \leqslant t \leqslant \frac{1}{u_p}\frac{L_g}{2L} \\ \left\{(\Lambda, \varphi, x_3) \,\middle|\, 0 \leqslant \varphi \leqslant 2\pi;\ 0 \leqslant x_3 \leqslant 1 + \frac{L_g}{2L} - u_p t\right\} \\ \quad \text{for } \frac{1}{u_p}\frac{L_g}{2L} \leqslant t \leqslant \frac{1}{u_p}\left(1 + \frac{L_g}{2L}\right) \end{array} \right. . \tag{22}$$

The set $(\partial\Omega_t)_I$ represents the melt boundary limited by the bottom of the ampoule for $0 \leqslant t \leqslant 1/u_p \cdot L_g/2L$ and for $1/u_p \cdot L_g/2L \leqslant t \leqslant 1/u_p \cdot (1 + L_g/2L)$ it is the precrystallization zone/crystal interface:

$$(\partial\Omega_t)_i = \left\{ \begin{array}{l} \{(r, \varphi, 1) \,|\, 0 \leqslant r \leqslant \Lambda;\ 0 \leqslant \varphi \leqslant 2\pi\} \text{ for } 0 \leqslant t \leqslant \frac{1}{u_p}\frac{L_g}{2L} \\ \left\{(r, \varphi, x_3) \,\middle|\, 0 \leqslant r \leqslant \Lambda;\ 0 \leqslant \varphi \leqslant 2\pi;\ x_3 = 1 + \frac{L_g}{2L} - u_p t\right\} \\ \quad \text{for } \frac{1}{u_p}\frac{L_g}{2L} \leqslant t \leqslant \frac{1}{u_p}\left(1 + \frac{L_g}{2L}\right) \end{array} \right. . \tag{23}$$

The initial condition (6) prescribes the velocity field in the melt. This is $\boldsymbol{u}_0(r, \varphi, x_3)$ and generally it will be considered equal to zero.

The boundary condition (7) is the no-flux condition for the heat on the top of the ampoule $(\partial\Omega_t)_{\text{top}}$.

Boundary condition (8) prescribes the value of the dimensionless temperature on the melt surface. According to [5], the heat exchange between the melt surface and the furnace is $Q = (T_H - T_C)(\theta_w - \theta_f)P/Rez \cdot Ac)$; where θ_w and θ_f are the (dimensionless) melt surface and furnace temperature, respectively; P/Ac is the geometry coefficient, equal to $2/R$, and Rez is the total heat resistance over the radial length between sample and furnace surfaces. If radiation is neglected, then the resistance is only a function of the heat conductivity K_a of the ampoule and wall thickness δ, $Rez = \delta/K_a$. Therefore, the heat exchange Q is given by

$$Q = (T_H - T_C))(\theta_w - \theta_f)\frac{2}{R}\frac{K_a}{\delta} \tag{24}$$

and θ_w is

$$\theta_w = \theta_f + \frac{Q}{(T_H - T_C)}\frac{R}{2}\frac{\delta}{K_a}. \tag{25}$$

If the heat exchange between the melt surface and furnace is negligible (the ampoule has negligible thermal mass [1]), then $\theta_w = \theta_f$ and is defined by

$$\theta_w = \theta_f = \begin{cases} 1; \text{ for } 0 \leqslant t \leqslant \frac{1}{u_p}\frac{L_g}{2L} \text{ and } 0 \leqslant x_3 \leqslant 1 - u_p t \\ -\frac{L_g}{L}(x_3 + u_p t) + 1 + \frac{L_g}{L}; \text{ for } 0 \leqslant t \leqslant \frac{1}{u_p}\frac{L_g}{2L} \text{ and } 1 - u_p t \leqslant x_3 \leqslant 1 \\ 1; \text{ for } \frac{1}{u_p}\frac{L_g}{2L} \leqslant t \leqslant \frac{1}{u_p} \text{ and } 0 \leqslant x_3 \leqslant 1 - u_p t \\ -\frac{L_g}{L}(x_3 + u_p t) + 1 + \frac{L_g}{L}; \text{ for } \frac{1}{u_p}\frac{L_g}{2L} \leqslant t \leqslant \frac{1}{u_p} \\ \qquad \text{and } 1 - u_p t \leqslant x_3 \leqslant 1 + \frac{L_g}{2L} - u_p t \\ 0; \text{ for } \frac{1}{u_p}\frac{L_g}{2L} \leqslant t \leqslant \frac{1}{u_p} \text{ and } 1 + \frac{L_g}{2L} - u_p t \leqslant x_3 \leqslant 1 \\ -\frac{L_g}{L}(x_3 + u_p t) + 1 + \frac{L_g}{L}; \text{ for } \frac{1}{u_p} \leqslant t \leqslant \frac{1}{u_p}\left(1 + \frac{L_g}{2L}\right) \\ \qquad \text{and } 0 \leqslant x_3 \leqslant 1 + \frac{L_g}{2L} - u_p t \\ 0; \text{ for } \frac{1}{u_p} \leqslant t \leqslant \frac{1}{u_p}\left(1 + \frac{L_g}{2L}\right) \text{ and } 1 + \frac{L_g}{L} - u_p t \leqslant x_3 \leqslant 1. \end{cases} \tag{26}$$

For ampoules with a small-diameter cross-section, in the one-dimensional approximation $\theta_w = \theta$, the boundary condition (8) disappears and in Eq. (3) θ_w is replaced by θ.

The boundary condition (9) prescribes the temperature on the bottom of the ampoule for $0 \leqslant t \leqslant 1/u_p \cdot L_g/2L$ and the temperature on the precrystallization zone/crystal interface for $1/u_p \cdot L_g/2L \leqslant t \leqslant 1/u_p \cdot (1 + L_g/2L)$. That is,

$$\theta_i = \begin{cases} -\frac{L}{L_g}(1 + u_p t) + 1 + \frac{L}{L_g}; \text{ for } 0 \leqslant t \leqslant \frac{1}{u_p}\frac{L_g}{2L}; \\ \qquad 0 \leqslant r \leqslant \Lambda;\ 0 \leqslant \varphi \leqslant 2\pi \text{ and } x_3 = 1 \\ \frac{1}{2}; \text{ for } \frac{1}{u_p}\frac{L_g}{2L} \leqslant t \leqslant \frac{1}{u_p}\left(1 + \frac{L_g}{2L}\right); \\ \qquad 0 \leqslant r \leqslant \Lambda;\ 0 \leqslant \varphi \leqslant 2\pi \text{ and } x_3 = 1 + \frac{L_g}{2L} - u_p t. \end{cases} \tag{27}$$

The initial condition (10) is the temperature distribution in the melt at the start; $\theta(0, r, \varphi, x_3) = \theta_0(r, \varphi, x_3)$. We shall consider the case when this distribution is uniform and equal to 1, i.e., $\theta_0(r, \varphi, x_3) \equiv 1$.

The boundary condition (11) expresses the no-flux condition of the dopant on the top of the ampoule $(\partial\Omega_t)_{\text{top}}$.

The boundary condition (12) is the no-flux condition of the dopant on the wall of the ampoule. For ampoules with a small-diameter cross-section, in the one-dimensional approximation, this boundary condition disappears.

Boundary condition (13) is the no-flux condition for the dopant on the bottom of the ampoule for $0 \leqslant t \leqslant 1/u_p \cdot L_g/2L$ and the flux of the dopant rejected into the melt for $1/u_p \cdot L_g/2L \leqslant\leqslant t \leqslant 1/u_p \cdot (1 + L_g/2L)$ and $x_3 = 1 + L_g/2L - u_p t$. That is,

$$\phi_i = \begin{cases} 0; \text{ for } 0 \leqslant t \leqslant \frac{1}{u_p}\frac{L_g}{2L};\ 0 \leqslant r \leqslant \Lambda;\ 0 \leqslant \varphi \leqslant 2\pi;\ x_3 = 1 \\ \frac{P_e}{D_*}(1-k)c + \sigma(T_h - T_c)c\frac{\partial\theta}{\partial x};\ \text{for } \frac{1}{u_p}\frac{L_g}{2L} \leqslant t \leqslant \frac{1}{u_p}\left(1 + \frac{L_g}{2L}\right); \\ \quad 0 \leqslant r \leqslant \Lambda;\ 0 \leqslant \varphi \leqslant 2\pi;\ x_3 = 1 + \frac{L_g}{2L} - u_p t. \end{cases} \tag{28}$$

3 TEMPERATURE AND DOPANT CONCENTRATION EVOLUTION IN THE MELT IN AMPOULES WITH A SMALL-DIAMETER CROSS-SECTION IN A SPACE EXPERIMENT IN STRICTLY ZERO GRAVITY

In a space experiment in strictly zero gravity, the Rayleigh number Ra is equal to zero ($Ra = 0$). Therefore, Eqs. (1) and (2) are completely decoupled. The solution of Eqs. (1) and (2), satisfying boundary condition (5) and initial condition (6), with $\boldsymbol{u}_0 = 0$, is the velocity field $\boldsymbol{u} \equiv 0$ on D_t.

For ampoules with a small-diameter cross-section we shall use the one-dimensional heat transport equation. In this particular case, the heat transport Eq. (3) becomes

$$b_\theta \frac{\partial\theta}{\partial t} = K_* \frac{\partial^2\theta}{\partial x^2} - 2\frac{K_a}{K}\frac{L}{\delta}\frac{L}{R}(\theta - \theta_f) \tag{3'}$$

in the domain D_t given by

$$D_t = \left\{(t,x)\,\middle|\, 0 \leqslant t \leqslant \frac{1}{u_p}\frac{L_g}{2L}; 0 \leqslant x \leqslant 1\right\} \cup \left\{(t,x)\,\middle|\, \frac{1}{u_p}\frac{L_g}{2L} \leqslant t \leqslant \frac{1}{u_p}\left(1 + \frac{L_g}{2L}\right); 0 \leqslant x \leqslant 1 + \frac{L_g}{2L} - u_p t\right\}. \tag{29}$$

If we assume that $\rho_s c_s = \rho_m c_m$, then the coefficient b_θ according to (17) is constant, equal to one, $b_\theta = 1$, on D_t.

The coefficient K_* is given by (17) in which the sets $D_t^1, D_t^{2,1}, D_t^{2,2}$, and $D_t^{2,3}$ are defined by (18).

The furnace temperature θ_f is given by formula (26) in which x_3 is replaced by x.

The boundary condition (7) in this case becomes

$$\left.\frac{\partial\theta}{\partial x}\right|_{x=0} = 0 \text{ for } 0 \leqslant t \leqslant \frac{1}{u_p}\left(1 + \frac{L_g}{2L}\right). \tag{7'}$$

The boundary condition (9) becomes

$$\theta_i(t) = \begin{cases} -\frac{L}{L_g}(1 + u_p t) + 1 + \frac{L}{L_g};\ \text{for } 0 \leqslant t \leqslant \frac{1}{u_p}\frac{L_g}{2L} \\ \frac{1}{2};\ \text{for } \frac{1}{u_p}\frac{L_g}{2L} \leqslant t \leqslant \frac{1}{u_p}\left(1 + \frac{L_g}{2L}\right). \end{cases} \tag{9'}$$

The initial condition (10) in this particular case is

$$\theta(0,x) = 1; \text{ for } 0 \leqslant x \leqslant 1. \tag{10'}$$

In the case of ampoules with a small-diameter cross-section for the dopant transport we can use also the one-dimensional mass transport equation. In this case, Eq. (4) becomes

$$b_c\frac{\partial c}{\partial t} = D_*\frac{\partial^2 c}{\partial x^2} + \sigma(T_H - T_C)\cdot\frac{\partial \theta}{\partial x}\cdot\frac{\partial c}{\partial x} + \sigma(T_H - T_C)\cdot\frac{\partial^2 \theta}{\partial x^2}c \tag{4'}$$

in the domain D_t given by (29).

The coefficients b_c and D_* are given by the formula (19).

The boundary condition (11) for the dopant concentration in this case is

$$\left.\frac{\partial c}{\partial x}\right|_{x=0} = 0 \text{ for } 0 \leqslant t \leqslant \frac{1}{u_p}\left(1 + \frac{L_g}{2L}\right). \tag{11'}$$

The boundary condition (12) disappears and the boundary condition (13) becomes

$$\frac{\partial c}{\partial x} = \phi_i(t) = \begin{cases} 0; \text{ for } 0 \leqslant t \leqslant \frac{1}{u_p}\frac{L_g}{2L}, \quad x = 1 \\ \frac{P_e}{D_*}(1-k)c + \sigma(T_h - T_c)c\frac{\partial\theta}{\partial x}; \text{ for } \frac{1}{u_p}\frac{L_g}{2L} \leqslant t \leqslant \frac{1}{u_p}\left(1 + \frac{L_g}{2L}\right). \\ \qquad x = 1 + \frac{L_g}{2L} - u_p t \end{cases} \tag{13'}$$

The initial dopant distribution in this case is given by

$$c(0,x) = c_0(x). \tag{14'}$$

For $c_0(x) = c_0 =$ const., the evolution of the temperature and the Ga concentration in the melt in this particular case was computed in [6].

Using the computed Ga concentration $c(t,x)$, it was possible to compute the amount of the dopant entering into the crystal at moment t. This is

$$c_{cr}(t) = k\cdot c(t, 1 + L_g/2L - u_p t); \;\; 1/u_p\cdot L_g/2L \leqslant t \leqslant 1/u_p\cdot(1 + L_g/2L).$$

From here it is possible to predict the dopant concentration in the crystal as a function of the solidified fraction $\xi = u_p t - L_g/2L$ for $1/u_p \cdot L_g/2L \leqslant t \leqslant 1/u_p \cdot (1 + L_g/2L)$:

$$c_{cr}(\xi) = k\cdot c((\xi + L_g/2L)/u_p, 1-\xi); 0 \leqslant \xi \leqslant 1.$$

This was computed in [6] and the result is plotted in Fig. 2a.

For $c_0(x) = 1 + 11.78\cdot\exp(-12.78x)$ the same computations were made in [6] and the dopant concentration in the crystal is plotted in Fig. 2b.

4 MODEL-BASED SIMULATION OF THE FLOW EVOLUTION, HEAT TRANSPORT, AND Ga DISPERSION IN THE MELT IN A LOW-GRAVITY EXPERIMENT

In order to compute flow, heat transport, and Ga dispersion in the model defined by Eqs. (1)–(14) we assume that at each moment the flow, the heat transport, and Ga distribution

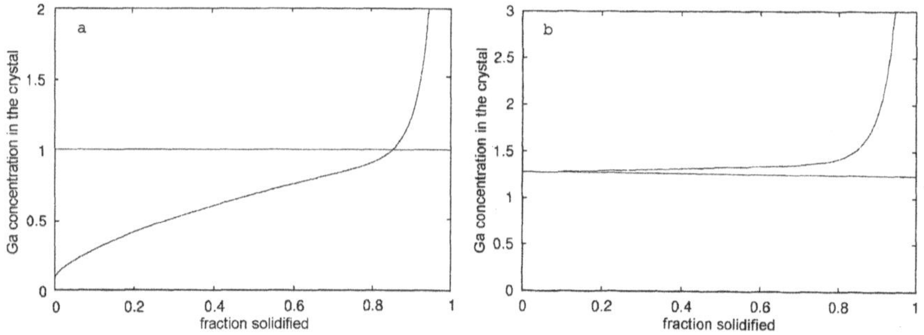

Figure 2 (a) Computed axial Ga distribution in the crystal due to the decrease of the melt in the ampoule and the Soret effect for initial Ga concentration $c_0(x) \equiv 1$. (b) Computed axial Ga distribution in the crystal due to the decrease of the melt in the ampoule and the Soret effect for initial Ga concentration $c_0(x) = 1 + 11.78 \cdot \exp(-12.78x)$.

are axisymmetric. This assumption is realistic because $Ra = 10^3$ and convenient because it permits reduction of the 3D problem to a 2D problem. We also assume that the ampoule has negligible thermal mass and therefore according to [1] $\theta_w = \theta_f$. Therefore in Eq. (3) the term $-2\frac{K_a}{K}\frac{L}{\delta}\frac{L}{R}(\theta_w - \theta_f)$ disappears. In our computation we also neglect the thermotransport and in Eq. (4) we take $\sigma = 0$. The dimensionless parameters and representative values used in the computations are given in Table 1. Computations were performed using the software Cosmos/M, which is a finite element system. For the initial Ga distribution $c_0(r, \varphi, x_3) \equiv 1$, the computed results obtained in [14] for the moments $t_1 = \frac{1}{u_p}\frac{L_g}{2L}$ and $t_2 = \frac{1}{u_p}\frac{L_g}{L}$ are plotted in Figs. 3 to 6.

5 MODEL-BASED PREDICTION OF THE AXIAL AND RADIAL SEGREGATION IN LOW-GRAVITY EXPERIMENTS

Using the computed Ga distribution in the melt $c(t, r, x_3)$ it is possible to predict the distribution of Ga in the crystal. For this it is convenient to introduce the solidified fraction $\xi = u_p t - L_g/2L$ for $1/u_p \cdot L_g/2L \leqslant t \leqslant 1/u_p \cdot (1 + L_g/2L)$ and to express the amount of Ga entering into the crystal at the moment $t : c_{cr}(t) = k \cdot c(t, r, 1 + L_g/2L - u_p t)$ as a function of the solidified fraction ξ:

$$c_{cr}(\xi, r) = k \cdot c((\xi + L_g/2L)/u_p, r, 1 - \xi);\ 0 \leqslant \xi \leqslant 1.$$

The computed values are plotted in Fig. 7.

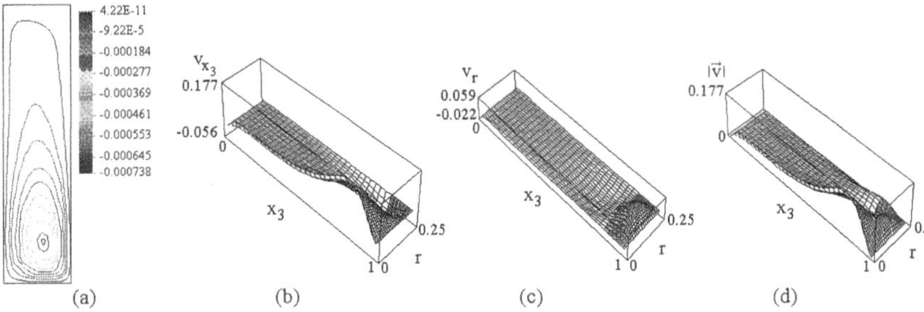

Figure 3 Flow field (a), axial velocity component (b), radial velocity component (c) and velocity magnitude (d) in the absence of the precrystallization zone for $t_1 = 1/u_p \cdot L_g/2L$ and $Ra = 10^3$.

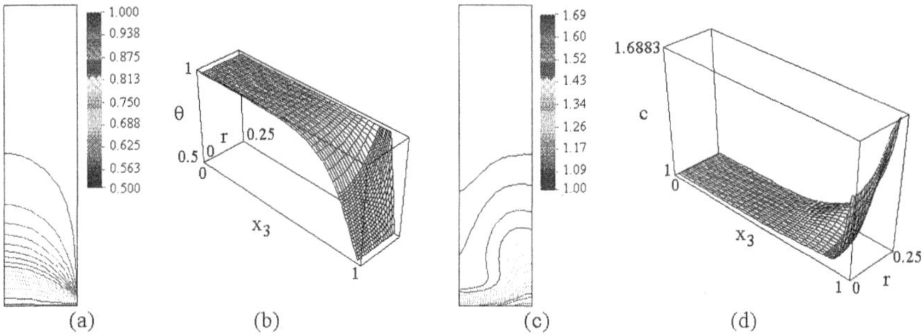

Figure 4 Temperature field ((a), (b)), and dopant concentration field ((c), (d)) in the absence of the precrystallization zone for $t_1 = 1/u_p \cdot L_g/2L$ and $Ra = 10^3$.

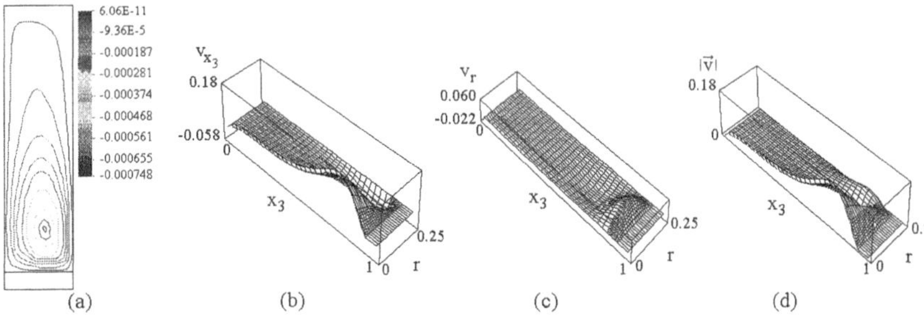

Figure 5 Flow field (a), axial velocity component (b), radial velocity component (c), and velocity magnitude (d) in the absence of the precrystallization zone for $t_2 = 1/u_p \cdot L_g/L$ and $Ra = 10^3$.

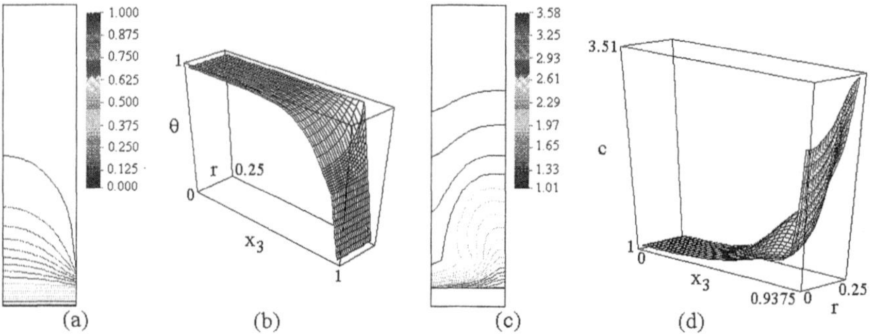

Figure 6 Temperature field ((a), (b)), and dopant concentration field ((c), (d)) in the absence of the precrystallization zone for $t_2 = 1/u_p \cdot L_g/L$ and $Ra = 10^3$.

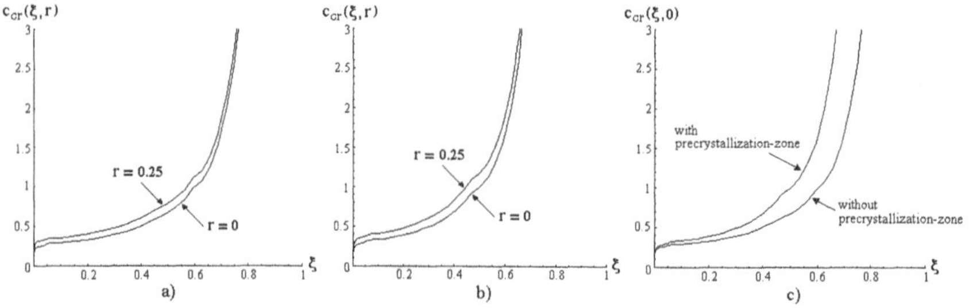

Figure 7 Ga concentration in the crystal function of the solidified fraction in the absence of the precrystallization zone (a) and in presence of the precrystallization zone (b). Plot (c) represents a comparison of the axial Ga concentration in the crystal in the absence/presence of the precrystallization zone.

REFERENCES

1. Ch.J. Chang and R.A. Brown, "Radial segregation induced by natural-convection and melt solid surface interface shape in vertical Bridgman growth," *J. Crystal Growth* 63 (1983), 343.
2. D.J. Larson Jr. and B.S. Dressler, "Shuttle mission 51-G, experiment MRS77F075 Flight sample characterization report," *NASA Report*, Re-753 (1988).
3. P.M. Adornato and R.A. Brown, "Convection and segregation in directional solidification of dilute and non-dilute binary alloys — effects of ampoule and furnace design," *J. Crystal Growth* 80 (1987), 155.
4. C. Wang, Ph.D. thesis, Dept. of Materials Science and Engineering, Massachussetts Institute of Technology, Cambridge (1984).
5. L.L. Zheng, D.J. Larson Jr., and H. Zhang, "Role of thermotransport (Soret effect) in macrosegregation during eutectic/off-eutectic directional solidification," *J. Crystal Growth* 191 (1998), 243.
6. A.M. Balint, E. Tulcan-Paulescu, and St. Balint, "The effect of the initial dopant distribution in the melt on the axial compositional uniformity of a thin doped crystal grown in strictly zero-gravity environment by Bridgman–Stockbarger method," *J. Crystal Growth* 247 (2003), 313.
7. S.R. Coriell, R.G. Boisvert, R.G. Rehm, and R.F. Sekerka, "Lateral solute segregation during unidirectional solidification of a binary alloy with curved solid–liquid interface. 2. Large departures from planarity," *J. Crystal Growth* 54 (1981), 167.
8. A.M. Balint, M.M. Mihailovici, D.G. Baltean, and St. Balint, "A modified Chang–Brown model for the determination of the dopant distribution in a Bridgman–Stockbarger semiconductor crystal growth system," *J. Crystal Growth* 230 (2001), 195.
9. J.A. Burton, R.C. Prim, and W.P. Slichter, "The distribution of solute in crystals grown from melt. 1. Theoretical," *J. Chem. Phys.* 21 (1953), 1987.
10. W.A. Tiller, K.A. Jackson, J.W. Rutter, et al., "The redistribution of solute atoms during the solidification of metals," *Acta Metall. Mater.* 1(4) (1953), 428.
11. A.F. Witt, H.C. Gatos, M. Lichtensteiger, et al., "Crystal growth and steady-state segregation under zero-gravity-INSB," *J. Electrochem. Soc.* 122(2) (1975), 276.
12. P.R. Griffin and S. Motakef, "Influence of non-steady gravity on natural convection during micro-gravity solidification of semiconductors. I. Time scale analysis," *Appl. Microgravity Tech.* 2(3) (1989), 121.
13. P.R. Griffin and S. Motakef, "Influence of non-steady gravity on natural convection during micro-gravity solidification of semiconductors. II. Implications for crystal growth experiments," *Appl. Microgravity Tech.* 2(3) (1989), 128.
14. H.M. Ettouney and R.A. Brown, "Effect of heat transfer on melt solid interface shape and solute segregation in edge-defined film-fed growth — finite element analysis," *J. Crystal Growth* 58 (1982), 313.
15. J.P. Kalejs, L.Y. Chin, and F.M. Carlson, "Interface shape studies for silicon ribbon growth by the EPG technique. 1. Transport phenomena modeling," *J. Crystal Growth* 61 (1983), 473.
16. M.M. Mihailovici, A.M. Balint, and St. Balint, "The axial and radial macrosegregation due to the thermo-convection, the decrease of the melt in the ampoule and the effect of the precrystallization zone in the semiconductor crystals grown in a Bridgman–Stockbarger system in a low gravity environment," *J. Crystal Growth* 237–239 (2002), 1752.

13

Identification of Stiffness Matrices of Structural and Mechanical Systems from Modal Data

Firdaus E. Udwadia

Aerospace and Mechanical Engineering, Civil Engineering, Mathematics, and Information and Operations Management, University of Southern California, Los Angeles, CA

In this chapter we present a simple method for the identification of stiffness matrices of structural and mechanical systems from information about some of their natural frequencies and corresponding mode shapes of vibration. The method is computationally efficient and is shown to perform well in the presence of measurement errors in the mode shapes of vibration. The method is applied to the identification of the stiffness distribution along the height of a simple vibrating structure. An example illustrating the method's efficacy in structural damage detection is also given. The efficiency and accuracy with which the method yields estimates of the system's stiffness make it worthy of further exploration for damage detection.

INTRODUCTION

Modal testing of structures is an extensive field in civil and mechanical engineering. It is generally used to understand/predict the dynamic behavior of a structure when subjected to low-amplitude vibrations. Often modal information is also used to identify/estimate the structural parameters of a system, under the assumption that it has classical normal modes of vibration [1]. Such identification leads to improved mathematical models that can be used in either predicting and/or controlling structural response to dynamic excitations.

Several different approaches to the parameter identification problem have appeared in the literature [2–10]. One approach is the so-called model updating method. Here a suitable analytical model of a structural system is developed using the equations of motion, and its numerical representation is obtained. Validation of the numerical model through modal testing is then sought. Such tests usually provide some of the frequencies of vibration (usually the lower frequencies) and the corresponding mode shapes. When these frequencies and mode shapes obtained from modal testing are compared with those obtained from the numerical model, they generally do not agree with one another. Discrepancies between the results from experimental testing and theoretical modeling arise due to a variety of reasons: simplifications used in developing the analytical model, uncertainties in things like material properties and boundary conditions in the analytical model, and experimental errors during modal testing. The problem of updating a numerical model so that it is as much as possible

0-415-30852-6/04/$0.00+$1.50

in conformity with experimental modal test data is referred to as the updating problem, and over the years it has received considerable attention.

In this paper we investigate a direct approach to structural identification through the use of modal test data. No *a priori* estimates are used. It should be pointed out that such experimental test data is seldom "complete," i.e., all the mode shapes of vibration and the corresponding natural frequencies are seldom available, for there is a practical limit to the range of frequencies that a structural or mechanical system can be tested for. Hence the idea is to obtain suitable models through the use of *incomplete* information, i.e., information on only a limited number of mode shapes and frequencies of vibration. We shall illustrate the method assuming that normal classical modes exist.

SYSTEM MODEL

Consider a structural system modeled by the linear differential equation

$$M\ddot{x} + \hat{C}\dot{x} + \hat{K}x = 0, \tag{1}$$

where, x is an n by 1 vector, and M is the n by n symmetric positive-definite mass matrix, $\hat{K}$ is the symmetric stiffness matrix, and $\hat{C}$ is the damping matrix. We shall assume that the elements of the mass matrix, M, are sufficiently well known, and that the system is classically damped. We could then rewrite Eq. (1) as

$$\ddot{y} + C\dot{y} + Ky = 0, \tag{2}$$

where $K = M^{-1/2}\hat{K}M^{-1/2}$, and $C = M^{-1/2}\hat{C}M^{-1/2}$.

Our intention is to investigate the identification of the stiffness matrix from a knowledge of the modal data corresponding to Eq. (2), i.e., from the eigenvectors and eigenvalues of the matrix K. We note that the eigenvectors $\hat{\varphi}_i$ of

$$\hat{K}\hat{\varphi}_i = \lambda_i M\hat{\varphi}_i \tag{3}$$

are related to the eigenvectors φ_i of

$$K\varphi_i = \lambda_i\varphi_i \tag{4}$$

by the relation $\hat{\varphi}_i = M^{-1/2}\varphi_i$, for the eigenvalue λ_i [1]. We shall assume that $\lambda_i \leqslant \lambda_{i+1}$, $i = 1, 2, ..., (n-1)$.

Often from the analytical model, the "structure" of the matrix K is known. Let us say that the matrix K is a function of the parameter s-vector $k := [k_1 k_2 ... k_s]^T$. Usually, because of the limited connectivity between the subassemblages of a structure, the number of parameters, s, that are required to be identified in the n by n symmetric matrix K is much less than $n(n+1)/2$.

We assume that each element of the matrix K is a linear combination of the parameters k_i, so that each of the n equations in the equation set $K\varphi_i = \lambda_i\varphi_i$ is linear in these parameters. Hence, each row of Eq. (4) is linear in both the parameters k_i and in the n components of the eigenvector φ_i. One can then rewrite equation (4) as

$$\Phi_i k = \lambda_i\varphi_i, \tag{5}$$

where the elements of the n by s matrix Φ_i are linear functions of the components of the n-vector φ_i. In what follows, we shall denote the jth component of the n-vector φ_i by φ_i^j.

Modal test data provides the *measured* frequencies and the *measured* mode shapes of vibration, yielding λ_i^m, $i = 1, 2, ..., r$, and the corresponding eigenvectors φ_i^m, $i = 1, 2, ..., r$, where some $r \leqslant n$. We denote experimental data by the superscript m. Were Eq. (5) to be true for the measured modal data, we would then obtain

$$\Phi_i^m \tilde{k} = \lambda_i^m \varphi_i^m, \quad i = 1, 2, ..., r, \tag{6}$$

giving us an estimate $\tilde{k}$ of the parameter s-vector k that contains the s parameters k_i, $i = 1, 2, ..., s$, of which the stiffness matrix K is a function. Equation (6) can be expressed in a more compact form, by stacking the measured data from each of the r modes, as

$$B\tilde{k}_r := \begin{bmatrix} \Phi_1^m \\ \Phi_2^m \\ . \\ . \\ \Phi_r^m \end{bmatrix} \tilde{k}_r = \begin{bmatrix} \lambda_1^m \varphi_1^m \\ \lambda_2^m \varphi_2^m \\ . \\ . \\ \lambda_r^m \varphi_r^m \end{bmatrix} = b_r, \tag{7}$$

where the matrix B is rn by s, and the vector b is an rn-vector.

The minimum-norm least-squares solution to this system of Eq. (7) is simply given by

$$\tilde{k}_r = B^+ b_r, \tag{8}$$

where B^+ stands for the Penrose generalized inverse of the matrix B; see Ref. [11] for details on the properties of generalized inverses of matrices, and their computation. The subscripts "r" indicate that measured modal data from r modes is "stacked" together in Eq. (7).

To illustrate the above equations, we consider a building structure modeled by a simple n-degree-of-freedom system (see Fig. 1) subjected to horizontal base motion. The mass matrix M is taken to be the identity matrix, and the structure is assumed to be lightly damped. Though effects like soil–structure interaction may be important in understanding the structural dynamics of such building structures, to illustrate our ideas we shall ignore soil-structure interaction and further assume that the structure is resting on a rigid base. We thus focus primarily on our ability to estimate the constant stiffness matrix of the structure from modal data. Furthermore, since numerous mechanical systems are often modeled by such a "chain" of springs and masses, and hence we use this as a prototypical system.

The equation of motion is described by Eq. (1), and the matrix $\hat{K} = K$ is tridiagonal and has the form

$$K = \begin{bmatrix} k_1 + k_2 & -k_2 & & & \\ -k_2 & k_2 + k_3 & -k_3 & & \\ & . & . & . & \\ & & . & . & . \\ & & & -k_n & k_n \end{bmatrix}. \tag{9}$$

One would like to estimate the parameter n-vector $k = [k_1 \; k_2 \; ... \; k_n]^T$ from modal test data.

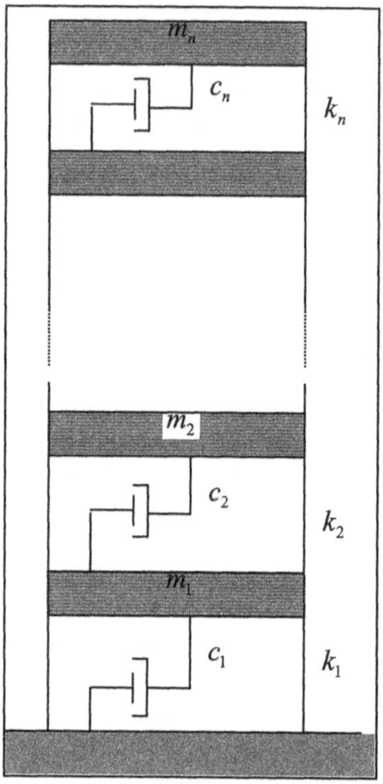

Figure 1 Model of a simple building structure.

The matrix Φ_i in Eq. (5) now becomes the upper triangular banded matrix

$$\Phi_i = \begin{bmatrix} \varphi_i^1 & -\varphi_i^{2,1} & & & & \\ & \varphi_i^{2,1} & -\varphi_i^{3,2} & & \text{O} & \\ & & \cdot & \cdot & & \\ \text{O} & & & \cdot & \cdot & \\ & & & & \varphi_i^{n-1,n-2} & -\varphi_i^{n.n-1} \\ & & & & & \varphi_i^{n.n-1} \end{bmatrix}, \tag{10}$$

where $\varphi_i^{p,q} = (\varphi_i^p - \varphi_i^q)$. In this special case one can explicitly obtain the inverse of the matrix Φ_i and solve Eq. (5) to yield

$$k = \lambda_i \Phi_i^{-1} \varphi_i,$$

where the upper triangular matrix Φ_i^{-1} is given by

$$\Phi_i^{-1} = \begin{bmatrix} 1/\varphi_i^1 & 1/\varphi_i^1 & 1/\varphi_i^1 & . & . & & 1/\varphi_i^1 \\ & & 1/\varphi_i^{2,1} & 1/\varphi_i^{2,1} & . & . & 1/\varphi_i^{2,1} \\ & & . & & . & . & . \\ & & & & . & . & . \\ & & & & & 1/\varphi_i^{n-1,n-2} & 1/\varphi_i^{n-1.n-2} \\ & & & & & & 1/\varphi_i^{n.n-1} \end{bmatrix}. \quad (11)$$

Were the eigenvalue λ_i obtained from measurements of the frequency of vibration of the ith normal mode, and the eigenvector φ_i obtained from experimental measurements of the ith mode shape of vibration of the building structure, we would obtain an *estimate* of the parameter n-vector k as

$$\tilde{k} = \lambda_i^m \left[\Phi_i^m\right]^{-1} \varphi_i^m, \quad (12)$$

where we have replaced all the φ_i^j's in (11) by their measured values, $(\varphi_i^j)^m$. While this estimate may be adequate when the mode shapes of vibration are assumed to be noise-free (a situation that never arises in practice), as shown below, it quickly deteriorates in the presence of measurement noise.

Were frequency and mode shape data obtained for modes $i = 1, 2, ..., r$, we would form the matrix B as shown in Eq. (7) and obtain the estimate of the parameters as indicated in Eq. (8). We illustrate this using the following numerical example.

A. NUMERICAL EXAMPLE

Consider a building structure modeled by a 9-degree-of-freedom system ($n = 9$) whose true (exact) story stiffnesses are taken to be (we assume the numerical data to be in consistent units):

$k_1 = 2000$; $k_2 = 2050$; $k_3 = 2090$; $k_4 = 2050$; $k_5 = 1950$; $k_6 = 1990$; $k_7 = 1920$; $k_8 = 1900$; and, $k_9 = 1900$.

The stiffness parameters are purposely chosen to have a small range of numerical values so that the power of the identification scheme in discriminating between these close stiffness values is assessed.

The fundamental frequencies of vibration corresponding to the 9 by 9 stiffness matrix shown in (9) are then: 1.1810, 3.4637, 5.6707, 7.7577, 9.5685, 11.1865, 12.4205, 13.3962, 14.0528 cycles/sec.

We begin by assuming that the experimentally determined frequencies and the mode-shapes of vibration from modal testing are accurate (noiseless). Thus, $\lambda_i = \lambda_i^m$, $\varphi_i = \varphi_i^m$, $i = 1, 2, ..., r$, where r is the total number of resonant frequencies and modeshapes obtained from the test data.

Figure 2(a) shows the percentage error in estimates of the parameter vector k whose components are the story stiffnesses. The error in the estimate, $\tilde{k}_i$, of the story stiffness k_i is defined as $(\tilde{k}_i - k_i)/k_i$. It is assumed that only the lowest four modes of vibration are measured, along with the lowest four modal frequencies. The figure shows that the stiffness distribution can be very accurately determined using Eq. (12) above, for i equal to 1 to 4. However, the addition of measurement noise alters the situation considerably.

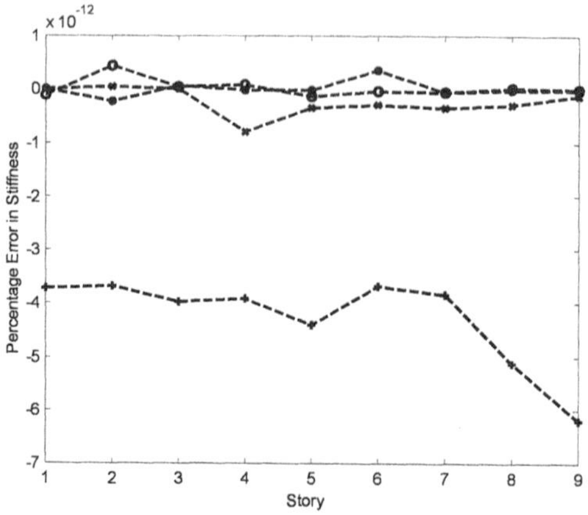

Figure 2 (a) Identification (ID) with noise-free data; + shows ID from first mode; x shows ID from second mode; * shows ID from third mode; o shows ID from fourth mode.

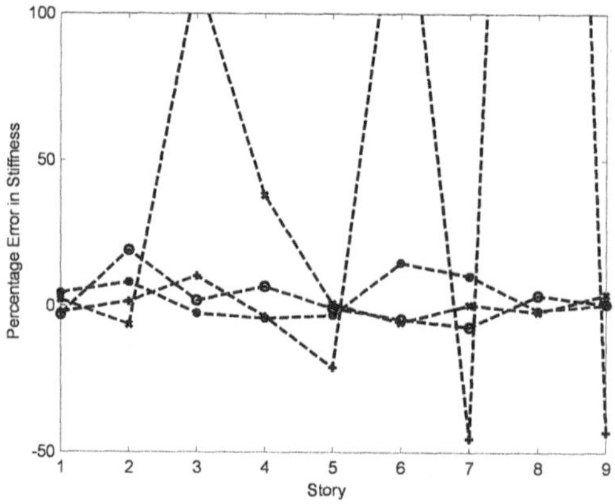

Figure 2 (b) Identification (ID) with 5% noise in modal amplitude data; + shows ID from first mode; x shows ID from second mode; * shows ID from third mode; o shows ID from fourth mode.

During modal testing it is customary to assume that the frequencies of vibration are accurately determined, and that it is in the determination of the amplitudes of the mode shapes that the experimental errors arise. This assumption is by and large valid because the frequency of shakers, even at resonance, can be quite accurately controlled.

Accordingly, the data is simulated by adding noise to each modal amplitude, and the "measured" component "j" of the ith-mode shape is taken to be

$$(\varphi_i^j)^m = \varphi_i^j(1 + \alpha_{noise}\xi), \tag{13}$$

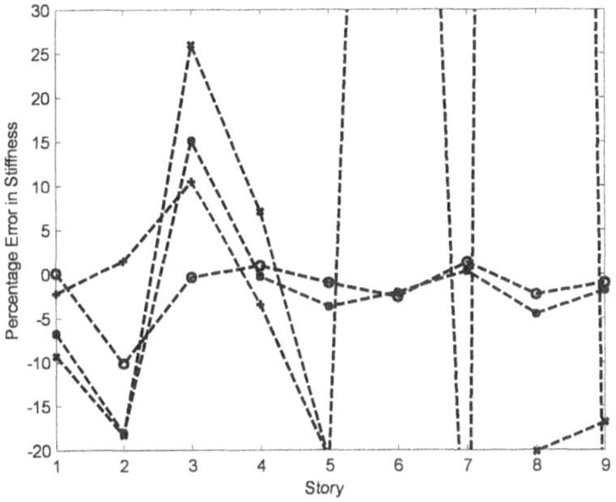

Figure 2 (c) Identification (ID) with 5% noise in modal amplitude data using generalized inverses; + shows ID using $r = 1$; x shows ID using $r = 2$; * shows ID using $r = 3$; o shows ID using $r = 4$.

where ξ is a uniformly distributed random number between -1 and $+1$. Figure 2(b) shows results with $\alpha_{noise} = 5\%$ indicating the dramatic increase in the error when Eq. (12) is used in estimating $\tilde{k}$ when using noisy mode shape data.

We next illustrate the improvement that is created by using Eq. (8) where the estimate is obtained by using the generalized inverse of the matrix B^+ for values of r ranging from 1 to 4. Figure 2(c) shows the progressive improvement in the estimates of the stiffness with the addition of information about each successive mode of vibration. The simultaneous use of data from all 4 modes ($r = 4$) shows that there is a substantial reduction in the percentage error in the estimation of the stiffness. The maximum percentage error is now less than 10%.

Iterative Improvements of the Stiffness Estimates in the Presence of Measurement Noise

Having obtained the minimum length least-squares estimate of the parameters by using the generalized inverse of the matrix B, we now attempt to improve this estimate when the mode shape information is corrupted by measurement noise, i.e., when we use the measured modal data, φ_i^m, $i = 1, 2, ..., r$. We assume, as is customary, that the measurement errors in determining the r resonant frequencies are negligible when compared to those incurred in the measurement of the mode shapes; hence $\lambda_i = \lambda_i^m$, $i = 1, 2, ..., r$. For convenience we will denote the minimum length least-squares estimate we have obtained so far by $\tilde{k}^{(0)}$.

This estimate, which is obtained from Eq. (8), gives us, in turn, an estimate, $\tilde{K}^{(1)}$, of the matrix K. Hence, a solution of the eigenvalue problem, $\tilde{K}^{(1)}\psi_j = \mu_j\psi_j$, provides: (1) an estimate of all the eigenvalues (frequencies), μ_j, $j = 1, 2, ..., n$ of vibration (estimates of the eigenvalues of K) and, (2) an estimate of all the modeshapes, ψ_j, $j = 1, 2, ..., n$ (estimates of the eigenvectors of K). Since $\tilde{K}^{(1)}$ is symmetric, ψ_j are (or can be chosen to be) orthogonal to one another, and can be normalized to have unit length.

In summary, through our estimate of the stiffness parameter vector $\tilde{k}^{(0)}$ at this point we have obtained estimates, μ_j, of the actual eigenvalues, λ_j, of K as well as estimates, ψ_j, of the actual mode shapes, φ_i. We next use this information to iteratively update the estimate $\tilde{k}^{(0)}$.

Since the eigenvectors of $\tilde{K}^{(1)}$ form a basis set in R^n, the "measured" noise-corrupted mode shape φ_i^m can be expanded in terms of the estimated mode shapes $\psi_j, j = 1, 2, ..., n$. We then have

$$\varphi_i^m = \sum_{j=1}^{n} \delta_j^i \psi_j, \quad i = 1, 2, ..., r, \tag{14}$$

where $\delta_j^i = \psi_j^T \varphi_i^m$. We then obtain

$$\tilde{K}^{(1)} \varphi_i^m = \tilde{K}^{(1)} \sum_{j=1}^{n} \delta_j^i \psi_j = \sum_{j=1}^{\delta_j^i} \tilde{K}^{(1)} \psi_j = \sum_{j=1}^{n} \delta_j^i \mu_j \psi_j, \quad i = 1, 2, ..., r. \tag{15}$$

As before, we can now express $\tilde{K}^{(1)} \varphi_i^m$ as $\Phi_i^m \tilde{k}^{(0)}$, and Eq. (15) yields

$$\Phi_i^m \tilde{k}^{(0)} = \sum_{j=1}^{n} \delta_j^i \mu_j \psi_j, \quad i = 1, 2, ..., r. \tag{16}$$

However, the measured r lowest frequencies are assumed to be accurately known and we can use this information in Eq. (16) by replacing the μ_j's by the known (i.e., accurately measured) eigenvalues λ_j for $j = 1, 2, ..., r$. This additional information when injected into Eq. (16) provides us with the opportunity to update our estimate of the stiffness parameter vector from $\tilde{k}^{(0)}$ to $\tilde{k}^{(1)}$, and we obtain

$$\Phi_i^m \tilde{k}^{(1)} = \sum_{j=1}^{r} \delta_j^i \lambda_j \psi_j + \sum_{j=r+1}^{n} \delta_j^i \mu_j \psi_j, \quad i = 1, 2, ..., r. \tag{17}$$

This forms the basis of our iterative improvement of the stiffness parameter vector $\tilde{k}^{(0)}$. We note that we have used: (1) our information from the r measured eigenvalues (frequencies of vibration) of the system, which we assume are accurate, (2) our best hereto available estimates, μ_j, of the remaining $(n - r)$ unmeasured eigenvalues (frequencies of vibration) of the system, and (3) our best hereto available estimates, ψ_j, $j = 1, 2, ..., n$ derived from the measured eigenvectors φ_j^m , $j = 1, 2, ..., r$, which are corrupted by measurement noise.

Equation (17) can be rewritten for convenience as

$$\Phi_i^m \tilde{k}^{(1)} = [\psi_1 \ \psi_2 ... \psi_{\text{n}}] \begin{bmatrix} \delta_1^i \lambda_1 \\ \delta_2^i \lambda_2 \\ \cdot \\ \cdot \\ \delta_r^i \lambda_r \\ \delta_{r+1}^i \mu_{r+1} \\ \cdot \\ \cdot \\ \delta_n^i \mu_n \end{bmatrix} := \Psi h_i, \quad i = 1, 2, ..., r. \tag{18}$$

Here Ψ is the n by n orthogonal matrix of eigenvectors of $\tilde{K}^{(1)}$. Lastly, since we have r measured mode shapes, we can, as before, stack the information again as

$$B\tilde{k}_r^{(1)} := \begin{bmatrix} \Phi_1^m \\ \Phi_2^m \\ . \\ . \\ \Phi_r^m \end{bmatrix} \tilde{k}_r^{(1)} = \begin{bmatrix} \Psi h_1 \\ \Psi h_2 \\ . \\ . \\ \Psi h_r \end{bmatrix} = b_r^{(1)}, \tag{19}$$

so that we again obtain the minimum-length least-squares solution for $\tilde{k}_r^{(1)}$ in Eq. (19) as

$$\tilde{k}_r^{(1)} = B^+ b_r^{(1)}. \tag{20}$$

It should be pointed out that the matrix B in Eq. (19) is the same as it was in Eq. (7); it is obtained from the r measured mode shapes.

The estimate $\tilde{k}_r^{(1)}$ is next used to create an estimate $\tilde{K}^{(2)}$ of the stiffness matrix, which in turn yields the new estimate $\tilde{k}_r^{(2)}$, and the iteration continues until the improvement in the vector $\tilde{k}_r^{(i)}$ is deemed to be negligible, or for a prefixed number of iterations.

The algorithm that we have developed can be summarized as follows:

1. Use the measured mode shape φ_i^m, to obtain the n by s matrices Φ_i^m, $i = 1, 2, ..., r$.
2. Stack the r matrices Φ_i^m's to obtain B.
3. Stack the vectors $\lambda_i^m \varphi_i^m$, $i = 1, 2, ..., r$, to obtain the vector $b_r^{(0)}$.
4. Calculate $\tilde{k}_r^{(0)} = B^+ b_r^{(0)}$ [Eq. (8)].
5. for $i = 1$ to N iterations, do:
   Obtain $\tilde{K}^{(i)}$, its eigenvalues μ_j, and its eigenvectors ψ_j, $j = 1, 2, ..., n$;
   Calculate $\delta_j^p = \psi_j^T \varphi_p^m$, $j = 1, 2, ..., n$; $p = 1, 2, ..., r$;
   Calculate the vectors Ψh_p, [Eq. 18] using measured λ_p^m, $p = 1, 2, ..., r$, and $\mu_p, p = r + 1, ..., n$;
   Stack the vectors Ψh_p, $p = 1,\ 2, \ldots, r$, to get $b_r^{(i)}$ [Eq. 19];
   Calculate $\tilde{k}_r^{(i)} = B^+ b_r^{(i)}$;
   If $\|\tilde{k}_r^{(i)} - \tilde{k}_r^{(i-1)}\| < \Delta$, exit do loop;
   end do loop.

B. NUMERICAL EXAMPLE (CONTINUED)

To illustrate the iterative improvement of the estimate of the stiffness parameter vector, we show in Fig. 2(d) the results from the same modal data shown in Figs. 2(b) and 2(c) by using the same realization as before of the random noise that corrupts the mode shape data. The solid line shows the identification results obtained after 1000 iterations, the dashed line is the result generated by direct use of the generalized inverse (i.e., Eq. (8)). We note in passing that negligible changes in the stiffness occur after 150 iterations.

Since the only experimental information that we assume here is that each component of the measured mode shape (corresponding to each natural frequency of vibration) is

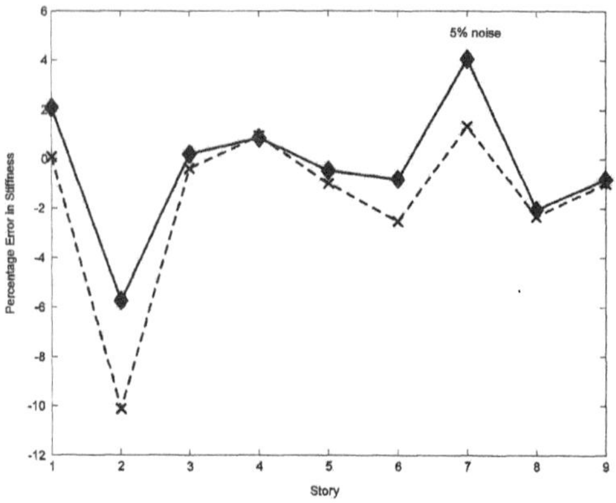

Figure 2 (d) Improvement in estimation caused by iterative scheme.

corrupted with zero mean, uniformly distributed white noise, what is of relevance are the expected values of the stiffness estimate obtained and its standard deviation. To find the expected value of the estimate, the simulation was carried out 1000 times with α_{noise} =5%, each time using a different set of randomly generated noise to corrupt the mode shape measurements. Figure 3(a) shows the expected value of the estimate of the stiffness at each story along with a 2-sigma band shown by the dashed lines to indicate the variance in the expected estimate. The true (exact) stiffness is shown by the shaded line and the expected value of the stiffness is shown by the solid dark line. As seen from the plot, the identification scheme appears to work remarkably well, and the expected estimate closely tracks the true stiffness.

To get a feeling for the expected error in the estimation of the stiffness at each story, i.e., $E\{\tilde{k} - k\}$, we show it in Fig. 3(b) along with its $\pm$1-sigma band. As seen from the figure, the estimation is quite accurate, and the expected error in the estimation of the stiffness is very small, the maximum of {mean error +1 sigma} being 5% at story number 2.

Lastly, we show in Fig. 3(c) the expected value (with a sample size of 1000) of the estimate along with the $\pm$1-sigma bands when only the generalized inverse is used *without* the iterative improvement in the estimates discussed in this section. Again the shaded line is the true stiffness, the dark solid line is the expected value of the stiffness, and dotted lines indicate the mean $\pm$1-sigma band. Comparing Fig. 3(a) with 3(c) we observe that the iteration discussed in this section (Eqs. (17)–(20)) has a definite advantage, both in terms of the expected value of the stiffness and in reducing the size of the $\pm$1-sigma band.

We observe that the estimates that we have obtained use information from only 4 modes of vibration of the structure. This is the usual situation, for complete information on all the modes and all the resonant frequencies of structures is seldom experimentally decipherable from measurement data. This is due to limitations in experimental testing. Also, higher modes "sample" smaller spatial domains and our models may be inadequate in representing these spatial heterogeneities.

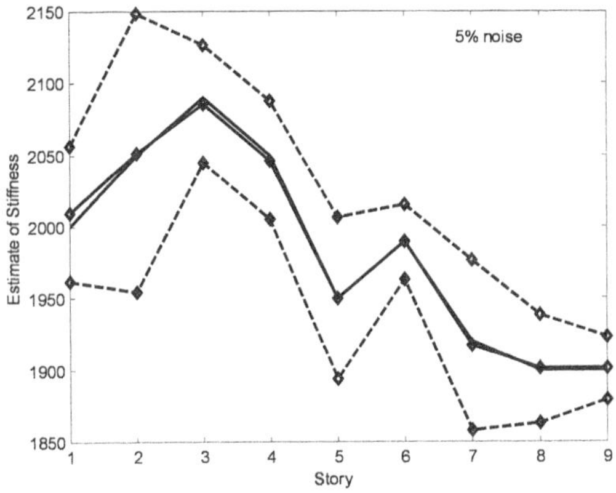

Figure 3 (a) Expected value of the identified estimate (dark line), true stiffness (shaded line), and the expected estimate ±standard deviation (dashed lines).

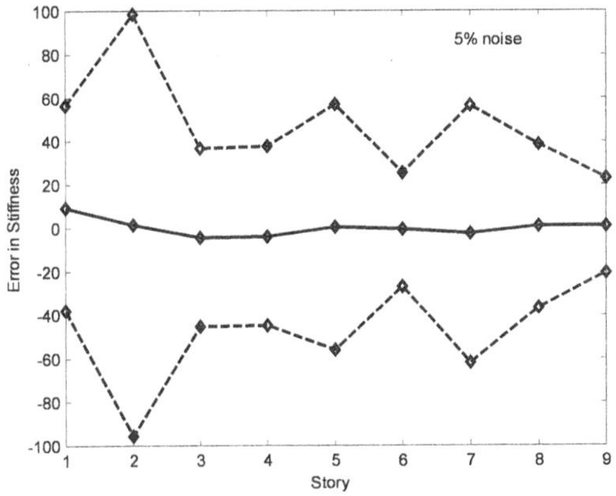

Figure 3 (b) Expected value of the error in the stiffness at each story and the ±1-sigma band around it (dashed lines).

We next show the influence of using more modal information on the identification results. We next use measured modal data from 6 modes instead of 4. Clearly, the additional information should improve the estimates of the stiffness distribution and reduce the uncertainty in the expected value of the estimates that are obtained. However, noise in the measurements of the modal amplitudes usually increases with increasing natural frequency, and so to simulate the noise-corrupted mode shape amplitude data we shall use α_{noise} =10% in Eq. (13), instead of 5% as we did when we used only 4 modes for the identification. We exhibit our results in Fig. 4.

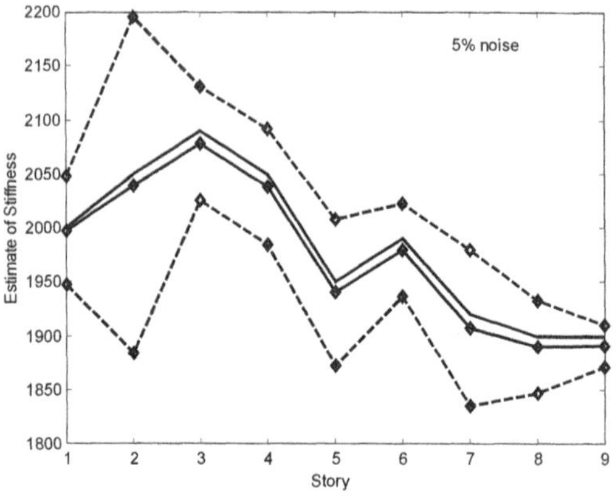

Figure 3 (c) Results with no iterative improvement. Expected value of stiffness at each story (dark line), true stiffness (shaded line), expected value ±1-sigma band (dashed line).

We observe that even with the sizably larger measurement noise of 10% in the mode shape measurements, the estimates of the 1-sigma band in the percentage error in the stiffness estimates cover only ±4%.

C. NUMERICAL EXAMPLE OF DAMAGE DETECTION

We now assume that the structural (or mechanical) system has, perhaps after being subjected to horizontal ground shaking, suffered deterioration along its height; and the stiffness at the third story has been reduced from its previous value by about 30%. Our aim is to estimate the new stiffness distribution, and investigate how well this drop in stiffness can be captured by modal testing, using again only 4 modes of vibration. We assume that the true stiffness of the damaged structure is given by

$$k_1 = 1950;\ k_2 = 1900;\ k_3 = 1460;\ k_4 = 1910;\ k_5 = 1910;\ k_6 = 1930;$$
$$k_7 = 1920; k_8 = 1900;\ \text{and,}\ k_9 = 1900.$$

Figure 5(a) shows that the weakened structure can be well identified, and the location of the drop in the stiffness is quite evident, and accurately determined using noise-corrupted modal information from only 4 modes. The solid line is the true stiffness, the dashed line is the stiffness after 1000 iterations, and the dash-dot line is the result of Eq. (8), with no iterative improvement. Figure 5(b) shows the percentage error in estimating the stiffness with (solid line) and without (dashed line) iterative improvements. With the iterative improvements, the percentage error in identification is seen to be less than 4%.

The expected value of the stiffness estimates are again obtained using 1000 realizations of noise-corrupted "measured" mode shapes. The solid line in Fig. 6(a) shows the expected value of the stiffness estimates at each story, the shaded solid line shows the true

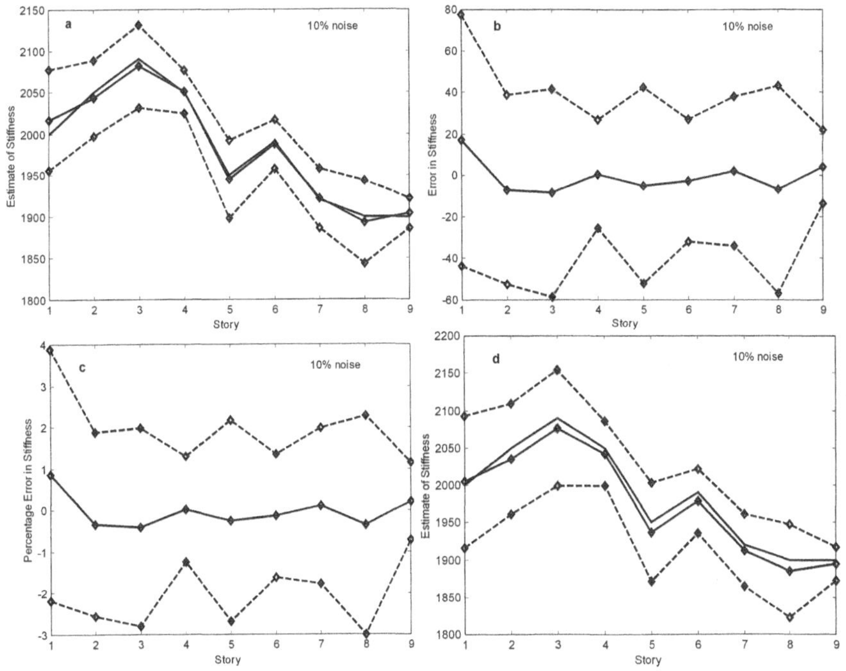

Figure 4 Identification using 6 modes. Figure 4(a): Expected value of the stiffness (dark line), true stiffness (shaded line), and the ±1-sigma band using the iterative improvement; Figure 4(b): Expected error in estimation of stiffness and ±1-sigma error band using the iterative improvement; Figure 4(c): Expected percentage error and ±1-sigma error band using iterative improvements; Figure 4(d): Expected value of the stiffness (dark line), true stiffness (shaded line), and the ±1-sigma band without iterative improvement.

stiffnesses, and the dashed lines show the ±1-sigma bands. Figure 6(b) shows the expected error in the stiffness estimates, and the ±1-sigma error bands. We see that the identification results are rather accurate and the ±1-sigma bands are far smaller than the drop in stiffness at the 3rd-story level, indicating that accurate damage detection can be accomplished in an efficient manner using the simple system identification approach describe here. Figure 6(c) shows the expected percentage error in estimation of the stiffness. The ±1-sigma bands show that changes of more than ±5% in the stiffness parameters can be rapidly identified with reasonable reliability. Lastly, Fig. 6(d) shows the expected values of the identification results without any iterative improvements. We note that though not as good as those obtained after iteration, these results too are considerably superior to those that are obtainable by other system identification techniques when using noise-corrupted data.

CONCLUSIONS AND DISCUSSION

We have proposed here a simple and efficient way of estimating the stiffness distribution in a structure from incomplete modal information. It relies on the Penrose generalized

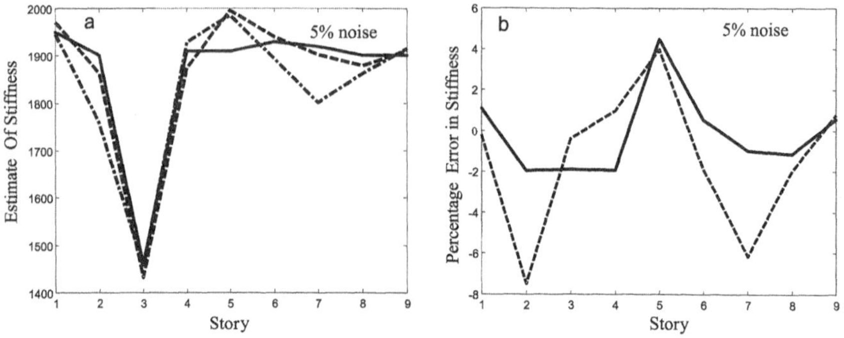

Figure 5 (a): Estimate of stiffness using one realization of noisy modal measurements; (b): Percentage error in stiffness estimates.

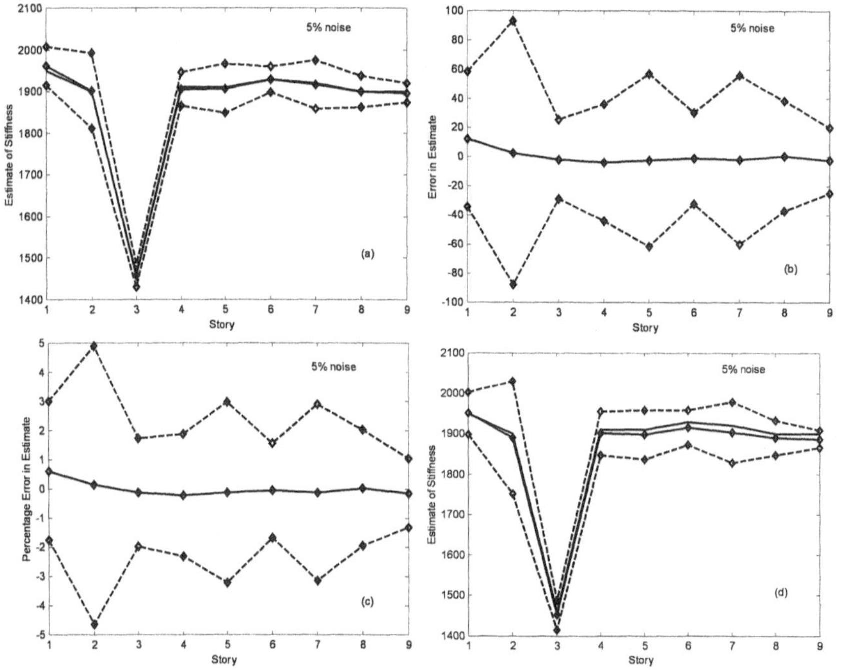

Figure 6 A case study of damage detection.

inverse of a matrix, which can be computed rapidly and efficiently. A subsequent iterative improvement of the estimates is then carried out. Our numerical computations show that this iterative improvement reduces the variance of the estimates and brings their expected values closer to the true (exact) values. The method appears to work well for estimating the stiffness distribution of simple building structures. We find that the method gives good results in the presence of noise-corrupted mode shape data, and is capable of tracking small changes in the stiffness parameters. Its accuracy in the presence of noise appears to surpass several other more sophisticated identification methods in common use (e.g., compared to results in [4, 8, 10]). This makes the method perhaps valuable for rapid damage detection in structures. Thus, we show that it can be used for real-time, on-line, accurate system identification and damage detection.

Though in this chapter we have used information from the lowest r mode shapes and frequencies of vibration, we could have chosen *any* r of the measured mode shapes and r corresponding frequencies of vibration for the identification. The identification method presented here can be extended to this situation in a simple and obvious manner.

It would be useful to compare the method proposed here with those commonly in use, especially the so-called updating methods. We note that in the method proposed herein:

1. The "structure" of the stiffness matrix is assumed to be known, and in addition, the way in which the unknown stiffness parameters enter each element of the stiffness matrix are assumed to be known.

2. The elements of the stiffness matrix are assumed to be linear functions of the parameters to be estimated; this is not as much of a restriction as may appear at first sight, for the "assemblage" of the stiffness matrix using, say, FEM models provides this sort of information in a more or less natural way.

3. We use generalized inverses to obtain the actual stiffness estimates, and no *a priori* stiffness estimates are required as in all the stiffness updating methods.

4. Unlike the usual updating methods that try to estimate each element of the stiffness matrix, thereby increasing the unknowns (usually to $n(n+1)/2$), the method proposed here uses the knowledge of the way in which the unknown parameters enter the stiffness matrix; this can reduce the number of unknowns significantly.

5. Our approach averts "connectivity" problems that require being addressed by the use of additional constraints when using some of the usual updating methods that employ minimization techniques.

6. Even when there is no measurement noise, the accuracy of the available updating methods is known to rely on the proper choice of "weighting matrices," and there appears to be no systematic method of choosing these matrices.

7. Updating methods are highly susceptible to measurement errors; we have shown that the method developed here can provide significant, high-quality information about the stiffness parameters even when the measurements of the mode shapes are incomplete, and corrupted by noise.

8. Unlike most updating methods, no attempt is made to orthonormalize the measured mode shapes.

9. The iterative procedure provides a significant advantage both in terms of reducing the variance of the stiffness estimates and in getting the expected estimates closer to the true values. This is achieved by expanding, at each iteration, the noise-corrupted

mode shapes in terms of the best available estimates of the mode shapes and eigenvalues, as well as the measured frequencies of vibration.

Despite the promise that this method shows, especially for rapid damage detection, and for providing *a priori* estimates that can then be used with other identification methods, the results presented herein should be considered as being preliminary. The way the method behaves when the number of unknown stiffness parameters becomes large, and when the structure is nonclassically damped still remains to be examined. Some of these issues will be presented in a future communication.

REFERENCES

1. Caughey T.K. and O'Kelley M.E.J., "General theory of vibration of damped linear dynamics systems," Dynamics Laboratory Report, California Institute of Technology, June 1963.
2. Baruch M. and Bar Itzhak Y., "Optimal weighted orthogonalization of measured modes," *AIAA Journ.*, 16 (4), pp. 346–351, 1978.
3. Kabe A.M., "Stiffness matrix adjustment using mode data," *AIAA Journ.*, 23 (9), pp. 1431–1436, 1985.
4. Kalaba R. and Udwadia F., "An associative memory approach to the rapid identification of nonlinear structural and mechanical systems," *Journ. Optim. Theory Appl.*, 76(2), pp. 207–223, 1993.
5. Kenigsbuch R. and Halevi Y., "Generalized reference basis model updating in structural dynamics," ASME Design Engineering Conference, Sacramento, CA, 1997.
6. Mottershead J.E. and Friswell M.I., "Model updating in structural dynamics: a survey," *Journ. Sound Vibration*, 16(2), pp. 347–375, 1993.
7. Udwadia F. and Ghodsi A., "Parameter identification problems in structural and geotechnical engineering," *Journ. Eng. Mech.*, 110, pp. 1409–1432, 1984.
8. Udwadia F. and Proskurowski, W., "A memory-matrix-based identification methodology for structural and mechanical systems," *Earthquake Eng. Struct. Dyn.*, 27, pp. 1465–1481, 1998.
9. Wei F.S., "Structural dynamic model modification using vibration data," *IMAC*, 7, pp. 562–567, 1989.
10. Koh C.G., Hong B., and Liaw C.-Y., "Parameter identification of large structural systems in the time domain," *Journ. Struct. Eng.*, 126(8), pp. 957–963, 2000.
11. Udwadia F., and Kalaba R., *Analytical Dynamics*, Chapter 2, Cambridge University Press, 1996.

14

A Survey of Applications in Structural–Acoustic Optimization for Passive Noise Control with a Focus on Optimal Shell Contour

Steffen Marburg
Institut für Festkörpermechanik Technische Universität, Dresden, Germany

This chapter reviews applications in structural–acoustic optimization. The author concentrates on problems in passive noise control. Several dozen papers on structural–acoustic optimization were found. An increasing community of authors indicates optimization of structures whereas a close-up confirms that just a few different variants were investigated. Herein, structural–acoustic optimization is understood as an autonomous (mostly iterative) search to minimize a certain objective function. In a brief survey of the literature, the articles are categorized into academic and more realistic applications. A few of them are discussed in more detail. These three applications have in common that a shell contour is optimized to obtain certain acoustic characteristics. The first application is the design of carillon bells, the second example reviews the design of a loudspeaker diaphragm, and the third application consists of the design of sedan body panels. It is emphasized that shell contour optimization accounts for an efficient tool to influence local structural stiffness without increasing the mass.

1 INTRODUCTION

The decrease of noise emission of machines, vehicles, domestic appliances, etc. will probably become one of the major challenges of engineering in the 21st century. Development of computers enabled a wide range of applications of numerical methods like the finite element method and the boundary element method during the past three decades. Further and further improvement of hardware systems and simulation methods encouraged researchers to employ methods of numerical optimization for structural–acoustic optimization over the past decade.

Herein, we focus on the field of structural–acoustic optimization for passive noise control. Structural–acoustic optimization is understood as the improvement of certain acoustic characteristics of a structure that emits sound or noise by mathematically controlled modification of the structure. That a multifield problem has to be solved is the reason that the subject is also considered as a problem of multidisciplinary optimization.

0-415-30852-6/04/$0.00+$1.50

Solution of multifield problems usually appears as a procedure that is computationally expensive. If structural–acoustic simulation covers a wide frequency range, the multifield problem has to be solved for a certain number of discrete frequencies. Optimization will then require an outer loop that contains the entire solution of the multifrequency multifield problem in a single step. This, however, considers analysis only. A number of additional technical problems is encountered if this type of multidisciplinary optimization is actually implemented. These difficulties may explain why comparatively few publications exist in the field of structural–acoustic optimization. We have not been able to find any paper on this subject for time–domain problems. A survey on analysis techniques in structural–acoustic optimization is given in the paper by Christensen et al. [1].

Several dozen papers have been published on the subject since 1990. However we found that often paper titles promise structural and/or acoustic optimization, but close examination reveals that only a number of different variants are compared while optimization in the sense of an (autonomous) iterative minimum search of an objective function (or cost function) is not reported, cf. Refs. [2–5]. Moreover, the field of structural–acoustic optimization is also important for problems in active noise control. This subject is excluded from our reflections for this chapter.

As the acoustic characteristics to be influenced, we can name the sound power, sound pressure level, directivity patterns, or still other measures. In practical applications, a number of very different objectives must be considered in the design process. Cases of multicriteria optimization are familiar because it often happens that two or more conflicting design criteria should be taken into account simultaneously. Then, we have to make a trade-off among them.

Structural modifications involve almost all possible changes of existing construction. To identify a few of them, modifications include plate and shell thicknesses, material data, size and location of added masses, damping characteristics, and shell curvature. In general, modifications must satisfy design and functionality requirements. Any further variation is subject to the engineer's experience and imagination. Many types of modifications necessitate certain constraints. These constraints typically comprise mass conditions but others like particular cost functions or even different definite conditions must be regarded.

Optimization involves a number of calculations of the objective function. Depending on the number of variables, the condition of the optimization problem, optimization method, and other features of the particular problem, this calculation is repeated between 10 and 10^6 times. The upper limit is an estimation and may be exceeded in some cases. This, however, demands efficient analysis techniques for calculation of the objective function. In most applications, the structural models are analyzed using finite element models. The variety of noise emission evaluation is greater. So, both finite element- and boundary element-based methods are used. Additionally, several methods based on the Rayleigh integral have been reported.

The optimization method will essentially influence success of the entire procedure. With respect to a high degree of nonlinearity of the objective function in terms of design parameters, only a numerical treatment like an iterative minimum search accomplishes the necessities of this problem. Deterministic and stochastic algorithms are appropriate for these purposes.

A number of the problems discussed above has been addressed in the books by Koopmann and Fahnline [6] and Kollmann [7]. One problem that was very little focused in these volumes is the technical completion for complex structural models. This includes mesh coupling between structure and fluid, parameterization, and data handling. For appreciation of optimization results, this point should not be neglected since it often requires major efforts of manual work. There are software tools for applications in structural–acoustic optimization. In most cases, however, one needs to combine several different tools or a particular code must be developed because general-purpose applications do not fit together with specific desiderata of the particular application.

In this chapter we will start with a brief review of the literature in the field of structural–acoustic optimization for passive noise control. Then, we will focus on optimization of shell curvature. The author has found three applications. First, there are some papers about bell design [8–13], second, one paper on the geometry optimization of a loudspeaker diaphragm was found [14], and, third, a number of papers were written on the optimization of metal sheets for sedan body noise, vibration, and harshness problems in the low-frequency range [15–22].

2 BRIEF REVIEW OF CONTRIBUTIONS IN STRUCTURAL–ACOUSTIC OPTIMIZATION

The relatively small number of contributions to this field can be categorized into academic examples and more realistic applications. It is the advantage of simple examples that one can experience and investigate many effects of a method. In some cases, however, particular adjustments become necessary if problems and models become more complex.

We start with the academic examples. These can be grouped into problems of beams, plates, shells, ducts, and boxes.

Few papers considered baffled beams, cf. Naghshineh et al. [23], Jog [24], and Fahnline and Koopmann [6]. The beam structure was likely chosen for simplicity. Some interesting effects could be demonstrated by investigating this example. Moreover, beam structures are suitable to test a method.

Plates have been investigated quite often [6, 14, 25–37]. However, the category of academic examples seems to be confined to either rectangular or circular ones. It was possible to find one article presenting the example of an engine cover plate [25] that will be discussed with the realistic applications. Choice of circular or rectangular plates could be critical since these structures contain symmetry properties that may have unfavorable effects on sound power radiation. For all of the plate examples, sound radiation problems were investigated. Usually, the sound power was to be minimized. Few examples on circular plates considered different targets. Fahnline and Koopmann [28] sought a circular plate of maximum power–conversion efficiency. Christensen et al. [14, 26] optimized the directivity pattern of sound radiation from a circular plate. Tinnsten [36] chose a circular plate to construct an example for experimental verification of structural–acoustic optimization. He measured the sound intensity above the plate. Wodtke and Lamancusa [37] optimized a circular baffled plate with an unconstrained damping layer for minimum sound power emission. In most of these examples on circular plates, the structure was excited by a point force [14, 26, 36, 37] or a ring load [37]. One example is based on modal properties and, therefore, independent of the excitation [28]. Rectangular plates were excited by point loads [6, 24, 25, 27, 30, 32–35] or by distributed loads [31, 34]. Lamancusa [30] optimized modal properties to achieve an optimal solution independent of a particular excitation. Plates were mostly clamped. Few examples of different boundary conditions are reported. Cases of clamped and free boundary conditions are compared in [36, 37].

Additionally, elastic support of edges is included in [37]. In [32], the example comprises a plate that is clamped at two edges while the other two edges are free. Ratle and Berry [34] investigated a simply supported rectangular plate. In some of these examples, the authors reported and compared both broad-band and single-frequency excitation [25, 33, 35]. Only one driving frequency or mode is considered (for the plate example) in [14, 26, 27]. Two or more frequencies or modes were included by [32, 34, 36, 37]. Broad-band sound radiation including many modes was minimized by Lamancusa and Eschenauer [30, 31] and by Hibinger [29]. Plates seem to be well suited for experimental verification as supplied in [6, 29, 32, 33, 36]. In addition to the axisymmetric examples discussed above, symmetry of vibration modes was assumed in some of the examples, cf. [27, 30, 31, 33, 35].

Structural–acoustic optimization of shell structures without an apparent application was addressed in [26, 27, 38, 39]. Hambric's [38, 40] example was a cylindrical shell with hemispherical end caps. In one test case the shell was simply supported, in another one these supports were omitted. The structure was excited by a ring load. Single-frequency and broad-band excitation cases were investigated for optimization. For this shell structure submerged in water, a full structure–fluid interaction became necessary as well as for the thin conical membrane that was considered by Christensen et al. [26]. This axisymmetric structure clamped at its edge was excited by a point load in the center. The directivity pattern was optimized for a single frequency. Constans et al. [27,39] chose the example of a half-cylindrical shell clamped along the two lower straight edges. The emitted sound power due to a driving force was minimized for the lowest five modes. Although structural and loading configuration were symmetric the optimal solution is asymmetric. An experimental verification was provided [39].

The simple example of a duct was investigated by Shii et al. The paper [41] is a reduced version of the Japanese paper [42]. The authors stated that the one-dimensional duct system with a coupled spring–mass-system should represent the rear window of a vehicle but complexity of vehicle structures is much higher than in their example. Sound pressure at a single point is calculated for a narrow-frequency band.

Boxes have been addressed for both, internal [43–46] and external [7, 29, 47] noise problems. Starting with the internal problem, a simple box model was investigated by Pal et al. [44–46]. A fully coupled structure–fluid interaction was included. The box was excited by a point load and analyzed for two discrete frequencies. The paper by Marburg et al. [43] addressed modeling techniques of a steel box structure that is reinforced by two frames. Although a frequency range containing about sixty modes was chosen, a reliable structural model was created for a range of about twenty modes. The root-mean-square sound pressure level over the frequency range was minimized for one test point. The paper by Tinnsten et al. [47] addressed sound radiation. A point load was applied for one or two discrete frequencies. The authors reported an uncoupled analysis. The box example of Hibinger [7, 29] was primarily focused on radiation problems. Sound radiation of this box was not computed, though. Minimization of the sum of mean-square velocities of the box surface accounted for the optimization target. A point load was applied. The frequency range contained the first 15 modes.

The more realistic applications of structural–acoustic optimization can be grouped into problems of sandwich panels, loudspeaker diaphragm, bells, a cylinder representing an aircraft fuselage, and vehicle noise problems. For the latter, we can further distinguish between body noise problems and others.

Dym et al. [48, 49] investigated optimization of sandwich panels. It was their target to maximize transmission loss of these plates. A large frequency range was analyzed and several, at that time new, findings were reported.

Although they belong to this category of realistic applications we will discuss examples on the design of carillon bells and a loudspeaker diaphragm in the next subsection.

A cylinder clamped at both ends accounted for a design problem similar to an aircraft fuselage in several papers by Cunefare, Engelstad et al. [50–52]. The structure was excited by a monopole source outside at a single frequency that coincided with an eigenfrequency of the original model. Optimization included thickness of shell rings [50], cross-section [51, 52] and location [52] of stiffening elements like stringers and frames. The sound pressure was evaluated at certain data recovery points inside the cylinder.

There are a few papers on the application to vehicle acoustics not primarily considering body vibrations. All of them address design and support of the engine or components. An engine cover plate accounts for the practical application in the paper by Belegundu et al. [25]. The plate was clamped with respect to rotations and supported on springs of high stiffness. Excitation forces of different phase angle were applied. In the optimization process, sound radiation was minimized with respect to the first three structural modes. The paper by Milstedt et al. [53] discussed the design and composition of an engine that consisted of several single substructures. Optimization was accomplished for the frequency range of 0 to 5000 Hz. Their model, however, is known as a concept model. Therefore, the objective of that study was to show the potential of structural optimization and to illustrate the concept. Real engine analysis requires finer models as the authors stated. La Civita and Sestieri [54] described the design of an optimal engine mounting system. They computed the sound pressure level at the driver's ear in the frequency range up to 120 Hz. The paper regards very different possibilities to construct engine supports, i.e., position and orientation, nonlinear spring characteristics, and, at the same time, they imposed different technical and geometrical constrains.

Most of the applied papers in structural–acoustic optimization addressed sedan body construction. These examples can be categorized in terms of the design variables. One category consists of seeking an optimal distribution of piecewise constant shell thickness [55–58]. These applications usually require many variables. The paper of Choi et al. [57] indicates use of 36,000 parameters. Hermans and Brughmans [58] composed the vehicle body as an assembly of several components being investigated independently in a first design stage. The other category utilizes modification of shell geometry that will be discussed in more detail later in this article [15, 18, 19, 21, 22].

All of these examples on sedan interior noise were aimed at pure reduction. Future work will likely design sound in cars as suggested by Freymann [59]. The importance and necessity of optimization in the body design process to achieve a "design right the first time" was emphasized in a paper on optimization of the noise, vibration, and harshness engineering process by Roesems [60].

There are a number of papers that do not actually address structural–acoustic optimization but are focused on structural–acoustic sensitivity analysis. Again, we start with the academic examples. Hahn and Ferri [61] demonstrated sensitivity analysis for a cylindrical shell submerged in water. Shepard and Cunefare [62] used a semi-infinite plate. Both papers mainly addressed modeling techniques including the question how coarse or fine certain details should be modeled. A number of papers were found on the interior noise of simple boxes [56, 63–69] and just one on the radiation of a box [70]. The latter

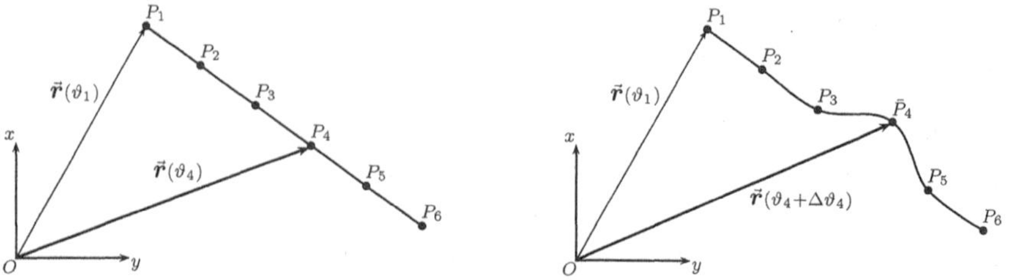

Figure 1 Geometry-based modeling: Shell contour modification by shifting a single keypoint.

also presented an example of a cylindrical shell. While most of the papers that consider the interior noise of boxes concentrated on coupled mode sensitivities, the contribution of Bitsie and Bernhard [67] is devoted to an unusual topic in the field, namely high-frequency noise problems. Applying energy finite element methods, they tried to reduce the energy level of structure and fluid due to excitation of either structure or fluid. Sensitivity analysis of vehicle body structures was accomplished in [56, 66, 71].

Some of the papers that consist of two parts discussed theoretical aspects in part one. The review paper of Christensen et al. [1] supplied detailed description of efficient analysis in structural acoustics. The author of this article described theoretical aspects of the optimization of vehicle noise transfer functions in [17]. In another recent paper [20], concepts of parametric description of shell structures were investigated. Geometry-based models were compared with a newly developed modification concept that enables a user to directly modify the shell geometry of a finite element mesh.

3 OPTIMAL SHELL GEOMETRY

3.1 Building and Parameterizing a Shell Model

In structural–acoustic optimization, two major problems are usually encountered when designing shell structure geometry parametrically: First, variable geometry may require modification of the mesh of the shell and, second, the volume where the sound is emitted changes its shape. For these two reasons, a finite-element model of the shell structure and a boundary-element model of the fluid were used for the three cases of shell contour optimization that will be reviewed in this section.

Apparently, axisymmetric shell models can be used for analysis of a bell or a loudspeaker diaphragm. In these cases, the structural model to be parameterized and meshed reduces to a line. This line may be defined by the position of a few keypoints. These keypoints are then connected by lines, i.e., by spline interpolation. Finite elements are created by dividing these lines into a certain number of line pieces accounting for axisymmetric shell elements.

This technique of modeling is known as geometry-based modeling. In a paper describing optimization of shell geometry for vehicle interior noise problems in general [17], the author proposed geometry-based modeling as a reasonable strategy of modeling and

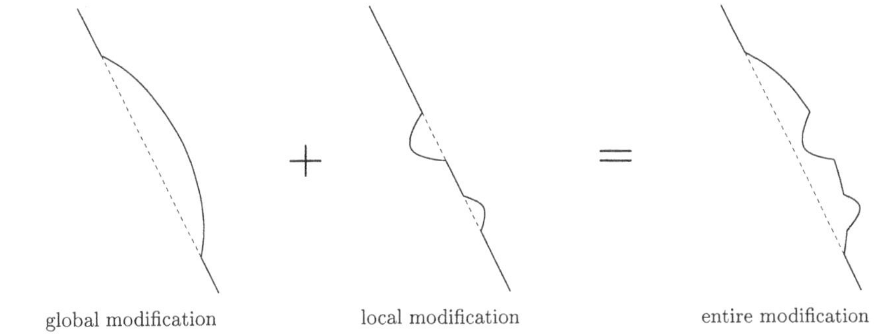

Figure 2 Direct modification of finite element mesh: Superposition of global and local modification yields complete modification of shell contour (mesh not visualized).

parameterizing the shell contour. Modification of one parameter affects a group of nodes and elements being shifted into normal direction, cf. Fig. 1.

An alternative technique to parameterize shell structures is based on direct modification of the geometry of a finite element model. This concept is explained in the author's paper [20]. It starts with definition of a modification domain Ω_m. Two types of modification are allowed on this domain: Global and local modification. We define both by so-called modification functions. A global modification function is a relatively complicated function being defined on the entire modification domain. A local modification function is defined on a local basis Ω_l as part of the modification domain. It is a very simple function but the local domain Ω_l is variable in size, position, and orientation. One example of both, global and local modification, is explained in Ref. [20]. Finally, global and local modification functions are superimposed and, if necessary, normalized to complete modification.

A major advantage of the alternative concept is that *a priori* existing detailed structures can be parameterized by a few variables. A geometry-based model would require many keypoints to describe detailed shell geometries. Another advantage is the simplified use of discrete acoustic influence coefficients [15, 17]. As they depend on this mesh topology, remeshing techniques require extraction of these coefficients for every computation of the objective function. For further discussion, we refer to Ref. [20].

3.2 Design of Carillon Bells

A comprehensive research project on the optimization of carillon bells, in particular the design of a major third one, was initiated and kept up for about two decades by Lehr and Schoofs. An early three-part paper gives an excellent and thorough survey of that problem [8–10]. Structural–acoustic optimization was incorporated by Roozen–Kroon [11]. Although still using a modal approach, she had added modal damping coefficients to the formulation. Another important step in bell design was completed by the dissertation of van Houten [12]. Schoofs and van Campen summarized these and other activities on bell design and optimization in a longer paper [13]. Finally, new generations of bells were designed by numeric optimization.

The objective function that was used for bell design in [11, 12] and in other papers is formulated as

$$F = w_D \left(\frac{f_D}{f_{D_{opt}}} - 1 \right)^2 + \sum_{i=1}^{N_f} w_{f,i} \left(\bar{f}_i - \bar{f}_{opt,i} \right)^2 + \sum_{i=1}^{N_f} w_{a,i} \left(\eta_{a_i} - \eta_{a_{opt,i}} \right)^2 . \tag{1}$$

In this function, $w_D, w_{f,i}$ and $w_{a,i}$ denote weighting factors for each single term. N_f is the number of structural eigenfrequencies being considered, $\bar{f}_i$ and η_{a_i} represent the eigenfrequencies and corresponding acoustic damping values of the current design while $\bar{f}_{opt,i}$ and $\eta_{a_{opt,i}}$ account for the target values. Acoustic damping has been defined as a frequency-scaled relation between the radiated sound power and the kinetic energy of the structure. The values of f_D and $f_{D_{opt}}$ correspond to a relation between the diameter of the bell and the shell thickness. For that, it was defined $f_D = D_l f_1$, the product of the lip diameter and the first eigenfrequency. Excluding this first term in Eq. (1), this objective function could also be utilized for autonomous model updating purposes where experimental and simulated data of a modal analysis are compared. In addition to the geometry, variable thickness was optimized.

In early papers on optimization of bells, optimization was performed by design of experiments [8,11]. This global approximation concept was used with respect to the computationally expensive analysis. In more recent papers, alternative approximation concepts have been favored, i.e., mid-range or local approximation. A thorough discussion on approximation concepts and applications in bell design has been presented by van Houten [12]. This work, however, took advantage of the axisymmetric boundary-element formulation for sound radiation developed by Kuijpers et al. [72].

3.3 Design of Loudspeaker Diaphragm

Christensen and Olhoff [14] optimized the geometry of the axisymmetric shell contour of a loudspeaker diaphragm. They presented three case studies, one for a clamped flat membrane with added ring masses, one for geometry optimization of the clamped membrane, and a third for geometry optimization combined with variable shell thickness in soft surrounds. A point force excitation was applied in the center of the membrane. Fully coupled fluid structure interaction was considered. The objective function was formulated as the squared deviation from a prescribed directivity pattern curve.

In the first case, a large mass in the driven center created a point source. Apparently, a point source guarantees a regular directivity pattern. Performance of the first case was improved in the second case. The axisymmetric geometry was defined by the variable positions of eight keypoints whereas the contour was created using B-spline interpolation of nine keypoints. (The position of one keypoint was arbitrarily chosen.) The best performance was found for the third case. The prescribed uniform directivity pattern was well adjusted for three different frequencies, 0.5, 10, and 15 kHz. Local differences of about 20 dB were observed for the initial design and less than about 1 dB for the optimal solution. In this case, the diaphragm was put into soft surround. This was necessary to achieve a uniform pattern at low frequencies. Constraint equations became necessary for the first case only. For that a mass condition was introduced to avoid an unlimited increase of the mass. Geometry optimization succeeded without any constraints.

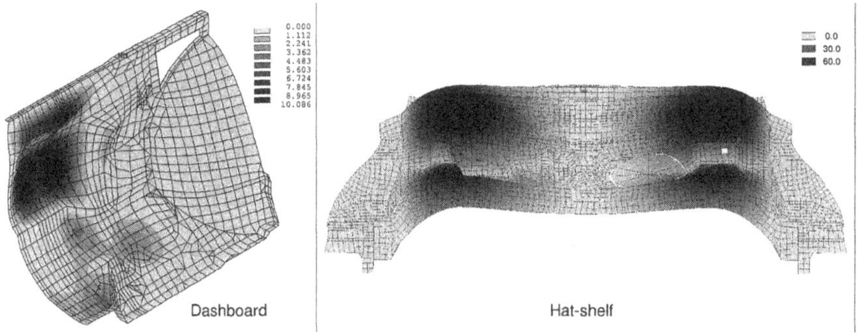

Figure 3 New design of sedan body panels: Optimized geometry as modification of the original geometry for symmetry half of a dashboard and hatshelf (scales in mm).

Although optimization was limited to three discrete frequencies only, the authors stated that uniform directivity patterns were also observed for other frequencies that they have checked. Finally, the authors emphasized suitability of optimization of the shell contour.

3.4 Design of Sedan Body Panels

In a number of papers on vehicle interior noise, the author of this chapter investigated several different issues of shell contour optimization for realistic applications, cf. Ref. [15–22]. These applications can be categorized into those that apply geometry-based models and those that use direct modification of finite-element meshes.

Geometry-based modeling was tested for a sedan roof [15, 16] and for a dashboard [17, 18]. Geometry of a symmetry half of the roof was defined by 15 keypoints, six of them with parametrically defined position. Simple support at six points and an excitation pressure from above were unrealistic conditions that essentially contributed to the success of this work. In that case, where the elastic modes occur at frequencies much lower than that for rigid body modes, the optimization strategy favors stiffening effects to decrease the average sound pressure level at the driver's ear. We observed a lowest eigenfrequency of 88 Hz and altogether 39 modes up to 200 Hz for the original structure. A 30-cm higher optimal roof had its first eigenfrequency at 209 Hz; in other cases it was even increased to about 250 Hz. Since optimization was limited to a frequency range of 0 to 200 or 0 to 300 Hz, amazing decreases of the objective function could be reported. Even for the case of only 10-mm modification significant decreases were observed. In that case, however, stiffness was not increased by maximizing the lowest eigenfrequency but by decreasing the number of modes in the frequency range under consideration. More generally, it became apparent that modes changed from global to local shapes with the tendency of extinguishing each other. This tendency was confirmed and utilized for optimization of a sedan dashboard [18]. Engine excitation, transmission through the body, and maximum feasible design modifications of 10 mm accounted for more realistic conditions. The geometry-based model consisted of several hundreds of keypoints, lines, and areas. Apparently, only

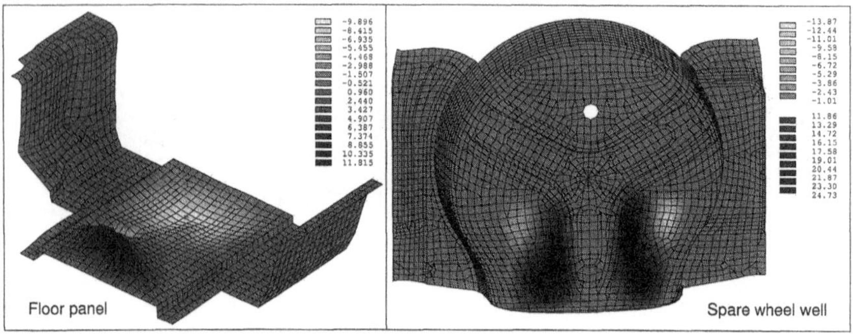

Figure 4 New design of sedan body panels: Optimized geometry as modification of the original geometry for part of a floor panel and spare wheel well (scales in mm).

few of them could be included in the optimization process. The objective function was more generally defined as [17]

$$F = \tilde{F}^{\frac{1}{n}}, \tag{2}$$

where $\tilde{F}$ is given by

$$\tilde{F} = \frac{1}{\omega_{max} - \omega_{min}} \int_{\omega_{min}}^{\omega_{max}} \Phi\{p_{\mathrm{L}}(\omega)\}\, d\omega. \tag{3}$$

The operator $\Phi\{\}$ applied to sound pressure level p_{L} represents a kind of a weighting function. A reasonable weighting function is

$$\Phi\{p_{\mathrm{L}}\} = \begin{cases} (p_{\mathrm{L}} - p_{\mathrm{Ref}})^n & \text{for} \quad p_{\mathrm{L}} > p_{\mathrm{Ref}} \\ 0 & \text{for} \quad p_{\mathrm{L}} \leq p_{\mathrm{Ref}}. \end{cases} \tag{4}$$

So, the objective function simply appears as a frequency-averaged sound pressure level and the exponent n controls the type of average. For $n = 1$, Eqs. (2)–(4) lead to the mean value where only values higher than a certain reference level p_{Ref} are taken into account. Similarly, this form provides the root-mean-square value for $n = 2$. The major advantage of this value is that high-level peaks are higher valued than low-level parts of the function. This helps to reduce these high-level peaks during optimization procedure and avoids deep valleys as compensation of high peaks. If n tends to infinity the objective function will be equal to the maximum value of the noise transfer function. In the case of the dashboard, 44 variables were optimized with the objective to decrease the root-mean-square sound pressure level at the driver's ear. New shapes showed well-known structures. So, a horizontal bead and two bulges could be identified, cf. Fig. 3.

The first application of direct modification of a finite-element mesh to find an optimal shell geometry for acoustics was reported for a sedan hat shelf [19]. In that case, an alternative to geometry-based modeling was required since the geometry-based model contained too many details. These details made it impossible to optimize keypoint positions. As a consequence, the mesh was directly modified by using just four design variables. Moreover, the case of stiff supports and excitation in the hat shelf, i.e., subwoofer, belt retractor, etc., made it possible to circumvent full structural–acoustic optimization and concentrate just on optimization of the lowest eigenfrequency. This was increased from 32 to 101 Hz, whereas the average sound pressure level at the driver's ear could decrease by between 4.4 and 13.9 dB, depending on the particular excitation and different parameters of the objective function, Eqs. (2)–(4). The new shape of the hat shelf is shown in Fig. 3. A full concept of direct modification was presented in [20] and applied to a sedan floor panel in [21] and a spare wheel well [22]. Similar to the dashboard case, excitation at engine supports and transmission through the body were considered. In both cases, we could not speak of stiffly supported structures. Consequently the mode spectrum remained the same, but vibration mode shapes varied. Both panels were stiffened, though. This became obvious for the floor panel at low frequencies, cf. new geometry in Fig. 4, and for the spare wheel well in the entire frequency range. A strong vibration loop at the bottom of the spare wheel well could be replaced by stiffening bead-like structures, cf. Fig. 4. Due to the weak support of both structures, comparatively small improvements of the objective function can be reported.

4 CONCLUSION

Herein we have reviewed contributions in structural–acoustic optimization. After a brief review of applications in the field, we have considered shell contour optimization in more detail. Based on the knowledge of the author, only three different applications were discussed. These were design of carillon bells, loudspeaker diaphragm, and sedan body panels. In all of these cases, it turned out that geometry optimization is an efficient tool for structural–acoustic design. It is one of the major advantages of geometry optimization that a mass condition is not required. Furthermore, it is advantageous compared to optimization of piecewise constant shell thickness that a large variety of geometries of metal sheets can be easily produced by conventional metal-forming techniques.

Finally, we emphasize again that optimization of the shell contour is an efficient tool to influence local stiffness of shells while its mass remains constant.

REFERENCES

1. S.T. Christensen, S.V. Sorokin, and N. Olhoff, "On analysis and optimization in structural acoustics – Part I: Problem formulation and solution techniques," *Structural Optimization*, vol. 16, pp. 83–95, 1998.
2. F. Dirschmid, H. Troidl, A. Kunert, S. Dillinger, O. von Estorff, E. Negele, and M. Stowasser, "Akustische Optimierung von Getriebegehäusen," in *Berechnung und Simulation im Fahrzeugbau*, pp. 633–651, VDI–Report 1283, 1996.
3. T. Bauer and G. Henneberger, "Three-dimensional calculation and optimization of the acoustic field of the induction furnace caused by electromagnetic forces," *IEEE Transactions on Magnetics*, vol. 35, pp. 1598–1601, May 1999.
4. N. Vincent, P. Bouvet, D.J. Thompson, and P.E. Gautier, "Theoretical optimization of track components to reduce rolling noise," *Journal of Sound and Vibration*, vol. 193, no. 1, pp. 161–171, 1996.
5. J.C.O. Nielsen, "Acoustic optimization of railway sleepers," *Journal of Sound and Vibration*, vol. 231, no. 3, pp. 753–764, 2000.

6. G.H. Koopmann and J.B. Fahnline, *Designing Quiet Structures: A Sound Power Minimization Approach*. San Diego, London: Academic Press, 1997.
7. F.G. Kollmann, *Maschinenakustik. Grundlagen, Messtechnik, Berechnung, Beeinflussung*. Berlin, Heidelberg: Springer Verlag, 1999.
8. A.J.G. Schoofs, F. van Asperen, P. Maas, and A. Lehr, "A carillon of major–third bells. Part I: Computation of bell profiles using structural optimization," *Music Perception*, vol. 4, no. 3, pp. 245–254, 1987.
9. A.J.M. Houtsma and H.J.G.M. Tholen, "A carillon of major–third bells. Part II: A perceptual evaluation," *Music Perception*, vol. 4, no. 3, pp. 255–266, 1987.
10. A. Lehr, "A carillon of major–third bells. Part III: From theory to practice," *Music Perception*, vol. 4, no. 3, pp. 267–280, 1987.
11. P.J.M. Roozen-Kroon, *Structural Optimization of Bells*. Dissertation, Technische Universiteit Eindhoven, 1992.
12. M.H. van Houten, *Function Approximation Concepts for Multidisciplinary Design Optimization*. Dissertation, Technische Universiteit Eindhoven, 1998.
13. A.J.G. Schoofs and D.H. van Campen, "Analysis and optimization of bells systems," in *11th Carillon World Congress* (Mechelen and Leuven, Belgium), August 1998.
14. S.T. Christensen and N. Olhoff, "Shape optimization of a loudspeaker diaphragm with respect to sound directivity properties," *Control and Cybernetics*, vol. 27, no. 2, pp. 177–198, 1998.
15. S. Marburg, H.-J. Hardtke, R. Schmidt, and D. Pawandenat, "An application of the concept of acoustic influence coefficients for the optimization of a vehicle roof," *Engineering Analysis with Boundary Elements*, vol. 20, no. 4, pp. 305–310, 1997.
16. S. Marburg, H.-J. Hardtke, R. Schmidt, and D. Pawandenat, "Design optimization of a vehicle panel with respect to cabin noise problems," in *Proceedings of the NAFEMS World Congress* (Stuttgart), pp. 885–896, 1997.
17. S. Marburg, "Efficient optimization of a noise transfer function by modification of a shell structure geometry. Part i: Theory." *Structural and Multidisciplinary Optimization*, vol. 24, no. 1, pp. 51–59, 2002.
18. S. Marburg and H.-J. Hardtke, "Efficient optimization of a noise transfer function by modification of a shell structure geometry. Part II: Application to a vehicle dashboard," *Structural and Multidisciplinary Optimization*, vol. 24, no. 1, pp. 60–71, 2002.
19. S. Marburg and H.-J. Hardtke, "Shape optimization of a vehicle hat–shelf: Improving acoustic properties for different load–cases by maximizing first eigenfrequency," *Computers and Structures*, vol. 79, no. 20–21, pp. 1943–1957, 2001.
20. S. Marburg, "A general concept for design modification of shell meshes in structural–acoustic optimization. Part i: Formulation of the concept," *Finite Elements in Analysis and Design*, vol. 38, no. 8, pp. 725–735, 2002.
21. S. Marburg and H.-J. Hardtke, "A general concept for design modification of shell meshes in structural–acoustic optimization. Part II: Application to a vehicle floor panel," *Finite Elements in Analysis and Design*, vol. 38, no. 8, pp. 737–754, 2002.
22. S. Marburg and H.-J. Hardtke, "Investigation and optimization of a spare wheel well to reduce vehicle interior noise," *International Journal for Acoustics and Vibration*, vol. 11, no. 3, 2003 (in press).
23. K. Naghshineh, G. H. Koopmann, and A. D. Belegundu, "Material tailoring of structures to achieve a minimum radiation condition," *Journal of the Acoustical Society of America*, vol. 92, no. 2, pp. 841–855, 1992.

24. C.S. Jog, "Reducing radiated sound power by minimizing the dynamic compliance," in *IUTAM–Symposium on Designing for Quietness*, December 2000 (M.L. Munjal, ed.). Dordrecht / Boston / London: Kluwer Academic Publishers, pp. 215–236, 2002.
25. A.D. Belegundu, R.R. Salagame, and G.H. Koopmann, "A general optimization strategy for sound power minimization," *Structural Optimization*, vol. 8, no. 2–3, pp. 113–119, 1994.
26. S.T. Christensen, S.V. Sorokin, and N. Olhoff, "On analysis and optimization in structural acoustics – Part II: Exemplifications for axisymmetric structures," *Structural Optimization*, vol. 16, pp. 96–107, 1998.
27. E.W. Constans, A.D. Belegundu, and G. H. Koopmann, "Design approach for minimizing sound power from vibrating shell structures," *AIAA Journal*, vol. 36, no. 2, pp. 134–139, 1998.
28. J.B. Fahnline and G.H. Koopmann, "Design for a high-efficiency sound source based on constrained optimization procedures," *Acoustical Physics*, vol. 41, no. 5, pp. 700–706, 1995.
29. F. Hibinger, *Numerische Strukturoptimierung in der Maschinenakustik.* Dissertation, Technische Universität Darmstadt, 1998.
30. J.S. Lamancusa, "Numerical optimization techniques for structural–acoustic design of rectangular panels," *Computers and Structures*, vol. 48, no. 4, pp. 661–675, 1993.
31. J.S. Lamancusa and H.A. Eschenauer, "Design optimization methods for rectangular panels with minimal sound radiation," *AIAA Journal*, vol. 32, no. 3, pp. 472–479, 1994.
32. K. Nagaya and L. Li, "Control of sound noise radiated from a plate using dynamic absorbers under the optimization by neural network," *Journal of Sound and Vibration*, vol. 208, no. 2, pp. 289–298, 1997.
33. R.L. St. Pierre Jr. and G.H. Koopmann, "A design method for minimizing the sound power radiated from plates by adding optimally sized, discrete masses," *Journal of Mechanical Design*, vol. 117, pp. 243–251, June 1995.
34. A. Ratle and A. Berry, "Use of genetic algorithms for the vibroacoustic optimization of a plate carrying point–masses," *Journal of the Acoustical Society of America*, vol. 104, no. 6, pp. 3385–3397, 1998.
35. R.R. Salagame, A.D. Belegundu, and G.H. Koopmann, "Analytical sensitivity of acoustic power radiated from plates," *Journal of Vibration and Acoustics*, vol. 117, pp. 43–48, January 1995.
36. M. Tinnsten, "Optimization of acoustic response — a numerical and experimental comparison," *Structural Optimization*, vol. 19, pp. 122–129, 2000.
37. H.W. Wodtke and J.S. Lamancusa, "Sound power minimization of circular plates through damping layer placement," *Journal of Sound and Vibration*, vol. 215, no. 5, pp. 1145–1163, 1998.
38. S.A. Hambric, "Approximation techniques for broad-band acoustic radiated noise design optimization problems," *Journal of Vibration and Acoustics*, vol. 117, pp. 136–144, January 1995.
39. E.W. Constans, G.H. Koopmann, and A.D. Belegundu, "The use of modal tailoring to minimize the radiated sound power of vibrating shells: Theory and experiment," *Journal of Sound and Vibration*, vol. 217, no. 2, pp. 335–350, 1998.
40. S.A. Hambric, "Sensitivity calculations for broad-band acoustic radiated noise design optimization problems," *Journal of Vibration and Acoustics*, vol. 118, pp. 529–532, July 1996.

41. Q. Shi, I. Hagiwara, A. Azetsu, and T. Ichkawa, "Holographic neural network approximations for acoustic optimization," *JSAE Review*, vol. 19, pp. 361–363, 1998.
42. Q. Shi, I. Hagiwara, S. Azetsu, and T. Ichikawa, "Optimization of acoustic problem using holographic neural network," *Transactions of the Society of Automotive Engineers of Japan*, vol. 29, pp. 93–97, July 1998 (in Japanese).
43. S. Marburg, R. Rennert, H.-J. Beer, J. Gier, H.-J. Hardtke, and F. Perret, "Experimental verification of structural–acoustic modeling and design optimization," *Journal of Sound and Vibration*, vol. 252, no. 4, pp. 591–615, 2002.
44. W. Kozukue, C. Pal, and I. Hagiwara, "Optimization of noise level reduction by truncated model coupled structural–acoustic sensitivity analysis," *Computers in Engineering (ASME)*, vol. 2, pp. 15–22, 1992.
45. C. Pal and I. Hagiwara, "Dynamic analysis of a coupled structural–acoustic problem. Simultaneous multi–modal reduction of vehicle interior noise level by combined optimization," *Finite Elements in Analysis and Design*, vol. 14, pp. 225–234, 1993.
46. C. Pal and I. Hagiwara, "Optimization of noise level reduction by truncated modal coupled structural–acoustic sensitivity analysis," *JSME International Journal, Series C*, vol. 37, no. 2, pp. 246–251, 1994.
47. M. Tinnsten, B. Esping, and M. Jonsson, "Optimization of acoustic response," *Structural Optimization*, vol. 18, no. 1, pp. 36–47, 1999.
48. M.A. Lang and C.L. Dym, "Optimal acoustic design of sandwich panels," *Journal of the Acoustical Society of America*, vol. 57, no. 6, pp. 1481–1487, 1975.
49. S.E. Makris, C.L. Dym, and J. MacGregor Smith, "Transmission loss optimization in acoustic sandwich panels," *Journal of the Acoustical Society of America*, vol. 92, no. 6, pp. 1833–1843, 1986.
50. S.P. Crane, K.A. Cunefare, S.P. Engelstad, and E.A. Powell, "Comparison of design optimization formulations for minimization of noise transmission in a cylinder," *Journal of Aircraft*, vol. 34, no. 2, pp. 236–243, 1997.
51. K. A. Cunefare, S.P. Crane, S.P. Engelstad, and E.A. Powell, "Design minimization of noise in stiffened cylinders due to tonal external excitation," *Journal of Aircraft*, vol. 36, no. 3, pp. 563–570, 1999.
52. S.P. Engelstad, K.A. Cunefare, E.A. Powell, and V. Biesel, "Stiffener shape design to minimize interior noise," *Journal of Aircraft*, vol. 37, no. 1, pp. 165–171, 2000.
53. M.G. Milsted, T. Zhang, and R.A. Hall, "A numerical method for noise optimization of engine structures," *Proceedings of the Institution of Mechanical Engineers / Part D: Journal of Automobile Engineering*, vol. 207, pp. 135–143, 1993.
54. M. la Civita and A. Sestieri, "Optimization of an engine mounting system for vibro–acoustic comfort improvement," *Proceedings — SPIE The International Society for Optical Engineering, Issue 3727/PT2*, pp. 1998–2004, 1999.
55. I. Hagiwara, Z.-D. Ma, A. Arai, and K. Nagabuchi, "Reduction of vehicle interior noise using structural–acoustic sensitivity analysis methods," *SAE Technical Paper Series No. 910208*, 1991.
56. Z.-D. Ma and I. Hagiwara, "Sensitivity analysis methods for coupled acoustic–structural systems, Part II: Direct frequency response and its sensitivities," *AIAA Journal*, vol. 29, no. 11, pp. 1796–1801, 1991.
57. K.K. Choi, I. Shim, and S. Wang, "Design sensitivity analysis of structure–induced noise and vibration," *Journal of Vibration and Acoustics*, vol. 119, pp. 173–179, April 1997.

58. L. Hermans and M. Brughmans, "Enabling vibro–acoustic optimization in a superelement environment: A case study," *Proceedings — SPIE The International Society for Optical Engineering, Issue 4062/PT2*, pp. 1146–1152, 2000.
59. R. Freymann, "Sounddesign und Akustikentwicklung im Automobilbau," in *Maschinenakustik '99 – Entwicklung lärm– und schwingungsarmer Produkte*, pp. 47–64, VDI–Report 1491, 1999.
60. D. Roesems, "A new methodology to support an optimized NVH engineering process," *Sound and Vibration*, vol. 31, no. 5, pp. 36–45, 1997.
61. S.R. Hahn and A.A. Ferri, "Sensitivity analysis of coupled structural–acoustic problems using perturbation techniques," *Journal of the Acoustical Society of America*, vol. 101, no. 2, pp. 918–924, 1997.
62. W.S. Shephard Jr. and K.A. Cunefare, "Sensitivity of structural acoustic response to attachment feature scales," *Journal of the Acoustical Society of America*, vol. 102, no. 3, pp. 1612–1619, 1997.
63. F. Scarpa and G. Curti, "A method for the parametric sensitivity of interior acousto–structural coupled systems," *Applied Acoustics*, vol. 58, no. 4, pp. 451–467, 1999.
64. F. Scarpa, "Parametric sensitivity analysis of coupled acoustic–structural systems," *Journal of Vibration and Acoustics*, vol. 122, pp. 109–115, April 2000.
65. Z.-D. Ma and I. Hagiwara, "Development of new mode–superposition technique for truncating lower and/or higher–frequency modes (application of eigenmode sensitivity analysis for systems with repeated eigenvalues)," *JSME International Journal, Series C*, vol. 37, no. 1, pp. 7–13, 1994.
66. Z.-D. Ma and I. Hagiwara, "Sensitivity analysis methods for coupled acoustic–structural systems, Part I: Modal sensitivities," *AIAA Journal*, vol. 29, no. 11, pp. 1787–1795, 1991.
67. F. Bitsie and R.J. Bernhard, "Sensitivity calculations for structural–acoustic efem predictions," *Noise Control Engineering Journal*, vol. 46, no. 3, pp. 91–96, 1998.
68. J. Luo and H.C. Gea, "Modal sensitivity analysis of coupled acoustic–structural systems," *Journal of Vibration and Acoustics*, vol. 119, pp. 545–550, October 1997.
69. K.K. Choi, I. Shim, J. Lee, and H.T. Kulkarni, "Design sensitivity analysis of dynamic frequency responses of acousto–elastic built-up structures," in *Optimization of Large Structural Systems* (G.I.N. Rozvany, ed.), Kluwer Academic Publishers, 1993, vol. 1, pp. 329–343.
70. K.A. Cunefare and G.H. Koopmann, "Acoustic design sensitivities for structural radiators," *Journal of Vibration and Acoustics*, vol. 114, pp. 178–186, April 1992.
71. S. Wang, "Design sensitivity analysis of noise, vibration, and harshness of vehicle body structure," *Mechanics of Structures and Machines*, vol. 27, no. 3, pp. 317–336, 1999.
72. A.H.W.M. Kuijpers, G. Verbeek, and J.W. Verheij, "An improved acoustic fourier boundary element formulation using fast fourier transform integration," *Journal of the Acoustical Society of America*, vol. 102, no. 3, pp. 1394–1401, 1997.

15

Intelligent Control of Aircraft Dynamic Systems with a New Hybrid Neuro-Fuzzy-Fractal Approach

Patricia Melin and Oscar Castillo
Department of Computer Science, Tijuana Institute of Technology, Chula Vista, CA

We describe in this chapter a hybrid method for adaptive model-based control of nonlinear dynamic systems using Neural Networks, Fuzzy Logic, and Fractal Theory. The new neuro-fuzzy-fractal method combines Soft Computing (SC) techniques with the concept of the fractal dimension for the domain of Nonlinear Dynamic System Control. The new method for adaptive model-based control has been implemented as a computer program to show that our neuro-fuzzy-fractal approach is a good alternative for controlling nonlinear dynamic systems. It is well known that chaotic and unstable behavior may occur for nonlinear systems. Normally, we will need to control this type of behavior to avoid structural problems with the system. We illustrate in this chapter our new methodology in the case of controlling aircraft dynamic systems. For this case, we use mathematical models for the simulation of aircraft dynamics during flight. The goal of constructing these models is to capture the dynamics of the aircraft, so as to have a way of controlling these dynamics to avoid dangerous behavior of the aircraft dynamic system.

1 INTRODUCTION

We describe in this chapter a new method for adaptive control of nonlinear dynamic systems based on the use of Neural Networks, Fuzzy Logic, and Fractal Theory. The dynamics of real-world systems are often highly nonlinear and difficult to control [4]. The problem of controlling them using conventional controllers has been widely studied [1]. Much of the complexity in controlling any process comes from the complexity of the process being controlled. This complexity can be described in several ways. Highly nonlinear systems are difficult to control, particularly when they have complex dynamics (such as instabilities to limit cycles and chaos [12]). Difficulties can often be presented by constraints, either on the control parameters or in the operating regime. Lack of exact knowledge of the process, of course, makes control more difficult. Optimal control of many processes also requires systems that make use of predictions of future behavior. The mathematical models for the dynamic systems are assumed to be differential equations. The goal of having these models is to capture the dynamics of nonlinear processes, so as to have a way of controlling these dynamics for industrial purpose [7].

0-415-30852-6/04/$0.00+$1.50

We need a mathematical model of the nonlinear dynamic system to understand the dynamics of the processes involved in the evolution of the system. For a specific case, this may require testing several models before obtaining the appropriate mathematical model for the process [9]. For real-world systems with complex dynamics, we may even need several models for different sets of parameter values to represent all of the possible behaviors of the system. A general mathematical model of a dynamic system can be expressed as follows:

$$\begin{aligned} dx/dt &= f_1(x, D, \alpha) - \beta f_2(x, D, \alpha) \\ dp/dt &= \beta f_2(x, D, \alpha) \end{aligned} \tag{1}$$

where $x \in R^n$ is a vector of state variables, $p \in R^m$ is a vector of outputs, $\beta \in R$ is a constant measuring the efficiency of the conversion process, $D \in (0, 3)$ is the fractal dimension of the process, and $\alpha \in R$ is a selection parameter. The fractal dimension is used to characterize the process, for example, in the case of biochemical reactors D represents the fractal dimension of the bacteria used for production [2, 9].

For a complex dynamical system it may be necessary to consider a set of mathematical models to represent adequately all possible dynamic behaviors of the system. In this case, we need a decision scheme to select the appropriate model to use according to the linguistic value of a selection parameter α. We use a new fuzzy inference system for differential equations to achieve fuzzy modeling [3]. We have fuzzy rules of the form:

$$\begin{array}{llllll} \textbf{IF} & \alpha \text{ is } A_1 & \textbf{AND} & D \text{ is } B_1 & \textbf{THEN} & M_1 \\ \dots & & \dots & & \dots & \\ \textbf{IF} & \alpha \text{ is } A_n & \textbf{AND} & D \text{ is } B_n & \textbf{THEN} & M_n \end{array} \tag{2}$$

where $A_1, ..., A_n$ are linguistic values for α, $B_1, ..., B_n$ are linguistic values for the fractal dimension D, and $M_1, ..., M_n$ are mathematical models of the form given by Eq. (1). The selection parameter α can be the temperature for biochemical processes, because temperature changes cause the presence of new bacteria in this case [9]. For the case of aircraft dynamic systems, α can be related to environment parameters.

We combine adaptive model-based control using neural networks with the method for modeling using fuzzy logic and fractal theory, to obtain a new hybrid neuro-fuzzy-fractal method for control of nonlinear dynamic systems. This general method combines the advantages of neural networks (ability for identification and control) with the advantages of fuzzy logic (ability for decision and use of expert knowledge) to achieve the goal of robust adaptive control of nonlinear dynamic systems. We also use the fractal dimension to characterize the processes in modeling these dynamical systems. We have developed intelligent control systems using this new method for adaptive control for several applications, to validate our new approach for control. We have obtained very good results in controlling biochemical reactors and chemical reactors with the hybrid approach for control [8]. In this chapter, we describe the application of our new method to the case of controlling aircraft dynamic systems.

2 FUZZY MODELING OF DYNAMICAL SYSTEMS

For a real-world dynamical system it may be necessary to consider a set of mathematical models to represent adequately all of the possible dynamic behaviors of the system [8, 9].

In this case, we need a fuzzy [15] decision procedure to select the appropriate model to use according to the value of a selection parameter vector α. To implement this decision procedure, we need a fuzzy inference system that can use differential equations as consequents. For this purpose, we have developed a new fuzzy inference system that can be considered as a generalization of Sugeno's inference system [13, 14], in which we are now using differential equations as consequents of the fuzzy rules, instead of simple polynomials like in the original Sugeno's method. Using this method, a fuzzy model for a general dynamical system can be expressed as follows:

$$\begin{array}{llllll}
\textbf{IF} \ \alpha_1 \text{ is } A_{11} & \textbf{AND} \ \alpha_2 \text{ is } A_{12} & \ldots & \textbf{AND} \ \alpha_m \text{ is } A_{1m} & \textbf{THEN} & \frac{dy}{dt} = f_1(y, \alpha) \\
\textbf{IF} \ \alpha_1 \text{ is } A_{21} & \textbf{AND} \ \alpha_2 \text{ is } A_{22} & \ldots & \textbf{AND} \ \alpha_m \text{ is } A_{2m} & \textbf{THEN} & \frac{dy}{dt} = f_2(y, \alpha) \\
\quad \vdots & \quad \vdots & & & & \\
\textbf{IF} \ \alpha_1 \text{ is } A_{n1} & \textbf{AND} \ \alpha_2 \text{ is } A_{n2} & \ldots & \textbf{AND} \ \alpha_m \text{ is } A_{nm} & \textbf{THEN} & \frac{dy}{dt} = f_n(y, \alpha)
\end{array} \tag{3}$$

where A_{ij} is the linguistic value of α_j for rule ith, $\alpha \in R^m$ and is defined by $\alpha = [\alpha_1, \ldots, \alpha_m]$, and $y \in R^p$ is the output obtained by the numerical solution of the corresponding differential equation. Of course, it is assumed that each differential equation in (3) locally approximates the real dynamical system over a neighborhood (or region) of R^m.

The numerical solution of the differential equations can be achieved by the standard Runge–Kutta type method:

$$\begin{aligned}
y_{n+1} &= RK(y_n) + 1/2(k_1 + k_2) \\
k_1 &= hf(y_n, t_n) \\
k_2 &= hf(y_n + k_1, t_{n+1}),
\end{aligned} \tag{4}$$

where h is the step size of the method and RK can be considered as the Runge–Kutta operator that transforms numerical solutions from time n to time $n+1$. Numerical solutions are then aggregated by weighted average with weights obtained by the minimum of the firing strengths of the inputs:

$$y = \frac{w_1 y_1 + w_2 y_2 + \ldots + w_n y_n}{w_1 + w_2 + \ldots + w_n} \tag{5}$$

where: $y_1 = RK(f_1(y, \alpha)), y_2 = RK(f_2(y, \alpha)), \ldots, y_n = RK(f_n(y, \alpha))$.

The new fuzzy inference system for differential equations can be illustrated as in Fig. 1, where a complex dynamical system is modeled by using four different mathematical models (M_1, M_2, M_3, and M_4). The decision scheme can be expressed as a single-input fuzzy model as follows:

$$\begin{array}{llll}
\textbf{IF} & \alpha \text{ is small} & \textbf{THEN} & dy/dt = f_1(y, \alpha) \\
\textbf{IF} & \alpha \text{ is regular} & \textbf{THEN} & dy/dt = f_1(y, \alpha) \\
\textbf{IF} & \alpha \text{ is medium} & \textbf{THEN} & dy/dt = f_1(y, \alpha) \\
\textbf{IF} & \alpha \text{ is large} & \textbf{THEN} & dy/dt = f_1(y, \alpha),
\end{array} \tag{6}$$

where the output y is obtained by the numerical solution of the corresponding differential equation.

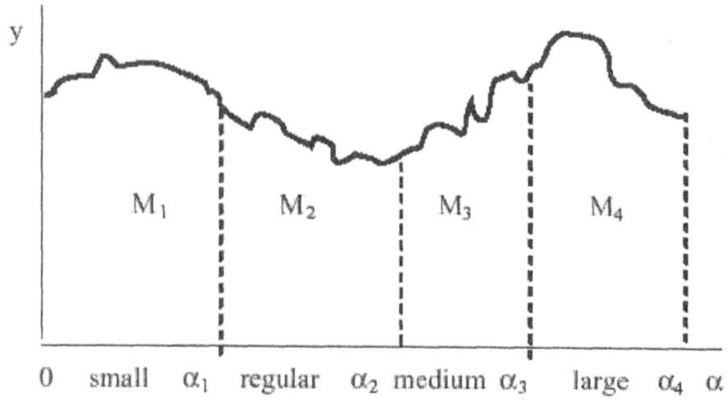

Figure 1 Modeling a complex dynamical system with the new fuzzy system.

3 NEURAL NETWORKS FOR CONTROL

Parametric Adaptive Control is the problem of controlling the output of a system with a known structure but unknown parameters. These parameters can be considered as the elements of a vector p. If p is known, the parameter vector θ of a controller can be chosen as θ^* so that the plant together with the fixed controller behaves like a reference model described by a difference (or differential) equation with constant coefficients [11]. If p is unknown, the vector $\theta(t)$ has to be adjusted on-line using all the available information concerning the system.

Two distinct approaches to the adaptive control of an unknown system are (i) direct control and (ii) indirect control. In direct control, the parameters of the controller are directly adjusted to reduce some norm of the output error. In indirect control, the parameters of the system are estimated as $p(t)$ at any time instant and the parameter vector $\theta(t)$ of the controller is chosen assuming that $p(t)$ represents the true value of the system parameter vector.

When indirect control is used to control a nonlinear system, the plant is parameterized using a mathematical model of the general form described in Section 1 and the parameters of the model are updated using the identification error. The controller parameters in turn are adjusted by backpropagating the error (between the identified model and the reference model outputs) through the identified model. A block diagram of such an adaptive system is shown in Fig. 2.

The overall structure of the adaptive system proposed in this paper to control a nonlinear dynamical system is the same as shown in Fig. 2 and is independent of the specific model used to identify the system. The delayed values of the system input and system output form the inputs to the neural network Nc, which generates the feedback control signal to the system. The parameters of the Neural Network Ni are adjusted by backpropagating the identification error e_i while those of the Neural Network Nc are adjusted by backpropagating the control error (between the output of the reference model and the identification model) through the identification model.

The mathematical model for the nonlinear dynamic system in the time domain is generated by the method of modeling (described in Section 2) using the real data that is measured on-line in the system. On the other hand, the fractal module is used to charac-

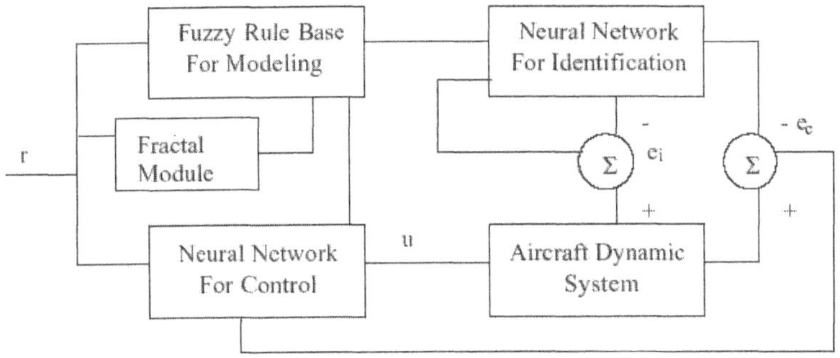

Figure 2 General architecture for adaptive neuro-fuzzy-fractal control.

terize the process and this information is used to specify the mathematical model in the time and space domain. This scheme enables dynamic changes of models according to the changes of on-line process identification. Our new method for adaptive model-based control combining Neural Networks, Fuzzy Logic and Fractal Theory differs from our previous approach of considering only the use of neural networks and models [5, 6].

4 ADAPTIVE CONTROL OF AIRCRAFT SYSTEMS

The mathematical models of aircraft systems can be represented as coupled nonlinear differential equations [10]. In this case, we can develop a fuzzy rule base for modeling that enables the use of the appropriate mathematical model according to the changing conditions of the aircraft and its environment. For example, we can use the following model of an airplane when wind velocity is relatively small:

$$p' = I_1(-q + l), \quad q' = I_2(p + m), \tag{7}$$

where I_1 and I_2 are the inertia moments of the airplane with respect to axis x and y, respectively, l and m are physical constants specific to the airplane, and p, q are the positions with respect to axis x and y, respectively. However, a more realistic model of an airplane in three-dimensional space is as follows:

$$p' = I_1(-qr + l), \quad q' = I_2(pr + m), \quad r' = I_3(-pq + n), \tag{8}$$

where now I_3 is the inertia moment of the airplane with respect to the z axis, n is a physical constant specific to the airplane, and r is the position along the z axis. Now considering wind disturbances in the model, we have the following equation:

$$p' = I_1(-qr + l) - u_g, \quad q' = I_2(pr + m), \quad r' = I_3(-pq + n), \tag{9}$$

where u_g is the wind velocity. The magnitude of wind velocity is dependent on the altitude of the airplane in the following form:

$$u_g = u_{wind510}\left(1 + \frac{\ln(h/510)}{\ln(51)}\right), \tag{10}$$

Table 1 Fuzzy rule base for modeling aircraft dynamic systems

IF		THEN	
Wind velocity	Inertia moment	Fractal dimension	Model
Small	Small	Low	M_1
Small	Small	Medium	M_2
Small	Large	Low	M_2
Small	Large	Medium	M_2
Large	Small	Medium	M_3
Large	Large	Medium	M_3
Large	Large	High	M_3

where $u_{wind510}$ is the wind speed at 510 ft altitude (typical value = 20 ft/sec).

If we use the models of Eqs. (7)–(9) for describing aircraft dynamics, we can formulate a set of rules that relate the models to the conditions of the aircraft and its environment. Let us assume that M_1 is given by Eq. (7), M_2 is given by Eq. (8), and M_3 is given by Eq. (9). Now using the wind velocity u_g and inertia moment I_1 as parameters, we can establish the fuzzy rule base for modeling as in Table 1.

In Table 1, we are assuming that the wind velocity u_g can have only two possible fuzzy values (small and large). This is sufficient to know if we have to use the mathematical model that takes into account the effect of wind (M_3) for u_g large or if we don't need to use it and simply the model M_2 is sufficient (for u_g small). Also, the inertia moment (I_1) helps in deciding between models M_1 and M_2 (or M_3).

5 EXPERIMENTAL RESULTS

To give an idea of the performance of our neuro-fuzzy-fractal approach for adaptive control, we show below simulation results for aircraft dynamic systems. First, we show in Fig. 3(a) the fuzzy rule base for a prototype intelligent system developed in the fuzzy logic toolbox of the MATLAB programming language. We show in Fig. 3(b) the nonlinear surface for the problem of aircraft dynamics using fractal dimension and wind velocity as input variables.

We show simulation results for an aircraft system obtained using our new method for modeling dynamical systems. In Figs. 4(a) and 4(b) we show results for an airplane with inertia moments: $I_1 = 1$, $I_2 = 0.4$, $I_3 = 0.05$ and the constants are: $l = m = n = 1$. The initial conditions are: $p(0) = 0$, $q(0) = 0$, $r(0) = 0$.

To give an idea of the performance of our neuro-fuzzy approach for adaptive model-based control of aircraft dynamics, we show below (Fig. 5) simulation results obtained for the case of controlling the altitude of an airplane for a flight of 6 hours. We assume that the airplane takes about one hour to achieve the cruising altitude 30,000 ft, then cruises along for about three hours at this altitude (with minor fluctuations), and finally descends for about two hours to its final landing point. We will consider the desired trajectory as

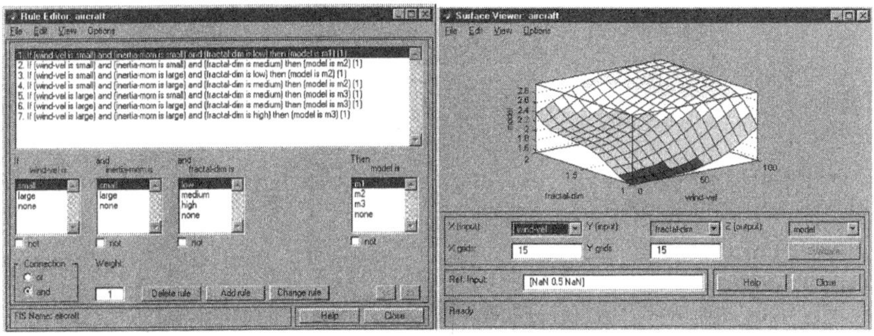

a b

Figure 3 (a) Fuzzy rule base. (b) Nonlinear surface for aircraft dynamics.

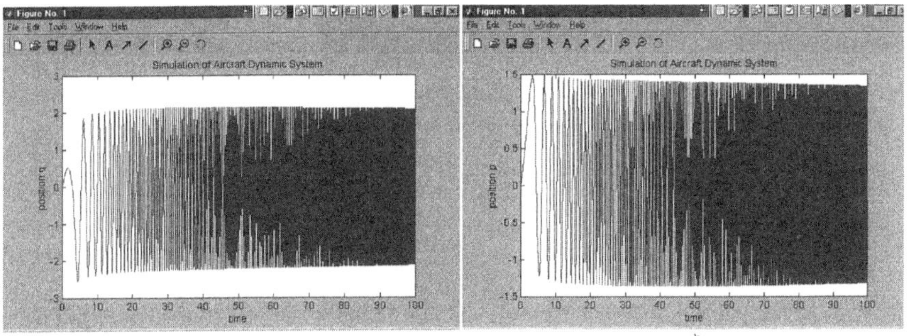

a b

Figure 4 (a) Simulation of position q. (b) Simulation of position p.

follows:

$$h_d = \begin{cases} 30t + \sin 2t & \text{for } 0 \leqslant t \leqslant 1 \\ 30 + 2\sin 10t & \text{for } 1 < t \leqslant 4 \\ 90 - 15t & \text{for } 4 < t \leqslant 6. \end{cases}$$

Of course, a complete desired trajectory for the airplane would have to include the positions for the airplane in the x and y directions. However we think that here for illustration purposes it is sufficient to show the control of the altitude h for the airplane.

We used three-layer neural networks (with 10 hidden neurons) with the Levenberg–Marquardt algorithm and hyperbolic tangent sigmoidal functions as the activation functions for the neurons. We show in Fig. 5 the function approximation achieved by the neural network for control after 800 epochs of training with a variable learning rate. The identification achieved by the neural network (after 800 epochs) can be considered very good because the error has been decreased to the order of 10^{-1}. Still, we can obtain a better approximation by using more hidden neurons or more layers. In any case, we can see clearly (from Fig. 5) how the neural network learns to control the aircraft, because it is able to

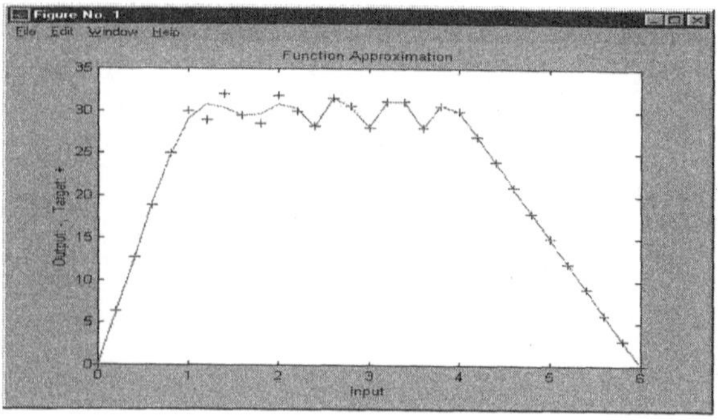

Figure 5 Function approximation of the neural network for control of an airplane.

follow the arbitrary desired trajectory.

We have to mention here that these simulation experiments for the case of a specific flight for a given airplane show very good results. We have also tried our approach for control with other types of flights and airplanes with good results.

6 CONCLUSIONS

We have developed a general method for adaptive model-based control of nonlinear dynamic systems using Neural Networks, Fuzzy Logic, and Fractal Theory. We illustrated our method for control by the case of controlling aircraft dynamics. In this case the models represent the aircraft dynamics during flight. We also described in this chapter an adaptive controller based on the use of neural networks and mathematical models for the system. The proposed adaptive controller performs rather well considering the complexity of the domain being considered in this research work. We have shown that our method can be used to control chaotic and unstable behavior in aircraft systems. Chaotic behavior has been associated with the "flutter" effect in real airplanes, and for this reason it is very important to avoid this kind of behavior. We can say that combining Neural Networks, Fuzzy Logic, and Fractal Theory, using the advantages that each of these methodologies has, can give good results for this kind of application. Also, we believe that our neuro-fuzzy-fractal approach is a good alternative for solving similar problems.

REFERENCES

1. P. Albertos, R. Strietzel, and N. Mart, *Control Engineering Solutions: A Practical Approach*, IEEE Computer Society Press, 1997.
2. O. Castillo and P. Melin, "Developing a new method for the identification of microorganisms for the food industry using the fractal dimension," *Journal of Fractals*, **2**, No. 3 (1994), 457–460.

3. O. Castillo and P. Melin, "A new fuzzy inference system for reasoning with multiple differential equations for modeling complex dynamical systems," in *Proceedings of CIMCA'99*, IOS Press, Vienna, Austria, 1999, pp. 224–229.
4. P. Melin and O. Castillo, "Modeling and simulation for bacteria growth control in the food industry using artificial intelligence," in *Proceedings of CESA'96*, Gerf EC Lille, Lille, France, 1996, pp. 676–681.
5. P. Melin and O. Castillo, "An Adaptive model-based neural network controller for biochemical reactors in the food industry," in *Proceedings of Control'97*, Acta Press, Canada, 1997, pp. 147–150.
6. P. Melin and O. Castillo, "An adaptive neural network system for bacteria growth control in the food industry using mathematical modeling and simulation," in *Proceedings of IMACS World Congress'97*, Berlin, Germany, 1997, Vol. 4, pp. 203–208.
7. P. Melin and O. Castillo, "Automated mathematical modeling and simulation for bacteria growth control in the food industry using artificial intelligence and fractal theory," *Journal of Systems, Analysis, Modeling and Simulation*, Gordon and Breach, 1997, pp. 189–206.
8. P. Melin and O. Castillo, "An adaptive model-based neuro-fuzzy-fractal controller for biochemical reactors in the food industry," in *Proceedings of IJCNN'98*, IEEE Computer Society Press, Anchorage, AK, 1998, Vol. 1 pp. 106–111.
9. P. Melin and O. Castillo, "A new method for adaptive model-based neuro-fuzzy-fractal control of nonlinear dynamic plants: the case of biochemical reactors," in *Proceedings of IPMU'98*, EDK Publishers, Paris, France, 1998, Vol.1, pp. 475–482.
10. P. Melin and O. Castillo, "A new method for adaptive model-based neuro-fuzzy-fractal control of nonlinear dynamical systems," in *Proceedings of the International Conference of Nonlinear Problems in Aviation and Aerospace'98*, European Publications, Daytona Beach, FL, 1998, Vol. 2, pp. 499-506.
11. K.S. Narendra and A.M. Annaswamy, *Stable Adaptive Systems*, Prentice-Hall, Englewood Cliffs, NJ, 1989.
12. S. Rasband, *Chaotic Dynamics of Nonlinear Systems*, John Wiley & Sons, New York, 1990.
13. M. Sugeno and G.T. Kang, "Structure identification of fuzzy model," *Fuzzy Sets and Systems*, **28** (1988), 15–33.
14. T. Takagi and M. Sugeno, "Fuzzy identification of systems and its applications to modeling and control," *IEEE Transactions on Systems, Man and Cybernetics*, **15** (1985), 116–132.
15. L.A. Zadeh, "The concept of a linguistic variable and its application to approximate reasoning," *Information Sciences*, **8** (1975), 43–80.

16

Closed-Form Solution of Three-Dimensional Ideal Proportional Navigation

Chi-Ching Yang† and Hsin-Yuan Chen*

† *Department of Electrical Engineering, Hsiuping Institute of Technology, Taichung, Taiwan*

* *Department of Automatic Control Engineering, Feng Chia University, Taichung, Taiwan*

The main goal of this research is to derive the complete closed-form solutions of three-dimensional ideal proportional navigation (3D IPN) for nonmaneuvering and maneuvering targets. IPN is a guidance law by which the commanded missile acceleration is applied in the direction perpendicular to the relative velocity between the missile and the target, and the acceleration magnitude is proportional to the product of line-of-sight (LOS) rate and relative speed. Three coupled nonlinear second-order dynamic equations modeling the relative motion are solved analytically without any linearization for performance and trajectory analysis. Properties of 3D IPN such that capture region, range-to-go, time-to-go, and two aspect angles within the spherical coordinates are all obtained in closed form.

I INTRODUCTION

Proportional navigation (PN) has been proved to be a useful guidance scheme in many air-to-air and surface-to-air homing systems for interception of airborne targets. Proportional navigation is so named because the magnitude of commanded acceleration is generated proportional to the measured rate of rotation of the missile-target line of sight (LOS). In traditional proportional navigation, pure proportional navigation (PPN) and true proportional navigation (TPN) have been widely discussed and thoroughly developed for homing systems of missiles. The earlier study started from Murtaugh and Criel [15]. During the three decades that followed, researchers such as Guelman [9, 10, 11, 12], Yang and Yeh [21–22], Yang and Yang [27], Shukla and Mahapatra [18, 19], Mahapatra and Shukla [14], Becker [2], Yuan and Chern [28–32], Dhar and Ghose [5], Ghose [6,7,8], Rao [16], Vathsal and Rao [19], have explored proportional navigation in many different ways, such as true proportional navigation (TPN), pure proportional navigation (PPN), generalized proportional navigation (GPN), realistic true proportional navigation (RTPN), and ideal proportional navigation (IPN), etc.

Because of the difficulties in analysis, most analytical investigations on PN guidance laws in the past used two-dimensional (2D) models, although practical pursuit-evasion motion occurred in a three-dimensional (3D) environment. As to 3D studies, earlier work was done by Adler [1]. The main viewpoint of his study was that the 3D relative motion can be

0-415-30852-6/04/$0.00+$1.50

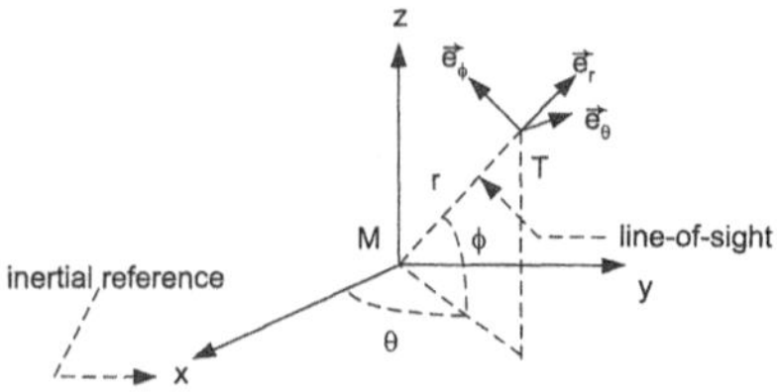

Figure 1 Three-dimensional pursuit geometry.

described by the principal plane defined by the instantaneous line of sight (LOS) and by the missile's velocity. Shinar et al. [17] used the linearized 3D model to analyze guidance laws. Guelman [13] extended the optimal guidance law in the plane to 3D models and obtained a preliminary version of 3D optimal guidance law. Cochran et al. [4], considered the 3D RTPN guidance problem, and revealed a partial solution. More recently, Yang and Yang [24–26] proposed a systematic solution technique for 3D guidance laws and the technique has been applied to generalized 3D PN, 3D RTPN, and 3D TPN problems, where complete solutions of the three coupled nonlinear differential equations within spherical coordinates were obtained without any linearization.

The discussion of 2D PN guidance law had used moving polar coordinates wherein the origin is fixed on the missile to characterize the relative motion. In 3D models, seeker measurements are spherical in nature [3] (range and angles), and are, therefore, nonlinear functions of the states in Cartesian coordinates. The nonlinear transformation of states could be avoided if the guidance laws were formulated in spherical coordinates. Thus, moving spherical coordintes seem to be most natural in characterizing 3D relative motions.

We will use spherical coordinates to describe the relative motion of a target with respect to the missile, which is guided by 3D IPN. The strategy of IPN guidance law is to turn the relative velocity in most efforts to the direction of LOS. In the IPN scheme the commanded acceleration is applied in the direction normal to the relative velocity between the missile and the target, and its magnitude is proportional to the product of the LOS rate and the relative velocity. Planar motion using IPN has been studied by Yuan and Chern (30). It was reported that with some more energy consumption, the IPN guidance law has a larger capture area and is much more effective than the other schemes.

The 3D relative motion resulting from the 3D IPN guidance law is governed by three second-order nonlinear differential equations whose solution has never been discussed before. Based on the systematic framework [24–26] for analyzing 3D guidance laws, we have made a breakthrough in obtaining the analytical solutions of these coupled nonlinear differential equations. The solutions are derived in terms of the unit angular momentum defined by the relative motion between the missile and the target. Unit angular momentum in the space is a performance index measuring the departure tendency of the 3D relative motion from a fixed plane. If the relative motion occurs wholly within a fixed plane, then the direction of the relative angular momentum will remain constant during the course of interception. Besides as an index for measuring 3D relative motions, unit angular momentum has a decoupling effect such that the relative distance and azimuths can be solved independently.

A systematic framework for analyzing 3D guidance laws is introduced in Section II.

3D IPN is defined in Section III where three nonlinear differential equations describing the relative motion with respect to a nonmaneuvering target are solved analytically, and the trajectory properties such as capture region, time-to-go, range-to-go, and aspect angles are all derived in closed forms. In Section IV, we solve the two-player-game problem wherein a missile guided by 3D IPN is designed to pursue a target guided by 3D RTPN, and vice versa. Numerical results and more physical insight about 3D IPN guidance law are discussed in Section V.

II GENERAL ANALYSIS OF 3D RELATIVE MOTION

Consider the spherical coordinates (r, θ, ϕ) with origin fixed at the missile, where r is the relative distance between the missile and the target, and θ and ϕ are azimuths. Let ($\vec{e}_r$, $\vec{e}_\theta$, $\vec{e}_\phi$) be unit vectors along the coordinate axes (see Fig. 1). According to the principles of kinematics, the three relative acceleration components (a_r, a_θ, a_ϕ) can be expressed by the following set of second-order nonlinear differential equations:

$$\ddot{r} - r\dot{\phi}^2 - r\dot{\theta}^2\cos^2\phi = a_{T_r} - a_{M_r} \equiv a_r \tag{1a}$$

$$r\ddot{\theta}\cos\phi + 2\dot{r}\dot{\theta}\cos\phi - 2r\dot{\phi}\dot{\theta}\sin\phi = a_{T_\theta} - a_{M_\theta} \equiv a_\theta \tag{1b}$$

$$r\ddot{\phi} + 2\dot{r}\dot{\phi} + r\dot{\theta}^2\cos\phi\sin\phi = a_{T_\phi} - a_{M_\phi} \equiv a_\phi, \tag{1c}$$

where a_{T_r},a_{T_θ}, and a_{T_ϕ} are the acceleration components of the target; a_{M_r},a_{M_θ}, and a_{M_ϕ} are the acceleration components of the missile. To analyze these coupled nonlinear equations, we find that the exploiting of the angular momentum of unit mass is very helpful. The unit angular momentum $\vec{h}$ for the missile-target relative motion is defined as

$$\vec{h} = \vec{r} \times \dot{\vec{r}}, \tag{2}$$

where $\vec{r}$ is the relative displacement along the line of sight (LOS), and $\dot{\vec{r}}$ is the relative velocity. Accordingly, if the relative motion occurs within a fixed plane, i.e., $\vec{r}$ and $\dot{\vec{r}}$ are in the same plane during the interception, then the direction of $\vec{h}$ will be constant, always being perpendicular to the plane spanned by $\vec{r}$ and $\dot{\vec{r}}$. Hence, the variation of the direction of $\vec{h}$ is a natural measure for departure tendency of the relative motion from a fixed plane.

$\vec{r}$ and $\dot{\vec{r}}$ can be expressed by the spherical unit vectors $(\vec{e}_r, \vec{e}_\theta, \vec{e}_\phi)$ as

$$\vec{r} = r\vec{e}_r \tag{3a}$$

$$\dot{\vec{r}} = \dot{r}\vec{e}_r + r\dot{\theta}\cos\phi\vec{e}_\theta + r\dot{\phi}\vec{e}_\phi. \tag{3b}$$

Substituting Eq. (3) into Eq. (2) yields the expression for $\vec{h}$:

$$\vec{h} = h\vec{e}_h = r^2(-\dot{\phi}\vec{e}_\theta + \dot{\theta}\cos\phi\vec{e}_\phi), \tag{4a}$$

where

$$h = r^2\sqrt{\dot{\phi}^2 + \dot{\theta}^2\cos^2\phi} \tag{4b}$$

is the magnitude of $\vec{h}$, and

$$\vec{e}_h = \frac{r^2}{h}(-\dot{\phi}\vec{e}_\theta + \dot{\theta}\cos\phi\vec{e}_\phi) \tag{4c}$$

is the unit vector along the direction of $\vec{h}$. According to the definition of $\vec{e}_h$, we know that $\vec{e}_h$ is perpendicular to $\vec{e}_r$. We can find another unit vector $\vec{e}_h^{\perp}$ that is perpendicular to both $\vec{e}_r$ and $\vec{e}_h$. It is straightforward to verify that

$$\vec{e}_h^{\perp} = \frac{r^2}{h}(\dot{\theta}\cos\phi\vec{e}_\theta + \dot{\phi}\vec{e}_\phi). \tag{5}$$

The set of unit vectors $(\vec{e}_r, \vec{e}_h, \vec{e}_h^{\perp})$ constitutes a new moving coordinate system that is more convenient for describing 3D guidance laws than the conventional spherical coordinates $(\vec{e}_r, \vec{e}_\theta, \vec{e}_\phi)$, as will be seen later.

We now proceed to derive the various physical components under the new coordinate system $(\vec{e}_r, \vec{e}_h, \vec{e}_h^{\perp})$. The differentiation of Eq. (2) gives

$$\frac{d\vec{h}}{dt} = \vec{r} \times \ddot{\vec{r}}, \tag{6}$$

where $\ddot{\vec{r}}$ is the relative acceleration defined by

$$\ddot{\vec{r}} = a_r\vec{e}_r + a_\theta\vec{e}_\theta + a_\phi\vec{e}_\phi. \tag{7}$$

Using Eq. (3a) and Eq. (7), the right-hand side of Eq. (6) can be rewritten as

$$\vec{r} \times \ddot{\vec{r}} = (-ra_\phi)\vec{e}_\theta + (ra_\theta)\vec{e}_\phi, \tag{8}$$

and the left-hand side of Eq. (6) becomes

$$\frac{d}{dt}(\vec{h}) = \dot{h}\vec{e}_h + h\dot{\vec{e}}_h, \tag{9}$$

where $\dot{h}$ and $\dot{\vec{e}}_h$ can be obtained from the differentiation of Eq. (4b) and Eq. (4c) with respect to time t. Upon differentiating h and $\vec{e}_h$, we need the relations :

$$\dot{\vec{e}}_r = \vec{\omega} \times \vec{e}_r, \quad \dot{\vec{e}}_\theta = \vec{\omega} \times \vec{e}_\theta, \quad \dot{\vec{e}}_\phi = \vec{\omega} \times \vec{e}_\phi,$$

where $\vec{\omega}$ is the angular velocity of the moving coordinates $(\vec{e}_r, \vec{e}_\theta, \vec{e}_\phi)$, which can be derived as

$$\vec{\omega} = \dot{\theta}\sin\phi\vec{e}_r - \dot{\phi}\vec{e}_\theta + \dot{\theta}\cos\phi\vec{e}_\phi. \tag{10}$$

$\dot{h}$ and $\dot{\vec{e}}_h$ are then derived as follows:

$$\dot{h} = \frac{2r\dot{r}(\dot{\phi}^2 + \dot{\theta}^2\cos^2\phi) + r^2(\dot{\theta}\ddot{\theta}\cos^2\phi - \dot{\phi}\dot{\theta}^2\sin\phi\cos\phi + \dot{\phi}\ddot{\phi})}{\sqrt{\dot{\phi}^2 + \dot{\theta}^2\cos^2\phi}} \tag{11a}$$

$$\begin{aligned}\dot{\vec{e}}_h = {} & \frac{1}{\sqrt{\dot{\phi}^2 + \dot{\theta}^2\cos^2\phi}} \\ & \times \left[\left(-\ddot{\phi} + \frac{\dot{\phi}(\dot{\phi}\ddot{\phi} + \dot{\theta}\ddot{\theta}\cos^2\phi - \dot{\theta}^2\dot{\phi}\sin\phi\cos\phi)}{\dot{\phi}^2 + \dot{\theta}^2\cos^2\phi} - \dot{\theta}^2\sin\phi\cos\phi\right)\vec{e}_\theta\right. \\ & \left. + \left(\ddot{\theta}\cos\phi - \dot{\theta}\dot{\phi}\sin\phi - \frac{\dot{\theta}\cos\phi(\dot{\phi}\ddot{\phi} + \dot{\theta}\ddot{\theta}\cos^2\phi - \dot{\theta}^2\dot{\phi}\sin\phi\cos\phi)}{\dot{\phi}^2 + \dot{\theta}^2\cos^2\phi} - \dot{\theta}\dot{\phi}\sin\phi\right)\vec{e}_\phi\right].\end{aligned} \tag{11b}$$

Equation (11) is substituted into Eq. (9) to obtain

$$\frac{d}{dt}(\vec{h}) = -r(r\ddot{\theta}\cos\phi + 2\dot{r}\dot{\theta}\cos\phi - 2r\dot{\phi}\dot{\theta}\sin\phi)\vec{e}_\theta + r(r\ddot{\phi} + 2\dot{r}\dot{\phi} + r\dot{\theta}^2\cos\phi\sin\phi)\vec{e}_\phi. \tag{12}$$

If we equate Eq. (8) with Eq. (12), we go back to the results of Eq. (1b) and Eq. (1c); however, this fact manifests that the roles of Eq. (1b) and Eq. (1c) can be replaced by the roles of Eq. (11) in describing the relative motion. Equation (11) can be further simplified to the following form:

$$\begin{aligned}\dot{h} &= \vec{r}\cdot(\ddot{\vec{r}}\times\vec{e}_h)\\ &= \frac{r^3}{h}\left(a_\phi\dot{\phi} + a_\theta\dot{\theta}\cos\phi\right) \qquad \text{(13a)}\\ \dot{\vec{e}}_h &= \frac{r}{h}(\ddot{\vec{r}}\cdot\vec{e}_h)\vec{e}_h^{\perp}\\ &= \frac{r^3}{h^2}(a_\theta\dot{\phi} - a_\phi\dot{\theta}\cos\phi)\vec{e}_h^{\perp}. \qquad \text{(13b)}\end{aligned}$$

The remaining equation describing the relative motion comes from Eq. (1a), which can be rewritten in terms of h as

$$\ddot{r} - \frac{h^2}{r^3} = a_r. \tag{13c}$$

Equation (13) is equivalent to Eq. (1), but Eq. (13) has the advantage of decoupling the radial motion from the tangential motion as can be seen in the following sections.

III 3D IPN WITH NONMANEUVERING TARGETS

Ideal proportional navigation is a guidance law whose commanded acceleration is applied in the direction normal to the relative velocity ($\dot{\vec{r}}$) between the missile and its target, and its magnitude is proportional to the product of the LOS rate and relative speed. A 2D version of IPN has been proposed by Yuan and Chern [26]. Commanded acceleration for a 3D IPN can be expressed in the following vector form:

$$\vec{a}_M = \lambda\dot{\vec{r}}\times\vec{\Omega}. \tag{14a}$$

The angular velocity $\vec{\Omega}$ of LOS can be found from the relation

$$\begin{aligned}\vec{\Omega} &= \frac{\vec{r}\times\dot{\vec{r}}}{r^2} = \frac{\vec{h}}{r^2}\\ &= -\dot{\phi}\vec{e}_\theta + \dot{\theta}\cos\phi\vec{e}_\phi, \qquad \text{(14b)}\end{aligned}$$

where Eq. (4a) has been used to obtain Eq. (14b). Therefore, missile acceleration becomes

$$\begin{aligned}\vec{a}_M &= \lambda\dot{\vec{r}}\times\vec{\Omega}\\ &= \lambda\dot{\vec{r}}\times\frac{\vec{h}}{r^2}\\ &= \lambda(\frac{h^2}{r^3}\vec{e}_r - \dot{r}\frac{h}{r^2}\vec{e}_h^{\perp}). \qquad \text{(15)}\end{aligned}$$

In this section a nonmaneuvering target is assumed, i.e., $\vec{a}_T = 0$. After substituting Eq. (15) into Eq. (13), we have

$$\ddot{r} - \frac{h^2}{r^3} = -\lambda \frac{h^2}{r^3} \tag{16a}$$

$$\dot{h} = \lambda h \frac{\dot{r}}{r} \tag{16b}$$

$$\dot{\vec{e}}_h = 0. \tag{16c}$$

Equation (16a) and Eq. (16b) corresponding to radial motion can be solved together, leading to the following results:

$$h = h_0 \left(\frac{r}{r_0} \right)^{\lambda} \tag{17}$$

$$\dot{r} = -\sqrt{\dot{r}_0^2 + \frac{h_0^2}{r_0^2} - \frac{h_0^2}{r_0^{2\lambda}} r^{2(\lambda-1)}} \,. \tag{18}$$

Combining Eq. (14b) and Eq. (17) yields

$$\Omega = \frac{h}{r^2} = \frac{h_0}{r_0^2} \left(\frac{r}{r_0} \right)^{\lambda-2} . \tag{19}$$

In order to maintain finite commanded acceleration, the navigation constant λ must be greater than 2. It is seen from Eq. (18) and Eq. (19) that the capture area for the IPN is simply $\lambda \geq 2$. That is, no matter what the initial condition of $\dot{r}$ is, the interception can always be achieved successfully when $\lambda \geq 2$.

Since the commanded acceleration (15) is normal to $\dot{\vec{r}}$, we can show that the relative speed $V(t) = |\dot{\vec{r}}|$ remains constant during the interception. This can be checked by using the definition of $\dot{\vec{r}}$ from Eq. (3b):

$$V(t) = |\dot{\vec{r}}| = \sqrt{\dot{r}^2 + \frac{h^2}{r^2}} .$$

Substituting Eq. (17) and Eq. (18) into the above equation, yields

$$V(t) = \sqrt{\dot{r}_0^2 + \frac{h_0^2}{r_0^2}} = V_0 = \text{constant}. \tag{20}$$

Equation (1) consists of three second-order differential equations, and six initial conditions shall be assigned to determine the solution uniquely. These six initial conditions are chosen as r_0, $\dot{r}_0$, θ_0, $\dot{\theta}_0$, ϕ_0, and $\dot{\phi}_0$. In general, we can choose the inertia reference line as the initial LOS such that $\theta_0 = \phi_0 = 0$. In order to have the analytical solution independent of the initial conditions, a nondimensionalized process is required before we proceed further. Dimensionless variables are defined as follows:

$$\rho = \frac{r}{r_0}, \quad \tau = \frac{t}{r_0/V_0}, \quad \bar{h} = \frac{h}{r_0 V_0}. \tag{21}$$

Using these dimensionless variables, Eq. (17) and Eq. (18) can be rewritten as

$$\bar{h} = \bar{h}_0 \rho^{\lambda} \tag{22}$$

$$d\tau = \frac{-d\rho}{\sqrt{1 - \bar{h}_0^2 \rho^{2\lambda-2}}}. \tag{23}$$

Equation (23) is integrated to obtain the time history of the relative distance ρ:

$$\tau = -\int_1^{\rho} \frac{d\rho}{\sqrt{1 - \bar{h}_0^2 \rho^{2\lambda-2}}}, \tag{24}$$

and the duration of interception can be found as

$$\tau_f = \int_0^{1} \frac{d\rho}{\sqrt{1 - \bar{h}_0^2 \rho^{2\lambda-2}}}. \tag{25}$$

The commanded missile acceleration can be calculated from Eq. (15) as

$$\begin{aligned} a_M &= \lambda \sqrt{\left(\frac{h^2}{r^3}\right)^2 + \left(\frac{\dot{r}h}{r^2}\right)^2} \\ &= \lambda \frac{h}{r^2} \sqrt{\frac{h^2}{r^2} + \dot{r}^2} \\ &= \lambda V_0 \frac{h}{r^2}. \end{aligned}$$

The dimensionless form becomes

$$\frac{a_M}{V_0^2 / r_0} = \lambda \bar{h}_0 \rho^{\lambda-2}. \tag{26}$$

The total cumulative velocity increment ΔV of the missile during interception is defined as

$$\Delta V = \int_0^{\tau_f} \left| \frac{a_M}{V_0^2 / r_0} \right| d\tau. \tag{27}$$

Substituting Eq. (23) and Eq. (27) into Eq. (28), we have

$$\Delta V = \int_0^{1} \frac{\lambda \bar{h}_0 \rho^{\lambda-2} d\rho}{\sqrt{1 - \bar{h}_0^2 \rho^{2(\lambda-1)}}}. \tag{28}$$

There are three time functions $\rho(t)$, $\theta(t)$, and $\phi(t)$ needed to describe the 3D relative motion within the spherical coordinates. Up to this stage, only relative distance $\rho(t)$ for radial motion is obtained. The following work is to find the azimuths $\theta(t)$ and $\phi(t)$ for tangential motion. This step is harder than the solution of r. The attack on the problem starts from Eq. (16c). Substituting Eq. (11b) into Eq. (16c) and using the transformation between the two coordinate systems $(\vec{e}_r, \vec{e}_\theta, \vec{e}_\phi)$ and $(\vec{e}_r, \vec{e}_h, \vec{e}_{h}^{\perp})$, we obtain the key equation characterizing the behavior of $\theta(t)$ and $\phi(t)$ as

$$\frac{\ddot{\theta}\cos\phi - \dot{\theta}\dot{\phi}\sin\phi}{-\dot{\phi}} + \frac{\dot{\theta}\cos\phi(\ddot{\theta}\dot{\theta}\cos^2\phi - \dot{\phi}\dot{\theta}^2\sin\phi\cos\phi + \dot{\phi}\ddot{\phi})}{\dot{\phi}(\dot{\phi}^2 + \dot{\theta}^2\cos^2\phi)} + \dot{\theta}\sin\phi = 0. \quad (29)$$

This equation can be simplified to a totally differentiable form:

$$\frac{d(\dot{\theta}\cos\phi)}{-\dot{\theta}\cos\phi} + \frac{d(\dot{\phi}^2 + \dot{\theta}^2\cos^2\phi)}{2(\dot{\phi}^2 + \dot{\theta}^2\cos^2\phi)} + \tan\phi d\phi = 0, \quad (30)$$

and the integration gives

$$\frac{\dot{\phi}^2 + \dot{\theta}^2\cos^2\phi}{\dot{\theta}^2\cos^4\phi} = \frac{\dot{\phi_0}^2 + \dot{\theta_0}^2\cos^2\phi_0}{\dot{\theta_0}^2\cos^4\phi_0} = l^2, \quad (31)$$

where l is an integration constant. The above equation can be solved for $\dot{\phi}^2$ in terms of $\dot{\theta}^2$:

$$\left(\frac{d\phi}{d\tau}\right)^2 = (l^2\cos^4\phi - \cos^2\phi)\left(\frac{d\theta}{d\tau}\right)^2. \quad (32)$$

The other relation between $d\phi/d\tau$ and $d\theta/d\tau$ comes from Eq. (4b):

$$\bar{h}^2 = \rho^4\left[\left(\frac{d\phi}{d\tau}\right)^2 + \left(\frac{d\theta}{d\tau}\right)^2\cos^2\phi\right]. \quad (33)$$

Substituting Eq. (22) and Eq. (32) into Eq. (33) yields

$$\left(\frac{d\theta}{d\tau}\right)^2 = \left(\frac{\bar{h}_0}{l}\right)^2\rho^{2\lambda}(\rho\cos\phi)^{-4}, \quad (34)$$

which, in turn, is substituted into Eq. (32) to obtain the expression for $d\phi/d\tau$ as

$$\frac{d\phi}{d\tau} = \mathrm{sign}(\dot{\phi}_0)\bar{h}_0\rho^{\lambda-2}\sqrt{1 - l^{-2}\cos^{-2}\phi}. \quad (35)$$

The relation between ϕ and ρ can be obtained by integrating Eq. (35) with the help of Eq. (23):

$$\int_0^\phi \frac{d\phi}{\sqrt{1 - l^{-2}\cos^{-2}\phi}} = \mathrm{sign}(\dot{\phi}_0)\int_\rho^1 \frac{\bar{h}_0\rho^{\lambda-2}d\rho}{\sqrt{1 - \bar{h}_0^2\rho^{2\lambda-2}}}. \quad (36)$$

The remaining step is to relate θ to ϕ by integrating Eq. (32):

$$\theta = \mathrm{sign}\left(\frac{\dot{\theta}_0}{\dot{\phi}_0}\right)\int_0^\phi \frac{d\phi}{\cos^2\phi\sqrt{l^2 - \cos^{-2}\phi}}. \quad (37)$$

It can be observed from Eq. (35) and Eq. (37) that both $\theta(\tau)$ and $\phi(\tau)$ are monotonically increasing when $\dot{\theta}_0 > 0$ and $\dot{\phi}_0 > 0$, while both $\theta(\tau)$ and $\phi(\tau)$ are monotonically decreasing when $\dot{\theta}_0 < 0$ and $\dot{\phi}_0 < 0$. Without loss of generality, we will assume $\dot{\theta}_0 > 0$ and $\dot{\phi}_0 > 0$ in the following discussion. These initial conditions can be naturally achieved by choosing the inertia reference line as the initial LOS and by defining the positive directions of θ and ϕ as the directions of $\dot{\theta}_0$ and $\dot{\phi}_0$.

To express the solutions more explicitly, we find that the introduction of the following auxiliary azimuths ψ and η is very helpful. The first auxiliary azimuth ψ is defined as

$$\frac{\cos\psi}{\cos\psi_0} = \rho^{\lambda-1}, \tag{38}$$

where $\psi_0 = \cos^{-1}\bar{h}_0$. It is noted that according to the definition

$$\bar{h}_0 = \frac{h_0}{r_0 V_0} = \frac{h_0}{\sqrt{h_0^2 + r_0^2 \dot{r}_0^2}}, \tag{39}$$

we have $0 \le \bar{h}_0 \le 1$. The second auxiliary azimuth η is defined as

$$\frac{\cos\eta}{\cos\eta_0} = \frac{1}{\cos\phi}, \tag{40}$$

where $\cos\eta_0 = 1/l = \dot{\theta}_0/\sqrt{\dot{\theta}_0^2 + \dot{\phi}_0^2} \le 1$. Instead of the time-to-go τ and the range-to-go ρ, we find that ψ and η are more appropriate to use as independent variables to describe the 3D relative motion. $\bar{h}$, ρ, τ, ϕ, and θ can all be expressed by ψ and η. From Eq. (22) and Eq. (38), we have

$$\bar{h} = \bar{h}_0 \left(\frac{\cos\psi}{\cos\psi_0}\right)^{\frac{\lambda}{\lambda-1}}. \tag{41}$$

The relation between time τ and ψ is established by substituting Eq. (38) into Eq. (24). After some manipulations, we have

$$\tau = \frac{1}{(\lambda-1)\cos\psi_0} \int_{\psi_0}^{\psi} \left(\frac{\cos\psi}{\cos\psi_0}\right)^{\frac{2-\lambda}{\lambda-1}} d\psi. \tag{42}$$

The required time of interception is obtained by setting $\psi = \pi/2$ (i.e., $\rho = 0$) in the upper limit of the above integration:

$$\tau_f = \frac{1}{(\lambda-1)\cos\psi_0} \int_{\psi_0}^{\pi/2} \left(\frac{\cos\psi}{\cos\psi_0}\right)^{\frac{2-\lambda}{\lambda-1}} d\psi. \tag{43}$$

When the initial closing rate $\dot{r}_0$ is zero, we have $\bar{h}_0 = 1$, i.e., $\psi_0 = 0$. In this case, τ_f has a cleaner expression:

$$\tau_f = \frac{\sqrt{\pi}}{2\lambda-2} \frac{\Gamma\left(\dfrac{1}{2\lambda-2}\right)}{\Gamma\left(\dfrac{\lambda}{2\lambda-2}\right)}, \tag{44}$$

where $\Gamma(\cdot)$ is gamma function. ϕ is related to ψ by substituting Eq. (38) into Eq. (36), leading to the following identity:

$$\psi - \psi_0 = (\lambda - 1) \int_0^{\phi} \frac{d\phi}{\sqrt{1 - l^{-2} \cos^{-2} \phi}}. \tag{45a}$$

The integration involved in Eq. (45a) can be evaluated in terms of elementary functions as

$$\phi = \sin^{-1} \left[\sqrt{1 - l^{-2}} \sin \left(\frac{\psi - \psi_0}{\lambda - 1} \right) \right]. \tag{45b}$$

On the other hand, θ is related to η by substituting Eq. (40) into Eq. (37).

$$\begin{aligned} \theta &= -\int_{\eta_0}^{\eta} \frac{\cos \eta d\eta}{\sqrt{\cos^2 \eta - l^{-2}}} \\ &= \cos^{-1} \left(\frac{\sin \eta}{\sin \eta_0} \right). \end{aligned} \tag{46}$$

Finally, the total cumulative velocity increment ΔV shown in Eq. (29) can be integrated to the following form:

$$\Delta V = \frac{\lambda}{\lambda - 1} \left(\frac{\pi}{2} - \psi_0 \right). \tag{47}$$

The auxiliary azimuth η has a concrete physical meaning. Combining Eq. (31) and Eq. (40), we can show that

$$\sin \eta = \frac{\dot{\phi}}{\sqrt{\dot{\phi}^2 + \dot{\theta}^2 \cos^2 \phi}}, \qquad \cos \eta = \frac{\dot{\theta} \cos \phi}{\sqrt{\dot{\phi}^2 + \dot{\theta}^2 \cos^2 \phi}}. \tag{48}$$

Substituting the above equation into Eq. (4c) yields

$$\vec{e}_h = -\sin \eta \, \vec{e}_\theta + \cos \eta \, \vec{e}_\phi. \tag{49}$$

This reveals that η is the angle between $\vec{e}_h$ and $\vec{e}_\phi$ as shown in Fig. 2. As mentioned earlier, $\vec{e}_h$ is normal to the instantaneous plane spanned by $\vec{r}$ and $\dot{\vec{r}}$. Consequently, when the relative motion remains within a fixed plane in the 3D space, $\vec{e}_h$ will not change. This is what we have derived in Eq. (16c). In the following we will show that when a nonmaneuvering target is pursued by a missile guided by the 3D IPN guidance law, the whole relative motion indeed occurs within a fixed plane, but this plane may not be limited to the horizontal $x - y$ plane. To show $\vec{e}_h$ is fixed in the 3D space, the most natural way is to express $\vec{e}_h$ in terms of the inertia coordinates $(\vec{i}, \vec{j}, \vec{k})$, and to show that the components of $\vec{e}_h$ in the $\vec{i}$, $\vec{j}$, and $\vec{k}$ directions are all constant.

Exploiting the transformation between the spherical coordinates and the inertia Cartesian coordinates,

$$\begin{aligned} \vec{e}_\theta &= -\sin \theta \, \vec{i} + \cos \theta \, \vec{j} \\ \vec{e}_\phi &= -\sin \phi \cos \theta \, \vec{i} - \sin \phi \sin \theta \, \vec{j} + \cos \phi \, \vec{k}, \end{aligned}$$

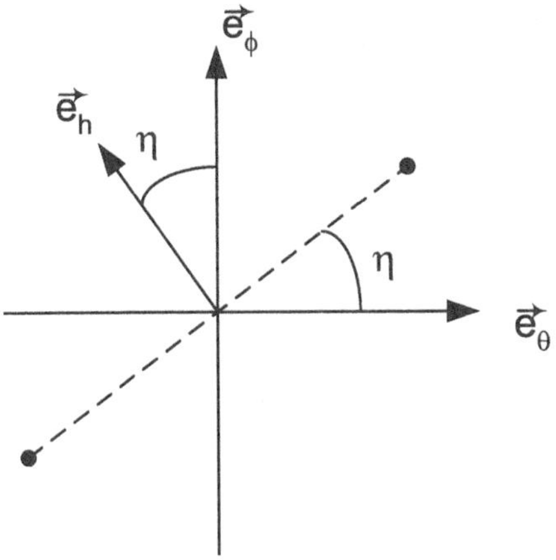

Figure 2 The direction of $\vec{e}_h$.

we can express $\vec{e}_h$ of Eq. (49) in terms of the inertia Cartesian coordinates as

$$\vec{e}_h = (\sin\eta \sin\theta - \cos\eta \sin\phi \cos\theta)\,\vec{\imath} + (-\sin\eta\cos\theta - \cos\eta \sin\phi \sin\theta)\,\vec{\jmath} + (\cos\eta\cos\phi)\,\vec{k}. \tag{50}$$

The simplification of Eq. (50) needs two relations coming from Eq. (40) and Eq. (46):

$$\cos\theta = \frac{\sin\eta}{\sin\eta_0}, \quad \sin\theta = \cot\eta_0 \tan\phi. \tag{51}$$

Using Eq. (40) and Eq. (51), we can establish the following identities:

$$\sin\eta\sin\theta - \cos\eta\sin\phi\cos\theta = 0 \tag{52a}$$

$$-\sin\eta\cos\theta - \cos\eta\sin\phi\sin\theta = -\sin\eta_0 \tag{52b}$$

$$\cos\eta\cos\phi = \cos\eta_0. \tag{52c}$$

Therefore, Eq. (50) becomes

$$\vec{e}_h = -\sin\eta_0\,\vec{\jmath} + \cos\eta_0\,\vec{k}. \tag{53}$$

As expected, $\vec{e}_h$ is fixed in the inertia coordinates and the orientation of the fixed plane where the relative motion occurs is determined by η_0, which is defined by the initial angular rates as

$$\eta_0 = \cos^{-1}\sqrt{\frac{\dot\theta_0^2}{\dot\theta_0^2 + \dot\phi_0^2}}.$$

Since we have chosen $\theta_0 = 0$ and $\phi_0 = 0$, the moving spherical coordinates coincide with the inertia Cartesian coordinates at the start of the interception, i.e., $\vec{e}_r = \vec{\imath}$, $\vec{e}_\theta = \vec{\jmath}$, and $\vec{e}_\phi = \vec{k}$. Hence, at the beginning of interception Eq. (49) is reduced to Eq. (53). As the

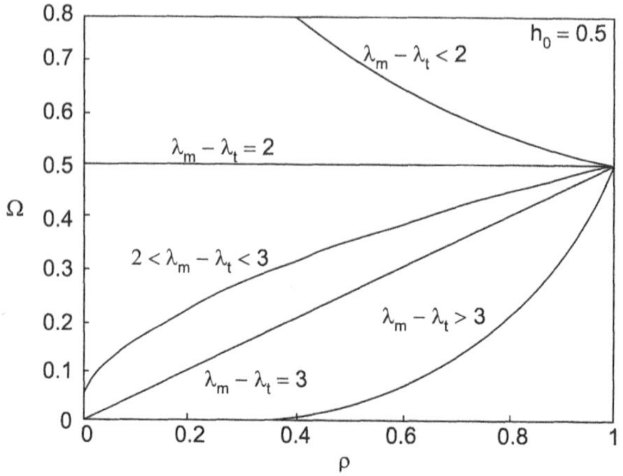

Figure 3 Angular rate for line of sight.

interception proceeds, η changes from η_0 and the spherical coordinates $(\vec{e}_r, \vec{e}_\theta, \vec{e}_\phi)$ rotate continuously in such a way that the orientation of $\vec{e}_h = -\sin\eta\,\vec{e}_\theta + \cos\eta\,\vec{e}_\phi$ keeps fixed in the space.

The physical meaning of the auxiliary azimuth ψ can be revealed by combining Eq. (19) and Eq. (38). It is easy to show that

$$\text{tangential speed} = V_0 \frac{\bar{h}}{\rho} = V_0 \cos\psi \tag{54a}$$

$$\text{radial speed} = V_0 \frac{d\rho}{d\tau} = V_0 \sin\psi. \tag{54b}$$

Therefore, ψ is the angle between the tangential relative velocity and the total relative velocity, as shown in Fig. 4.

IV NUMERICAL RESULTS

The analysis of the 3D IPN guidance law can be greatly simplified by using the above explicit formulae. Some inherent properties of the 3D IPN, gained from numerical results, are discussed in the following.

Influence of Navigation Constant

Figures 3 to 6 show the variations of Ω, $d\rho/d\tau$, ϕ, and θ with respect to the dimensionless relative distance ρ for various values of navigation constant λ_M and for various initial conditions. The relative motions can be divided into three categories according to the value of λ_M:

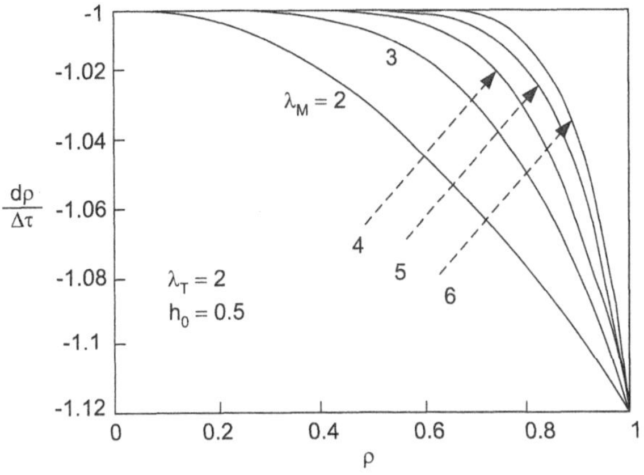

Figure 4 Closing speed for 3D IPN

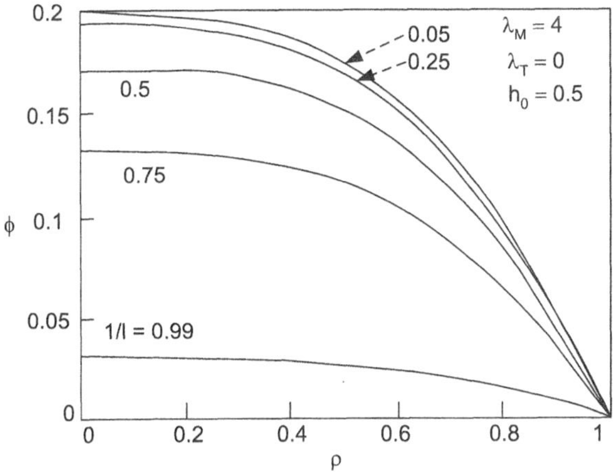

Figure 5 Azimuth ϕ for 3D IPN with varying l.

- (1) Divergent relative motion: When $\lambda_M < \lambda_T + 2$, the angular rate of LOS, the azimuths θ and ϕ, and the maximum required missile acceleration all grow without bound, indicating a failure of interception.

- (2) Steady-state relative motion: When $\lambda_M = \lambda_T + 2$, , we have $\Omega = \Omega_0$=constant. In this case, the missile approaches the target steadily, and the required interception time is evaluated as

$$\tau_f = \frac{\pi/2 - \psi_0}{\sqrt{1 - \lambda_T h_0^2 \cos \psi_0}}.$$

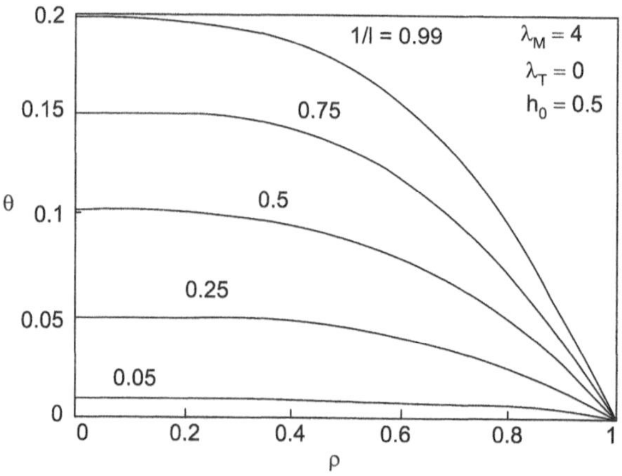

Figure 6 Azimuth θ for 3D IPN with varying l.

- (3) Convergent relative motion : When $\lambda_M > \lambda_T + 2$, all the physical variables converge as the collision point approaches, and the converging speed of the relative motion increases as λ_M is increased.

V CONCLUSIONS

This paper is devoted to the study of three-dimensional pursuit-evasion motion. Three second-order nonlinear differential equations which describe the motion of missiles guided by three-dimensional ideal proportional navigation (3D IPN) are solved analytically without any linearization. Trajectory and performance analysis are described in closed form for nonmaneuvering targets. The general analysis proposed here can be applied to other 3D relative motions using guidance laws other than IPN in unified framework.

Acknowledgment

This research was supported by the National Science Council under Grant No. NSC 90-2213-E-164-002.

REFERENCES

1. Adler, F.P., "Missile guidance by three-dimensional proportional navigation," *Journal of Applied Physics*, 27 (1956) 500–507.
2. Becker, K., "Closed-form solution of pure proportional navigation," *IEEE Trans. Aerospace and Electronic Systems*, 26 (1990) 526–533.

3. Cloutier, J.R., Evers, J.H., and Feeley, J.J., "Assessment of air-to-air missile guidance and control technology," *IEEE Control Systems Magazine*, October (1989) 27–34.
4. Cochran, J.E., No, T.S., and Thaxton, D.G., "Analytical solutions to a guidance problem," *Journal of Guidance, Control, and Dynamics*, 14 (1991) 117–122.
5. Dhar, A. and Ghose, D., "Capture region for a realistic TPN guidance law," *IEEE Trans. Aerospace and Electronic Systems*, 29 (1993) 995–1003.
6. Ghose, D., "True proportional navigation with maneuvering target," *IEEE Trans. Aerospace and Electronic Systems*, 30 (1994) 229–237.
7. Ghose, D., "Capture region for true proportional navigation guidance with nonzero miss-distance," *Journal of Guidance, Control, and Dynamics*, 17 (1994) 627–628.
8. Ghose, D., "On the generalization of true proportional navigation," *IEEE Trans. Aerospace and Electronic Systems*, 30 (1994) 545–555.
9. Guelman, M., "A qualitative study of proportional navigation," *IEEE Trans. Aerospace and Electronic Systems*, 7 (1971) 638–643.
10. Guelman, M., "Proportional navigation with a maneuvering target," *IEEE Trans. Aerospace and Electronic Systems*, 8 (1972) 364–371.
11. Guelman, M., "Missile acceleration in proportional navigation," *IEEE Trans. Aerospace and Electronic Systems*, 9 (1973) 462–463.
12. Guelman, M., "The closed form solution of true proportional navigation," *IEEE Trans. Aerospace and Electronic Systems*, 12 (1976) 472–482.
13. Guelman, M. and Shinar, J., "Optimal guidance law in the plane," *Journal of Guidance, Control, and Dynamics*, 7 (1984) 471–476.
14. Mahapatra, P.R. and Shukla, U.S., "Accurate solution of proportional navigation for maneuvering targets," *IEEE Trans. Aerospace and Electronic Systems*, 25 (1989) 81–89.
15. Murtaugh, S.A. and Criel, H.E., "Fundamental of proportional navigation," *IEEE Spectrum*, 3 (1966) 75–85.
16. Rao, M.N., "New analytical solutions for proportional navigation," *Journal of Guidance, Control, and Dynamics*, 16 (1993) 591–594.
17. Shinar, J., Rotsztein, Y., and Bezner, E., "Analysis of three-dimensional optimal evasion with linearized kinematics," *Journal of Guidance, Control, and Dynamics*, 2 (1979) 353–360.
18. Shukla, U.S. and Mahapatra, P.R., "Optimalization of biased proportional navigation," *IEEE Trans. Aerospace and Electronic Systems*, 25 (1989) 73–80.
19. Shukla, U.S. and Mahapatra, P.R., "The proportional navigation dilemma — pure or true?" *IEEE Trans. Aerospace and Electronic Systems*, 26 (1990) 382–392.
20. Vathsal, S. and Rao, M.N., "Analysis of generalized laws for homing missiles," *IEEE Trans. Aerospace and Electronic Systems*, 31 (1995) 514–521.
21. Yang, C.D., Yeh, F.B., and Chen, C.H., "The closed-form solution of generalized proportional navigation," *Journal of Guidance, Control, and Dynamics*, 10 (1987) 216–218.
22. Yang, C.D. and Yeh, F.B., "Closed-form solution of a class of guidance laws," *Journal of Guidance, Control, and Dynamics*, 10 (1987) 412–415.
23. Yang, C.D., Hsiao, F.B., and Yeh, F.B., "Generalized guidance law for homing missiles," *IEEE Trans. Aerospace and Electronic Systems*, 25 (1989) 197–212.
24. Yang, C.D. and Yang, C.C., "Analytical solution of generalized three-dimensional

proportional navigation," *Journal of Guidance, Control, and Dynamics*, 19 (1996) 721–724.

25. Yang, C.D. and Yang, C.C., "Analytical solution of three-dimensional realistic true proportional navigation," *Journal of Guidance, Control, and Dynamics*, 19 (1996) 569–577.
26. Yang, C.D. and Yang, C.C., "Analytical solution of three-dimensional true proportional navigation," *IEEE Trans. Aerospace and Electronic Systems*, 32 (1997) 1509–1522.
27. Yang, C.D. and Yang, C.C., "A unifed approach to proportional navigation," *IEEE Trans. Aerospace and Electronic Systems*, 33 (1997) 557–567.
28. Yuan, P.J. and Chern, J.S., "Analytic study of biased proprotional navigation," *Journal of Guidance, Control, and Dynamics*, 15 (1992) 185–190.
29. Yuan, P.J. and Chern, J.S., "Solutions of true proportional navigation for maneuvering and non-maneuvering targets," *Journal of Guidance, Control, and Dynamics*, 15 (1992) 268–271.
30. Yuan, P.J. and Chern, J.S., "Ideal proportional navigation," *Journal of Guidance, Control, and Dynamics*, 15 (1992) 1161–1165.
31. Yuan, P.J. and Hsu, S.C., "Exact closed-form solution of generalized proportional navigation," *Journal of Guidance, Control, and Dynamics*, 16 (1993) 963–966.
32. Yuan, P.J. and Hsu, S.C., "Solutions of generalized proportional navigation with maneuvering and nonmaneuvering targets," *IEEE Trans. Aerospace and Electronic Systems*, 31 (1995) 469–474.

17

Guidance Design with Nonlinear H_2/H_∞ Control

Hsin-Yuan Chen

Department of Automatic Control Engineering, Feng Chia University, Taichung, Taiwan

A mixed H_2/H_∞ optimal control design and its application to a missile guidance problem of homing phase are studied. The problem consists of combining the performance requirements of quadratic optimal controllers with the robustness properties of H_∞ controllers. Our approach has five features. (1) The complete nonlinear kinematics of the pursuit-evasion motion are considered; neither linearization nor small angle assumption is made here. (2) The nonlinear H_2/H_∞ guidance law is derived analytically and expressed in a very simple form; neither iterative approximation nor complicated numerical computation is required. (3) Unlike adaptive and neural guidance laws, the implementation of the proposed robust H_2/H_∞ law does not need information on target acceleration. (4) Under appropriate constraints, the pair of cross-coupled Hamilton-Jacobi Partial Inequality (HJPDI) can be easily solved. (5) The derived nonlinear H_2/H_∞ guidance law exhibits strong robustness and excellent performance against maneuvering targets.

I INTRODUCTION

The objective of optimal control design is minimizing or maximizing some index performance by which various designs can be compared. The traditional H_2 optimization attempts to minimize the energy of the system output when the system is faced with white noise inputs. The result is a controller adept at handling the problem with white noises but potentially weak in robustness characteristics and tracking performance. The H_∞ optimization, on the other hand, attempts to minimize the system output energy to unknown but bounded energy inputs. The optimal H_∞ optimal controller results in a highly robust system but one that can be notably deficient in handling the performance of the system. The design method explored in this chapter is the mixed H_2/H_∞ optimization [1] that handles performance and bounded energy input simultaneously and establishes a link between H_2 and H_∞ optimization. Through this method, the designer can determine the tradeoff between the performance (H_2) and robust stability characteristics (H_∞).

Recently, there has been a great deal of interest in formulating a mixed H_2/H_∞ optimal control problem. The mixed H_2/H_∞ optimal control has been studied for linear systems [2–7]. Although this problem can be stated and motivated quite easily, solving it has turned out to be difficult. So, very powerful algorithms and associated theory have recently been developed for convex optimization in Refs. [8] and [9]. As a result of this

0-415-30852-6/04/$0.00+$1.50

development, many convex optimization problems for which no traditional analytic solutions are known can be solved rapidly. In particular, the first application of the theory of Nash games to H_2/H_∞ control problems can be found in Ref. [10]. Based on this development, the mixed H_2/H_∞ controllers have been extensively applied to aircraft control in Ref. [11]. Reference [12] contains examples of gain scheduling and H_2/H_∞ optimal control of aircraft and missiles. In Ref. [13], it is shown that robotic systems can be robustly controlled by the H_2/H_∞ performance even if the different noises are causally dependent. Additionally, in many filtering problems [14,15], the applicability of this H_2/H_∞ control theory is illustrated by designing a filter to estimate the states of an aircraft flying through a downburst and flexible structures. To our knowledge, however, this design method has not yet been discussed for nonlinear systems and even not applied to tear practical (physical) missile systems. The design objective is achieve the optimal H_2 optimal performance under a desired H_∞ disturbance rejection constraint.

In this study, a mixed nonlinear H_2/H_∞ control design is proposed for the homing missile system under unpredictable accelerations of the target (unpredictable disturbances). This design scheme is useful for robust performance design of the system under external disturbances. The solution to this nonlinear mixed H_2/H_∞ control problem is explicitly presented by combining dissipative theory [16] with the LQ optimal control method. A pair of coupled nonlinear Hamilton-Jacobi Partial Inequality (HJPDI) must first be solved. This subsequently leads to a sufficient condition for the solvability of the missile system of the maneuvering targets via a mixed H_2/H_∞ control method. Next, two nonlinear coupled HJPDIs can be transformed to corresponding algebraic inequalities via an adequate choice of the solution of HJPDI. Finally, these two coupled HJPDI can be easily solved and a simplified mixed H_2/H_∞ guidance law is ultimately obtained for missile guidance systems.

The robust H_2/H_∞ guidance law obtained from the solution of HJPDI exhibits robustness properties against target maneuvers [17] and in the meanwhile, maintains some special performances. Comparisons with conventional proportional navigation [18] schemes show that the conventional PN schemes do perform very well in the absence of the target's maneuvers. However, when target maneuvers are present, their performance may degrade dramatically. On the other hand, the H_2/H_∞ robust guidance law does not only demonstrate strong performance robustness under the variation of target maneuvers, but also gives a better performance than the H_∞ robust guidance law [19].

This chapter is organized as follows. In Sec. II, we briefly survey some preliminaries of the nonlinear H_2/H_∞ control theory. In Sec. III, we formulate the missile guidance problem as a nonlinear disturbance attenuation H_2/H_∞ control problem. Then we solve analytically the associated HJPDIs in Sec. IV. Finally, comparison with PN schemes, robust H_∞ guidance law and the robustness and performance of the H_2/H_∞ guidance laws against target maneuvers are illustrated numerically in Sec. V.

II DISSIPATION AND THE NONLINEAR H_2/H_∞ CONTROL PROBLEM

In this section, we apply dissipative theory [6] to derive the nonlinear mixed H_2/H_∞ control theory for later use. Consider a nonlinear control system governed by a set of differential equations of the form

$$\dot{\boldsymbol{x}} = f(\boldsymbol{x}) + g_1(\boldsymbol{x})\boldsymbol{w} + g_2(\boldsymbol{x})\boldsymbol{u} \tag{1a}$$

$$\boldsymbol{z} = \begin{bmatrix} h_1(\boldsymbol{x}) \\ \rho\boldsymbol{u} \end{bmatrix}, \tag{1b}$$

where $\boldsymbol{x} = [x_1 \ \cdots \ x_n]^T$ consists of the n independent generalized coordinates describing the motion of the system; $\boldsymbol{w}(t)$ is the input vector; $\boldsymbol{z}(t)$ is the penalized output vector, and $\boldsymbol{u}$ is the control to be designed. System (1) is said to be dissipative with respect to the supply rate $r(\cdot,\cdot)$, if there exists an energy storage function $V(\boldsymbol{x}) > 0$, $\forall \boldsymbol{x} \neq 0$, satisfying the following inequality:

$$0 \leq V(\boldsymbol{x}(t)) \leq \int_0^t r(\boldsymbol{w}(\xi), \boldsymbol{z}(\xi))d\xi, \tag{2}$$

for all t and for all $\boldsymbol{x}(\cdot)$, $\boldsymbol{z}(\cdot)$, and $\boldsymbol{w}(\cdot)$ satisfying Eq. (1). For a differentiable function $V(\boldsymbol{x})$, an equivalent statement of dissipative condition (2) is

$$\dot{D} = r(\boldsymbol{w}(t), \boldsymbol{z}(t)) - \dot{V}(\boldsymbol{x}(t)) \geq 0. \tag{3}$$

The energy dissipative rate $\dot{D}$ is recognized as the energy supply rate r minus the energy storage rate $\dot{V}$. The term $dV(\boldsymbol{x}(t))/dt = \frac{\partial V}{\partial \boldsymbol{x}}\dot{\boldsymbol{x}} = \left[\frac{\partial V}{\partial x_1}\ \frac{\partial V}{\partial x_2} \cdots \frac{\partial V}{\partial x_n}\right]\dot{\boldsymbol{x}}$ denotes the total time derivative of $V(\boldsymbol{x}(t))$ along the state trajectory $\boldsymbol{x}(t)$.

The supply rate r_∞ relating to H_∞ performance is defined as

$$r_\infty(\boldsymbol{w}(t), \boldsymbol{z}(t)) = \gamma_\infty^2 \boldsymbol{w}^T\boldsymbol{w} - \boldsymbol{z}^T\boldsymbol{z}. \tag{4}$$

Using r_∞ in Eq. (2), we have

$$\int_0^T \boldsymbol{z}(t)^T\boldsymbol{z}(t)dt \leq \gamma_\infty^2 \int_0^T \boldsymbol{w}^T(t)\boldsymbol{w}(t)dt, \quad \forall \boldsymbol{w} \in \boldsymbol{L}_2, \tag{5}$$

where $\int_0^T \boldsymbol{z}^T\boldsymbol{z}dt / \int_0^T \boldsymbol{w}^T\boldsymbol{w}dt$ is known as the L_2 gain of the system. Equation (5) implies that system (1) has a L_2 gain lower than or equal to γ_∞. A special case of Eq. (1) is the familiar linear system:

$$\dot{\boldsymbol{x}} = A\boldsymbol{x} + B_1\boldsymbol{w} + B_2\boldsymbol{u} \tag{6a}$$

$$\boldsymbol{z} = \begin{bmatrix} C_1\boldsymbol{x} \\ \rho\boldsymbol{u} \end{bmatrix}. \tag{6b}$$

For linear systems, it can be shown that the L_2-gain constraint (5) is equivalent to the ∞-norm constraint:

$$\|H_{zw}(s)\|_\infty \leq \gamma_\infty, \tag{7}$$

where $\|H_{zw}(s)\|_\infty$ is the ∞-norm of the transfer function from the input $\boldsymbol{w}$ to output $\boldsymbol{z}$. The equivalence between Eq. (3) and Eq. (5) indicates that a system with L_2-gain $\leq \gamma_\infty$ is a dissipative system, and vice versa. Hence, with respect to dissipation, we can say that the nonlinear H_∞ control is an essential means to make a nonlinear system dissipative. In terms of an energy storage function $V = W$ and the supply rate $r = r_\infty$, the H_∞ constraint (5) can be rewritten by using Eq. (3) as

$$\dot{D}_\infty(\boldsymbol{w},\boldsymbol{u}) = \left[\gamma_\infty^2 \boldsymbol{w}^T\boldsymbol{w} - \frac{\partial W}{\partial \boldsymbol{x}} g_1(\boldsymbol{x})\boldsymbol{w}\right] - \left[\rho \boldsymbol{u}^T\boldsymbol{u} + \frac{\partial W}{\partial \boldsymbol{x}} g_2(\boldsymbol{x})\boldsymbol{u}\right] - \frac{\partial W}{\partial \boldsymbol{x}} f(\boldsymbol{x}) - h_1(\boldsymbol{x})^T h_1(\boldsymbol{x}) \geq 0, \quad \forall \boldsymbol{w} \in \boldsymbol{L_2}. \tag{8}$$

On the other hand, the supply rate r_2 relating to H_2 performance is defined as

$$r_2(\boldsymbol{w}(t), \boldsymbol{z}(t)) = \frac{\gamma_2^2}{T} - \boldsymbol{z}^T\boldsymbol{z}. \tag{9}$$

Substituting this r in Eq. (2), we have

$$\int_0^T \boldsymbol{z}(t)^T \boldsymbol{z}(t)dt \leq \gamma_2^2. \tag{10}$$

For the linear system (6), Eq. (10) reduces to the familiar quadratic performance index:

$$\int_0^T (\boldsymbol{x}^T Q\boldsymbol{x} + \boldsymbol{u}^T R\boldsymbol{u})dt \leq \gamma_2^2, \tag{11}$$

where $Q = C_1^T C_1$ and $R = \rho^2 I$. In terms of energy storage function $V = U$ and supply rate $r = r_2$, the quadratic constraint (10) can be rewritten by using Eq. (3) as

$$\dot{D}_2(\boldsymbol{w},\boldsymbol{u}) = \frac{\gamma_2^2}{T} - \frac{\partial U}{\partial \boldsymbol{x}} g_1 w - \left[\rho^2 \boldsymbol{u}^T\boldsymbol{u} + \frac{\partial U}{\partial \boldsymbol{x}} g_2(\boldsymbol{x})\boldsymbol{u}\right] - \frac{\partial U}{\partial \boldsymbol{x}} f(\boldsymbol{x}) - h_1(\boldsymbol{x})^T h_1(\boldsymbol{x}) \geq 0. \tag{12}$$

The purpose of the nonlinear mixed H_2/H_∞ control is to find the control $\boldsymbol{u}$ such that the H_∞ criterion (5) and the H_2 criterion (10) are satisfied simultaneously. In terms of the dissipative condition (3), the solvability of the mixed H_2/H_∞ controller relies on the existence of a controller $\boldsymbol{u}_0$ and two positive energy storage functions W and U satisfying Eq. (8) and Eq. (12), i.e.,

$$\dot{D}_\infty(\boldsymbol{w},\boldsymbol{u}_0) \geq 0, \quad \forall \boldsymbol{w} \in \boldsymbol{L}_2 \tag{13a}$$

$$\dot{D}_2(\boldsymbol{w},\boldsymbol{u}_0) \geq 0, \quad \forall \boldsymbol{w} \in \boldsymbol{L}_2. \tag{13b}$$

Conditions (13a) and (13b) are known as robust H_∞ constraint and robust H_2 constraint, respectively. Condition (13a) can be met by requiring $\boldsymbol{u}_0$ to satisfy

$$\dot{D}_\infty(\boldsymbol{w}^*,\boldsymbol{u}_0) = \min_{\boldsymbol{w}} \dot{D}_\infty(\boldsymbol{w},\boldsymbol{u}_0) \geq 0, \tag{14}$$

where $\boldsymbol{w}^*$ is called the worst-case disturbance. Using this worst-case disturbance $\boldsymbol{w}^*$ in $\dot{D}_2$ results in the worst-case H_2 performance $\dot{D}_2(\boldsymbol{w}^*,\boldsymbol{u}_0)$. In conjunction with condition (13b), controller $\boldsymbol{u}_0$ need be chosen to make $\dot{D}_2(\boldsymbol{w}^*, \boldsymbol{u}_0)$ as large as possible. The maximum value that can be achieved by $\boldsymbol{u}_0$ is $max_{\boldsymbol{u}_0} \dot{D}_2(\boldsymbol{w}^*, \boldsymbol{u}_0)$. Hence, condition (13b) requires

$$\dot{D}_2(\boldsymbol{w}^*, \boldsymbol{u}^*) = \max_{\boldsymbol{u}_0} \dot{D}_2(\boldsymbol{w}^*, \boldsymbol{u}_0) \geq 0. \tag{15}$$

Combining Eq. (14) with Eq. (15), we can replace conditions (13) by

$$\dot{D}_\infty(\boldsymbol{w}^*, \boldsymbol{u}^*) \geq 0 \tag{16a}$$

$$\dot{D}_2(\boldsymbol{w}^*, \boldsymbol{u}^*) \geq 0, \tag{16b}$$

where the extreme functions $\boldsymbol{w}^*$ and $\boldsymbol{u}^*$ are defined as

$$\boldsymbol{w}^* = \arg\min_{\boldsymbol{w}} \dot{D}_\infty(\boldsymbol{w}, \boldsymbol{u}), \text{ for fixed } \boldsymbol{u} \tag{17a}$$

$$\boldsymbol{u}^* = \arg\max_{\boldsymbol{u}} \dot{D}_\infty(\boldsymbol{w}^*, \boldsymbol{u}). \tag{17b}$$

In the following step, we will express the worst-case disturbance $\boldsymbol{w}^*$ and the optimal control $\boldsymbol{u}^*$ in terms of the energy function W and U. From Eq. (8) it can be seen that $\dot{D}_\infty$ is a quadratic function in $\boldsymbol{w}$, and $\dot{D}_\infty$ has a global minimum at

$$\boldsymbol{w}^*(\boldsymbol{x}) = \frac{1}{2\gamma_\infty^2} g_1^T(\boldsymbol{x}) \left(\frac{\partial W}{\partial \boldsymbol{x}} \right)^T. \tag{18}$$

By substituting this $\boldsymbol{w}^*$ into Eq. (12), $\dot{D}_2(\boldsymbol{w}^*, \boldsymbol{u})$ becomes

$$\begin{aligned} \dot{D}_2(\boldsymbol{w}^*, \boldsymbol{u}) =& \frac{\gamma_2^2}{T} - \frac{1}{2\gamma_\infty^2} \frac{\partial U}{\partial \boldsymbol{x}} g_1 g_1^T \left(\frac{\partial W}{\partial \boldsymbol{x}} \right)^T - \left[\rho^2 \boldsymbol{u}^T \boldsymbol{u} + \frac{\partial U}{\partial \boldsymbol{x}} g_2(\boldsymbol{x}) \boldsymbol{u} \right] \\ & - \frac{\partial U}{\partial \boldsymbol{x}} f(\boldsymbol{x}) - h_1(\boldsymbol{x})^T h_1(\boldsymbol{x}) \geq 0, \end{aligned} \tag{19}$$

which has a global maximum at

$$\boldsymbol{u}^*(\boldsymbol{x}) = -\frac{1}{2\rho^2} g_2^T(\boldsymbol{x}) \left(\frac{\partial U}{\partial \boldsymbol{x}} \right)^T. \tag{20}$$

Substituting u^* and w^* into Eq. (8) and Eq. (12), we can express the conditions (16) in terms of the two unknown functions $W(\boldsymbol{x})$ and $U(\boldsymbol{x})$ as

$$\begin{aligned} \left(\frac{\partial W}{\partial \boldsymbol{x}} \right) f - \frac{1}{2\rho^2} \left(\frac{\partial W}{\partial \boldsymbol{x}} \right) g_2 g_2^T \left(\frac{\partial U}{\partial \boldsymbol{x}} \right)^T + \frac{1}{4\rho^2} \left(\frac{\partial U}{\partial \boldsymbol{x}} \right) g_2 g_2^T \left(\frac{\partial U}{\partial \boldsymbol{x}} \right)^T & \\ + \frac{1}{4\gamma_\infty^2} \left(\frac{\partial W}{\partial \boldsymbol{x}} \right) g_1 g_1^T \left(\frac{\partial W}{\partial \boldsymbol{x}} \right)^T + h_1^T h_1 \leq 0 & \end{aligned} \tag{21a}$$

$$\begin{aligned} \left(\frac{\partial U}{\partial \boldsymbol{x}} \right) f + \frac{1}{2\gamma_\infty^2} \left(\frac{\partial U}{\partial \boldsymbol{x}} \right) g_1 g_1^T \left(\frac{\partial W}{\partial \boldsymbol{x}} \right)^T - \frac{1}{4\rho^2} \left(\frac{\partial U}{\partial \boldsymbol{x}} \right) g_2 g_2^T \left(\frac{\partial U}{\partial \boldsymbol{x}} \right)^T & \\ + h_1^T h_1 - \frac{\gamma_2^2}{T} \leq 0. & \end{aligned} \tag{21b}$$

The above two inequalities are known as HJPDIs whose solutions characterize the admissible storage functions $W(\boldsymbol{x})$ and $U(\boldsymbol{x})$. Once W and U are obtained, the mixed H_2/H_∞

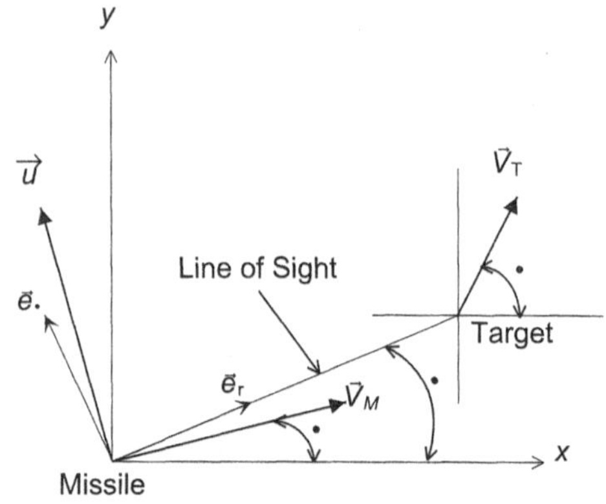

Figure 1 Planar pursuit geometry.

control law $\boldsymbol{u}^*$, and the worst-case disturbance $\boldsymbol{w}^*$ can then be determined according to Eq. (20) and Eq. (18). The corresponding HJPDIs relating to the missile guidance problem will be derived in the next section. Before closing this section, it is worth noting that for the linear system (6) the two coupled HJPDIs have a simple solution in quadratic form:

$$W(\boldsymbol{x}) = \boldsymbol{x}^T P \boldsymbol{x}, \quad U(\boldsymbol{x}) = \boldsymbol{x}^T Q \boldsymbol{x}, \tag{22}$$

where P and Q are two constant positive definite matrices to be determined. Substituting Eq. (22) into Eq. (21), and replacing $f(\boldsymbol{x})$, $g_1(\boldsymbol{x})$, $g_2(\boldsymbol{x})$, $h_1(\boldsymbol{x})$ with $A\boldsymbol{x}$, B_1, B_2, $C_1\boldsymbol{x}$, respectively, we have

$$PA + A^T P - \frac{1}{2\rho^2}\left(PB_2B_2^TQ + QB_2B_2^TP - QB_2B_2^TQ\right) + \frac{1}{2\gamma_\infty^2}PB_1B_1^TP + 2C_1C_1 \leq 0 \tag{23a}$$

$$QA + A^T Q + \frac{1}{2\gamma_\infty^2}\left(PB_1B_1^TQ + QB_1B_1^TP\right) - \frac{1}{2\rho^2}QB_2B_2^TQ + 2C_1C_1 - 2\gamma_2^2/T \leq 0, \tag{23b}$$

which are recognized as the coupled Riccati inequalities that appeared in the linear mixed H_2/H_∞ control theory.

III FORMULATION OF H_2/H_∞ GUIDANCE PROBLEMS

The pursuit geometry of relative motion between missile and target is described by the polar coordinate (r, θ) (refer to Fig. 1). Now, we consider the relation between missile and target as point mass models, and only kinematics are considered. The governing equation in Fig. 1 can be derived as

$$\ddot{r} - r\dot{\theta}^2 = w_r - u_r \tag{24a}$$

$$r\ddot{\theta} + 2\dot{r}\dot{\theta} = w_\theta - u_\theta, \tag{24b}$$

where r is the relative distance between missile and target; θ is the aspect angle of the line of sight (LOS) with respect to an inertial reference line; w_r and w_θ are the target's acceleration components, which are assumed to be an unpredictable disturbance; u_r and u_θ are the missile's acceleration components that are to be found. Now to transfer the governing equation (Eq. (24)) into the standard state-space equation as Eq. (1) by selecting the state variables r, V_r, and V_θ, and the transformed equations as

$$\dot{\boldsymbol{x}} = f(\boldsymbol{x}) + g_1(\boldsymbol{x})\boldsymbol{w} + g_2(\boldsymbol{x})\boldsymbol{u}, \tag{25}$$

where

$$\boldsymbol{x} = \begin{bmatrix} r \\ V_r \\ V_\theta \end{bmatrix}, \quad f(\boldsymbol{x}) = \begin{bmatrix} V_r \\ \frac{V_\theta^2}{r} \\ \frac{-V_r V_\theta}{r} \end{bmatrix}, \quad g_1(\boldsymbol{x}) = -g_2(\boldsymbol{x}) = \begin{bmatrix} 0 & 0 \\ 1 & 0 \\ 0 & 1 \end{bmatrix}. \tag{26}$$

The penalized output $\boldsymbol{z}$ is chosen as

$$\boldsymbol{z} = \begin{bmatrix} h_1(r, V_\theta) \\ \rho\boldsymbol{u} \end{bmatrix}, \tag{27}$$

where

$$h_1(r, V_\theta) = \frac{1}{r}V_\theta^2 = r\dot{\theta}^2 \tag{28}$$

is a measure of guidance performance, and ρ is a weighting coefficient concerning the tradeoff between guidance performance and acceleration command. It can be seen from Eq. (29) that the missile is close to the target (small r), and/or the LOS angular rates $\dot{\theta}$ are small if h_1 can be kept small. By choosing the weighting coefficient ρ properly, it is possible to obtain an acceptable small h_1 without consuming a lot of acceleration $\boldsymbol{u}$.

The problem H_2/H_∞ robust guidance law design can be stated as follows: To find the missile acceleration command $\boldsymbol{u} = [u_r u_\theta]^T$ such that the following mathematical expressions can be satisfied:

$$\frac{\int_0^T \boldsymbol{z}^T\boldsymbol{z}dt}{\int_0^T \boldsymbol{w}^T\boldsymbol{w}dt} = \frac{\int_0^T (h_1{}^2 + \rho^2\boldsymbol{u}^T\boldsymbol{u})dt}{\int_0^T (w_r^2 + w_\theta^2)dt} \leq \gamma_\infty^2, \quad \forall \boldsymbol{w} \in \boldsymbol{L_2} \tag{29}$$

$$\int_0^T \boldsymbol{z}^T\boldsymbol{z}dt = \int_0^T (h_1{}^2 + \rho^2\boldsymbol{u}^T\boldsymbol{u})dt \leq \gamma_2^2. \tag{30}$$

IV SOLUTIONS OF THE HAMILTON-JACOBI PDI

In this section, the associated HJPDIs for the missile guidance problem will be derived, and the corresponding solutions are derived analytically. Now, to substitute $f(\boldsymbol{x})$, $g_1(\boldsymbol{x})$,

$g_2(\boldsymbol{x})$, and $h_1(\boldsymbol{x})$ from Eq. (26), Eq. (27), and Eq. (28) into Eq. (21), resulting in

$$V_r\frac{\partial W}{\partial r}+\frac{1}{r}V_\theta^2\frac{\partial W}{\partial V_r}-\frac{V_rV_\theta}{r}\frac{\partial W}{\partial V_\theta}-\frac{1}{2\rho^2}\left[\left(\frac{\partial W}{\partial V_r}\right)\left(\frac{\partial U}{\partial V_r}\right)+\left(\frac{\partial W}{\partial V_\theta}\right)\left(\frac{\partial U}{\partial V_\theta}\right)\right]$$
$$+\frac{1}{4\rho^2}\left[\left(\frac{\partial U}{\partial V_r}\right)^2+\left(\frac{\partial U}{\partial V_\theta}\right)^2\right]+\frac{1}{4\gamma_\infty^2}\left[\left(\frac{\partial W}{\partial V_r}\right)^2+\left(\frac{\partial W}{\partial V_\theta}\right)^2\right]+\frac{1}{r^2}V_\theta^4\leq 0. \tag{31}$$

The second set of HJPDI can be obtained by substituting $f(\boldsymbol{x})$, $g_1(\boldsymbol{x})$, $g_2(\boldsymbol{x})$, and $h_1(\boldsymbol{x})$ from Eq. (26), Eq. (27), and Eq. (28) into Eq. (21), resulting in

$$V_r\frac{\partial U}{\partial r}+\frac{1}{r}V_\theta^2\frac{\partial U}{\partial V_r}+\frac{-V_rV_\theta}{r}\frac{\partial U}{\partial V_\theta}+\frac{1}{2\gamma_\infty^2}\left[\left(\frac{\partial U}{\partial V_r}\right)\left(\frac{\partial W}{\partial V_r}\right)+\left(\frac{\partial U}{\partial V_\theta}\right)\left(\frac{\partial W}{\partial V_\theta}\right)\right]$$
$$-\frac{1}{4\rho^2}\left[\left(\frac{\partial U}{\partial V_r}\right)^2+\left(\frac{\partial U}{\partial V_\theta}\right)^2\right]+\frac{1}{r^2}V_\theta^4-\gamma_2^2/T\leq 0. \tag{32}$$

Both are nonlinear first-order partial differential inequalities with the unknown function $U(r,V_r,V_\theta)$ and $W(r,V_r,V_\theta)$. Once qualified U and W are determined, the nonlinear H_2/H_∞ robust guidance law can be given by Eq. (18) as

$$\boldsymbol{u}(r,V_r,V_\theta)=-\frac{1}{\rho^2}g_2^T\left(\frac{\partial U}{\partial x}\right)^T=\frac{1}{\rho^2}\begin{bmatrix}\partial U/\partial V_r\\ \partial U/\partial V_\theta\end{bmatrix}. \tag{33}$$

And $u_r=\frac{1}{\rho^2}\partial U/\partial V_r$, and $u_\theta=\frac{1}{\rho^2}\partial U/\partial V_\theta$. Here, we assume the optimal pursuing decision confronts the optimal escaping decision, i.e., assuming $U=W$. Both can be simplified as

$$V_r\frac{\partial U}{\partial r}+\frac{1}{r}V_\theta^2\frac{\partial U}{\partial V_r}-\frac{V_rV_\theta}{r}\frac{\partial U}{\partial V_\theta}-\frac{1}{4\rho^2}\left[\left(\frac{\partial U}{\partial V_r}\right)^2+\left(\frac{\partial U}{\partial V_\theta}\right)^2\right]$$
$$+\frac{1}{4\gamma_\infty^2}\left[\left(\frac{\partial U}{\partial V_r}\right)^2+\left(\frac{\partial U}{\partial V_\theta}\right)^2\right]+\frac{1}{r^2}V_\theta^4\leq 0 \tag{34}$$

and

$$V_r\frac{\partial U}{\partial r}+\frac{1}{r}V_\theta^2\frac{\partial U}{\partial V_r}-\frac{V_rV_\theta}{r}\frac{\partial U}{\partial V_\theta}+\frac{1}{2\gamma_\infty^2}\left[\left(\frac{\partial U}{\partial V_r}\right)^2+\left(\frac{\partial U}{\partial V_\theta}\right)^2\right]$$
$$-\frac{1}{4\rho^2}\left[\left(\frac{\partial U}{\partial V_r}\right)^2+\left(\frac{\partial U}{\partial V_\theta}\right)^2\right]+\frac{1}{r^2}V_\theta^4-\gamma_2^2/T\leq 0. \tag{35}$$

It is noted that a function U satisfying Eq. (34) and Eq. (35) is also a solution of the HJPDIS in Eq. (31) and Eq. (32), but it is not unique. Now, we can see these HJPDIs clearly, and find that the solution of HJPDIs can be substituted by the solution of the following inequality:

$$V_r\frac{\partial U}{\partial r}+\frac{1}{r}V_\theta^2\frac{\partial U}{\partial V_r}-\frac{V_rV_\theta}{r}\frac{\partial U}{\partial V_\theta}+\frac{1}{4}\left(\frac{2}{\gamma_\infty^2}-\frac{1}{\rho^2}\right)\left[\left(\frac{\partial U}{\partial V_r}\right)^2+\left(\frac{\partial U}{\partial V_\theta}\right)^2\right]+\frac{1}{r^2}V_\theta^4\leq 0. \tag{36}$$

This solution is developed by the dissipative rate concept. This solution of Eq. (34) and Eq. (35) can be found readily as

$$U(r, V_r, V_\theta) = W = \frac{\lambda V_r V_\theta^2}{r}; \quad \gamma_2 \geq 0. \tag{37}$$

When $V_r < 0$, $\lambda < 0$, and $V_r > 0$, $\lambda > 0$. To check whether the U given by Eq. (37) is feasible or not, we substitute Eq. (37) into the HJPDI (36) to obtain the expression

(1) $V_r < 0$

$$\left[\lambda^2\left(\frac{2}{\gamma_\infty^2} - \frac{1}{\rho^2}\right) - 3\lambda\right] V_r^2 + \left[\frac{\lambda^2}{4}\left(\frac{2}{\gamma_\infty^2} - \frac{1}{\rho^2}\right) + \lambda + 1\right] V_\theta^2 \leq 0. \tag{38}$$

This condition is satisfied for arbitrary $V_r(t)$, and $V_\theta(t)$, if each leading coefficient of the two terms in Eq. (38) is negative, i.e.,

$$\lambda^2\left(\frac{2}{\gamma_\infty^2} - \frac{1}{\rho^2}\right) - 3\lambda \leq 0 \tag{39a}$$

$$\frac{\lambda^2}{4}\left(\frac{2}{\gamma_\infty^2} - \frac{1}{\rho^2}\right) + \lambda + 1 \leq 0. \tag{39b}$$

We find that the intersection of the two inequalities exists if and only if $\gamma_\infty > \sqrt{2}\rho$. Under this condition, the intersection gives two admissible ranges for λ:

$$\text{(A) } \lambda \leq \frac{2 - 2\sqrt{1 + \rho^{-2} - 2\gamma_\infty^{-2}}}{\rho^{-2} - 2\gamma_\infty^{-2}}, \text{ when } \gamma_\infty \geq \sqrt{2}\left(\rho^{-2} - 21/4\right)^{-1/2}, \ \rho < \sqrt{4/21} \tag{40a}$$

$$\text{(B) } \lambda \leq -\frac{3}{\left(\rho^{-2} - 2\gamma_\infty^{-2}\right)}, \text{ when } \sqrt{2}\rho < \gamma_\infty < \sqrt{2}\left(\rho^{-2} - 21/4\right)^{-1/2}, \ \rho < \sqrt{4/21}. \tag{40b}$$

(2) $V_r > 0$ Under this condition, the range will become as follows:

$$0 \leq \lambda \leq \frac{2 + 2\sqrt{1 + \rho^{-2} - 2\gamma_\infty^{-2}}}{\rho^{-2} - 2\gamma_\infty^{-2}} \quad \text{when} \quad \gamma_\infty \geq \sqrt{2}\rho. \tag{41}$$

As mentioned earlier, γ_∞ is an index for the disturbance attenuation level. Hence, a smaller γ_∞ means that the interceptive performance is less sensitive to target maneuvers. From Eq. (38) we can see that the γ_∞ achieved by using the navigation gain (A) is larger than that by using the navigation gain (B). Up to now, we have proved that the functions U and W in Eq. (29) and Eq. (30), respectively, are truly solutions of the HJPDIs. Once the storage functions are determined, the resulting H_2/H_∞ guidance law can be obtained from Eq. (35) as

$$u_r = \lambda\frac{V_\theta^2}{r} \tag{42}$$

$$u_\theta = \lambda\frac{2}{\rho^2}\frac{V_r V_\theta}{r}, \tag{43}$$

and then from Eq. (16), the worst-case maneuvering target turned out to be at

$$w_r = \frac{1}{\gamma_\infty^2}\left(\frac{\partial U}{\partial V_r}\right) = \frac{\lambda V_\theta^2}{\gamma_\infty^2 r} \tag{44}$$

$$w_\theta = \frac{1}{\gamma_\infty^2}\left(\frac{\partial U}{\partial V_\theta}\right) = \frac{2\lambda V_r V_\theta}{\gamma_\infty^2 r}. \tag{45}$$

γ_∞ is not only an index of disturbance attenuation level but also an important parameter describing the worst-case target's maneuvering ability. It can be seen that as $\gamma_i \longrightarrow \infty$, the effect of disturbance attenuation entirely disappears, and the worst-case target that the missile can handle becomes a nonmaneuvering target, i.e., $w_r = w_\theta = w_\phi = 0$. Decreasing γ_∞ means that the missile's robustness is increased and the allowable target acceleration that can be captured by the missile is also increased.

V NUMERICAL RESULTS

In this section, maneuvering targets are used, and the mean value of this target is used to test the robustness and robust performance of the nonlinear 2D H_2/H_∞ robust guidance law. In this simulation, we need to normalize the governing equations to their dimensionless forms.

(1) The relative velocities V_r and V_θ are normalized to $\overline{V}_r = V_r/V_0$, and $\overline{V}_\theta = V_\theta/V_0$, respectively, where $V_0 = \sqrt{V_{r_0}^2 + V_{\theta_0}^2}$.

(2) The acceleration components u_r and u_θ are normalized to $\overline{u}_r = u_r/(V_0^2/r_0)$, and $\overline{u}_\theta = u_\theta/(V_0^2/r_0)$, respectively. Now the performance and robustness of four missile guidance laws will be tested and compared. Their acceleration commands are listed as below.

(M1) H_∞ Guidance Law [18] $\overline{u}_r = \lambda\overline{V}_\theta^2/(\rho^2\overline{r})$, and $\overline{u}_\theta = 2\lambda\overline{V}_r\overline{V}_\theta/(\rho^2\overline{r})$, where λ is a constrained variable.

(M2) Pure proportional navigation [19] (PPN) guidance law: $\overline{u}_r = -\lambda\overline{V}_M(d\theta/d\tau)\sin(k\theta+\alpha_0)$, $\overline{u}_\theta = \lambda\overline{V}_M(d\theta/d\tau)\cos(k\theta+\alpha_0)$, where $\overline{V}_M$ is the normalized missile's velocity; $k = \lambda - 1$ with λ being the navigation gain; α_0 is the initial aspect angle of $\vec{V}_M$ with respect to the inertial reference line (see Fig. 1).

(M3) Nonlinear H_2/H_∞ Guidance Law: $\overline{u}_r = \lambda\overline{V}_\theta^2/(\rho^2\overline{r})$, and $\overline{u}_\theta = 2\lambda\overline{V}_r\overline{V}_\theta/(\rho^2\overline{r})$.

The target commands are the following:

(T1) Sinusoidal target: $\overline{w}_r = \overline{w}_\theta = \lambda_T \sin(\tau d\theta/d\tau)$

(T2) Ramp target: $\overline{w}_r = \overline{w}_\theta = \lambda_T \tau$

(T3) Step target: $\overline{w}_r = 0, \overline{w}_\theta = \lambda_T$

The robustness and desired performance of the nonlinear H_2/H_∞ guidance law can be illustrated in the following aspects:

1. Robustness properties
2. Comparison of the quadratic performance

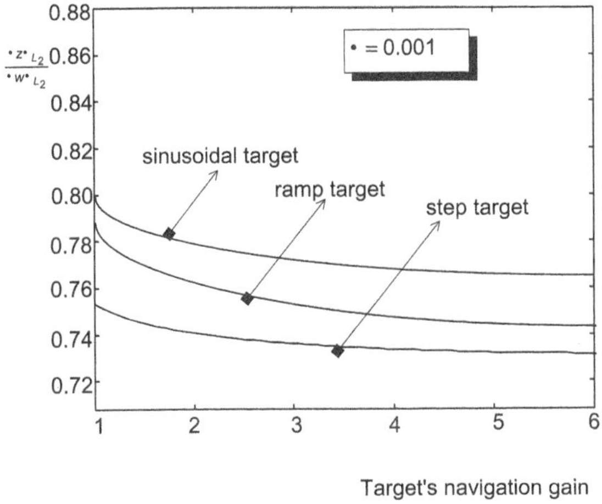

Figure 2 Robustness of L_2-gain for nonlinear H_2/H_∞ guidance law.

(1) Robustness properties

Depending on the definition of the system's L_2 gain, we know that γ_∞ is small, and indicated that good robustness can be preserved in the presence of unknown target maneuvers. We find the H_∞ guidance law has better robustness than the H_2/H_∞ guidance law, but the H_2/H_∞ has better expression. In Fig. 2, when the navigation of the target's acceleration, its associated initial conditions (Fig. 3) increase, H_2/H_∞ guidance law (M3) shows its robustness against the maneuvering target, which can keep the L_2 gain smaller than γ_∞.

(2) Comparison of the quadratic performance

Here, the quadratic performance is used as an index to reflect the interceptive performance. It is shown in Fig. 4 that as the navigation of the target's acceleration increases, the H_2/H_∞ guidance law maintains excellent performance with output energy (i.e., $\|z\|_2$) smaller than 0.4, but the performance of the PPN becomes worse as the variation of target acceleration changes.

VI CONCLUSIONS

A nonlinear mixed H_2/H_∞ control has been proposed in this chapter for the first time for robustness of missile guidance design under maneuvering targets. This nonlinear mixed H_2/H_∞ robust control design problem must first solve two coupled nonlinear Hamilton-Jacobi Partial Inequalities. In order to avoid the difficulty of solving these two coupled nonlinear HJPDIs, some adequate assumptions are also made for the solution of coupled

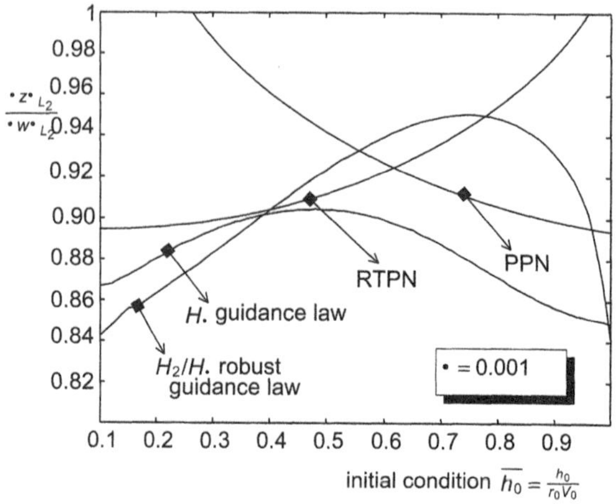

Figure 3 System L_2-gain vs. $\bar{h}_0$ for four guidance laws.

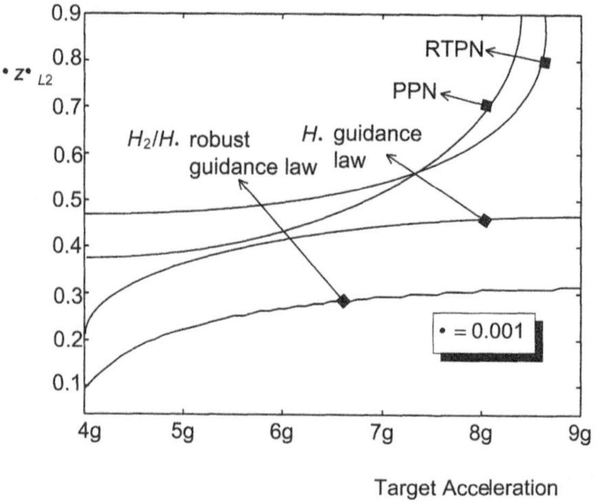

Figure 4 Comparison of performances for four guidance laws.

HJPDIs to obtain the mixed H_2/H_∞ guidance law. This solution is optimal and of closed form, and a simple design procedure is also proposed.

Acknowledgment

I sincerely thank the National Scientific Council, Taiwan, for its kind support (NSC 89-2218-E-035-022).

REFERENCES

1. Lin, W., "Mixed H_2/H_∞ control via state feedback for nonlinear systems," *International Journal of Control*, 1996, Vol. 5, pp. 899–922.
2. Berstein, D.S., and Haddad, W.M., "LQG control with an H_∞ performance bound: a Riccati equation approach," *IEEE Transactions on Automatic Control*, 1989, Vol. 34, pp. 293–305.
3. Doyle, J., Glover, K., Khargonekar, P.P., and Francis, B.A., "State-space solutions to standard H_2 and H_∞ control problem," *IEEE Transactions on Automatic Control*, 1989, Vol. 34, pp. 831–847.
4. Doyle, J., Zhou, K., Khargonekar, P.P., and Francis, B.A., "Mixed H_2 and H_∞ performance objectives II: optimal control," *IEEE Transactions and Automatic Control*, 1994, Vol. 39, pp. 1575–1587.
5. Khargonekar, P.P. and Rotea, M.A., "Mixed H_2/H_∞ control: a convex optimization approach," *IEEE Transactions on Automatic Control*, Vol. 36, 1991, pp. 824–837.
6. Mosca, E., Casavola, A., and Giarre, L., "Minimax LQ stochastic tracking and servo problems," *IEEE Transactions on Automatic Control*, 1990, Vol. 35, pp. 95–97.
7. Zhou, K., Glover, K., Bodenheimer, B., and Doyle, J., "Mixed H_2 and H_∞ performance objectives I: robust performance analysis," *IEEE Transactions on Automatic Control*, 1994, Vol. 39, pp. 1564-1574.
8. Hardt, M., Helton, J.W., and Delgado, K.K., "Numerical solution of nonlinear H_2 and H_∞ control problems with application to jet engine control," *Proceedings the 36th Conference on Decision and Control*, 1997, Vol. 3, pp. 2317–2322.
9. Fan, Y., Cliff, M.M., Lutze, F.H., and Anderson, M.R., "Mixed H_2/H_∞ optimal control for an elastic aircraft," *Journal of Guidance, Control, and Dynamics*, 1996, Vol. 19, pp. 650–655.
10. Limebeer, D.N., Anderson, B.D.O., and Handel, B., "Nash game approach to mixed H_2/H_∞ control," *IEEE Transactions on Automatic Control*, 1994, Vol. 39, pp. 69–82.
11. Ozbay, H. and Bachmann, G.R., "H_2/H_∞ controller design for a two-dimensional thin airfoil flutter suppression," *Journal of Guidance, Control, and Dynamics*, 1994, Vol. 4, pp. 722–728.
12. Shue, S.-P., Sawan, M.E., and Rokhsaz, K., "Mixed H_2/H_∞ method suitable for gain scheduled aircraft control," *Journal of Guidance, Control, and Dynamics*, 1997, Vol. 4, pp. 699–706.
13. Chen, B.-S. and Chang, Y.-C., "Nonlinear mixed H_2/H_∞ control for robust tracking design of robotic systems," *International Journal of Control*, 1997, Vol. 6, pp. 837–857.
14. Rotstein, H., Sznaier, M., and Idan, M., "H_2/H_∞ filtering theory and an aerospace application," *International Journal of Robust and Nonlinear Control*, 1996, Vol. 6, pp. 347–366.

15. Gawronski, W. and Lim, K.B., "Frequency weighting for the H_∞ and H_2 control design of flexible structures," *Journal of Guidance, Control, and Dynamics*, 1998, Vol. 21, pp. 664–666.
16. Willems, J.C., "Dissipative dynamical systems part I: general theory," *Arch. Rational Mechanics and Analysis*, 1972, Vol. 5, pp. 321–351.
17. Austin, F., Carbone, G., Falco, M., and Hinz, H., "Game theory for automated maneuvering during air-to-air combat," *Journal of Guidance, Control, and Dynamics*, 1990, Vol. 6, pp. 1143–1149.
18. Yang, C.D. and Yang, C.C., "A unified approach to proportional navigation," *IEEE Transactions on Aerospace and Electronic Systems*, 1997, Vol. 2, pp. 557–567.
19. Yang, C.D. and Chen, H.-Y., "Nonlinear H_∞ robust guidance law for homing missiles," *Journal of Guidance, Control, and Dynamics*, 1998, Vol. 6, pp. 882–890.

18

A Control Algorithm for Nonlinear Multiconnected Objects

V.Yu. Rutkovsky, S.D. Zemlyakov, V.M. Sukhanov, and V.M. Glumov
Trapeznikov Institute of Control Sciences of Russian Academy of Sciences, Moscow, Russia

The control of mechanical objects is considered here. A mathematical model of such objects is described by nonlinear Lagrange equations of the second type. The control algorithms, which under the conditions of restricted control actions ensure the separate control for each of the generalized coordinates, monotonic character and guaranteed response speed of the transient processes are synthesized. Such severe requirements for the motion of a nonlinear multiconnected control system are fulfilled when the mathematical model complies with some conditions. But these conditions take place for the considered sets of space vehicles.

1 THE PROBLEM MOTIVATION

Airplanes, space vehicles, above-water and submarine aids, means of transport, robots-manipulators, and others as the control objects are mechanical systems. As a rule, a mathematical model (MM) of such objects is derived on the basis of Lagrange equations of the second type:

$$\frac{d}{dt}\frac{\partial T}{\partial \dot{q}} - \frac{\partial T}{\partial q} = Q, \tag{1}$$

where $q, Q = Q(q, \dot{q}, t) \in R^n$ are the vectors of generalized coordinates and generalized forces; $T = T(q, \dot{q})$ is the kinetic energy of the mechanical system. In particular the equation of free-flying space robotic module [1, 2] is as follows:

$$A(q)\ddot{q} + \sum_{s=1}^{n} \left(\dot{q}^T D_s(q)\dot{q}\right) e^s = Q(q, \dot{q}, t), \tag{2}$$

where $A(q)$ is the positive definite matrix; $D_s(q)$ is the matrix of Christoffel's deltas of the first type; ${e^s}^T = (e_1, e_2, ..., e_n)$ – vector with zero components with the exception of only one component $e_s = 1$; T is the symbol of transposition.

Equation (2) represents the nonlinear multiconnected, in a general case, nonstationary MM of a mechanical system. The control of this object is realized with the help of purposeful action on the vector of generalized forces $Q(q, \dot{q}, t)$.

0-415-30852-6/04/$0.00+$1.50

For many mechanical objects, for example, which were mentioned above, a high guaranteed accuracy of control is required. Thus the free-flying space robotic module that is intended for servicing the manned orbital station must realize all operations not only with very high accuracy but with the predicted dynamics (predicted dynamics is the preassigned form of the transient processes for all variable coefficients of the module's nonstationary mathematical model). This requirement is dictated both by the safety of the manned orbital station, other space objects, and by the safety of the module itself.

In this chapter the task of the design of the control algorithms for the object with a mathematical model (2) is considered. The algorithms must realize monotonic tracking with guaranteed response speed of each generalized coordinate $q_i = q_i(t)$ $(i = \overline{1,n})$ after the prescribed function $q_i^{pr}(t)$. Moreover this tracking must be executed at the restricted control action and independently on the presence or the absence of the control actions for the other generalized coordinates.

It is clear that this task is very difficult for such a complicated object. For the solution of this task, mathematical model (2) will be accepted with the following simplifications:

- It is assumed that the system has for each coordinate q_i $(i = \overline{1,n})$ the independent control action M_i and this control action is restricted, i.e., $|M_i| \leqslant M_i^{\max}$, $M_i^{\max} = const > 0$.
- It is assumed that the velocities $\dot{q}_i$ $(i = \overline{1,n})$ are small and the second term on the left-hand side of Eq. (2) can be neglected (in many mechanical systems the special automatic limiters are introduced for the restriction of the velocities $\dot{q}_i$).
- It is assumed that the object moves under the conditions of the weightlessness and the damping forces' absence in the airless space (for example, free-flying space robotic module [2]).
- Let us introduce a further assumption. In this chapter the external disturbances will not be taken into account. This assumption is not a matter of principle and is introduced only for the brevity of the mathematical calculations (for the free-flying space robotic module these disturbances are very small).

The aforementioned assumptions allow one to simplify the mathematical model (2) of the mechanical system rather essentially and to represent it as

$$A(q)\ddot{q} = D(q)M, \tag{3}$$

where M is the vector of control actions, which are created by the control system, i.e.,

$$M^T = (M_1, M_2, \ldots, M_n), \tag{4}$$

$D(q)$ is the matrix, which defines the interconnection of the generalized coordinates' motion by means of control actions.

Let us denote by $M^{\max}$ the vector of the restrictions on the control actions, i.e.,

$$M^{\max^T} = (M_1^{\max}, M_2^{\max}, \ldots, M_n^{\max}). \tag{5}$$

It is important to note that the mechanical system with the model (3) in spite of the simplification is nonlinear and multiconnected.

2 FORMULATION OF THE PROBLEM

It will be supposed that the components of the vectors $q(t)$ and $\dot{q} = \dot{q}(t)$ are measured.

The control law $M = M(q, \dot{q}, t)$ must be synthesized so that it could realize for each coordinate monotonic motion with guaranteed response speed and these requirements must be fulfilled by properly choosing only one control action, i.e.,

$$M_i = M_i(q_i(t), \dot{q}_i(t), t). \tag{6}$$

Definition 1. The motion $q_i = q_i(t)\ (i = \overline{1,n})$ is called to be monotonic if the coordinate q_i moves from the initial point $(q_i(t_0) = q_{i0}, \dot{q}_i(t_0) = 0)$ at $t = t_0$ into the new preassigned position $(q_i(t_1) = q_{i1}, \dot{q}_i(t_1) = 0)$ at $t = t_1\ (t_1 > t_0)$ without changing the sign of the coordinate's velocity $\dot{q}_i(t)$ on the time interval $[t_0, t_1]$.

Definition 2. The motion $q_i(t)$ is called as one with guaranteed response speed if the coordinate $q_i\ (i = \overline{1,n})$ gets over from the initial point $(q_i(t_0) = q_{i0}, \dot{q}_i(t_0) = 0)$ at $t = t_0$ into the new preassigned position $(q_i(t_1) = q_{i1}, \dot{q}_i(t_1) = 0)$ at $t = t_1\ (t_1 > t_0)$ during the time $T_i(\Delta q_i) \leqslant \lambda_i \sqrt{\Delta q_i}$, where $\Delta q_i = |q_i(t_1) - q_i(t_0)|, \lambda_i = const > 0$ is the preassigned constant value.

3 AUTONOMOUS TECHNICAL CONTROLLABILITY OF AN OBJECT

The solving of the formulated task is essentially based on the concept of autonomous technical controllability (AT controllability).

Definition 3. The control object with the mathematical model (3) is called autonomous technical controllable (AT controllable) with respect to the coordinate $q_i = q_i(t)\ (i = \overline{1,n})$ at the position $q = q^*$ of phase space q, if with the applying at any moment $t = t^*$ the control action M_i, which is equal to its maximum value, i.e., $|M_i(t^*)| = M_i^{\max}$, the second derivative $\ddot{q}_i$ of the coordinate q$_i$ obtains the value $|\ \ddot{q}_i(t^*)\ | \geqslant \rho_i^0, \rho_i^0 = const > 0$ and the sign of the $\ddot{q}_i(t^*)$ is the same as the sign of the control action $M_i(t^*)$ independently of the presence or absence of the control actions applying to the other coordinates.

Let us find out the conditions at which the object (3) is AT controllable.

Beforehand we will introduce the following.

Definition 4. Quadratic matrix B is called the matrix with diagonal prevalence if its elements $b_{ij}\ (i, j = \overline{1,m})$ are real and possess the following properties: 1) $b_{ii} > 0$; $b_{ij} \leqslant 0$ $(j = \overline{1,m};\ j \neq i)$; 2) There exists the positive vector $\gamma \in R^m$, $\gamma^T = (\gamma_1, \gamma_2, \ldots, \gamma_m)$, $\gamma > 0$ such that $\gamma_i b_{ii} > \sum\limits_{j=1;\ (j \neq i)}^{m} \gamma_j |b_{ij}|$.

In this chapter we consider any matrix $B = (b_{ij})\ (i = \overline{1,m};\ j = \overline{1,n})$ with the real elements to be positive (nonnegative), if $b_{ij} > 0\ \ (b_{ij} \geqslant 0)$ [3].

On the basis of definition 4 the following lemma can be formulated.

Lemma. The matrix $B = (b_{ij})\ \ (i, j = \overline{1,m})$ with the diagonal prevalence possesses an inverse matrix and $B^{-1} \geqslant 0$.

The lemma is proved in the Appendix.

Let us represent model (3) in the following form:

$$\ddot{q} = R(q)M, \tag{7}$$

where $R(q) = A^{-1}(q)D(q),\ R(q) = (r_{ij}(q))\ (i, j = \overline{1,n})$.

Let us consider the matrix $S(q) = (s_{ij}(q))$ with elements $s_{ii}(q) = r_{ii}(q),\ s_{ij}(q) = -|r_{ij}(q)|\ (j = \overline{1,n};\ j \neq i)$. According to definition 3, a necessary condition of AT controllability is the requirement of $r_{ii}(q) > 0\ (i = \overline{1,n})$ but the elements $r_{ij}(q)$ $(j = \overline{1,n};\ j \neq i)$ can be positive, negative, and zero.

It is evident that the requirement of the object's AT controllability is reduced to the following relation:

$$S(q^*)M^{\max} \geqslant \rho^0, \tag{8}$$

i.e., to the existence of the positive vector $M^{\max} > 0$ for each positive vector ρ^0 at which the inequality (8) occurs.

On the basis of the lemma the following theorem can be proved.

Theorem. The necessary and sufficient conditions for AT controllability of the object with the mathematical model (3) in the position $q = q^*$ is the belonging of the matrix $S(q^*)$, which is obtained from the matrix $R(q^*)$ to the set of the matrices with diagonal prevalence.

The theorem is proved in the Appendix.

From the theorem the importance for practice consequence can be obtained:

Consequence. Among all the vectors $M^{\max}$ complying with the condition of AT controllability of the object with mathematical model (3) at the position $q = q^*$, the vector $M^{\max}$ defined by the equality $M^{\max} = S^{-1}(q^*)\,\rho^0$ is minimal with respect to the norm.

4 WORKING DOMAIN OF THE OBJECT'S MOTION

According to the assumption on the restrictions of the coordinates' velocities,

$$|\dot{q}_i| \leqslant \dot{q}_{i0},\ \dot{q}_{i0} = const > 0 \qquad (i = \overline{1,n}), \tag{9}$$

the domain of the definition for the object with mathematical model (3) in the space $\{q, \dot{q}\}$ is limited by a cylindrical one:

$$G(q, \dot{q}) = \left\{ q, \dot{q} \;\middle|\; |\dot{q}_i| \leqslant \dot{q}_{i0},\ i = \overline{1,n} \right\}. \tag{10}$$

A formulated theorem enables one for every position $q = q^*$, $q^* \in G(q, \dot{q})$ to answer the question: Is the object at this position AT controllable or not?

Let us notice that according to the theorem the object's AT controllability in the position $q^* \in G(q, \dot{q})$ does not depend on the control actions $M_i = M_i(q, \dot{q}, t)$. It is defined only by the object's constructive parameters, i.e., the concrete object is AT controllable in the position q^* or it is not AT controllable. In the latter case it is impossible to make the object to be AT controllable by the variations of the vector $M^{\max}$ of control restrictions. For this purpose it is necessary to change the constructive parameters of the object. If the object is AT controllable we can get any desired vector ρ^0 (it is the vector of AT controllability degree) by changing of vector's $M^{\max}$ control restrictions.

Thus in the domain of the definition of the object's motion $G(q,\dot{q})$, it is possible to select a domain $G_0(q,\dot{q}) \subset G(q,\dot{q})$ in which the object is AT controllable in each position $q^* \in G_0(q,\dot{q})$. But in each position $q^* \in G_0(q,\dot{q})$ the different vector $M^{\max}$ of control restrictions (5) exists guaranteeing the desired degree of AT controllability.

From the practice point of view an inverse task can be more important: for the object with the mathematical model (3) the certain vector $M^{\max} = M_o^{\max}$ of control restrictions and the certain vector of AT controllability degree ρ^0 are given. Can the preassigned vector $M_o^{\max}$ guarantee the preassigned degree ρ^0 of AT controllability in every position $q* \in G_0(q,\dot{q})$? If not then does a domain $G_W(q,\dot{q}) \subset G_0(q,\dot{q})$ with the aforementioned property exist?

Definition 5. We denote as the working domain $G_w(q,\dot{q})$ for the object with the mathematical model (3) a set of points of the space $\{q,\dot{q}\}$ including the point (0,0) for which the same constant vector $M^{\max} = M_o^{\max}$, $M_o^{\max} > 0$ of control restrictions guaranteeing the preassigned degree of AT controllability exists, i.e., in any position $q^* \in G_w(q,\dot{q})$ the degree of AT controllability $\rho(q^*, M_0^{\max}) \geqslant \rho^0$, where $\rho^0 > 0$ is the preassigned constant vector.

In this chapter we will not consider the very interesting and rather nontrivial question about the existence and the methods of finding out the object's working domains $G_w(q,\dot{q})$, but we will assume that the working domain exists.

5 THE SOLVING OF THE FORMULATED PROBLEM

The solving of the problem formulated in Section 2 is given by the following.

Statement. The object with the mathematical model (3) and with the control law:

$$M_i = M_i^{\max}\text{sign}\,(u_i); \quad u_i = [(q_i^0 - q_i) - \frac{(\dot{q}_i)^2}{2\rho_i^0}\text{sign}\,(\dot{q}_i)]\ (i = \overline{1,n}), \tag{11}$$

where q_i^0 = const is the required value of the coordinate q_i, under the condition that the points of the interval $[q_i(t_0),\ q_i^0]$ belong to the object's working domain $G_w(q,\dot{q})$, has monotonic transient processes for each coordinate q_i with the guaranteed time $T_i(\Delta q_i)$ of motion from the position $(q_i(t_0) = q_{i0};\ \dot{q}_i(t_0) = 0)$ into the position $(q_i(t_1) = q_i^0;\ \dot{q}_i(t_1) = 0)\ (t_1 > t_0)$, which is defined by inequality $T_i(\Delta q_i) \leqslant T_i^0(\Delta q_i)$, $\Delta q_i = |\,q_i^0 - q_i(t_0)\,|$, $T_i^0(\Delta q_i) = \sqrt{4\Delta q_i/\rho_i^0}$.

In the opinion of the authors the proof of this statement includes some interesting facts. So we will consider it in more detail.

Let us write the motion equations defining the control system of the object (7) in the following form:

$$\begin{aligned} \ddot{q}_i &= \sum_{j=1}^{n} r_{ij}(q)M_j \quad (i = \overline{1,n}), \\ M_j &= M_j^{\max}\text{sign}\,(u_j), \\ u_j &= (q_j^0 - q_j) - \frac{(\dot{q}_j)^2}{2\rho_j^0}\text{sign}\,(\dot{q}_j). \end{aligned} \tag{12}$$

Let us introduce the notation

$$x_j = q_j^0 - q_j \quad (j = \overline{1,n}), \tag{13}$$

and taking into account $q_j^0 = const$ we rewrite the system of Eq. (12) in the following form:

$$\ddot{x}_i = -r_{ii}(q)M_i^{\max}\text{sign}\,(u_i) - \sum_{\substack{j=1 \\ (j \neq i)}}^{n} r_{ij}(q)M_j^{\max}\text{sign}\,(u_j) \quad (i = \overline{1,n}), \tag{14}$$

or in the other form:

$$\ddot{x}_i = -\left[r_{ii}(q)M_i^{\max} + \sum_{\substack{j=1 \\ (j \neq i)}}^{n} r_{ij}(q)M_j^{\max}\text{sign}\,(u_j)\text{sign}\,(u_i) \right] \cdot \text{sign}\,(u_i) \quad (i = \overline{1,n}). \tag{15}$$

Let us introduce the notation

$$\rho_i(t) = r_{ii}(q)M_i^{\max} + \sum_{\substack{j=1 \\ (j \neq i)}}^{n} r_{ij}(q)M_j^{\max}\text{sign}\,(u_j)\text{sign}\,(u_i) \quad (i = \overline{1,n}), \tag{16}$$

and again we rewrite the system (15) as

$$\begin{aligned} \ddot{x}_i &= -\rho_i(t)\text{sign}\,(u_i), \\ u_i &= x_i + \frac{(\dot{x}_i)^2}{2\rho_i^0}\text{sign}\,(\dot{x}_i) \ (i = \overline{1,n}). \end{aligned} \tag{17}$$

In accordance with the assumption that the motion of the control system described by Eq. (15) takes place in the working domain $G_w\,(q, \dot{q})$, the following inequality occurs:

$$\rho_i(t) \geqslant \rho_i^0 \ (i = \overline{1,n}). \tag{18}$$

Let us note a very important result: nonlinear multiconnected system (12) from the mathematical point of view is decomposed on separate subsystems for each generalized coordinate q_i $(i = \overline{1,n})$ and moreover each subsystem is described by an analogous nonlinear and nonstationary equation from the system of Eq. (17).

Further we will prove that in each subsystem with the equation of motion of the type (17) under condition (18), the transient processes are monotonic with preassigned response speed.

Let us consider the case:

$$\rho_i(t) \equiv \rho_i^0 \ (i = \overline{1,n}), \tag{19}$$

and represent for this case the system (17) as follows:

$$\begin{aligned} \dot{x}_{i1} &= x_{i2}, \\ \dot{x}_{i2} &= -\rho_i^0\text{sign}\,(u_i), \\ u_i &= x_{i1} + \frac{x_{i2}^2}{2\rho_i^0}\text{sign}\,(x_{i2}), \end{aligned} \tag{20}$$

where the notations $x_i = x_{i1}$, $\dot{x}_i = x_{i2}$ are introduced.

On the phase plane with coordinates x_{i1}, x_{i2} the representative point of the system (20) is moving from the position $x_{i1}(0) = x_{i1}^0$, $x_{i2}(0) = 0$ into the origin of coordinates $x_{i1} = 0$, $x_{i2} = 0$ along the optimal trajectory with minimal control time [4]:

$$T_i^0(x_{i1}^0) = \sqrt{\frac{4|x_{i1}^0|}{\rho_i^0}}, \tag{21}$$

and the transient process is monotonic.

Now let us consider the case

$$\rho_i(t) \not\equiv 0, \quad \rho_i(t) \geqslant \rho_i^0 . \tag{22}$$

For this case we will prove that the transient process is monotonic and the control time is

$$T_i(x_{i1}^0) \leqslant T_i^0(x_{i1}^0). \tag{23}$$

According to the works [5, 6] for the system,

$$\begin{aligned} \dot{x}_{i1} &= x_{i2}, \\ \dot{x}_{i2} &= -\rho_i(t)\,\mathrm{sign}\,(u_i), \\ u_i &= x_{i1} + \frac{x_{i2}^2}{2\rho_i^0}\,\mathrm{sign}\,(x_{i2}); \end{aligned} \tag{24}$$

we will consider Lyapunov's function

$$V_i(x_{i1}, x_{i2}) = \frac{1}{2} S_i^2(x_{i1}, x_{i2}), \tag{25}$$

where

$$S_i(x_{i1}, x_{i2}) = x_{i1} + \frac{x_{i2}^2}{2\rho_i^0}\,\mathrm{sign}\,(x_{i2}). \tag{26}$$

Lyapunov's function is positive at any point of the phase plane $\{x_{i1}, x_{i2}\}$ with the exception of the points of the switching line:

$$S_i(x_{i1}, x_{i2}) = 0, \tag{27}$$

in which it is zero.

The derivative of Lyapunov's function (25) by virtue of the system (24) for any point-of-phase plane $\{x_{i1}, x_{i2}\}$ with the exception of the points of the switching line (27) is as follows:

$$\frac{d}{dt} V_i(x_{i1}, x_{i2}) = S_i(x_{i1}, x_{i2}) \frac{dS_i(x_{i1}, x_{i2})}{dt}. \tag{28}$$

At first let us consider the half-plane

$$x_{i2} > 0. \tag{29}$$

For the derivative $\frac{dS_i(x_{i1},x_{i2})}{dt}$ we will have the correlation

$$\frac{dS_i(x_{i1}, x_{i2})}{dt} = x_{i1} \left(1 - \frac{\rho_i(t)}{\rho_i^0}\,\mathrm{sign}\,(S_i(x_{i1}, x_{i2}))\right). \tag{30}$$

Taking into account the condition (29) and the inequality

$$\frac{\rho_i(t)}{\rho_i^0} > 1, \tag{31}$$

from the correlation (30) we obtain two inequalities:

$$\begin{aligned} \frac{dS}{dt} &< 0 \text{ for } S > 0, \\ \frac{dS}{dt} &> 0 \text{ for } S < 0. \end{aligned} \tag{32}$$

Further we will consider the half-plane

$$x_{i2} < 0. \tag{33}$$

On this half-plane the following correlation takes place:

$$\frac{dS_i(x_{i1}, x_{i2})}{dt} = x_{i2}\left(1 + \frac{\rho_i(t)}{\rho_i^0}\operatorname{sign}\left(S_i(x_{i1}, x_{i2})\right)\right). \tag{34}$$

For this case under conditions (31) and (33) from equality (34) we also can obtain two inequalities of form (32).

Thus for any point-of-phase plane $\{x_{i1}, x_{i2}\}$ with the exception of the points on the switching line (27) the inequalities (32) take place. These inequalities are the conditions of arising of the sliding mode on the switching line (27). Since the instant when the representative point gets on this switching line the motion goes along the line (27) into the origin of the coordinate, which is the stable equilibrium.

Because the trajectories of the system (24) on the phase plane $\{x_{i1}, x_{i2}\}$ having the beginning in any initial point $(x_{i1}(0) = x_{i1}^0, x_{i2}(0) = 0)$ by all means intersect the switching line (27) without changing the coordinate's x_{i2} sign, the monotonic character of the transient processes is proved.

Now let us prove that the transient processes have the guaranteed response speed. As the transient processes are monotonic the representative points of the systems (24) and (20) overcome the equal distance x_{i1}^0 at their motion from the initial position $x_{i1}(0) = x_{i1}^0, x_{i2}(0) = 0$ to the origin of coordinates. But it is obvious that under condition (31) the module of the velocity, $|x_{i2}|$, of the system's (24) representative point in any moment of the transient process is more than the module of the velocity, $|x_{i2}|$, of the representative point of the system (20), i.e., the system (24) overcomes the required distance more quickly. However for system (20) the control time is determined by expression (21). So we have proved that the control time in system (24) satisfies condition (23).

Thus the statement is proved completely.

6 CONCLUSION

For nonlinear multiconnected objects with mathematical model (3) and with the restricted control actions the control law of form (11) is suggested. Under certain conditions formulated in this chapter the system with such control law is separated into some subsystems with mathematical models of form (17). Transient processes in each subsystem are monotonic and have a guaranteed response speed. We would like to note another result that is important for practice: giving vector $M^{\max}$ of control restrictions we can point out the guaranteed response speed of all the subsystems and vice versa being given by the desired response speed of the subsystems we can find the minimal values of the control vector elements restrictions at which this response speed will take place.

APPENDIX

The proof of the lemma. Let us introduce the matrices Γ and L, $\Gamma = \text{diag}(\gamma_1, \gamma_2, \ldots, \gamma_m)$ and $L = \text{diag}\ (b_{11}, b_{22}, \ldots, b_{mm})$ where b_{ii} $(i = \overline{1,m})$ are the diagonal elements of the matrix B, and write the matrix product in the form $B\Gamma = \Gamma L(E - Q)$, where E is the identity matrix, $Q = (q_{ij})\ (i,j = \overline{1,m}\ )$, $q_{ii} = 0$; $q_{ij} = \gamma_j |b_{ij}| / \gamma_i b_{ii}$ $(j = \overline{1,m};\ j \neq i)$. Matrix Q is nonnegative and $\sum_{j=1}^{m} q_{ij} < 1$ for any $i = \overline{1,m}$. In this case according to the Perron–Frobenius theorem [3] the matrix $(E - Q)^{-1}$ exists, and moreover $(E - Q)^{-1} \geqslant 0$. From the correlation $B\Gamma = \Gamma L(E - Q)$ we obtain $B^{-1} = \Gamma(E - Q)^{-1}(\Gamma L)^{-1}, B^{-1} \geqslant 0$. □

The proof of the theorem. The proof of the necessity. Let the object with a mathematical model of form (3) be AT controllable in the position $q = q^*$. Given vector $M^{\max}$ as the vector γ in definition 4 we obtain that the matrix $S(q^*)$ is relative to the class of the matrices with diagonal prevalence.

The proof of the sufficiency. Let matrix $S(q^*)$ be relative to the class of the matrices with diagonal prevalence. Then according to the lemma the matrix $S^{-1}(q^*) \geqslant 0$ exists. In such case for any $\rho^0 > 0$ the vector $M^{\max} > 0$ exists, determined by the correlation $M^{\max} = S^{-1}(q^*)\,\rho^0$, which satisfies the condition (8) of AT controllability.

The theorem is proved. □

Acknowledgments

The work is supported by the Department of Energetics, Machine-building, Mechanics and Control Processes of RAS (project 14) and by the Russian Foundation for Basic Research (project 03-01-00062) for which the authors are grateful.

REFERENCES

1. Kulakov F.M. *Supervisory Control of Manipulator Robots*, Moscow: Nauka, 1980 (in Russian).
2. Rutkovsky V.Yu. and Sukhanov V.M. “A dynamic model of a free-flying space robotic engineering module,” *Automation and Remote Control*, Vol 61, N 5, Part 1, 2000, pp. 749–767.
3. Gantmakher F.R. *Matrices' Theory*, Moscow: Nauka, 1967 (in Russian).
4. Boltjansky V.G. *Mathematical Methods of Optimal Control*, Moscow: Nauka, 1973 (in Russian).

5. Emelyanov S.V. *Automatic Control Systems with Variable Structure*, Moscow: Nauka, 1967 (in Russian).
6. *The Theory of the System with Variable Structure*, Ed. S.V. Emelyanov. Moscow: Nauka, 1970 (in Russian).

19

Optimal Control and Differential Riccati Equations under Singular Estimates for $e^{At}B$ in the Absence of Analyticity

Irena Lasiecka and Roberto Triggiani
Department of Mathematics, University of Virginia, Charlottesville, VA

In this chapter, we study the quadratic optimal control problem over a finite time horizon in the case where the free dynamics operator A and the control operator B yield a singular estimate for $e^{At}B$. Here, e^{At} is the corresponding s.c. semigroup which, by assumption, is *not* analytic. The resulting abstract model covers systems of coupled Partial Differential Equations, which possess an analytic *component*, but which are *not themselves analytic.* Several applications are given to hyperbolic/parabolic structural acoustic problems, to thermoelastic problems, and to sandwich beam problems. As to structural acoustic problems, we have that a hyperbolic PDE (a wave equation within an acoustic chamber) is coupled with a parabolic PDE (the flexible wall), which is either modeled by an elastic equation with structural damping [1], [5], or else by a thermoelastic equation with no rotational inertia [18–21], [23], [16, Vol. I], or else by a sandwich beam [11], [28].

1 MATHEMATICAL SETTING AND FORMULATION OF THE CONTROL PROBLEM

Dynamical model

Let U (control), Y (state) be separable Hilbert spaces. In this chapter, we consider the following abstract state equation:

$$\dot{y}(t) = Ay(t) + Bu(t) + w(t) \text{ on, say, } [\mathcal{D}(A^*)]'; \tag{1.1}$$

subject to the following assumptions, to be maintained throughout the chapter.

(H.1) $A : Y \supset \mathcal{D}(A) \to Y$ is the infinitesimal generator of a strongly continuous (s.c.) semigroup e^{At} on Y. Without loss of generality, at the price of replacing A with a suitable translation of A, we may assume that $A^{-1} \in \mathcal{L}(Y)$.

(H.2) B is a linear operator $U = \mathcal{D}(B) \to [\mathcal{D}(A^*)]'$, the dual space of the domain $\mathcal{D}(A^*)$, with respect to the pivot space Y. Here A^* is the adjoint of A in Y. Thus, e^{At} can be extended as an s.c. semigroup on $[\mathcal{D}(A^*)]'$ as well.

0-415-30852-6/04/$0.00+$1.50

(H.3) There exist constants $0 \le \gamma < 1$ and $T > 0$, such that the following singular estimate holds true:

$$\left\|e^{At}B\right\|_{\mathcal{L}(U;Y)} = \left\|B^*e^{A^*t}\right\|_{\mathcal{L}(Y;U)} \le \frac{c_T}{t^\gamma}, \quad \forall\, 0 < t \le T, \tag{1.2}$$

where $(Bu, v)_Y = (u, B^*v)_U$, $u \in U$, $v \in \mathcal{D}(B^*) \supset \mathcal{D}(A^*)$.

(H.4) The function w is a deterministic disturbance, satisfying

$$w \in L_2(0, \infty; Y) \tag{1.3}$$

to be kept fixed throughout.

(H.5)

$$R \in \mathcal{L}(Y; Z). \tag{1.4}$$

Optimal Control Problem on $[0, T]$

With the dynamics (1.1), we associate the following quadratic cost functional over a finite time horizon:

$$J_w(u, y) = \int_0^T [\|Ry(t)\|_Z^2 + \|u(t)\|_U^2]dt, \tag{1.5}$$

and the corresponding Optimal Control Problem is:

$$\begin{cases} \text{For a fixed } w \in L_2(0, T; Y), \text{ minimize } J_w(u, y) \text{ over all } u \in \\ L_2(0, T; U), \text{ where } y \text{ is the solution of (1.1) due to } u \text{ (and } w). \end{cases} \tag{1.6}$$

Remark 1.1. What makes the above optimal control problem different from those studied in the literature [4], [15], [16], is, of course, the new set of assumptions imposed on model (1.1); in particular, the presence of the singular estimate of hypothesis (H.3) = (1.2), while, however, the semigroup e^{At} is only assumed to be strongly continuous. Explicitly, e^{At} is *not* assumed to be analytic. In [16, Chapters 1, 2, 6], the singular estimate (H.3) = (1.2) was also available in the treatment of those chapters; however, it was so *a posteriori* as a *consequence* of the original assumption that the s.c. semigroup e^{At} be, moreover, analytic. Thus, writing $e^{At}B = (-A)^\gamma e^{At}(-A)^{-\gamma}B$ would at once yield estimate (1.2) under the two assumptions in those aforementioned chapters: (i) analyticity of the semigroup of e^{At} on Y, and (ii) the property $(-A)^{-\gamma}B \in \mathcal{L}(U; Y)$. In the present paper, by contrast, the s.c. semigroup e^{At} is *not* assumed to be analytic. The above set of assumptions on (1.1) are motivated by the structural acoustic problem, with hyperbolic/parabolic coupling, where these assumptions are, in fact, properties of the coupled dynamics: see the examples in Section 4 below, with both $\gamma < \frac{1}{2}$ (the easier case: Example 4.1 of Section 4.1) and $\frac{1}{2} < \gamma < 1$ (the more challenging case of Example 4.2 in Section 4.1; and of the examples in Sections 4.2–4.4). Similarly, in line with the structural acoustic problem, we are taking a pure distributed L_2-disturbance w (that is G = Identity in the notation of [16, Eq. (6.1.1.1) of Chapter 6]. The corresponding optimal control problem over an infinite time horizon,

$T = \infty$, was studied in [13], [17] (definite cost) and [26] (nondefinite cost and min-max problem). □

Preliminaries. As in previous work [16], we shall have to consider the corresponding Optimal Control Problem over the interval $[s, T]$, $0 \leq s < T$, rather than just over $[0, T]$; thus with s, rather than 0, as the initial time. Let s, $0 \leq s < T$ be fixed throughout as an initial time with initial datum $y_0 \in Y$. The corresponding solution to (1.1) is then

$$y(t, s; y_0) = e^{A(t-s)} y_0 + (L_s u)(t) + (W_s w)(t); \tag{1.7}$$

$$(L_s u)(t) = \int_s^t e^{A(t-\tau)} Bu(\tau) d\tau, \qquad s \leq t \leq T, \tag{1.8a}$$

:continuous $L_2(s, T; U) \to L_2(s, T; Y)$ (1.8b)

:continuous $C([s, T]; U) \to C([s, T]; Y)$, with operator norms that are uniform with respect to s, $0 \leq s \leq T$; (1.8c)

$$(W_s w)(t) = \int_s^t e^{A(t-\tau)} w(\tau) d\tau \tag{1.9a}$$

:continuous $L_2(s, T; Y) \to C([s, T]; Y)$, with operator norms that are uniform with respect to s, $0 \leq s \leq T$. (1.9b)

Again, the above regularity properties are due to the Young's inequality [25] and (H.3) = (1.2). They are formalized in Proposition 2.1.2 below. The L_2-adjoints of L_s and W_s are

$$(L_s^* f)(t) = \int_t^T B^* e^{A^*(\tau - t)} f(\tau) d\tau, \qquad s \leq t \leq T; \tag{1.10a}$$

:continuous $L_2(s, T; Y) \to L_2(s, T; U)$ (1.10b)

:continuous $C([s, T]; Y) \to C([s, T]; U)$, with operator norms that are uniform with respect to s, $0 \leq s \leq T$; (1.10c)

$$(W_s^* v)(t) = \int_t^T e^{A^*(\tau - t)} v(\tau) d\tau, \quad s \leq t \leq T \tag{1.11a}$$

:continuous $L_2(s, T; Y) \to C([s, T]; Y)$, with operator norms that are uniform with respect to s, $0 \leq s \leq T$. (1.11b)

Henceforth, for $s = 0$, we *shall use the notation* L, W, L^*, W^* *instead of* L_0, W_0, L_0^*, W_0^*.

Statement of main results

The main result of this chapter is the following counterpart of [21, Theorem 2.1] as it pertains to the finite horizon case.

Theorem 1.1. Assume hypotheses (H.1)–(H.5) above. Then:

(a1) for each $y_0 \in Y$ and $w \in L_2(0,T;Y)$ fixed, there exists a unique optimal pair $\{u^0_w(\,\cdot\,,0;y_0), y^0_w(\,\cdot\,,0;y_0)\}$ of the optimal control problem (1.5), (1.6), for the dynamics (1.1), or (1.7), which satisfies the following properties:

$$u^0_w(\,\cdot\,,0;y_0) \in C([0,T];Y); \quad y^0_w(\,\cdot\,,0;y_0) \in C([0,T];Y). \tag{1.12}$$

(a2) The operator $\Phi(t,\tau) \in \mathcal{L}(Y)$ defined by

$$\Phi(t,\tau)x = y^0_{w=0}(t,\tau;x) \in C([0,T];Y), \quad x \in Y \tag{1.13a}$$

:solution to optimal control problem (1.1), (1.5), (1.6), with $w = 0$, over the interval $[\tau, T]$, $0 \le \tau < T$, (1.13b)

is an evolution operator:

$$\Phi(t,t) = I, \quad \Phi(t,s) = \Phi(t,\tau), \Phi(\tau,s),\ 0 \le s \le \tau \le t \le T. \tag{1.13c}$$

(a3) The bounded operator $P(t) \in \mathcal{L}(Y)$ defined on Y by

$$P(t)x = \int_t^T e^{A^*(\tau-t)} R^* R y^0_{w=0}(\tau,t;x)d\tau \tag{1.14a}$$

$$= \int_t^T e^{A^*(\tau-t)} R^* R\Phi(\tau,t)x, \quad x \in Y \tag{1.14b}$$

$$: \text{continuous } Y \to C([0,T];Y), \tag{1.14c}$$

is nonnegative, self-adjoint on $Y : P(t) = P^*(t)$,

$$(P(t)x_1, x_2)_Y = \int_t^T \Big[(Ry^0_{w=0}(\tau,t;x_1), Ry^0_{w=0}(\tau,t;x_2))_Z$$

$$+ \ (u^0_{w=0}(\tau,t;x_1), u^0_{w=0}(\tau,t;x_2))_U \Big] d\tau. \tag{1.15}$$

In particular,

$$J^0_{w=0}(t,y_0) = J(u^0_{w=0}(\,\cdot\,,t;y_0), y^0_{w=0}(\,\cdot\,,t;y_0)) = (P(t)y_0, y_0)_Y. \tag{1.16}$$

(a4) The gain operator $B^*P(t)$ is bounded $Y \to U$ at each t and, in fact,

$$B^*P(t): \ \text{continuous } Y \to C([0,T];U). \tag{1.17}$$

(a5) The operator

$$A_P(t) = A - BB^*P(t), \quad \mathcal{D}(A_P(t)) = \{x \in Y : [I - A^{-1}B(B^*P(t))] \in \mathcal{D}(A)\}, \tag{1.18}$$

satisfies for any $x \in Y$ and any t, $s < t < T$:

$$\frac{\partial \Phi(t,s)x}{\partial t} = A_P(t)\Phi(t,s)x \in C([0,T];[\mathcal{D}(A^*)]'); \tag{1.19}$$

$$\frac{\partial \Phi(t,s)x}{\partial s} = -\Phi(t,s)A_P(s)x \in {}_\gamma C([s,T];Y), \; x \in \mathcal{D}(A). \tag{1.20}$$

(a6) The following singular estimate holds true for the evolution operator $\Phi(t,s)$:

$$\|\Phi(t,s)B\|_{\mathcal{L}(U;Y)} = \|B^*\Phi^*(t,s)\|_{\mathcal{L}(Y;U)} \leq \frac{c_T}{(t-s)^\gamma}, \quad 0 \leq s < t \leq T, \tag{1.21}$$

where $\gamma < 1$ is the constant in assumption (H.3) = (1.2).

(a7) The operator $P(t)$ in (1.14) satisfies the following Differential Riccati Equation:

$$\begin{cases} (\dot{P}(t)x, z)_Y &= -(Rx, Rz)_Z - (P(t)x, Az)_Y - (P(t)Ax, y)_Y \\ & \quad + (B^*P(t)x, B^*P(t)z)_U \\ P(T) &= 0, \qquad \text{for all } x, z \in \mathcal{D}(A). \end{cases} \tag{1.22}$$

(a8) The optimal pair $\{u_w^0(t,s;y_0), y_w^0(t,s;y_0)\}$ satisfies

$$\begin{aligned} u_w^0(t,s;y_0) &= -B^*p_w(t,s;y_0) = -B^*[Py_w^0(t,s;y_0) + r_w(t,s)] \\ &= -B^*P(t)y_w^0(t,s;y_0) - B^*r_w(t,s); \end{aligned} \tag{1.23}$$

$$B^*P(t)y_w^0(t,s;y_0) \in C([s,T];U); \quad B^*r_w(t,s) \in C([s,T];U). \tag{1.24}$$

Here, $r_w(t,s)$ is the function defined by

$$r_w(t,s) \equiv p_w(t,s;y_0 = 0) - P(t)y_w^0(t,s;y_0 = 0) \in C([s,T];Y), \tag{1.25}$$

(a9) such function satisfies the equation

$$\dot{r}_w(t,s) = -A_P^*(t)r_w(t,s) - P(t)w(t) \in [\mathcal{D}(A^*)]' \text{ a.e. in } t, \tag{1.26}$$

so that $r_w(t,s)$ is also explicitly given by

$$r_w(t,s) = \int_t^T \Phi^*(\tau,t)P(\tau)w(\tau)d\tau. \qquad \Box \tag{1.27}$$

Dedication. This chapter is dedicated to our friend, Professor A. V. Balakrishnan, who spurred our interest in optimal control problems for PDEs with boundary control, through his pioneering work [2, Sect. 4.12], [3], which we gladly acknowledge in details in [16, Vol. 1, pp. 420–424].

2 PROOF OF THEOREM 1.1

2.1 Existence of a Unique Optimal Pair, Characterization, and Regularity Properties: Proof of (a1)

The following result is the perfect counterpart of [21, Proposition 3.1.1] in the case $T = \infty$, and it is proved in the same way.

Proposition 2.1.1. Assume (H.1)–(H.5). Let $y_0 \in Y$. Then:

(i) for any $w \in L_2(s,T;Y)$ fixed, the optimal control problem (1.5), (1.6), over $[s,T]$ rather than $[0,T]$, for the dynamics (1.1), or (1.7), admits a unique optimal pair $\{u_w^0(\cdot\,,s;y_0), y_w^0(\cdot\,,s;y_0)\}$ satisfying the optimality condition

$$u_w^0(\cdot\,,s;y_0) = -L_s^* R^* R y_w^0(\cdot\,,s;y_0) \in L_2(s,T;U). \tag{2.1.1}$$

(ii) Moreover, the optimal pair is given explicitly by the following formulae:

$$u_w^0(\cdot\,,s;y_0) = -[I_s + L_s^* R^* R L_s]^{-1} L_s^* R^* R [e^{A(\cdot\,-s)} y_0 + W_s w] \in L_2(s,T;U) \tag{2.1.2a}$$

$$= u_{w=0}^0(\cdot\,,s;y_0) + u_w^0(\cdot\,,s;y_0 = 0) \tag{2.1.2b}$$

$$y_w^0(\cdot\,,s;y_0) = [I_s + L_s L_s^* R^* R]^{-1} [e^{A(\cdot\,-s)} y_0 + W_s w] \in L_2(s,T;Y) \tag{2.1.3a}$$

$$= \{I_s - L_s [I_s + L_s^* R^* R L_s]^{-1} L_s^* R^* R\} [e^{A(\cdot\,-s)} y_0 + W_s w] \tag{2.1.3b}$$

$$= y_{w=0}^0(\cdot\,,s;y_0) + y_w^0(\cdot\,,s;y_0 = 0) \tag{2.1.3c}$$

$$R y_w^0(\cdot\,,s;y_0) = [I_s + R L_s L_s^* R^*]^{-1} R [e^{A(\cdot\,-s)} y_0 + W_s w] \in L_2(s,T;Z) \tag{2.1.4a}$$

$$= R y_{w=0}^0(\cdot\,,s;y_0) + R y_w^0(\cdot\,,s;y_0 = 0), \tag{2.1.4b}$$

where, as usual, the inverse operators in the above formulas are well defined as bounded operators on all of $L_2(s,T;\,\cdot\,)$ [16, Chapter 2, Appendix 2A for (2.1.3a)]. Moreover, with $y_0 \in Y$, the corresponding optimal dynamics is

$$y_w^0(t,s;y_0) = e^{A(t-s)} y_0 + \{L_s u_w^0(\cdot\,,s;y_0)\}(t) + \{W_s w\}(t) \in L_2(s,T;Y). \tag{2.1.5}$$

(iii) For $y_0 \in Y$, the optimal cost satisfies the following relations:

$$J_w^0(s;y_0) = J_w(u_w^0(\cdot\,,s;y_0), y_w^0(\cdot\,,s;y_0))$$

$$= J_{w=0}^0(s;y_0) + J_w^0(s;y_0 = 0) + \mathcal{X}_w(s;y_0); \tag{2.1.6}$$

$$J_{w=0}^0(s;y_0) = (R e^{A(\cdot\,-s)} y_0, [I_s + R L_s L_s^* R^*]^{-1} R e^{A(\cdot\,-s)} y_0)_{L_2(s,T;Z)}; \tag{2.1.7}$$

$$J_w^0(s;y_0 = 0) = (w, W_s^* R^* [I_s + R L_s L_s^* R^*]^{-1} R W_s w)_{L_2(s,T;Y)}; \tag{2.1.8}$$

$$\mathcal{X}_w(s;y_0) = 2(R e^{A(\cdot\,-s)} y_0, [I_s + R L_s L_s^* R]^{-1} R W_s w)_{L_2(s,T;Z)}$$

$$= \text{linear in } w. \quad \Box \tag{2.1.9}$$

The next result is the perfect counterpart of [21, Proposition 3.1.2] in the case $T = \infty$.

Proposition 2.1.2. Assume (H.1), (H.2), (H.3), (H.5). Then, with reference to the operators L_s and L_s^* in (1.8) and (1.10), we have the following boundedness (continuity) results, all uniformly in the operator norm, with respect to the parameter s, $0 \le s \le T$, but with bounds that will generally depend on γ and T:

(i)

$$L_s : \text{continuous } L_2(s,T;U) \to L_2(s,T;Y) \quad (2.1.10a)$$
$$: \text{continuous } C([s,T];U) \to C([s,T];Y); \quad (2.1.10b)$$

(ii)

$$L_s^* : \text{continuous } L_2(s,T;Y) \to L_2(s,T;U) \quad (2.1.11a)$$
$$: \text{continuous } C([s,T];Y) \to C([s,T];U); \quad (2.1.11b)$$

(iii)

$$[I_s + L_s^* R^* R L_s]^{-1} \in \mathcal{L}(L_2(s,T;U)) \cap \mathcal{L}(C([s,T];U)); \quad (2.1.12)$$

(iv)

$$[I_s + L_s L_s^* R^* R]^{-1} \in \mathcal{L}(L_2(s,T;Y)) \cap \mathcal{L}(C([s,T];Y)); \quad (2.1.13)$$

(v)

$$L_s : \text{ continuous } L_2(s,T;U) \to C([s,T];Y), \text{ if } \gamma < \frac{1}{2}. \quad (2.1.14)$$

(vi) Generally, let $r_0 = 2$, and let r_n, $n = 1, 2, \ldots$ be arbitrary positive numbers such that

$$2 < r_1 < \cdots < r_n < \frac{2}{2n\gamma - (2n-1)}, \quad n = 1, 2, \ldots \text{ for } \frac{2n-1}{2n} \le \gamma < 1. \quad (2.1.15)$$

Then, for $n = 0, 2, 4, \ldots$ we have

$$L_s : \text{ continuous } L_{r_n}(s,T;U) \to L_{r_{n+1}}(s,T;Y), \quad (2.1.16)$$

where for $0 \le \gamma < [2(n+1) - 1]/[2(n+1)]$ we may take $r_{n+1} = \infty$; and

$$L_s^* : \text{ continuous } L_{r_{n+1}}(s,T;Y) \to L_{r_{n+2}}(s,T;U), \quad (2.1.17)$$

where r_{n+1} in (2.1.17) is the same as in (2.1.16), and where for $0 \le \gamma < [2(n+2) - 1]/[2(n+2)]$ we may take $r_{n+2} = \infty$.

(vii) for $p > 1/(1-\gamma)$,

$$L_s : \text{ continuous } L_p(s,T;U) \to C([s,T];Y). \quad (2.1.18)$$

(viii) Thus, *a fortiori*, there exists a positive integer $n_0 = n_0(\gamma)$, depending on γ, such that for all positive integers $n \ge n_0(\gamma)$, we have

$$(L_s^* R^* R L_s)^n : \text{ continuous } L_2(s,T;U) \to C([s,T];Y). \quad \square \quad (2.1.19)$$

The next result is the present counterpart of [21, Proposition 3.1.3] in the case $T = \infty$.

Proposition 2.1.3. Assume (H.1)–(H.5). Then, for any $y_0 \in Y$ and any $w \in L_2(0,T;Y)$ (fixed), the optimal pair established in Proposition 2.1.1 satisfies the additional regularity properties:

$$u_w^0(\,\cdot\,,s;y_0) \in C([s,T];U); \quad y_w^0(\,\cdot\,,s;y_0) \in C([s,T];Y), \tag{2.1.20}$$

continuously in $y_0 \in Y$, and uniformly in s. $\square$

Thus, the proof of property (a1) of Theorem 1.1 is complete.

2.2 The Evolution Operator $\Phi(t,s)$; the Functions $p_w(t,s;y_0)$, $r_w(t,s)$; the Operator $P(t) \in \mathcal{L}(Y)$. Proof of (a2), (a3), (a4), (a8)

The operator $\Phi(t,s)$. Lemma 2.2.1. Assume (H.1), (H.2), (H.3), (H.5). The operator $\Phi(t,s)$, defined in (1.13a), has the following properties:

$$\Phi(t,s)y_0 = y_{w=0}^0(t,s;y_0) = \{[I_s + L_s L_s^* R^* R]^{-1} e^{A(\,\cdot\,-s)} y_0\}(t) \tag{2.2.1a}$$

$$\underset{(\text{in } t)}{\in} C([s,T];Y) \text{ uniformly in } s,\ 0 \le s \le T \tag{2.2.1b}$$

$$\Phi(t,s)y_0 \underset{(\text{in } s)}{\in} C([0,t];Y); \ \|\Phi(t,s)\|_{\mathcal{L}(Y)} \le c_T,\ 0 \le s \le t \le T. \tag{2.2.1c}$$

(ii) Moreover, $\Phi(t,s)$ is an evolution operator

$$\Phi(t,t) = I, \quad \Phi(t,s)x = \Phi(t,\tau)\Phi(\tau,s)x, \quad x \in Y,\ 0 \le s \le \tau \le t \le T. \tag{2.2.2}$$

Proof. As in [16, Lemma 1.4.6.2 of Chapter 1, p. 59]. The regularity noted in (2.2.1b): $t \to \Phi(t,s)y_0$ continuous in Y, for s fixed, uniformly in s, $s \le t \le T$, follows from Proposition 2.1.3, Eq. (2.1.20). Then, such continuity in the first variable implies continuity in the second variable: $s \to \Phi(t,s)y_0$ for t fixed, $0 \le s \le t$, as seen in the proof of [16, Lemma 1.4.6.2]. $\square$

The function $p_w(t,s;y_0)$, and the operator $P(t) \in \mathcal{L}(Y)$. For $y_0 \in Y$, we define [in line with, say, [16, Eq. (6.2.21) of Chapter 6]:

$$p_w(t,s;y_0) = \int_t^T e^{A^*(\tau-t)} R^* R y_w^0(\tau,s;y_0)\,d\tau \in C([s,T];Y), \tag{2.2.3}$$

which is the unique solution of the problem:

$$\dot p_w(t,s;y_0) = -A^* p_w(t,s;y_0) - R^* R y_w^0(t,s;y_0) \in [\mathcal{D}(A)]' \tag{2.2.4a}$$

$$]p_w(T,s;y_0) = 0, \tag{2.2.4b}$$

with zero initial condition at $t = T$. Moreover, we introduce the operator $P(t) \in \mathcal{L}(Y)$, as in, say, [16, (6.2.25) of Chapter 6], by setting

$$P(t)x = \int_t^T e^{A^*(\tau-t)} R^* R \Phi(\tau,t)x\,d\tau, \quad x \in Y \tag{2.2.5a}$$

$$: \text{continuous } Y \to C([0,T];Y),\ P(T) = 0, \tag{2.2.5b}$$

where the claimed regularity in (2.2.5b) stems, via property (2.2.1c), as in obtaining [16, Eq. (1.2.25) of Chapter 1].

The next result is the present counterpart of [21, Lemma 3.2.3] for $T = \infty$.

Lemma 2.2.2. Assume (H.1)–(H.5). Then, for $y_0 \in Y$, with reference to (2.2.3) and (2.2.5), we have

(i)

$$p_w(t, s; y_0) = p_{w=0}(t, s; y_0) + p_w(t, s; y_0 = 0) \in C([s, T]; Y); \tag{2.2.6}$$

$$p_{w=0}(t, s; y_0) = \int_t^T e^{A^*(\tau - t)} R^* R y^0_{w=0}(\tau, s; y_0) d\tau \tag{2.2.7}$$

$$= \int_t^T e^{A^*(\tau - t)} R^* R \Phi(\tau, s) y_0 d\tau$$

$$= \int_t^T e^{A^*(\tau - t)} R^* R \Phi(\tau, t) \Phi(t, s) y_0 d\tau$$

$$= P(t)\Phi(t, s) y_0 \in C([s, T]; Y); \tag{2.2.8}$$

(ii)

$$-u^0_w(t, s; y_0) = \{L^*_s R^* R y^0_w(\cdot\, , s; y_0)\}(t)$$

$$= \int_t^T B^* e^{A^*(\tau - t)} R^* R y^0_w(\tau, s; y_0) d\tau \tag{2.2.9}$$

$$= B^* p_w(t, s; y_0) \in C([s, T]; U); \tag{2.2.10}$$

(iii) $P(t) \in \mathcal{L}(Y)$, and moreover,

$$B^* P(t) : \text{ continuous } Y \to C([0, T]; U), \tag{2.2.11}$$

$$B^* P(t) x = \int_t^T B^* e^{A^*(\tau - t)} R^* R \Phi(\tau, t) x \, d\tau; \tag{2.2.12}$$

(iv)

$$-u^0_{w=0}(t, s; y_0) = B^* p_{w=0}(t, s; y_0) = B^* P(t) \Phi(t, s) y_0 \in C([s, T]; U). \tag{2.2.13}$$

Proof. (i) The decomposition of p_w in (2.2.6) is a consequence of using the decomposition of y^0_w in (2.1.3c) in the definition of (2.2.6). The steps from (2.2.7) to (2.2.8) use: (2.2.1a) for Φ, its evolution property (2.2.2), and definition (2.2.5a) for $P(t)$.

(ii) The step from (2.2.9) to (2.2.10) is self-explanatory, once one recalls the optimality condition (2.1.1), and then (1.10) for L^*_s and (2.2.3) for p_w.

(iii) Equation (2.2.12) follows from (2.2.5a): it is here that one uses assumption (H.3) = (1.2) on the singular estimate of $B^* e^{A^* t}$ to obtain $B^* P(t) \in \mathcal{L}(Y)$ at each t, as in

[21, (3.2.15)]. Then, property (2.2.1c) yields, in fact, (2.2.11) as in, say, [16, Proposition 1.4.6.5].

(iv) Equation (2.2.13) follows from (2.2.10) and (2.2.8). □

The function $r_w(t,s)$. Proof of (a8). Next, the present counterpart of [164, Lemma 6.2.2.1, Chapter 6, p. 565], or of [21, Proposition 3.6.1] is:

Proposition 2.2.3. Assume (H.1)–(H.5).

(i) With reference to p_w in (2.2.3), the following identity holds true:

$$p_w(t,s;y_0) = P(t)\Phi(t,s)y_0 + p_w(t,s;y_0=0) = P(t)y^0_{w=0}(t,s;y_0) + p_w(t,s;y_0=0) \tag{2.2.14}$$

$$= P(t)y^0_w(t,s;y_0) + r_w(t,s), \tag{2.2.15}$$

where the function $r_w(t,s)$ is defined by

$$r_w(t,s) \equiv p_w(t,s;y_0=0) - P(t)y^0_w(t,s;y_0=0). \tag{2.2.16}$$

(ii) For $y_0 \in Y$, the optimal control is written in feedback synthesis as

$$-u^0_w(t,s;y_0) = B^*p_w(t,s;y_0) = B^*[P(t)y^0_w(t;s;y_0) + r_w(t,s)] \in C([s,T];U) \tag{2.2.17}$$

$$= B^*P(t)y^0_w(t,s;y_0) + B^*r_w(t,s) \in C([s,T];U); \tag{2.2.18}$$

$$B^*P(t)y^0_w(t,s;y_0) \text{ and } B^*r_w(t,s) \in C([s,T];U). \tag{2.2.19}$$

(iii) The optimal dynamics may thus be rewritten as

$$\begin{cases} \dot{y}^0_w(t,s;y_0) &=& A_P(t)y^0_w(t,s;y_0) - BB^*r_w(t,s) + w(t) \in [\mathcal{D}(A^*)]' \\ y^0_w(s,s;y_0) &=& y_0 \in Y; \end{cases} \tag{2.2.20}$$

$$A_P(t) = A - BB^*P(t) \tag{2.2.21a}$$

$$\mathcal{D}(A_P(t)) = \{x \in Y : [I - A^{-1}BB^*P(t)]x \in \mathcal{D}(A)\}, \tag{2.2.21b}$$

in the sense that, for all $x \in \mathcal{D}(A^*)$:

$$\begin{aligned}(\dot{y}^0_w(t,s;y_0),x)_Y &= ([I - A^{-1}BB^*P(t)]y^0_w(t,s;y_0),A^*x)_Y \\ &\quad - (B^*r_w(t,s),B^*A^{*-1}A^*x)_U + (w(t),x)_Y;\end{aligned} \tag{2.2.22}$$

$$\begin{aligned}y^0_w(t,s;y_0) &= e^{A(t-s)}y_0 - \int_s^t \Phi(\tau,s)BB^*r_w(\tau,s)d\tau \\ &\quad + \int_s^t \Phi(\tau,s)w(\tau)d\tau \in C([s,T];Y).\end{aligned} \tag{2.2.23}$$

Proof. (i) Using the decomposition (2.1.3c) in definition (2.2.3), and recalling (2.2.1a), (2.2.2), (2.2.5a), (2.2.3), (2.1.3c), yields

$$p_w(t,s;y_0) = \int_t^T e^{A^*(\tau-t)} R^* R[y^0_{w=0}(\tau,s;y_0) + y^0_w(\tau,s;y_0=0)]d\tau \quad (2.2.24)$$

$$\text{(by (2.2.1), (2.2.2))} = \int_t^T e^{A^*(\tau-t)} R^* R\Phi(\tau,t)\Phi(t,s)y_0\, d\tau$$

$$= \int_t^T e^{A^*(\tau-t)} R^* R y^0_w(\tau,s;y_0=0)]d\tau$$

$$\text{(by (2.2.5a), (2.2.3))} = P(t)\Phi(t,s)y_0 + p_w(t,s;y_0=0)$$

$$\text{(by (2.1.3c))} = P(t)[y^0_w(t,s;y_0) - y^0_w(t,s;y_0=0)] + p_w(t,s;y_0=0)$$

$$= P(t)y^0_w(t,s;y_0)$$

$$+ [p_w(t,s;y_0=0) - P(t)y^0_w(t,s;y_0=0)], \quad (2.2.25)$$

and (2.2.25) yields (2.2.15) via the definition (2.2.16) for $r_w(t,s)$.

(ii) To get (2.2.17), we use (2.2.15) in (2.2.10). The regularity of $B^*P(t)y^0_w(t,s;y_0)$ in (2.2.19) follows by combining (2.2.11) for $B^*P(t)$ with (2.1.20) for $y^0_w(t,s;y_0)$. Then, recalling likewise (2.2.18) for $y^0_w(t,s;y_0)$, we obtain

$$B^* r_w(t,s) = -u^0_w(t,s;y_0) - B^*P(t)y^0_w(t,s;y_0) \in C([s,T];U), \quad (2.2.26)$$

where we could, equally well, set $y_0 = 0$ in (2.2.26) by (2.2.16).

2.3 Singular Estimate for $\Phi(t,s)B$: Proof of (a5), (a6)

Orientation. The present section is the counterpart of [21, Section 3.3] in the case $T < \infty$. We shall transfer the assumed singularity for $e^{At}B$ in (H.3) = (1.2) into the same singularity for $\Phi(t,s)B$. We shall make critical use of [21, Proposition 3.3.1], which is repeated below for convenience. Briefly, as in the proof of, say, [16, Theorem 6.26.3.1], in order to derive that the operator $P(t)$ defined in (2.2.5) satisfies the Differential Riccati Equation (1.22) on $\mathcal{D}(A)$, we need to differentiate strongly $\Phi(t,s)$ on $\mathcal{D}(A)$. This, in turn, is accomplished if we can establish the same singular estimate for $\Phi(t,s)B$ that holds true under (H.3) for $e^{At}B$. The resulting Theorem 2.3.2 below is a delicate new point of the present development, which was not explicitly needed in our treatment of the abstract analytic or parabolic case of [16, Chapters 1, 2, and 6]. □

We begin by collecting [21, Proposition 3.3.1].

Proposition 2.3.1. Assume (H.1) through (H.5). With reference to the operators L and L^* in (1.8) and (1.10), and recalling that $0 \le \gamma < 1$ as postulated in assumption (H.3), we have

(i) Let $0 < r < 1$. Then for any $0 < T \le \infty$,

$$L: \text{ continuous } {}_rC([0,T];U) \to_{(r+\gamma-1)} C([0,T];Y). \quad (2.3.1)$$

(ii) Let $r > 0$, and $\epsilon > 0$ arbitrary,

$$L^* : \text{ continuous } {}_rC([0,T];Y) \to_{(r+\gamma-1+\epsilon)} C([0,T];U). \tag{2.3.2}$$

(iii) Let $0 < r < 1$. Then, there exists a positive integer $m = m(r)$ such that

$$(L^*R^*RL)^m : \text{ continuous } {}_rC([0,T];U) \to C([0,T];U). \tag{2.3.3}$$

(iv)

$$[I + L^*R^*RL]^{-1} \in \mathcal{L}({}_\gamma C([0,T];Y)). \tag{2.3.4}$$

(v)

$$[I + LL^*R^*R]^{-1} = I - L[I + L^*R^*RL]^{-1}L^*R^*R \in \mathcal{L}({}_\gamma C([0,T];Y)), \tag{2.3.5}$$

where we recall that, if X is a Banach space, then

$${}_rC([0,T];X) \equiv \left\{ f(t) \in C((0,T];X) : \|f\|_{{}_rC([0,T];X)} = \sup_{0<t\leq T} t^r \|f(t)\|_X < \infty \right\}, \tag{2.3.6}$$

see [16, p. 3 and p. 46 of Chapter 1].

Theorem 2.3.2. Assume (H.1), (H.2), (H.3), (H.5).
(i) With reference to (2.2.1a) we have

$$\begin{cases} \Phi(t,s)Bu \in {}_\gamma C([s,T];Y) \text{ continuously in } u \in U, \text{ so that} & (2.3.7a) \\ \|\Phi(t,s)B\|_{\mathcal{L}(U;Y)} \leq c_T \dfrac{1}{(t-s)^\gamma}, \quad 0 \leq s < t \leq T, & (2.3.7b) \end{cases}$$

where $\gamma < 1$ is the constant of assumption (H.3) = (1.2).

(ii) Let $x \in \mathcal{D}(A)$. Then, $\Phi(t,s)x$ is strongly differentiable in Y in the s variable for $0 \leq s < t \leq T$, and

$$\frac{\partial \Phi(0,s)x}{\partial s} = -[I_s + L_sL_s^*R^*R]^{-1}\left\{ e^{A(\,\cdot\,-s)}[Ax + Bu^0_{w=0}(s,s;x)] \right\}$$
$$\in {}_\gamma C([s,T];Y), \; x \in \mathcal{D}(A) \tag{2.3.8}$$

$$\frac{\partial \Phi(t,s)x}{\partial s} = -\Phi(t,s)[Ax + Bu^0_{w=0}(s,s;x)] \tag{2.3.9}$$
$$= -\Phi(t,s)[A - BB^*P(s)]x = -\Phi(t,s)A_P(s)x$$
$$\in {}_\gamma C([s,T];Y), \; x \in \mathcal{D}(A), \tag{2.3.10}$$

where the operator $A_P(t)$ is defined by (2.2.21),

$$\left\| \frac{\partial \Phi(t,s)x}{\partial s} \right\|_Y \leq c_T \left[\|Ax\|_Y + \frac{1}{(t-s)^\gamma}\|x\|_Y \right], \; 0 \leq s < t \leq T, \; x \in \mathcal{D}(A), \tag{2.3.11}$$

and the singularity in (2.3.11) is integrable, since $0 \leq \gamma < 1$.

Proof. (i) From (2.2.1a), we obtain, continuously, in $u \in U$:

$$\Phi(t,s)Bu = [I_s + L_sL_s^*R^*R]^{-1}e^{A(\,\cdot\,-s)}Bu \in {}_\gamma C([s,T];Y), \tag{2.3.12}$$

since $e^{A(\cdot - s)}Bu \in {}_\gamma C([s,T];Y)$, continuously in $u \in U$, by assumption (H.3) = (1.2), and by application of Proposition 2.3.1(v), Eq. (2.3.5) above [applied with s in place of 0]. Then, (2.3.12) yields (2.3.1).

(ii) The following proof is similar in spirit to that of [16, Section 1.4.5, Chapter 1].

Let $x \in Y$. By (2.2.1a) we have

$$\Phi(\cdot, s)x = [I_s + L_s L_s^* R^* R]^{-1} e^{A(\cdot - s)}x; \text{ or } \{[I_s + L_s L_s^* R^* R]\Phi(\cdot, s)x\}(t) = e^{A(t-s)}x, \tag{2.3.13}$$

or explicitly via (1.8), (1.10),

$$\Phi(t,s)x + \int_s^t e^{A(t-\tau)}BB^* \int_\tau^T e^{A^*(\sigma-\tau)}R^*R\Phi(\sigma,s)x\,d\sigma\,d\tau = e^{A(t-s)}x. \tag{2.3.14}$$

Specialize now to $x \in \mathcal{D}(A)$. Differentiating (2.3.14) in s, we obtain

$$\begin{aligned}\frac{\partial\Phi(t,s)x}{\partial s} &+ \int_s^t e^{A(t-\tau)}BB^* \int_\tau^T e^{A^*(\sigma-\tau)}R^*R\frac{\partial\Phi(\sigma,s)x}{\partial s}\,d\sigma\,d\tau \\ &- e^{A(t-s)}BB^*\int_s^T e^{A^*(\sigma-s)}R^*R\Phi(\sigma,s)x\,d\sigma = -e^{A(t-s)}Ax, \ s < t \le T,\end{aligned} \tag{2.3.15}$$

or, again via (1.8), (1.10):

$$[I_s + L_s L_s^* R^* R]\frac{\partial\Phi(\cdot,s)x}{\partial s} = e^{A(\cdot-s)}B\{L_s^*R^*R\Phi(\cdot,s)x\}(s) - e^{A(\cdot-s)}Ax \tag{2.3.16}$$

$$= -e^{A(\cdot-s)}Bu_{w=0}^0(s,s;x) - e^{A(\cdot-s)}x, \tag{2.3.17}$$

where, by (2.1.1), and (2.1.20) for u_w^0:

$$\{L_s^*R^*R\Phi(\cdot,s)x\}(s) = \{L_s^*R^*Ry_{w=0}^0(\cdot,s;x)\}(s) = -u_{w=0}^0(s,s;x) \in U, \tag{2.3.18}$$

so that by (H.3) = (1.2) and (2.3.18),

$$e^{A(\cdot-s)}Bu_{w=0}^0(s,s;x) + e^{A(\cdot-s)}Ax \in {}_\gamma C([s,T];Y), \ x \in \mathcal{D}(A). \tag{2.3.19}$$

Invoking once more the above Proposition 2.3.1(v), Eq. (2.3.5), we obtain from (2.3.17) and (2.3.19) that (2.3.2) holds true, as a well-defined element in ${}_\gamma C([s,T];Y)$:

$$\begin{aligned}\frac{\partial\Phi(\cdot,s)x}{\partial s} &= -[I_s + L_s L_s^* R^* R]^{-1}\left\{e^{A(\cdot-s)}[Ax + Bu_{w=0}^0(s,s;x)]\right\} \\ &\in {}_\gamma C([s,T];Y), \quad x \in \mathcal{D}(A).\end{aligned} \tag{2.3.20}$$

Next, recalling (2.2.1a), we rewrite (2.3.20) as

$$\frac{\partial\Phi(t,s)x}{\partial s} = -\Phi(t,s)[Ax + Bu_{w=0}^0(s,s;x)] \in {}_\gamma C([s,T];Y), \quad x \in \mathcal{D}(A), \tag{2.3.21}$$

and (2.3.3) is established. Finally, we return to (2.2.13), with $t = s$ and obtain

$$-u^0_{w=0}(s, s; x) = B^*P(s)\Phi(s, s)x = B^*P(s)x, \tag{2.3.22}$$

and (2.3.22) inserted in (2.3.21) yields (2.3.10), as desired.

Finally, estimate (2.3.12) follows at once from (2.3.9), or (2.3.10), via estimate (2.3.7) and the continuity

$$\|u^0_{w=0}(s, s; x)\|_U = \|B^*P(s)x\|_U \leq c_T\|x\|_Y, \quad 0 \leq s \leq T, \tag{2.3.23}$$

due to (2.1.20), or (2.2.11). Theorem 2.3.2 is proved. □

2.4 The Operator $P(t)$ Satisfies the DRE on $\mathcal{D}(A)$: Proof of (a7)

We proceed as, say, in the proof of [16, Chapter 1, Section 4.6], [21, Section 3.5] in the present circumstances, by critically using Theorem 2.3.1(ii), to differentiate $\Phi(\ ,\ )x$ strongly in the second variable, for $x \in \mathcal{D}(A)$.

Theorem 2.4.1. Assume (H.1)–(H.5). Then, the operator $P(t) \in \mathcal{L}(Y)$ defined by (2.2.5) — which was noted in property (a3) to be nonnegative, self-adjoint: $P(t) = P^*(t) \geq 0$ — satisfies the following Differential Riccati Equation on $\mathcal{D}(A)$:

$$\begin{aligned}(\dot{P}(t)x, z)_Y &= -(Rx, Ry)_Z - (P(t)x, Az)_Y - (P(t)Ax, y)_Y \\ &\quad + (B^*P(t)x, B^*P(t)z)_U, \quad \forall\, x, z \in \mathcal{D}(A).\end{aligned} \tag{2.4.1}$$

Proof. Let $x, z \in Y$. By (2.2.5a),

$$(P(t)x, z)_Y = \int_t^T (R\Phi(\tau, t)x, Re^{A(\tau-t)}z)_Z d\tau. \tag{2.4.2}$$

We next specialize to $x, z \in \mathcal{D}(A)$, take the time derivative of (2.4.2), and use Theorem 2.3.1(ii), Eq. (2.3.4) to obtain

$$\begin{aligned}(\dot{P}(t)x, z)_Y &= -\int_t^T (R\Phi(\tau, t)x, Re^{A(\tau-t)}Ax)_Z d\tau - (Rx, Rz)_Z \\ &\quad + \int_t^T \left(R\frac{\partial\Phi(\tau, t)x}{\partial t}, Re^{A(\tau-t)}z\right)_Z d\tau \end{aligned} \tag{2.4.3}$$

$$\begin{aligned}\text{(by (2.4.2), (2.3.4))} &= -(P(t)x, Az)_Y - (Rx, Rz)_Z \\ &\quad - \int_t^T (R\Phi(\tau, t)A_P(t)x, Re^{A(\tau-t)}z)_Z d\tau\end{aligned} \tag{2.4.4}$$

$$
\begin{aligned}
\text{(by (2.4.2))} &= -(P(t)x, Az)_Y - (Rx, Rz)_Z - (P(t)A_P(t)x, z)_Y \\
\text{(by (2.2.21))} &= -(P(t)x, Az)_Y - (Rx, Rz)_Z - (P(t)[A - BB^*P(t)]x, z)_Y \qquad (2.4.5)\\
&= -(P(t)x, Az)_Y - (P(t)Ax, z) - (Rx, Rz) \\
&+ (B^*P(t)x, B^*P(t)z)_U, \quad x, z \in \mathcal{D}(A). \qquad (2.4.6)
\end{aligned}
$$

We recall that the singularity in (2.3.4) at s is integrable as $\gamma < 1$. Above, we have invoked (2.4.2), (2.3.4) to pass from (2.4.3) to (2.4.4); and then (2.4.2) and (2.2.21) to finally obtain (2.4.6), where $B^*P(t) \in \mathcal{L}(Y; U)$ by (2.2.11).

Remark 2.4.1. In the present section on a finite time horizon, the property in Theorem 2.3.1, that $\Phi(t,s)B$ inherits the same degree γ of singularity at $t = s$, from the original $e^{At}B$, is critical in (obtaining Eq. (2.3.3), hence in) deriving the Differential Riccati Equation in Theorem 2.4.1. By contast, in the infinite time horizon of [21], where the regularity properties of [21, Proposition 3.4.1, in particular (3.4.2)], are available for the Riccati operator P, the singular estimate inherited by $e^{A_P t}B$ from $e^{At}B$, with the same degree of singularity, is *not* critical in deriving the Algebraic Riccati Equation in [21, Theorem 3.5.1]. The counterpart of [21, Proposition 3.4.1] does not appear to be available for $T < \infty$. □

3 THE FUNCTION r_w AND ITS DIFFERENTIAL EQUATION. PROOF OF (A9)

In this section we prove property (a9) for r_w.

Theorem 3.1. Assume (H.1)–(H.5). Then the function $r_w(t,s)$ defined in (2.2.16) satisfies

$$\frac{\partial r_w}{\partial t}(t,s) = -A_P^*(t) r_w(t,s) - P(t)w(t) \in [\mathcal{D}(A)]' \text{ a.e. in } t, \qquad (3.1)$$

and is therefore given by

$$r_w(t,s) = \int_t^T \Phi^*(\tau, s)P(\tau)w(\tau)d\tau \in C([s,T]; Y), \qquad (3.2)$$

from which one re-proves that

$$B^*r_w(t,s) = \int_t^T B^*\Phi^*(\tau, s)P(\tau)w(\tau)d\tau \in C([s,T]; U), \qquad (3.3)$$

in agreement with (2.2.19).

Proof. The key step is the following result, whose proof is in the same spirit as the proof of Theorem 2.3.2.

Step 1. Proposition 3.2. Assume (H.1)–(H.5). Then, writing for simplicity, $y_w^0(s,\xi;0)$ for $y_w^0(s,\xi;y_0 = 0)$, $\xi \le s < t \le T$, throughout this Proposition, we have

$$
\begin{aligned}
\frac{\partial}{\partial s}[\Phi(\cdot, s)y_w^0(s,\xi;0)] &= -[\Phi(\,\cdot\,, s)B]B^*r_w(s,\xi) + \Phi(\,\cdot\,, s)w(s) \\
&\in {}_\gamma C([s,T]; Y) \text{ a.e. in } s. \qquad (3.4)
\end{aligned}
$$

Proof. This is similar to the proof of Theorem 2.3.2. We return to (2.2.1a) and write

$$\Phi(\,\cdot\,,s)y_w^0(s,\xi;0) = [I_s + L_s L_s^* R^* R]^{-1} e^{A(\,\cdot\,-s)} y_w^0(s,\xi;0), \tag{3.5}$$

or

$$[I_s + L_s L_s^* R^* R]\Phi(\,\cdot\,,s)y_w^0(s,\xi;0) = e^{A(\,\cdot\,-s)} y_w^0(s,\xi;0), \tag{3.6}$$

in the $L_2(s,T;Y)$-sense; or explicitly, pointwise in t, by (1.8a), (1.10a),

$$\Phi(t,s)y_w^0(s,\xi;0) + \int_s^t e^{A(t-\tau)} BB^* \int_\tau^T e^{A^*(\sigma-\tau)} R^* R[\Phi(\sigma,s)y_w^0(s,\xi;0)]d\sigma\, d\tau$$
$$= e^{A(t-s)} y_w^0(s,\xi;0). \tag{3.7}$$

We next differentiate (3.7) in s. We use

$$\frac{\partial}{\partial s}\left[e^{A(t-s)} y_w^0(s,\xi;0)\right] = -Ae^{A(t-s)} y_w^0(s,\xi;0) + e^{A(t-s)} \frac{\partial y_w^0}{\partial s}(s,\xi;0) \tag{3.8}$$

(by (2.2.20))

$$= e^{A(t-s)}[-Ay_w^0(s,\xi;0) + A_P(s)y_w^0(s,\xi;0)$$
$$- BB^* r_w(s,\xi) + w(s)] \tag{3.9}$$

(by (2.2.21))

$$= e^{A(t-s)}\{[(\not{A} - BB^*P(s) - \not{A}]y_w^0(s,\xi;0)$$
$$- BB^* r_w(s,\xi) + w(s)\}, \tag{3.10}$$

or

$$\frac{\partial}{\partial s} e^{A(t-s)} y_w^0(s,\xi;0)] = -e^{A(t-s)} B[B^*P(s)y_w^0(s,\xi;0) + B^* r_w(s,\xi)] + e^{A(t-s)} w(s), \tag{3.11}$$

where, with reference to the right side of (3.11), we have

(i) For each s,

$$e^{A(t-s)} B[B^*P(s)y_w^0(s,\xi;0) + B^* r_w(s,\xi)] \in {}_\gamma C([s,T];Y). \tag{3.12}$$

The regularity noted in (3.12) follows by invoking Theorem 2.3.2(i) on $e^{A\,\cdot}\, B$; Eq. (2.2.11) on $B^*P(s)$; Eq. (2.1.20) on $y_w^0(s,\xi;0) \in Y$; and Eq. (2.2.19) on $B^* r_w(s,\xi) \in U$.

(ii) Moreover, a.e. in s, by (H.4) = (1.3),

$$e^{A(t-s)} w(s) \in C([s,T];Y). \tag{3.13}$$

We note explicitly that the term $Ay_w^0(s,\xi;0) \in [\mathcal{D}(A^*)]'$, so that cancellation of this term in (3.10) takes place in $[\mathcal{D}(A^*)]'$. All the other terms in the right side of (3.11) are well defined.

Thus, differentiating (3.7) in s, and using (3.11) on its right side yields

$$\frac{\partial}{\partial s}[\Phi(t,s)y_w^0(s,\xi;0)] + \int_s^t e^{A(t-\tau)}BB^* \int_\tau^T e^{A^*(\sigma-\tau)}R^*R\frac{\partial}{\partial s}[\Phi(\sigma,s)y_w^0(s,\xi;0)]d\sigma\, d\tau - e^{A(t-s)}BB^*\int_s^T e^{A^*(\sigma-s)}R^*R[\Phi(\sigma,s)y_w^0(s,\xi;0)]d\sigma$$

$$= -Ae^{A(t-s)}y_w^0(s,\xi;0) + e^{A(t-s)}\frac{\partial}{\partial s}y_w^0(s,\xi;0)$$

$$= -e^{A(t-s)}B[B^*P(s)y_w^0(s,\xi;0) + B^*r_w(s,\xi)] + e^{A(t-s)}w(s). \tag{3.14}$$

As to the third term on the left side of (3.14) we rewrite it as

$$-e^{A(t-s)}BB^*\int_s^T e^{A^*(\sigma-s)}R^*R[\Phi(\sigma,s)y_w^0(s,\xi;0)]d\sigma = -e^{A(t-s)}BB^*P(s)y_w^0(s,\xi;0) \in {}_\gamma C([s,T];Y), \tag{3.15}$$

by recalling the definition of $P(\ )$ in (2.2.5a), and hence it cancels the same first term on the right side of (3.14). The regularity noted in (3.15) is obtained by invoking assumption (H.3) = (1.2), as well as (2.2.11) and (2.1.20). In conclusion, after said cancellation, (3.14) yields, recalling again (1.8a), (1.10a):

$$[I_s + L_sL_s^*R^*R]\frac{\partial}{\partial s}[\Phi(\,\cdot\,,s)y_w^0(s,\xi;0)] = -[e^{A(t-s)}B]B^*r_w(s,\xi) + e^{A(t-s)}w(s), \tag{3.16}$$

where, for each s,

$$[e^{A(t-s)}B]B^*r_w(s,\xi) \in {}_\gamma C([s,T];Y), \tag{3.17}$$

due to (H.3) = (1.2), as well as (2.2.19). Because of the available regularity (3.16) and (3.13) on the right side of (3.15), we can invoke property (2.3.5) (for s, rather than $s=0$) and property (2.1.13) for $[I_s + L_sL_s^*R^*R]^{-1}$ and conclude from (3.16) that

$$\frac{\partial}{\partial s}[\Phi(\,\cdot\,,s)y_w^0(s,\xi;0)] = [I_s + L_sL_s^*R^*R]^{-1}\{-e^{A(\,\cdot\,-s)}BB^*r_w(s,\xi) + e^{A(\,\cdot\,-s)}w(s)\} \tag{3.18}$$

(by (2.2.1a))
$$= -\Phi(\,\cdot\,,s)BB^*r_w(s,\xi) + \Phi(\,\cdot\,,s)w(s) \in_\gamma C([s,T];Y) \text{ a.e. in } s, \tag{3.19}$$

where the regularity claimed in (3.19) follows this time by Theorem 2.3.2, Eq. (2.3.7), as well as $B^*r_w(s,\xi) \in U$ by (2.2.19). Thus, (3.19) proves (3.4), as desired. □

Step 2. We return to the definition of r_w in (2.2.16), rewritten explicitly by (2.2.5a) as

$$r_w(t,s) = p_w(t,s;y_0=0) - P(t)y_w^0(t,s;y_0=0) \tag{3.20}$$

$$\text{(by (2.2.5a))} \quad = p_w(t,s;y_0=0) - \int_t^T e^{A^*(\tau-t)}R^*R\Phi(\tau,t)y_w^0(t,s;y_0=0)d\tau. \tag{3.21}$$

We next differentiate (3.21) in t obtaining via (2.2.4) in $[\mathcal{D}(A)]'$:

$$\begin{aligned}\dot{r}_w(t,s) &= \dot{p}_w(t,s;y_0=0) + R^*Ry_w^0(t,s;y_0=0)\\ &+ A^*\int_t^T e^{A^*(\tau-t)}R^*R\Phi(\tau,t)y_w^0(t,s;y_0=0)d\tau\\ &- \int_t^T e^{A^*(\tau-t)}R^*R\frac{\partial}{\partial t}[\Phi(\tau,t)y_w^0(t,s;y_0=0)]d\tau,\end{aligned} \tag{3.22}$$

(by (2.2.4), (2.2.5), (3.4))

$$\begin{aligned}&= -A^*p_w(t,s;y_0=0) - R^*Ry_w^0(t,s;y_0=0)\\ &+ R^*Ry_w^0(t,s;y_0=0) + A^*P(t)y_w^0(t,s;y_0=0)\\ &- \int_t^T e^{A^*(\tau-t)}R^*R\Phi(\tau,t)[BB^*r_w(t,s)+w(t)]d\tau.\end{aligned} \tag{3.23}$$

To pass from (3.21) to (3.22), we have used (2.2.4), (2.2.5), (3.4). Notice that each term in (3.23) is well defined in $[\mathcal{D}(A)]'$ (that is, as an inner product with $x \in \mathcal{D}(A)$) and that the singularity in (3.4) is integrable since $\gamma < 1$. Thus, from (6.23), we obtain

$$\begin{aligned}\dot{r}_w(t,s) &= -A^*p_w(t,s;y_0=0) + A^*P(t)y_w^0(t,s;y_0=0)\\ &- P(t)[BB^*r_w(t,s)+w(t)] \in [\mathcal{D}(A)]' \text{ a.e. in } t\end{aligned} \tag{3.24}$$

$$\begin{aligned}&= -A^*[p_w(t,s;y_0=0) - P(t)y_w^0(t,s;y_0=0)]\\ &- P(t)BB^*r_w(t,s) - P(t)w(t)\end{aligned} \tag{3.25}$$

$$= -A^*r_w(t,s) - P(t)BB^*r_w(t,s) - P(t)w(t) \in [\mathcal{D}(A)]' \text{ a.e. in } t. \tag{3.26}$$

Finally, (3.25) can be rewritten as

$$\dot{r}_w(t,s) = -(A - BB^*P(t))r_w(t,s) - P(t)w(t) \in [\mathcal{D}(A)]' \text{ a.e. in } t, \tag{3.27}$$

and (3.1) is proved, as desired, via (2.2.21). Then (3.3) follows, again via Theorem 2.3.2, Eq. (2.3.7) and (1.3). The proof of Theorem 3.1 is complete. □

Remark 3.1. The above proof of Theorem 3.1 has the advantage of *avoiding* the use of the DRE for $\dot{P}(t)$, as in other proofs [see, e.g., the proof of [16, Proposition 6.2.3.3, p. 568 in Chapter 6]. The "traditional" proof based on the use of the DRE (2.4.1) for $P(t)$ is only formal in the present case. It involves, as usual, the *cancellation* of the term $P(t)A_p(t)y_w^0(t,s;y_0=0)$ or $P(t)Ay_w^0(t,s;y_0=0)$; but it is not clear where this term is well-defined, as the counterpart of [21, Proposition 3.4.1] for the algebraic Riccati operator P, in the infinite time horizon case, does not appear to be available for the time-varying $P(t)$ in the finite horizon case. □

4 ILLUSTRATIONS. STRUCTURAL ACOUSTIC PROBLEMS SATISFYING (H.1), (H.2), (H.3) WITH $\gamma = \frac{3}{8} + \epsilon$ OR $\gamma = \frac{3}{4} + \epsilon$

In this section, divided in various parts, we provide several nontrivial and physically motivated PDEs illustrations. In Subsection 4.1 we provide two *structural acoustic models* with elastic or thermoelastic flexible walls, which, once written abstractly as in (1.1), satisfy all the required assumptions (H.1), (H.2), (H.3) = (1.2), the latter one with either $0 < \gamma < \frac{1}{2}$, typically $\gamma = \frac{3}{8} + \epsilon$ (Example 4.1), or else $\gamma = \frac{3}{4} + \epsilon > \frac{1}{2}$ (Example 4.2).

In Subsection 4.2 we provide a *thermoelastic plate model* with boundary control and $\gamma = \frac{3}{4} + \epsilon$, $\epsilon > 0$. Furthermore, in Sections 4.3, 4.4, we provide a *sandwich beam* model with boundary control and $\gamma = \frac{3}{4} + \epsilon$, $\epsilon > 0$. In our present treatment, which is after [27], and unlike [11], the sandwich beam is reduced to the thermoelastic model of Subsection 4.2. This explains why we obtain the same value of γ. This reduction, then, permits consideration of the structural acoustic chamber of Example 4.2, and replace the thermoelastic flexible wall with a sandwich beam. This is noted in Subsection 4.5. Naturally, we obtain the same value $\gamma = \frac{3}{4} + \epsilon$ as in Example 4.2.

4.1 Two classes of structural acoustic problems

Example 4.1: A class of structural acoustic problems with constant $0 < \gamma < \frac{1}{2}$

The model. We consider the following class of structural acoustic problems, where Ω is an acoustic chamber with a flexible (elastic) wall Γ_0, an assumed *flat* and rigid wall Γ_1. Thus, let $\Omega \subset \mathbb{R}^n$, $n = 2, 3$, be an open bounded domain with boundary $\Gamma = \overline{\Gamma_0 \cup \Gamma_1}$, where Γ_0 and Γ_1 are open, connected, and disjoint parts, $\Gamma_0 \cap \Gamma_1 = \phi$, in $\mathbb{R}^{n-1}$, and Γ_0 is flat. We allow either Γ to be sufficiently smooth (say, C^2), or else Ω to be convex: this assumption will then guarantee that solutions to classical elliptic equations with $L_2(\Omega)$-nonhomogeneous terms be in $H^2(\Omega)$ [7]. Let z denote the velocity potential of the acoustic medium within the chamber. For simplicity of notation, we take equal to 1 both the density of the fluid and the speed of sound in the fluid. Then z_t is the acoustic pressure. Let v denote the displacement of the flat flexible wall Γ_0, modeled by an elastic beam or plate equation ($n = 2$, or $n = 3$). The structural acoustic model considered here is as follows:

$$\begin{cases} \text{acoustic chamber:} & \begin{cases} z_{tt} = \Delta z - d_1 z_t + f & \text{in } (0,T] \times \Omega, & (4.1.1a) \\ \dfrac{\partial z}{\partial \nu} + d_2 z_t = 0 & \text{in } (0,T] \times \Gamma_1, & (4.1.1b) \\ \dfrac{\partial z}{\partial \nu} + d_3 z_t = \pm v_t & \text{in } (0,T] \times \Gamma_0, & (4.1.1c) \end{cases} \\ \text{elastic wall} & \mathcal{M}_k v_{tt} + \mathcal{A} v_t + \mathcal{A} v \pm z_t = \mathcal{B} u \text{ in } (0,T] \times [\mathcal{D}(\mathcal{A}^{\frac{1}{2}})]' \quad (4.1.1d) \\ z(0,\cdot) = z_0,\ z_t(0,\cdot) = z_1 \text{ in } \Omega; & \\ v(0,\cdot) = v_0,\ v_t(0,\cdot) = v_0 \text{ in } \Gamma_0. & (4.1.1e) \end{cases}$$

Assumptions. In (4.1.1a–c), $f \in L_2(0,T; L_2(\Omega))$ denotes the deterministic external noise within the chamber; and the nonnegative constant d_i, when positive, introduce interior/boundary damping in the model. Equation (4.1.1d) is an abstract version encompassing several "concrete" elastic models, as documented below. At the abstract level, we make the following assumptions:

(a1)

$$\begin{cases} \mathcal{M}_k : L_2(\Gamma_0) \supset \mathcal{D}(\mathcal{M}_k) \to L_2(\Gamma_0); \quad \mathcal{A} : L_2(\Gamma_0) \supset \mathcal{D}(\mathcal{A}) \to L_2(\Gamma_0) \\ \text{are two positive, self-adjoint operators} \end{cases} \tag{4.1.2}$$

(the stiffness operator, and the elastic operator, respectively, the first depending on a non-negative parameter $k \geq 0$; in concrete situations, if $k > 0$, the elastic model on Γ_0 accounts for rotational forces);

(a2)

$$\mathcal{D}(\mathcal{A}^{\frac{1}{2}}) \subset \mathcal{D}(\mathcal{M}_k^{\frac{1}{2}}); \tag{4.1.3}$$

(a3) there is a positive constant $\rho < \frac{1}{2}$ such that

$$\mathcal{A}^{-\rho}\mathcal{B} \in \mathcal{L}(\mathcal{U}; L_2(\Gamma_0)), \text{ equivalently } \mathcal{B}: \text{ continuous } \mathcal{U} \to [\mathcal{D}(\mathcal{A}^{\rho})]', \tag{4.1.4}$$

where $[\ \ ]'$ denotes duality with respect to $L_2(\Gamma_0)$ as a pivot space, and $\mathcal{U}$ is the control Hilbert space.

(a4) Moreover, the parameter ρ in (4.1.4) satisfies either one of the following additional assumptions:

(a4i) either

$$\begin{cases} \rho \leq \dfrac{5}{12}, & \text{if } \Omega \text{ is a smooth domain,} \\ \rho \leq \dfrac{7}{16}, & \text{if } \Omega \text{ is a parallelopiped,} \end{cases} \tag{4.1.5a}$$

(a4ii) or else

$$\mathcal{D}(\mathcal{M}_k^{\frac{1}{2}}) \subset H^{\frac{1}{3}}(\Gamma_0). \tag{4.1.5b}$$

Remark 4.1.1 (on the control operator $\mathcal{B}$). In concrete PDE examples of the structural acoustic problems, such as they arise in smart material technology, all of the above assumptions are satisfied. First, in this case, the control operator $\mathcal{B}$ is given by

$$\mathcal{B}u = \sum_{j=1}^{J} a_i u_i \delta'_{\xi_j}, \quad u = [u_1, \ldots, u_J] \in \mathbb{R}^J = \mathcal{U}, \tag{4.1.6}$$

where: (i) if dim $\Gamma_0 = 1$ (dim $\Omega = 2$), then ξ_j are points on Γ_0, a_j are constants, and δ'_{ξ_j} are derivatives of the Dirac distribution supported at ξ_j; (ii) if dim $\Gamma_0 = 2$ (dim $\Omega = 3$), then ξ_j denote closed regular curves on Γ_0, a_j are smooth functions and δ'_{ξ_j} denotes the normal derivative supported at ξ_j:

$$(\delta'_{\xi_j}, \phi)_{L_2(\Gamma_0)} = \begin{cases} -\phi'(\xi_j), \ \dim \Gamma_0 = 1 & \text{(4.1.7a)} \\ -\displaystyle\int_{\xi_j} \nabla\phi \cdot \nu \, d\xi_j, \ \dim \Gamma_0 = 2; & \text{(4.1.7b)} \end{cases} \quad \forall \phi \in H^{\frac{3}{2}+\epsilon}(\Gamma_0),$$

where ν is the unit outer normal vector to the closed curve ξ_j, and $\epsilon > 0$ is arbitrary. Thus, from (4.1.7) we have that, *a fortiori*, $(\delta'_{\xi_j}, \mathcal{A}^{-(\frac{3}{8}+\frac{\epsilon}{4})}\psi)_{L_2(\Gamma_0)}$ is well defined $\forall\, \psi \in L_2(\Gamma_0)$, since $\mathcal{A}^{-(\frac{3}{8}+\frac{\epsilon}{4})}\psi \in \mathcal{D}(\mathcal{A}^{\frac{3}{8}+\frac{\epsilon}{4}}) \subset H^{\frac{3}{2}+\epsilon}(\Gamma_0)$. *Hence the operator $\mathcal{B}$ defined by (4.1.6) satisfies assumption (a3) = (4.1.4) with $\rho = \frac{3}{8} + \epsilon$, $\forall\, \epsilon > 0$ small* [13], [16, vol. 2, p. 907]. Then, *such $\rho = \frac{3}{8} + \epsilon$ satisfies both conditions (a4i) = (4.1.5a) as well.*

Remark 4.1.2 (on the stiffness operator $\mathcal{M}_k$). In concrete PDE examples where the parameter $k > 0$ (and so the elastic beam/plate-model on Γ_0 accounts for rotational forces), $\mathcal{M}_k$ is the translation by the identity of the realization of $(-\Delta)$ on Γ_0 subject to appropriate boundary conditions. Thus, $\mathcal{D}(\mathcal{M}_k^{\frac{1}{2}}) \subset H^1(\Gamma_0) \subset H^{\frac{1}{3}}(\Gamma_0)$ and assumption (a4ii) = (4.1.5b) is satisfied as well. □

The above structural acoustic model (4.1.1), subject to assumptions (a1) through (a4), satisfies assumption (H.3) = (1.2) with $\gamma = \rho < \frac{1}{2}$. The following claims are shown in [13], [1]. The structural acoustic problem (4.1.1) can be rewritten in the abstract form (1.1), with operators A and B explicitly identified, and $w = [0, f, 0, 0]$. Moreover, the operator A is the generator of an s.c. contraction semigroup e^{At} on an appropriate finite energy space Y_k given by

$$Y_k \equiv H^1(\Omega) \times L_2(\Omega) \times \mathcal{D}(\mathcal{A}^{\frac{1}{2}}) \times \mathcal{D}(\mathcal{M}_k^{\frac{1}{2}}), \tag{4.1.8}$$

for the variables $[z(t), z_t(t), v(t), v_t(t)] = e^{At}[z_0, z_1, v_0, v_1]$.

Finally, such operators A and B do satisfy assumption (H.3) = (1.2) for $\gamma = \rho < \frac{1}{2}$, the constant in assumptions (a3) and (a4). In particular, if the control operator $\mathcal{B}$ is defined by (4.1.6), then we have $\gamma = \rho = \frac{3}{8} + \epsilon$, $\forall\, \epsilon > 0$.

Remark 4.1.3 (on the uniform stability of problem (4.1.1)). There are several configurations — that is, choices of the damping constants d_i in (4.1.1a–c) and corresponding geometrical conditions — which ultimately yield uniform stability on Y_k of the associated s.c. semigroup e^{At} [12]. They include the following cases: (1) $d_2 = d_3 = 0$, $d_1 > 0$ (viscous damping); (2) $d_1 = d_3 = 0$, $d_2 > 0$ (boundary damping on rigid wall Γ_1), with no geometrical conditions; (3) $d_1 = d_2 = 0$, $d_3 > 0$ (boundary damping on flexible wall Γ_0) under the geometrical condition that Ω is convex and there exists a point $x_0 \in \mathbb{R}^n$ such that $(x - x_0) \cdot \nu(x) \leq 0$, $\forall\, x \in \Gamma_1$ [22, Appendices]. Additional cases are also possible [13].

Remark 4.1.4. At the price of introducing heavy notation, it would be possible to include into problem (4.1.1), also the case where the elastic wall Γ_0 is *curved* and, accordingly, modeled by a shell equation to be written abstractly as in (4.1.1d), see [1.3]. □

Concrete illustrations of the abstract elastic equation (4.1.1d). As canonical illustrations of the abstract elastic equation (4.1.1d) — say, with no coupling term z_t and with no control: $u \equiv 0$ — we may take the classical Euler-Bernoulli equation on Γ_0 ($k = 0$) or the corresponding Kirchhoff equation on Γ_0 ($k > 0$):

$$v_{tt} - k\Delta v_{tt} + \Delta^2 v + \Delta^2 v_t = 0 \quad \text{on } (0, T] \times \Gamma_0, \tag{4.1.9}$$

under a variety of B.C. on $(0, T] \times \partial\Gamma_0$: hinged, clamped, free B.C., etc. [1], [13], [21]. Then, $\mathcal{A}$ is the realization of Δ^2 on $L_2(\Gamma_0)$ subject to the appropriate B.C. Finally, $\mathcal{M}_k = I + k\mathcal{A}_1$ where $\mathcal{A}_1$ is the realization of $(-\Delta)$ on $L_2(\Gamma_0)$ under suitable B.C.

Conclusion. Under the above assumptions, including those of Remark 4.1.3, model (4.1.1) is covered by the abstract theory of Sections 1 and 2, with $\gamma < \frac{1}{2}$.

Remark 4.1.5. Example 4.1 can be extended to replace, in Eq. (4.1.1d), the damping term $\mathcal{A}v_t$ with a damping term $\mathcal{A}^\alpha v_t$, $\frac{1}{2} \leq \alpha \leq 1$, the range of analyticity of the free v-problem. New interesting phenomena arise, which are analyzed in detail in [5]. In particular, the singular estimate for $e^{At}B$ continues to hold true, with a parameter γ that depends on α. □

Example 4.2: A class of structural acoustic problems with constant $\frac{1}{2} < \gamma < 1$

The model. We use, when possible, the same notation as in Example 4.1. We consider again an acoustic chamber Ω endowed with a rigid wall Γ_1 and a flexible wall Γ_0, where, now, however, we introduce two main changes over Example 4.1: (i) the wave equation in z displays a "strong" damping on the wall Γ_0 (much stronger than damping on z_t in (4.1.1c) of Example 4.1: see operator D in (4.1.10c) below); (ii) the flexible wall Γ_0 now also accounts for thermal effects, and is therefore modeled by a thermoelastic beam or plate, where w and θ denote displacement and temperature. Accordingly, the new model is now given by

$$\left\{\begin{array}{ll}
\text{acoustic chamber:} & \left\{\begin{array}{lll}
z_{tt} = \Delta z - d_1 z + f & \text{in } (0,T] \times \Omega, & \text{(4.1.10a)} \\
\alpha \dfrac{\partial z}{\partial \nu} + z = 0 & \text{in } (0,T] \times \Gamma_1, & \text{(4.1.10b)} \\
\alpha = 0: \text{ Dirichlet B.C.}; & & \\
\alpha > 0: \text{ Robin B.C.} & & \\
\dfrac{\partial z}{\partial \nu} + D z_t = v_t & \text{in } (0,T] \times \Gamma_0, & \text{(4.1.10c)}
\end{array}\right. \\
\text{thermo-elastic wall} & \left\{\begin{array}{lll}
v_{tt} - k\Delta v_{tt} + \Delta^2 v + \Delta\theta + z_t = \mathcal{B}u & \text{in } (0,T] \times \Gamma_0, & \text{(4.1.10d)} \\
\theta_t - \Delta\theta - \Delta v_t = 0 & \text{in } (0,T] \times \Gamma_0, & \text{(4.1.10e)} \\
\text{plus Boundary Conditions} & &
\end{array}\right. \\
z(0,\cdot) = z_0,\ z_t(0,\cdot) = z_1 \text{ in } \Omega; & \\
v(0,\cdot) = v_0,\ v_t(0,\cdot) = v_1 \text{ in } \Gamma_0. & \text{(4.1.10f)}
\end{array}\right.$$

where

$$z_0 \in H^1_{\Gamma_1}(\Omega) \equiv \{\phi \in H^1(\Omega) : \phi = 0 \text{ on } \Gamma_1\};\ z_1 \in L_2(\Omega); \tag{4.1.11a}$$

$$v_0 \in H^2(\Gamma_0);\ v_1 \in H^1(\Gamma_0; k) \equiv \begin{cases} L_2(\Gamma_0) & \text{if } k = 0 \\ H^1(\Gamma_0) & \text{if } k > 0 \end{cases};\ \theta_0 \in L_2(\Gamma_0), \tag{4.1.11b}$$

where, for $k > 0$, $H^1(\Gamma_0; k)$ is topologized by

$$\|\psi\|^2_{H^1(\Gamma_0;k)} \equiv \|\psi\|^2_{L_2(\Gamma_0)} + k\|\|\nabla\psi\|\|^2_{L_2(\Gamma_0)}. \tag{4.1.11c}$$

In (4.1.10b–c), ν is the unit outward vector to the boundary $\Gamma = \partial\Omega$. By contrast, below, when $\dim \Omega = 3$ and so $\dim \Gamma_0 = 2$, we shall let $\tilde{\nu}$ be the unit outward normal to $\partial\Gamma_0$ as the boundary of Γ_0; and, $\tilde{\tau}$ be the unit tangent vector to $\partial\Gamma_0$, oriented counterclockwise. System (4.1.10) is supplemented with Boundary Conditions (B.C.). We shall explicitly consider three sets of B.C. for the thermoelastic component:

Hinged B.C.:

$$v = \Delta v = \theta = 0 \text{ on } (0, T] \times \partial\Gamma_0; \tag{4.1.12}$$

Clamped B.C.:

$$v = \frac{\partial v}{\partial \tilde{\nu}} = \theta = 0 \text{ on } (0, T] \times \partial\Gamma_0. \tag{4.1.13}$$

Free B.C. when $\dim \Omega = 3$, $\dim \Gamma_0 = 2$:

$$\begin{cases} \Delta v + B_1 v + \theta = 0 & & \text{(4.1.14a)} \\ \dfrac{\partial}{\partial \tilde{\nu}} \Delta v + B_2 v - \gamma \dfrac{\partial}{\partial \tilde{\nu}} v_{tt} + \dfrac{\partial \theta}{\partial \tilde{\nu}} = 0 & \text{on } (0, T] \times \partial\Gamma_0 & \text{(4.1.14b)} \\ \dfrac{\partial \theta}{\partial \tilde{\nu}} + \lambda\theta = 0, \ \lambda > 0, & & \text{(4.1.14c)} \end{cases}$$

where, with constant $0 < \mu < 1$, we have

$$B_1 v = (1-\mu)(2\nu_1\nu_2 v_{xy} - \nu_1^2 v_{yy} - \nu_2^2 v_{xx}), \quad \tilde{\nu} = [\nu_1, \nu_2]; \tag{4.1.14d}$$

$$B_2 v = (1-\mu)\left\{\frac{\partial}{\partial \tilde{\tau}}[(\nu_1^2 - \nu_2^2)v_{xy} + \nu_1\nu_2(v_{yy} - v_{xx})] + \ell v\right\}, \ \ell > 0. \tag{4.1.14e}$$

Remark 4.1.6. The parameter $k \geq 0$ in (4.1.10d) is critical in describing the character of the dynamics of the uncoupled free thermoelastic system (4.1.10d–e) [that is, with no coupling term z_t and with $u \equiv 0$]: for $k = 0$, such a thermoelastic problem generates an s.c. analytic semigroup ("parabolic" case) [16, Chapter 3, Appendices 3E-3I], [18–20], while for $k > 0$ the corresponding s.c. semigroup is "hyperbolic-dominated" in a technical sense [17]. □

Throughout this example, we let

$$\mathcal{A} = \text{realization in } L_2(\Gamma_0) \text{ of } \Delta^2 \text{ subject to hinged, or clamped, or free homogeneous B.C.} \tag{4.1.15}$$

Regarding the control operator $\mathcal{B}$ in (4.1.10d), we shall assume the same hypothesis (A3) = (4.1.4), here restated as

(h1) there exists a positive constant $\rho < \frac{1}{2}$, such that

$$\mathcal{A}^{-\rho}\mathcal{B} \in \mathcal{L}(\mathcal{U}; L_2(\Gamma)), \text{ equivalently } \mathcal{B}: \text{ continuous } \mathcal{U} \to [\mathcal{D}(\mathcal{A}^{\rho})]' \tag{4.1.16}$$

[an assumption satisfied with $\rho = \frac{3}{8} + \epsilon$, if $\mathcal{B}$ is the operator defined in (4.1.6)]. In addition, we make the following assumption on the tangential positive self-adjoint operator D occurring in (4.1.10c):

(h2) With ρ as in (4.1.16), there exist positive constants δ_1, δ_2, such that $\mathcal{D}(D^{\frac{1}{2}}) = \mathcal{D}(\mathcal{A}^{\rho_0})$ and

$$\delta_1 \|z\|^2_{\mathcal{D}(\mathcal{A}^{\rho_0})} \leq (Dz, z)_{L_2(\Gamma_0)} \leq \delta_2 \|z\|^2_{\mathcal{D}(\mathcal{A}^{\rho_0})}, \quad \forall\, z \in \mathcal{D}(\mathcal{A}^{\rho_0}) \subset \mathcal{D}(D^{\frac{1}{2}}), \tag{4.1.17}$$

where

$$\begin{cases} \text{if } \frac{1}{4} \le \rho < \frac{1}{2}, & \text{then: } \rho - \frac{1}{4} \le \rho_0 \le \frac{1}{4}; \quad (4.1.18a) \\ \text{if } \rho < \frac{1}{4}, & \text{then: } \rho_0 = 0. \quad (4.1.18b) \end{cases}$$

A typical example of such tangential operator D is a realization of the Laplace-Beltrami operator on $L_2(\Gamma_0)$.

Remark 4.1.7 (on assumption (h2)). (a) If the constant ρ in (4.1.16) satisfies $\rho < \frac{1}{4}$, then the damping operator D may be taken to be the identity operator on $L_2(\Gamma_0)$.

(b) If, however, $\frac{1}{4} \le \rho < \frac{1}{2}$ [as in the case of the control operator $\mathcal{B}$ in (4.1.6)], then a stronger, unbounded damping operator D is needed. More precisely:

(b1) Let $\frac{1}{4} \le \rho \le \frac{3}{8}$ so that $\rho - \frac{1}{4} \le \frac{1}{8}$ and we can take $\rho_0 = \frac{1}{8}$. Then $4\rho_0 - \frac{1}{2} = 0$ and then $\mathcal{D}(\mathcal{A}^{\rho_0}) = \mathcal{D}(\mathcal{A}^{\frac{1}{8}})$ is topologically equivalent to $H^{\frac{1}{2}}(\Gamma_0)$ subject to appropriate B.C. [e.g., in the case of hinged or clamped B.C., then $\mathcal{D}(\mathcal{A}^{\frac{1}{8}}) = H_{00}^{\frac{1}{2}}(\Gamma_0)$].

(b2) Let $\frac{3}{8} < \rho < \frac{1}{2}$ so that $\frac{1}{8} = \frac{3}{8} - \frac{1}{4} < \rho - \frac{1}{4} < \frac{1}{4}$, and we can take ρ_0 satisfying $\frac{1}{8} < \rho_0 \le \frac{1}{4}$. Then, (*): $0 < 4\rho_0 - \frac{1}{2} \le \frac{1}{2}$. Then, the following two subcases need to be considered.

(b2i) Assume either hinged or clamped B.C. for the operator $\mathcal{A}$ in (4.1.15), see (4.1.12) or (4.1.13). Then (*) above implies [6] that

$$\mathcal{D}(\mathcal{A}^{\rho_0}) = H_0^{4\rho_0}(\Gamma_0) \subset H^1(\Gamma_0). \quad (4.1.19)$$

Thus, in this case, Dz_t well-defined requires, by assumptions (4.1.17) and (4.1.19), that $z|_{\partial\Gamma_0} \equiv 0$. To ensure this, we then take $\alpha = 0$ in (4.1.10b), so that the z-problem is endowed with Dirichlet, rather than Robin, B.C.

(b2ii) Assume now free B.C. for the operator $\mathcal{A}$ in (4.1.15). Then, (*) above implies that

$$\mathcal{D}(\mathcal{A}^{\rho_0}) = H^{4\rho_0}(\Gamma_0) \subset H^1(\Gamma_0), \quad (4.1.20)$$

and then we can allow $\alpha \ge 0$ in (4.1.10b): that is, either Robin or Dirichlet B.C.

The above structural acoustic model (4.1.10), subject to assumptions (h1), (h2), and $k = 0$ *(no rotational forces accounted for) satisfies assumption (H.3) = (1.3)* with $\gamma = 2\rho$ [13]. Thus, if $\mathcal{B}$ is the control operator defined by (4.1.6), then $\frac{1}{2} < \gamma = 2(\frac{3}{8} + \epsilon) = \frac{3}{4} + 2\epsilon < 1$.

When $k = 0$ in (4.1.10d) [Euler-Bernoulli rather than Kirchhoff equation] and assumptions (h1), (h2) above are in force, then the structural acoustic model (4.1.10) can be rewritten in the abstract form (1.1), with operators A and B explicitly identified. Moreover, the operator A is the generator of an s.c. contraction semigroup e^{At} on an appropriate finite energy space Y given by

$$Y \equiv H^1_{\Gamma_1}(\Omega) \times L_2(\Omega) \times \mathcal{D}(\mathcal{A}^{\frac{1}{2}}) \times L_2(\Gamma_0) \times L_2(\Gamma_0) \quad (4.1.21)$$

for the variables $[z(t), z_t(t), v(t), v_t(t), \theta(t)]$. The abstract deterministic disturbance w in Eq. (1.1) is now $w = [0, f, 0, 0, 0]$, with f the disturbance in (4.1.10a). Finally, the s.c. semigroup e^{At} is uniformly stable on Y [16, Chapter 3, Sections 3.11–3.13]. (In the case of free B.C., this is due to the term $(1-\mu)\ell v$ in (4.1.14e) with coefficient $(1-\mu)\ell < 0$).

Conclusion. Under the above assumptions, model (4.1.10) is covered by the abstract theory of Sections 1–3.

4.2 A class of thermoelastic plates with thermal boundary control and $\gamma = \frac{3}{4} + \epsilon$, $\forall\, \epsilon > 0$

In this subsection we consider a class of thermoelastic plate problems with rotational inertia and boundary control in the thermal component. More precisely, let $\Omega \subset \mathbb{R}^2$ be a smooth bounded domain with boundary Γ. [The following model is the counterpart of the thermoelastic component modeling the elastic wall in the structural acoustic problem of Example 4.2.] If w denotes the vertical displacement of the plate and θ denotes the temperature (with respect to the stress-free state), the thermoelastic plate model with boundary thermal control u considered throughout this subsection is

$$\begin{cases} w_{tt} - \rho\Delta w_{tt} + \Delta^2 w - \Delta\theta = 0 & \text{in } (0, T] \times \Omega \equiv Q; & (4.2.1\text{a}) \\ \theta_t - \Delta\theta + \Delta w_t = 0 & \text{in } Q; & (4.2.1\text{b}) \\ w|_\Sigma \equiv 0,\ \Delta w|_\Sigma \equiv 0,\ \theta|_\Sigma = u & \text{in } (0, T] \times \Gamma \equiv \Sigma; & (4.2.1\text{c}) \\ w(0, \cdot) = w_0,\ w_t(0, \cdot) = w_1,\ \theta(0, \cdot) = \theta_0 & \text{in } \Omega. & (4.2.1\text{d}) \end{cases}$$

The boundary control u acts in the Dirichlet boundary conditions of the thermal component θ. The boundary conditions of the mechanical variable w are "hinged" B.C. The positive constant $\rho > 0$ is proportional to the square of the thickness of the plate and accounts for the rotational forces on the model. It is known [17] that for $\rho > 0$, the free dynamics (3.2) with $u \equiv 0$ is hyperbolic-dominated, in the sense that the resulting thermoelastic s.c. semigroup is a compact perturbation, at each $t > 0$, of a damped s.c. group [17].

Consistently with (4.1.15), we define the operator $\mathcal{A}$ by

$$\mathcal{A}f = \Delta^2 f, \quad \mathcal{D}(\mathcal{A}) \equiv H^4(\Omega) \cap H_0^2(\Omega). \tag{4.2.2}$$

Then, with equivalent norms:

$$\mathcal{D}(\mathcal{A}^{\frac{1}{2}}) \equiv H_0^2(\Omega), \qquad \mathcal{D}(\mathcal{A}^{\frac{1}{4}}) \equiv H_0^1(\Omega). \tag{4.2.3}$$

The state space of problem (4.2.1) (with $\rho > 0$) is

$$Y_\rho \equiv \mathcal{D}(\mathcal{A}^{\frac{1}{2}}) \times \mathcal{D}(\mathcal{A}^{\frac{1}{4}}) \times L_2(\Omega). \tag{4.2.4}$$

It is possible to *rewrite problem (4.2.1) in the form of the abstract model (1.1)*, with operators A and B explicitly identified ([1], [13], [16, Chapter 3], [17]). Moreover, in the same references, it is shown that the resulting operator A is *the infinitesimal* generator of an s.c. semigroup e^{At} of contractions on the space Y_ρ above (4.2.2), so that assumption (H.1) holds true. Further, the resulting operator B satisfies assumption (H.2). The key issue is, of course, assumption (H.3): This was established to hold true in [14], with $\gamma = \frac{3}{4} + \epsilon$, with respect to A, B, Y as above, and $U = L_2(\Gamma)$.

Conclusion. Model (4.2.1) satisfies the abstract assumptions (H.1), (H.2), (H.3) of Section 1 and the abstract theory of Sections 1–3 applies. □

4.3 A class of composite beams models

In this subsection we focus on a model, due to [9], of a sandwich beam consisting of a thin, compliant middle layer and two identical stiff outer layers. In this simplified model, the middle layer resists shear, but not bending, and the thickness is assumed to be small enough, so that the mass may be neglected or included in the outer layers. Damping is included in the middle layer in such a way that shear motions are resisted by a force proportional to the rate of shear. The outer layers obey the usual assumptions of the Euler-Bernoulli beam theory.

The original model. For $0 < x < 1$ and $0 < t < T$, let $w(t,x)$ = transverse displacement; and $\xi(t,x)$ = effective rotation angle of the beam. Then the following mathematical model of a sandwich beam is proposed in [9], where $\Omega = (0,1)$:

$$\left\{\begin{array}{lr} w_{tt} - \rho w_{ttxx} - (d\xi_x)_{xx} = 0 \quad \text{in } (0,T] \times \Omega; & (4.3.1a) \\ \frac{\beta}{3}(\xi_t + w_{tx}) - d\left(\xi_x + \frac{1}{4} w_{xx}\right)_x + \frac{\delta}{3}(\xi + w_x) = 0 \quad \text{in } (0,T] \times \Omega; & (4.3.1b) \\ w(0,\cdot) = w_0,\ w_t(0,\cdot) = w_1,\ \xi(0,\cdot) = \xi_0 \quad \text{in } \Omega, & (4.3.1c) \end{array}\right.$$

along with appropriate Boundary Conditions (B.C.) at $x = 0$ and $x = 1$. We shall consider two cases:

CASE 1: *Hinged/Neumann B.C. for* $\{w, \xi\}$. These are

$$w|_\Sigma \equiv 0,\ w_{xx}|_\Sigma \equiv 0;\ \xi_x|_\Sigma \equiv 0, \quad \Sigma \equiv (0,T] \times \Gamma, \tag{4.3.2}$$

where throughout this section, $\Gamma = \{0,1\}$ consists of the two end points $x = 0$ and $x = 1$.

CASE 2: *Clamped/Dirichlet B.C. for* $\{w, \xi\}$. These are

$$w|_\Sigma \equiv 0,\ w_x|_\Sigma \equiv 0;\ \xi|_\Sigma \equiv 0,\ \Sigma \equiv (0,T] \times \Gamma. \tag{4.3.3}$$

In (4.3.1a-b), we have that: γ represents a rotational moment of inertia; $D > 0$ is the stiffness of the beam; $\delta > 0$ is the stiffness of the middle layer; and β is the damping parameter for the middle layer.

A first transformation of model (4.1). We introduce a new variable [11]

$$s(t,x) \equiv \xi(t,x) + w_x(t,x) \tag{4.3.4}$$

in terms of the effective rotation angle ξ and the transverse displacement w. Physically, s is proportional to the *shear* of the middle layer. Then, model (1.1) in the variables $\{w, \xi\}$ is transformed into the following system in the variables $\{w, s\}$, by virtue of (4.3.4):

$$\left\{\begin{array}{lr} w_{tt} - \rho w_{ttxx} + d w_{xxxx} - d s_{xxx} \equiv 0 \quad \text{in } (0,T] \times \Omega; & (4.3.5a) \\ \beta s_t - 3d\left[s_{xx} - \frac{3}{4} w_{xxx}\right] + \delta s \equiv 0 \quad \text{in } (0,T] \times \Omega; & (4.3.5b) \\ w(0,\cdot) = w_0,\ w_t(0,\cdot) = w_1; \quad \text{in } \Omega, & \\ s(0,\cdot) = s_0 = \xi_0 + w_x(0,\cdot) \quad \text{in } \Omega, & (4.3.5c) \end{array}\right.$$

along with the following Boundary Conditions at $x = 0$ and $x = 1$.

CASE 1: *Hinged/Neumann B.C. for* $\{w, \xi\}$ *as in (4.3.2) yield Hinged/Neumann B.C. for* $\{w, s\}$. In fact, direct use of (4.3.2) and (4.3.4) yields

$$w|_\Sigma \equiv 0,\ w_{xx}|_\Sigma \equiv 0;\ s_x|_\Sigma = [\xi_x + w_{xx}]_\Sigma \equiv 0, \quad \Sigma = (0,T] \times \Gamma. \tag{4.3.6}$$

CASE 2: *Clamped/Dirichlet B.C. for* $\{w, \xi\}$ *as in (4.3.3) yield Clamped/Dirichlet B.C. for* $\{w, s\}$. In fact, direct use of (4.3.2) and (4.3.4) yields

$$w|_\Sigma \equiv 0,\ w_x|_\Sigma \equiv 0;\ s|_\Sigma = [\xi + w_x]_\Sigma \equiv 0, \quad \Sigma \equiv (0,T] \times \Gamma. \tag{4.3.7}$$

A final transformation of model (4.1) into a thermoelastic model. With reference to (4.3.4) we introduce a new variable [28]

$$\theta(t,x) \equiv s_x(t,x) - \frac{3}{4} w_{xx}(t,x) \equiv \xi_x(t,x) + \frac{1}{4} w_{xx}(t,x). \tag{4.3.8}$$

Then, model (4.3.1) in the variables $\{w, \xi\}$, or model (4.3.5) in the variables $\{w, s\}$ are transformed into the following system in the variables $\{w, \theta\}$, by virtue of (4.3.8):

$$\begin{cases} w_{tt} - \rho w_{ttxx} + \frac{d}{4} w_{xxxx} - d\theta_{xx} = 0 \quad \text{in } (0,T] \times \Omega; & \text{(4.3.9a)} \\ \beta\theta_t - 3d\theta_{xx} + \delta\theta + \dfrac{3}{4}\beta w_{xxt} + \dfrac{3\delta}{4} w_{xx} \quad \text{in } (0,T] \times \Omega; & \text{(4.3.9b)} \\ w(0,\cdot) = w_0,\ w_t(0,\cdot) = w_1; \\ \theta(0,\cdot) = \theta_0 = \xi_x(0,\cdot) + \frac{1}{4} w_{xx}(0,\cdot) \quad \text{in } \Omega. & \text{(4.3.9c)} \end{cases}$$

Indeed, for instance, starting from (4.3.5): to obtain (4.3.9a) we replace $s_{xxx} = \theta_{xx} + \frac{3}{4} w_{xxxx}$ (obtained from (4.3.8)) into (4.3.5a). To obtain (4.3.9b), we rewrite (4.3.5b) as $\beta s_t = 3d\theta_x - \delta s$ by (4.3.8) and substitute $\beta s_{tx} = 3d\theta_{xx} - \delta s_x = 3d\theta_{xx} - \delta\left[\theta + \frac{3}{4} w_{xx}\right]$ into $\beta\theta_t = \beta s_{xt} - \frac{3}{4}\beta w_{xxt}$ obtained from (4.3.8). With (4.3.9) we associate appropriate B.C. at $x = 0$ and $x = 1$.

CASE 1: *Hinged/Neumann B.C. (for* $\{w, \xi\}$, *hence) for* $\{w, s\}$ *yields Hinged/Dirichlet B.C. for* $\{w, \theta\}$. In fact, direct use of (4.3.6), (4.3.2), and (4.3.8), yields

$$w|_\Sigma \equiv 0,\ w_{xx}|_\Sigma \equiv 0;\ \theta|_\Sigma = \left[s_x - \frac{3}{4} w_{xx}\right]_\Sigma = \left[\xi_x + \frac{1}{4} w_{xx}\right]_\Sigma \equiv 0, \qquad \Sigma = (0,T] \times \Gamma. \tag{4.3.10}$$

CASE 2: *Clamped/Dirichlet B.C. (for* $\{w, \xi\}$, *hence) for* $\{w, s\}$ *yields Clamped/Neumann B.C. for* $\{w, \theta\}$. In fact, direct use of (4.3.7) with $s|_\Sigma = 0$, hence of $s_t|_\Sigma \equiv 0$, yields

$$w|_\Sigma \equiv 0,\ w_x|_\Sigma \equiv 0;\ \theta_x|_\Sigma = \left[s_{xx} - \frac{3}{4} w_{xxx}\right]_\Sigma \equiv 0;\ \left[\xi_{xx} + \frac{1}{4} w_{xxx}\right] \equiv 0, \tag{4.3.11}$$

by virtue also of (4.3.8) and the equation in (4.3.5b).

Thus, we recognize the $\{w, w_t, \theta\}$-problem in (4.3.9a-b-c) along with either the Hinged/Dirichlet B.C. (4.3.10), or else the clamped/Neumann B.C. (4.3.11), to be a thermoelastic problem of the type of Section 4.2. In Eq. (4.3.9b), the terms θ and w_{xx} which are premultiplied by δ are lower-order terms.

Energy function and state space. With reference to model (4.3.1), we associate its natural energy function, which is given by

$$E_\rho(t) \equiv \int_0^1 \Big[w_t^2(t,x) + \rho w_{tx}^2(t,x) + d\xi_x^2(t,x)$$

$$+\frac{d}{3}|\xi_x(t,x) + w_{xx}(t,x)|^2 + \frac{4\delta}{9}|\xi(t,x) + w_x(t,x)|^2\Big]\, dx. \quad (4.3.12)$$

We readily see that the energy function (4.3.12) is well-defined for $t \geq 0$, provided that the variables $\{w(t), w_t(t), \xi(t)\}$ belong to the following state space, where $\Omega = (0,1)$,

$$\begin{cases} Y_\rho \equiv [H^2(\Omega) \cap H_0^1(\Omega)] \times \begin{matrix} H_0^1(\Omega) \\ L_2(\Omega) \end{matrix} \times H^1(\Omega) & \begin{matrix} \text{for } \rho > 0, \\ \text{for } \rho = 0, \end{matrix} \\ \text{in the case of Hinged/Neumann B.C. for } \{w,\xi\} \text{ in (4.3.2);} \end{cases} \quad (4.3.13)$$

$$\begin{cases} Y_\rho \equiv H^2(\Omega) \times \begin{matrix} H_0^1(\Omega) \\ L_2(\Omega) \end{matrix} \times H_0^1(\Omega) & \begin{matrix} \text{for } \rho > 0, \\ \text{for } \rho = 0, \end{matrix} \\ \text{in the case of Clamped/Dirichlet B.C. for } \{w,\xi\} \text{ in (4.3.3).} \end{cases} \quad (4.3.14)$$

In turn, with reference to problem (4.3.5), we see that the variable s defined in (4.3.4) satisfies: $s = \xi + w_x \in H^1(\Omega)$ for $\{w,\xi\}$ as in (4.3.13), (4.3.14); and hence, finally, the variable θ in (4.3.8) satisfies then $\theta = s_x - \frac{3}{4}w_{xx} \in L_2(\Omega)$. We collect these results:

$$s = \xi + w_x \in H^1(\Omega);\ \theta = s_x - \frac{3}{4}w_{xx} \in L_2(\Omega). \quad (4.3.15)$$

4.4 Composite beams models under homogeneous hinged mechanical B.C. and boundary control on $\xi_x|_\Sigma$. Well-posedness. Singular estimates

Nonhomogeneous boundary problem. In this subsection, we continue the study of the sandwich beam model of Section 4.3, originally derived in the original variables $\{w,\xi\}$ as in (4.3.1a–c); hence in the variables $\{w,s\}$ as in (4.3.5a–c); finally, in the "thermoelastic" variables $\{w,\theta\}$ as in (4.3.9a–c), under the following nonhomogeneous B.C.:

$$w|_\Sigma \equiv 0, \qquad w_{xx}|_\Sigma \equiv 0; \qquad \xi_x|_\Sigma \equiv u, \quad (4.4.1a)$$

or

$$s_x|_\Sigma \equiv [\xi_x + w_{xx}]_\Sigma \equiv u; \quad \text{or} \quad \theta|_\Sigma \equiv \left[\xi_x + \frac{1}{4}w_{xx}\right]_\Sigma \equiv \left[s_x - \frac{3}{4}w_{xx}\right]_\Sigma \equiv u. \quad (4.4.1b)$$

Following [27], we shall concentrate on the reduction of $\{w,\xi\}$ to the thermoelastic variables $\{w,\theta\}$.

Well-posedness of $\{w,\theta\}$-problem. In this subsection, we shall exploit the reduction of the original $\{w,\xi\}$-system to its thermoelastic version of $\{w,\theta\}$ with hinged B.C. in the mechanical variables w, and thermal boundary control u in the Dirichlet B.C. of θ. To this end, we introduce the following operators, where $\rho > 0$ throughout:

$$\mathcal{A}f = -\Delta f, \quad \mathcal{D}(\mathcal{A}) = H^2(\Omega) \cap H_0^1(\Omega); \; \mathcal{A}_\rho = (I + \rho\mathcal{A}); \tag{4.4.2}$$

$$\mathcal{A}^2 f = \Delta^2 f, \quad \mathcal{D}(\mathcal{A}^2) = \{f \in H^4(\Omega) : \; f|_\Gamma = \Delta f|_\Gamma = 0\}, \tag{4.4.3}$$

and the space

$$Y \equiv Y_\rho \equiv X_1 \times X_2 \times X_3; \tag{4.4.4}$$

$$X_1 \equiv \mathcal{D}_{\sqrt{D/4}}(\mathcal{A}); \quad X_2 \equiv \mathcal{D}(\mathcal{A}_\rho^{\frac{1}{2}}); \quad X_3 \equiv L_{2;(\sqrt{4/3})}(\Omega); \tag{4.4.5}$$

$$\|x_1\|_{X_1}^2 \;\; \equiv \;\; \frac{D}{4}\|\mathcal{A}x_1\|_{L_2(\Omega)}^2; \quad \|x_2\|_{X_2}^2 = \|\mathcal{A}_\rho^{\frac{1}{2}} x_2\|_{L_2(\Omega)}^2; \tag{4.4.6}$$

$$\|x_3\|_{X_3}^2 \;\; = \;\; \frac{4}{3}\|x_3\|_{L_2(\Omega)}^2. \tag{4.4.7}$$

Thus, for $f \equiv [f_1, f_2, f_3] \in Y_\rho, \quad g = [g_1, g_2, g_3] \in Y_\rho$, the Y_ρ-inner product is

$$(f, g)_{Y_\rho} = \frac{D}{4}(\mathcal{A}f_1, \mathcal{A}g_1)_{L_2(\Omega)} + ((I + \rho\mathcal{A})f_2, g_2)_{L_2(\Omega)} + \frac{4}{3}(f_3, g_3)_{L_2(\Omega)}. \tag{4.4.8}$$

Further, we introduce the operator

$$A_\rho = \begin{bmatrix} 0 & I & 0 \\ -\frac{D}{4}\mathcal{A}_\rho^{-1}\mathcal{A}^2 & 0 & -\mathcal{A}_\rho^{-1}\mathcal{A} \\ \frac{3\delta}{4\beta}\mathcal{A} & \frac{3}{4}\mathcal{A} & -\frac{3D}{\beta}\mathcal{A} - \frac{\delta}{\beta} \end{bmatrix} : \; Y_\rho \supset \mathcal{D}(A_\rho) \to Y_\rho, \tag{4.4.9}$$

with its natural domain. The following result then follows from thermoelastic system [1], [17].

Lemma 4.4.1. Let $\rho > 0$. (i) The operators

$$\begin{bmatrix} 0 & I \\ -\frac{d}{4}\mathcal{A}_\rho^{-1}\mathcal{A}^2 & 0 \end{bmatrix}, \quad \begin{bmatrix} 0 & I & 0 \\ -\frac{d}{4}\mathcal{A}_\rho^{-1}\mathcal{A}^2 & 0 & -\mathcal{A}_\rho^{-1}\mathcal{A} \\ 0 & \frac{3}{4}\mathcal{A} & 0 \end{bmatrix}, \tag{4.4.10}$$

with their natural domains, are *unitary* operators on $X_1 \times X_2$, and $Y_\rho = X_1 \times X_2 \times X_3$, respectively.

(ii) For $\delta = 0$, the operator A_ρ in (4.4.9) is maximal dissipative, hence the generator of an s.c. contraction semigroup on Y_ρ. Similarly for its Y_ρ-adjoint A_ρ^*.

(iii) Let $\delta > 0$. The operator obtained from A_ρ in (4.4.9) by removing the left bottom corner entry $\frac{3\delta}{4\beta}\mathcal{A}$ is maximal dissipative, and hence the generator of an s.c. contraction semigroup on Y_ρ. What is left is the operator

$$\begin{bmatrix} 0 & 0 \\ \frac{4}{3}\frac{\delta}{\beta}\mathcal{A} & 0 \end{bmatrix} \in \mathcal{L}(Y_\rho), \tag{4.4.11}$$

which is bounded on Y_ρ.

(iv) Thus, the operator A_ρ in (4.4.9), being a bounded perturbation of the generator of an s.c. contraction semigroup on Y_ρ, is itself the generator of an s.c. semigroup on Y_ρ (noncontraction, however), which is denoted by $e^{A_\rho t}$, $t \geq 0$.

(v) System (4.3.9a-b-c) in $\{w, \theta\}$ with hinged/Dirichlet homogeneous B.C. (4.3.10) (Case 1), can be rewritten abstractly as

$$\dot{y}(t) = A_\rho y(t), \quad y(t) = \{w(t), w_t(t), \theta(t)], \tag{4.4.12}$$

with A_ρ defined by (4.4.9). Thus, for any I.C. $y_0 = [w_0, w_1, \theta_0] \in Y_\rho$, we have

$$\begin{aligned} y(t) \equiv [w(t), w_t(t), \theta(t)] &= e^{A_\rho t} y_0 = e^{A_\rho t}[w_0, w_1, \theta_0] \\ &\in C([0,T]; Y_\rho). \end{aligned} \tag{4.4.13}$$

(vi) System (4.3.9a-b-c) on $\{w, \theta\}$ with nonhomogeneous B.C. $w|_\Sigma = w_{xx}|_\Sigma \equiv 0$, $\theta|_\Sigma = u$ in (4.4.1), can be rewritten abstractly as

$$\dot{y}(t) = A_\rho y(t) + Bu \in [\mathcal{D}(A_\rho^*)]', \tag{4.4.14}$$

where A_ρ is defined in (4.4.9), and B is defined by

$$Bu = \begin{bmatrix} 0 \\ A_\rho^{-1} ADu \\ ADu \end{bmatrix}; \quad D : L_2(\Gamma) \to H^{\frac{1}{2}}(\Omega); \tag{4.4.15}$$

$$\begin{cases} \Delta(Dg) = 0 & \text{in } \Omega; \\ Dg = g & \text{on } \Gamma, \end{cases}$$

where D is the Dirichlet map. □

Lemma 4.4.1 verifies assumptions (H.1), (H.2) of Section 1.

Singular estimate for $e^{A_\rho t}B$. We now come to the verification of the critical assumption (H.3) = (1.2) of Section 1.

Proposition 4.4.2. With reference to system (4.3.9a-b-c) in $\{w, \theta\}$ with nonhomogeneous B.C. (4.4.1), or its abstract version (4.4.13), we have:

(i) Let $\delta = 0$ in (4.3.9b) or (4.4.9). Then, with corresponding $e^{A_\rho t}B$ satisfies the singular estimate:

$$\|e^{A_\rho t} B\|_{\mathcal{L}(Y_\rho)} \leq \frac{C_T}{t^\gamma}, \quad 0 < t \leq T, \ \gamma = \frac{3}{4} + \epsilon, \ \forall \epsilon > 0. \tag{4.4.16}$$

(ii) Now let $\delta > 0$. Then, the corresponding $e^{A_\rho t}B$ satisfies the same singular estimate as in (4.4.16). Explicitly, this means the following: Consider system (4.3.9a-b) in $\{w, \theta\}$ with homogeneous B.C. (4.3.10): $w|_\Sigma \equiv w_{xx}|_\Sigma \equiv \theta|_\Sigma \equiv 0$, and Initial Conditions (see 4.4.15)):

$$w_0 = 0, \quad w_1 = A_\rho^{-1} ADu, \quad \theta_0 = ADu, \ u \in \mathbb{R}^2. \tag{4.4.17}$$

Then, the corresponding solution $\{w(t), w_t(t), \theta(t)\}$ satisfies the following singular estimate, recalling (4.4.4)–(4.4.8):

$$\|w(t)\|_{H^2(\Omega)} + \|w_t(t)\|_{H^1(\Omega)} + \|\theta(t)\|_{L_2(\Omega)} \leq \frac{C_T}{t^\gamma}|u|, \quad 0 < t \leq T, \qquad \gamma = \frac{3}{4} + \epsilon, \tag{4.4.18}$$

or, in terms of ξ, recalling (4.3.8),

$$\|\xi_x(t)\|_{L_2(\Omega)} \leq \|\theta(t)\|_{L_2(\Omega)} + \frac{1}{4}\|w_{xx}(t)\|_{L_2(\Omega)} \leq \frac{C_T}{t^\gamma}, \; 0 < t \leq T. \tag{4.4.19}$$

Proof. (i) For $\delta = 0$ (canonical thermoelastic case), the singular estimate (4.4.16) is established in [14], using techniques similar to those used in obtaining singular estimates in Sections 4.1 and 4.2.

(ii) For $\delta > 0$, the singular estimate (4.4.16) is retained, since the case $\delta > 0$ yields only a *bounded* perturbation of the generator (4.4.9) over the case $\delta = 0$. This is true because of the result in Appendix, Proposition A.2. Then, recalling that the space Y_ρ in (4.4.5) is topologically equivalent to $H^2(\Omega) \times H^1(\Omega) \times L_2(\Omega)$, we readily see that the singular estimate (4.4.16) means precisely the estimate (4.4.18) for problem (4.3.9a-b-c) with initial condition Bu given by (4.4.17), in view of (4.4.15). Then estimate (4.4.19) follows by (4.4.18) and $\xi_x = \theta - \frac{1}{4}w_{xx}$ by (4.3.8). □

The corresponding cost functional. We consider the optimal control problem of Section 1, for the boundary control problem (4.3.9a-b-c), (4.4.1), with cost functional given by

$$J(u; \{w, w_t, \xi\}) = \int_0^T \{\|w(t)\|^2_{H^2(\Omega)} + \|w_t(t)\|^2_{H^1(\Omega)} + \|\xi(t)\|^2_{H^1(\Omega)} + |u(t)|^2\}dt. \tag{4.4.20}$$

Conclusion. The boundary control system (4.3.9a-b-c), (4.4.1) describing a sandwich beam with boundary control and cost functional (4.4.20) satisfies the setting of Section 1, in particular, assumptions (H.1), (H.2), (H.3) with $Y \equiv Y_\rho$ given by (4.4.4), $A = A_\rho$ and B given by (4.4.9) and (4.4.15), and cost given by (4.4.20). Accordingly, the results of Sections 1–3 apply to it.

4.5 A structural acoustic 2-D model with a composite (sandwich) beam as flexible wall

In this subsection we return to the 2-D model of a structural acoustic chamber of Section 4.1 and replace the elastic flexible wall of Example 4.1, or the thermoelastic flexible wall of Example 4.2 with a composite (sandwich) beam, as in Section 4.3. The resulting coupled PDE system is

$$
\begin{cases}
\begin{cases}
z_{tt} = \Delta z - d_1 z_t + f & \text{in } (0,T] \times \Omega, & \text{(4.5.1a)}\\
\frac{\partial z}{\partial \nu} + d_2 z_t = 0 & \text{in } (0,T] \times \Gamma_1, & \text{(4.5.1b)}\\
\frac{\partial z}{\partial \nu} + d_3 z_t = \pm v_t & \text{in } (0,T] \times \Gamma_0, & \text{(4.5.1c)}
\end{cases}\\
\begin{cases}
v_{tt} - \rho v_{ttxx} - (d\xi_x)_{xx} \pm z_t = 0 & \text{in } (0,T] \times \Gamma_0, & \text{(4.5.1d)}\\
\frac{\beta}{3}(\xi_t + v_{tx}) - d(\xi_x + \frac{1}{4} v_{xx})_x & \text{in } (0,T] \times \Gamma_0, & \text{(4.5.1e)}\\
\quad + \frac{\delta}{3}(\xi + v_x) = 0 & & \\
v| = 0, \quad v_{xx}| = 0, \quad \xi_x| = u & \text{on } (0,T] \times \partial\Gamma_0, & \text{(4.5.1f)}
\end{cases}\\
z(0, \cdot) = z_0, \; z_t(0, \cdot) = z_1 \qquad \text{in } \Omega, & & \text{(4.5.1g)}\\
v(0, \cdot) = v_0, \; v_t(0, \cdot) = v_1 \qquad \text{in } \Gamma_0. & & \text{(4.5.1h)}
\end{cases}
$$

where the first system refers to the acoustic chamber, and the second system to the composite beam.

Problem (4.5) satisfies the abstract assumptions (H.1), (H.2), (H.3) of Section 1, on the state space

$$Y = H^1(\Omega) \times L_2(\Omega) \times X_1 \times X_2 \times X_3, \tag{4.5.2}$$

for

$$\left\{ z(t), z_t(t), v(t), v_t(t), \theta(t) = \xi_x(t) + \frac{1}{4} v_{xx}(t) \right\}$$

with X_i, $i = 1, 2, 3$ defined by (4.4.5)–(4.4.7); where $\gamma = \frac{3}{4} + \epsilon$, $\epsilon > 0$ as in Section 4.2. All this follows by using Sections 4.3, 4.4 in the analysis of Example 4.2.

APPENDIX: TWO RESULTS ON SINGULAR ESTIMATES

In this appendix we collect two results of interest regarding the singular estimate condition (H.3) = 1.2.

Proposition A.1. Assume (H.1), with, say, $\|e^{At}\| \leq M e^{\omega_0 t}$, $t \geq 0$, as well as (H.2), and (H.3), so that [21, Lemma 1.1], the following singular estimate holds true:

$$\|e^{At} B\|_{\mathcal{L}(U;Y)} \leq \frac{M e^{\omega t}}{t^\gamma}, \quad 0 < t; \; 0 < \gamma < 1, \tag{A.1}$$

for a constant $\omega > \omega_0$. [If the s.c. semigroup e^{At} is, moreover, exponentially stable, then the singular estimate (H.3) on $0 < t \leq T$ implies that the singular estimate (A.1) on $0 < t$

with some $\omega < 0$]. Then, for all $\lambda \in \mathbb{C}$ with Re $\lambda > \omega$, we have

$$n! \|R^{n+1}(\lambda, A)Bu\|_{\mathcal{L}(Y)} = \left\| \frac{d^n R(\lambda, A)Bu}{d\lambda^n} \right\|_{\mathcal{L}(Y)} \tag{A.2}$$

$$\leq \frac{M\Gamma(1-\gamma+n)}{(\text{Re}\,\lambda - \omega)^{1-\gamma+n}} \|u\|_U, \; n = 0, 1, 2, \ldots, \; \text{Re}\,\lambda > \omega, \tag{A.3}$$

where $\Gamma(\,\cdot\,)$ is the gamma function.

(ii) Conversely, suppose that estimates (A.3) hold true for all $\lambda = \text{Re}\,\lambda$ (real) $> \omega$, and $0 < \gamma < 1$. Then, we obtain

$$\|e^{At}B\|_{\mathcal{L}(U;Y)} \leq \frac{Me^{\omega t}}{t^\gamma}, \quad 0 < t, \; 0 < \gamma < 1. \tag{A.4}$$

Proof. First, (A.1) follows from (H.3) as in the elementary proof of [21, Lemma 1.1]. Next,

$$R(\lambda, A)Bu = \int_0^\infty e^{-\lambda t} e^{At} Bu\, dt, \tag{A.5}$$

where $u \in U$, and thus interpreted at first in $[\mathcal{D}(A^*)]'$, we obtain by (A.1) by recalling known results [8, p. 55] on the Laplace transform, with Re $\lambda > \omega$, $\gamma < 1$,

$$\frac{d^n}{d\lambda^n} R(\lambda, A)Bu = \int_0^\infty (-1)^n \frac{t^n e^{-\lambda t} e^{At} Bu}{t^\gamma}\, dt; \tag{A.6}$$

$$\left\| \frac{d^n}{d\lambda^n} R(\lambda, A)Bu \right\|_{\mathcal{L}(Y)} \leq M \int_0^\infty e^{-(\text{Re}\,\lambda - \omega)t} t^{n-\gamma} dt \|u\|_U$$

$$= \frac{M\Gamma(1-\gamma+n)}{(\text{Re}\,\lambda - \omega)^{1-\gamma+n}} \|u\|_U, \tag{A.7}$$

since $n - \gamma > -1$, as required, $n = 0, 1, \ldots$. Thus (A.7) establishes (A.2) since

$$\frac{d^n}{d\lambda^n} R(\lambda, A) = (-1)^n n!\, R^{n+1}(\lambda, A), \; \lambda \in \rho(A). \tag{A.8}$$

(ii) *Case* $\omega = 0$. Conversely, it suffices to restrict, at first, to the case $\omega = 0$. Thus, assume the validity of estimates (A.3) for, say, $\lambda = \text{Re}\,\lambda > \omega = 0$. We then seek to establish conclusion (A.4) with $\omega = 0$.

Step 1. We shall use the well-known formula of Hille-Phillips:

$$e^{At}Bu = \lim_{k \to \infty} \left[\frac{k}{t} R\left(\frac{k}{t}, A \right) \right]^k Bu, \quad t > 0, \; u \in U, \tag{A.9}$$

again interpreted, at first, in $[\mathcal{D}(A^*)]'$. We estimate

$$\left\| \left[\frac{k}{t} R\left(\frac{k}{t}, A\right)\right]^k Bu \right\|_{\mathcal{L}(Y)} \leq \frac{M}{t^\gamma} k^\gamma \frac{\Gamma(k-\gamma)}{(k-1)!}, \quad k = 1, 2, \ldots. \tag{A.10}$$

In fact,

$$\left\| \left[\frac{k}{t} R\left(\frac{k}{t}, A\right)\right]^k Bu \right\|_{\mathcal{L}(Y)} = \frac{k^k}{t^k} \left\| R^k\left(\frac{k}{t}, A\right) Bu \right\|_{\mathcal{L}(Y)} \tag{A.11}$$

$$\text{(by (A.3))} \qquad \leq \frac{k^k}{t^k} \frac{M}{(k-1)!} \frac{\Gamma(k-\gamma)}{\left(\frac{k}{t}\right)^{k-\gamma}} \|u\|_U \tag{A.12}$$

$$= \frac{M}{t^\gamma} \frac{k^\gamma \Gamma(k-\gamma)}{(k-1)!} \|u\|_U, \quad t > 0, \tag{A.13}$$

where, in (A.12), we have recalled assumption (A.3) with $\omega = 0$ and Re $\lambda = \lambda = \frac{k}{t}$, as assumed.

Step 2. Next, with $0 \leq \gamma < 1$, we shall show that

$$\frac{k^\gamma \Gamma(k-\gamma)}{(k-1)!} \leq \text{const.}, \qquad \forall\, k = 1, 2, \ldots, \tag{A.14}$$

as required, after which we then obtain by (A.13), (A.14),

$$\left\| \left[\frac{k}{t} R\left(\frac{k}{t}, A\right)\right]^k Bu \right\|_{\mathcal{L}(Y)} \leq \frac{C}{t^\gamma} \|u\|_U, \text{ for all } 0 < t, \tag{A.15}$$

as desired.

To prove (A.14), we set $x = k - 1 - \gamma = K - \gamma$, $k - 1 = K$, and recall the Stirling formula [Ru, p. 194], to obtain

$$\begin{aligned} \Gamma(k-\gamma) &= \Gamma(x+1) \sim \left(\frac{x}{e}\right)^x \sqrt{2\pi x} \\ &= \frac{(K-\gamma)^{K-\gamma}}{e^{K-\gamma}} \sqrt{2\pi(K-\gamma)}. \end{aligned} \tag{A.16}$$

Thus,

$$\frac{k^\gamma \Gamma(k-\gamma)}{(k-1)!} \sim \frac{(K+1)^\gamma}{K!} \frac{(K-\gamma)^{K-\gamma}}{e^{K-\gamma}} \sqrt{2\pi(K-\gamma)} \tag{A.17}$$

$$= \left(\frac{K+1}{K-\gamma}\right)^\gamma e^\gamma \frac{(K-\gamma)^K}{K! e^K} \sqrt{2\pi(K-\gamma)} \tag{A.18}$$

$$\sim \left(\frac{K+1}{K-\gamma}\right)^\gamma e^\gamma \frac{(K-\gamma)^K \sqrt{2\pi(K-\gamma)}}{K^K \sqrt{2\pi K}}, \tag{A.19}$$

where in the last step we have again invoked the Stirling formula for

$$K!\, e^K \sim K^K \sqrt{2\pi K}. \tag{A.20}$$

Then (A.19) establishes (A.14), as desired, at least for $\omega = 0$.

Case $\neq 0$. Finally, let (A.3) hold true with $\omega \neq 0$ and $\lambda = \text{Re}\,\lambda > \omega$. Define: $\tilde{A} \equiv A - \omega I$, $\tilde{\lambda} = \lambda - \omega > 0$, so that $\tilde{\lambda} I - \tilde{A} = \lambda I - A$ and $R^n(\tilde{\lambda}, \tilde{A}) \equiv R^n(\lambda, A)$. Thus

$$\|R^n(\tilde{\lambda}, A)B\|_{\mathcal{L}(U;Y)} = \|R^n(\lambda, A)B\|_{\mathcal{L}(U,Y)} \leq \frac{M\Gamma(1-\gamma+n)}{\tilde{\lambda}^{(1-\gamma+n)}}, \ \tilde{\lambda} > 0. \tag{A.21}$$

Then the previous part with $\omega = 0$ applies to $\tilde{A}$ and we obtain

$$e^{-\omega t}\|e^{At}B\|_{\mathcal{L}(U;Y)} = \|e^{\tilde{A}t}B\|_{\mathcal{L}(U;Y)} \leq \frac{C_T}{t^\gamma}, \quad 0 < t;\ 0 < \gamma < 1, \tag{A.22}$$

and the desired conclusion (A.4) follows for $e^{At}B$. The proof of Proposition A.1 is complete. □

The next Proposition shows that the singular estimate (H.3) = (1.2) for $e^{At}B$ is preserved under a bounded perturbation Π of A.

Proposition A.2. Assume (H.1), (H.2), (H.3) = (1.2). Let $\Pi \in \mathcal{L}(Y)$. Then, $e^{(A+\Pi)t}B$ satisfies the same singular estimate (1.2),

$$\|e^{(A+\Pi)t}B\|_{\mathcal{L}(U;Y)} \leq \frac{C}{t^\gamma}, \quad 0 < t \leq T,\ 0 < \gamma < 1. \tag{A.23}$$

Proof. Let $y_0 \in Y$. Setting

$$y(t) \equiv e^{At}y_0, \text{ or } \dot{y}(t) + (A+\Pi)y(t) - \Pi y(t), \tag{A.24}$$

we may write by the variation of the parameter formula,

$$y(t) \equiv e^{At}y_0 = e^{(A+\Pi)t}y_0 - \int_0^t e^{(A+\Pi)(t-\tau)}\Pi e^{A\tau}y_0\, d\tau. \tag{A.25}$$

Thus, from

$$e^{(A+\Pi)t}Bu = e^{At}Bu + \int_0^t e^{(A+\Pi)(t-\tau)}\Pi e^{A\tau}Bu\, d\tau, \tag{A.26}$$

with $u \in U$, and so interpreted, at first, in $[\mathcal{D}(A^*)]'$, we estimate for $t > 0$, $u \in U$, in the $\mathcal{L}(Y)$-norm:

$$\|e^{(A+\Pi)t}Bu\| \leq \|e^{At}Bu\| + \int_0^T C_T\|\Pi\|\ \|e^{A\tau}Bu\|d\tau \tag{A.27}$$

$$\leq \left\{\frac{C_T}{t^\gamma} + M_T\int_0^T \frac{1}{\tau^\gamma}\,d\tau\right\}\|u\|_U \tag{A.28}$$

$$= \left\{\frac{C_T}{t^\gamma} + M_T\frac{T^{1-\gamma}}{1-\gamma}\right\|\ \|u\|_U, \tag{A.29}$$

and (A.29) establishes (A.23). □

REFERENCES

1. G. Avalos and I. Lasiecka, "Differential Riccati equations for the active control of a problem in structural acoustics," *JOTA* 91 (1996), 695–728.
2. A. V. Balakrishnan, *Applied Functional Analysis*, Springer-Verlag, 1981.
3. A. V. Balakrishnan, "Boundary control of parabolic equations, L-Q-R theory," *Proc. V International Summer School*, Control Inst. Math. Mech. Acad. Sci., GDR, Berlin (1977).
4. A. Bensoussan, G. Da Prato, M. Delfour, and S. Mitter, *Representation and Control of Infinite Dimensional Systems*, vols. 1 and 2, Birkhäuser, 1993.
5. F. Bucci, I. Lasiecka, and R. Triggiani, "Singular estimates and uniform stability of coupled systems of hyperbolic/parabolic PDEs," *Abstract and Applied Analysis*, vol. 7(4) (2002), 169–237.
6. P. Grisvard, "Characterisation del quelques espaces d'interpolation," *Arch. Rational Mech. Anal.* 25 (1967), 40–63.
7. P. Grisvard, *Elliptic Problems in Nonsmooth Domains*, Monographs and Studies in Mathematics, 24, Pitman, 1985.
8. W. Grobner and N. Hofreiter, *Integraltafeln*, Zweiter Teil, *Bestimmte Integral*, Springer-Verlag, 1973.
9. S. Hansen, "A model for a two-layered plate with interfacial slip," in *Control and Estimation of Distributed Parameter Systems: Non-linear Phenomena*, Vol. 1SNA118, Editors: W. Desch, F. Kappel, K. Kunish, Birkhäuser, 143–170.
10. S. Hansen, "Modeling and analysis of multilayer laminated plates," *ESAIM Proc.* 4 (1998), 117–135.
11. S. Hansen and I. Lasiecka, "Analyticity, hyperbolicity and uniform stability of semigroups arising in models of composite beams," *Mathematical Models and Methods in Applied Sciences*, vol. 10 (2000), 555–580.
12. I. Lasiecka, "Optimization problems for structural acoustic models with thermoelasticity and smart materials," *Discussiones Mathematicae, Differential Inclusions, Control and Optimization*, vol. 20 (2000), 113–140.
13. I. Lasiecka, *Mathematical Control Theory of Coupled PDEs Systems*, SIAM NSF-CMBS Lecture Notes, N. 75, 2002.
14. I. Lasiecka, "Well-posedness of optimal control problems for systems with unbounded controls and partially analytic generators," *Control and Cybernetics* 31 (2002), 751–777.
15. I. Lasiecka and R. Triggiani, *Differential and Algebraic Riccati Equations with Applications to Boundary/Point Control Problems: Continuous and Approximation Theory*, LNICS, Springer Verlag, 1991.
16. I. Lasiecka and R. Triggiani, *Control Theory for Partial Differential Equations*, vol. I: Abstract Parabolic Equations; Vol II: Abstract hyperbolic-like systems over a finite time horizon. Cambridge University Press, Encyclopedia of Mathematics and its Applications, January 2000.
17. I. Lasiecka and R. Triggiani, "Structural decomposition of thermo-elastic semigroups with rotational forces," *Semigroup Forum* 60 (2000), 16–66.
18. I. Lasiecka and R. Triggiani, "Two direct proofs on the analyticity of the s.c. semigroup arising in abstract thermo-elastic equations," *Advances Diff. Eqns.* 3(3) (May 1998), 387–416.

19. I. Lasiecka and R. Triggiani, "Analyticity of thermo-elastic semigroups with coupled hinged/Neumann B.C.," *Abstract Appl. Anal.* 3(1–2) (1998), 153–169.
20. I. Lasiecka and R. Triggiani, "Analyticity of thermo-elastic semigroups with free B.C.," *Annali Scuola Normale Superiore*, Pisa, Cl. Sci. (4), vol. 37 (1998), 457–482.
21. I. Lasiecka and R. Triggiani, "Optimal control and algebraic riccati equations under singular estimates for $e^{At}B$ in the absence of analyticity. Part I: The stable case," in *Differential Equations and Control Theory*, Lecture Notes in Pure and Applied Mathematics, Editors: S. Aizicovici and N. Pavel, vol. 225 (2001), Marcel Dekker, 198–219.
22. I. Lasiecka, R. Triggiani, and X. Zhang, "Nonconservative wave equations with unobserved Neumann B.C.: Global uniqueness and observability in one shot," *Contemporary Mathematics*, vol. 268, AMS (2000), 227–326.
23. K. Liu and M. Renardy, "A note on equations of a thermoelastic plates," *Appl. Math. Letters* 8 (1995), 1–6.
24. W. Rudin, *Principles of Mathematical Analysis*, 3rd edition, McGraw-Hill, New York, 1976.
25. C. Sadovsky, *Interpolationn of Operators and Singular Integrals*, Monographs in Pure and Applied Mathematics, Marcel Dekker, 1979.
26. R. Triggiani, "Min-max game theory and optimal control with indefinite cost under a singular estimate for $e^{At}B$ in the absence of analyticity." Invited paper in a special volume: *Evolution Equations, Semigroups and Functional Analysis*, Editors: A. Lorenzi and B. Ruf, Birkhäuser Verlag, 2002, 353–380. (In memory of B. Terreni.)
27. R. Triggiani, "Backward uniqueness of the s.c. semigroup arising in the coupled PDE-system of structural acoustic," *Advances Diff. Eqns.*, to appear.
28. R. Triggiani, "The coupled PDE system of a composite (sandwich) beam revisited," *Discrete & Continuous Dynamical Systems–Series B*; vol. 3(2) (May 2003), 285–298.

20

Nonlinear Problems of Spacecraft Fault Tolerant Control Systems

V.M. Matrosov*† and Ye.I. Somov*‡

*Stability and Nonlinear Dynamics Research Center, Mechanical Engineering Research Institute, Russian Academy of Sciences, Moscow, Russia

†Moscow Aviation Institute (State Technical University), Moscow, Russia

‡Research Institute of Mechanical Systems Reliability, Samara State Technical University, Samara, Russia

A brief overview and new results on nonlinear methods for research of logic-dynamic fault-tolerant control systems and their applications to precise gyromoment attitude control of information spacecraft are presented in this chapter.

1 INTRODUCTION

During the last 25 years the authors have accumulated substantial experience in research of the spacecraft attitude control systems (ACSs) with high fault tolerance, survivability, and autonomy at the expense of using functional excessibility.

The dynamic requirements of ACS of contemporary spacecraft for communication, navigation, and geodetics are: (i) continuous precision 3-axis orientation of the spacecraft body, which requires only a minimum number of measurements under possible failures of the ACS onboard equipment, disturbances on optical devices from the Sun, the Moon, etc., and also in executing the spacecraft orbit correction; (ii) the possibility of the spacecraft body reorientation for its orbit correction, as well as autonomous orientation of the solar array panels (SAPs) and each high-gain receiving-transmitting antenna (RTA) with respect to the spacecraft body; (iii) robustness to variations of the spacecraft inertial and rigidity characteristics under minimum mass, size, and power expenditures.

For remote sensing spacecraft there is the need to orient the line of sight to a predetermined part of the Earth's surface with the scan in a designated direction and to compensate an image motion at the onboard optical telescope's focal plane. Moreover, for the low-orbit remote-sensing spacecraft these requirements are expressed by rapid angular maneuvering and spatial compensative motion with a variable vector of angular rate. Increased requirements of such information spacecraft — exactness of spatial rotation maneuvers (SRMs) with the effective damping of the spacecraft flexible construction oscillations, efficiency, fault tolerance, reliability, as well as reasonable mass, size, and energy characteristics, has motivated intensive development the ACS with executive devices in the form of moment gyrocomplex (MGC) based on an excessive number of gyrodynes — single-gimbal control moment gyroscopes (CMGs).

0-415-30852-6/04/$0.00+$1.50

Mathematical aspects of nonlinear spacecraft attitude control [1] were represented in a number of research works [2]. The recent results of B.R. Hoelscher and S.R. Vadali [3] on nonlinear dynamics of gyromoment ACSs are based on methods of optimization and *Lyapunov functions*, whereas the *exact feedback linearization* (EFL) technique, suggested by A. Isidori [4], also was applied to this problem by S.N. Singh and T. Bosart [5].

This chapter discusses our newest results on nonlinear logic-dynamic fault-tolerant gyromoment ACSs for considered classes of information spacecraft.

2 NONLINEAR FAULT-TOLERANT FEEDBACK CONTROL

Let there be the nonlinear generalized controlled object $\boldsymbol{\mathcal{O}}$:

$$D^{+}\mathrm{x}(t) = \mathcal{F}(\mathrm{x}(t),\ \mathrm{u},\ \mathrm{p}(t,\mathrm{x}),\ \gamma_{\nu}^{f}(t)),\ \ \mathrm{x}(t_0) = \mathrm{x}_0; \tag{1}$$

$$\mathrm{y}(t) = \psi^{o}(\mathrm{x}(t),\ \gamma_{\nu}^{f}(t));\quad \mathrm{z}^{o}(t) = \phi^{o}(\mathrm{x}(t),\ \mathrm{y}(t),\ \mathrm{p}(t,\mathrm{x})), \tag{2}$$

where $\mathrm{x}(t) \in \mathcal{H} \subset \mathbb{R}^{n_\nu}$; $\mathrm{x}_0 \in \mathcal{H}_0 \subseteq \mathcal{H}$; $\mathrm{y}(t) \in \mathbb{R}^{r_\nu^s}$ is an output vector for measurement and diagnosis of an object's state, and $\mathrm{z}^{o}(t) \in \mathbb{R}^{r_\nu^f}$ is a vector for description of its failure conditions; $\mathrm{u} = \{u_j\} \in \mathrm{U} \subset \mathbb{R}^{r_\nu^c}$ is a control vector, and $\mathrm{p}(t,\mathrm{x}) \in \mathcal{P}$ is the vector-function of disturbances in class $\mathcal{P}$; D^{+} is the symbol of a right derivative with respect to time $t \in \mathrm{T}_{t_0} = [t_0, \infty)$; a vector $\gamma_{\nu}^{f}(t) \in \mathcal{B}^m \equiv \mathcal{B}\times\mathcal{B}\cdots\times\mathcal{B}$, where $\mathcal{B} = \{0,1\}$, is a vector of logic variables, which are outputs of a "fault" asynchronous logic automaton (ALA) $\boldsymbol{\mathcal{A}}^{f}$:

$$\gamma_{\nu}^{f} = \delta^{f}(\kappa_{\nu}^{f}, \mathrm{l}_{\nu}^{f});\ \kappa_{\nu+1}^{f} = \lambda^{f}(\kappa_{\nu}^{f}, \mathrm{l}_{\nu}^{f}),\quad \kappa_0^{f} = \kappa^{f}(0),\ \ \nu \in \mathrm{N}_0 = [0,1,2,\ldots), \tag{3}$$

with memory, logic vectors of the object's state $\kappa_{\nu}^{f} = \kappa^{f}(\nu)$ and input $\mathrm{l}_{\nu}^{f} = \mathrm{l}^{f}(\nu) = \mathrm{g}^{f}(\mathrm{z}^{o}(t_{\nu}^{f}))$, which are used for representing fault occurrences and damage development depending on the automaton time ν, bound up with the continuous time as $t = t_{\nu}^{f} + (\tau^{f} - t_{\nu}^{f})$; $\tau^{f} \in \tau_{\nu}^{f} = [t_{\nu}^{f}, t_{\nu+1}^{f})$, $\nu \in \mathrm{N}_0$. Moreover $\mathrm{l}_{\nu}^{f}(t) = \mathrm{const}\ \forall t \in \tau_{\nu}^{f}$ and change of the logic vector γ_{ν}^{f} in general leads to variation of dimensions for vectors $\mathrm{x}(t)$ and $\mathrm{y}(t)$ under mappings in time moments $t = t_{\nu}^{f}$:

$$\mathrm{x}(t_{\nu+}^{f}) = \mathcal{P}_{\nu}^{x}(\mathrm{x}(t_{\nu-}^{f}));\ \ \mathrm{y}(t_{\nu+}^{f}) = \mathcal{P}_{\nu}^{y}(\mathrm{y}(t_{\nu-}^{f})).$$

Let there also be a given subsystem for discrete measurement of the object's state and digital filtering:

- for diagnostics of the structural state of object $\boldsymbol{\mathcal{O}}$:

$$\mathrm{y}_s^{d} = \psi^{d}(\mathrm{y}_s);\quad \mathrm{z}_k^{\mathrm{df}} = \mathcal{F}_{\mathrm{T}_u}(\mathrm{y}_s^{d});\ \ k, s \in \mathrm{N}_0; \tag{4}$$

- for forming the control loop and its reconfiguration:

$$\mathrm{y}_s^{u} = \psi^{u}(\mathrm{y}_s);\ \mathrm{y}_k^{\mathrm{f}} = \mathcal{F}_{\mathrm{T}_u}(\mathrm{y}_s^{u});\ \mathrm{z}_{\mu}^{\mathrm{f}} = \mathcal{F}_{\mathrm{T}_r}(\mathrm{z}_k^{\mathrm{df}}),\ \mathrm{T}_r = t_{\mu+1} - t_{\mu};\ \mu, k, s \in \mathrm{N}_0. \tag{5}$$

We use denotations $x_k = x(t_k)$; $t_k = k\mathrm{T}_u$, $t_s = s\mathrm{T}_q$, $k, s \in \mathbb{N}_0$; $x_k^{\mathrm{f}} = \mathcal{F}_{\mathrm{T}_u}(x_s)$, where T_u and $\mathrm{T}_q \leq \mathrm{T}_u$ are fixed sampling periods of control and state measurement; moreover, multiplicity conditions must be satisfied for the periods $\mathrm{T}_q, \mathrm{T}_u, \mathrm{T}_r$; a variable x_k^{f} is the value of a variable x_s measured with the sampling period T_q, which is filtered out at time $t = t_k$; $\mathcal{F}_{\mathrm{T}_y}(\cdot)$ is the digital filtering operator with the sampling period $\mathrm{T}_y, y = u, r$.

Principal problems are contained in the *synthesis* of:

- the synchronous logic automaton (SLA) $\boldsymbol{\mathcal{A}}^d$ for the flowing structural state diagnosis:

$$\gamma_k^d = \delta^d(\kappa_k^d, \mathrm{l}_k^d); \quad \kappa_{k+1}^d = \lambda^d(\kappa_k^d, \mathrm{l}_k^d), \; \kappa_0^d = \kappa^d(t_0); \; k \in \mathbb{N}_0, \tag{6}$$

  with memory, logic vectors of state κ_k^d, input $\mathrm{l}_k^d = \mathrm{g}^d(\mathrm{z}_k^{\mathrm{df}})$ and output γ_k^d;

- the SLA $\boldsymbol{\mathcal{A}}^r$, also with memory, for the description of damage blockkeeping and the reconfiguration sequence depending on the automaton time μ and sampling time $t_\mu = \mu\mathrm{T}_r$:

$$\gamma_\mu^r = \delta^r(\kappa_\mu^r, \mathrm{l}_\mu^r); \quad \kappa_{\mu+1}^r = \lambda^r(\kappa_\mu^r, \mathrm{l}_\mu^r), \; \kappa_0^r = \kappa^r(t_0); \; \mu \in \mathbb{N}_0, \tag{7}$$

  with logic vectors of state κ_μ^r, input $\mathrm{l}_\mu^r = \mathrm{g}^r(\mathrm{z}_\mu^{\mathrm{f}}, \gamma_\mu^{\mathrm{df}})$ with $\gamma_\mu^{\mathrm{df}} = \mathcal{F}_{\mathrm{T}_r}(\gamma_k^d)$ and output γ_μ^r;

- the nonlinear control law (NCL) with its reconfigurations due to the SLA $\boldsymbol{\mathcal{A}}^r$ routine:

$$\mathrm{u}_k = \mathcal{U}(\hat{\mathrm{x}}_{e\,k}, \mathrm{y}_{e\,k}^{\mathrm{f}}, \mathrm{y}_{o\,k}, \gamma_\mu^r); \; \hat{\mathrm{x}}_{e\,k+1} = \hat{\mathcal{F}}_e(\hat{\mathrm{x}}_{e\,k}, \mathrm{y}_{e\,k}^{\mathrm{f}}, \mathrm{y}_k^o, \mathrm{u}_k, \gamma_k^d, \gamma_\mu^r), \; \hat{\mathrm{x}}_{e\,0} = \hat{\mathrm{x}}_e(t_0), \tag{8}$$

  where $k, \mu \in \mathbb{N}_0$; $\mathrm{y}_{e\,k}^{\mathrm{f}} = \mathcal{F}_{\mathrm{T}_u}(\psi_e^u(\mathrm{y}_{e\,s}))$; $\mathrm{y}_{e\,s} = \psi_e^o(\mathrm{x}_{e\,s}, \gamma_k^d)$; $\mathrm{x}_{e\,s} = \mathrm{x}_e(t_s) \in \mathbb{R}^{n_\mu^e}$ is the state vector of a simplified discrete object's model:

$$\mathrm{x}_{e\,s+1} = \mathcal{F}_e(\mathrm{x}_{e\,s}, \mathrm{u}_k, \gamma_k^d, \gamma_\mu^r), \; \mathrm{x}_{e\,0} = \mathrm{x}_e(t_0); \quad s, k, \mu \in \mathbb{N}_0, \tag{9}$$

  and $\hat{\mathrm{x}}_{e\,k} = \hat{\mathrm{x}}_e(t_k) \in \mathbb{R}^{n_\mu^e}$ is its estimation; $n_\mu^e \leq n = \max\{n_\nu\}$ and y_k^o is a programmed vector.

Feedback loops (4)–(9) are intended for fault-tolerant control of objects (1)–(3).

Over the last two decades, the basic research on fault diagnostics, the isolation of faults, and reconfiguration of control systems has received much attention: see survey [6]. The model-based approach to this problem is now recognized as an important method for linear control systems; moreover there are trends in extending that methodology to nonlinear systems [7]. Therefore now we present in detail only our approach to *synthesis* of *nonlinear control systems* when their structural state is known.

Along with different methods for nonlinear dynamic investigations as the primary approach, we use the method of *vector Lyapunov functions* [8, 9, 10, 11, 12], which has a strong mathematical basis for the analysis of stability and other dynamical properties of various nonlinear interconnected and logic-dynamic systems with the *discontinuous right-hand side*, depending on *bounded control*, in cooperation with *exact feedback linearization* technique [4]. Let there be a nonlinear controlled object with a certain sequence of the structural state $\gamma_\nu^f(t)$ and, for simplicity, with a complete and continuous measurement of its state:

$$D^+\mathrm{x}(t) = \mathcal{F}(\mathrm{x}(t), \mathrm{u}, \mathrm{p}^f(t,\mathrm{x})); \; \mathrm{p}^f(t,\mathrm{x}) \equiv \{\mathrm{p}(t,\mathrm{x}), \gamma_\nu^f(t)\}; \; \mathrm{x}(t_0) = \mathrm{x}_0; \; \mathrm{x} \in \mathbb{R}^n; t \in \mathrm{T}_{t_0},$$

and also let *vector norms* $\rho(\mathrm{x}) \in \overline{\mathbb{R}}_{+}^{l}$ and $\rho^{0}(\mathrm{x}_0) \in \overline{\mathbb{R}}_{+}^{l_0}$ be *given*. For any control law $\mathrm{u} = \mathcal{U}(\mathrm{x})$ this model has the form

$$D^{+}\mathrm{x}(t) = \mathcal{F}(\mathrm{x}, \mathcal{U}(\mathrm{x}), \mathrm{p}^{f}(t, \mathrm{x})) \equiv \mathcal{X}(t, \mathrm{x}); \;\; \mathrm{x}(t_0) = \mathrm{x}_0, \tag{10}$$

where $\mathcal{X} : \mathrm{T}_{t_0} \times \mathcal{H} \to \mathcal{H}$ is a *discontinuous* operator.

Assuming the existence and the nonlocal continuability of the *right-sided* solution $x(t) \equiv \mathrm{x}(t_0, \mathrm{x}_0; t)$ of the closed-loop control system (10) for its *extended definition* in the aspect of physics (engineering), the most important dynamic property is obtained, that is, *$\rho\rho^0$-exponential invariance* of the solution $x(t) = 0$ under the *desired* $\gamma \in \overline{\mathbb{R}}_{+}^{l}$:

$$\begin{aligned} &(\exists \alpha \in \mathbb{R}_{+}) \; (\exists \mathcal{B} \in \overline{\mathbb{R}}_{+}^{l \times l_0}) \; (\exists \delta \in \mathbb{R}_{+}^{l_0}) \; (\forall \rho^{0}(\mathrm{x}_0) < \delta) \\ &\rho\,(x(t)) \leq \gamma + \mathcal{B}\,\rho^{0}\,(\mathrm{x}_0) \cdot \exp(-\alpha(t - t_0)) \qquad \forall t \in \mathrm{T}_{t_0}. \end{aligned} \tag{11}$$

For the *vector Lyapunov function* (VLF) $\upsilon : \mathcal{H} \to \overline{\mathbb{R}}_{+}^{k}$ with the components $\upsilon^{s}(\mathrm{x}) \geq 0$, $\upsilon^{s}(0) = 0, \; s = 1 : k$ and the norm $\|\upsilon(\mathrm{x})\| = \max\,\{\upsilon^{s}(\mathrm{x}), \; s = 1 : k\}$, defined are the scalar function $\overline{\upsilon}(\mathrm{x}) = \max\{\upsilon^{s}(\mathrm{x}), \; s = 1 : l_k, \; 1 \leq l_k \leq k\}$ and the *upper-right derivative* of the VLF $\upsilon(\mathrm{x})$ with respect to (10) as

$$\upsilon'(\mathrm{x}) = \overline{\lim_{\delta t \to 0+}} \; \{(\upsilon(\mathrm{x} + \delta t \cdot \mathcal{X}(t, \mathrm{x})) - \upsilon(\mathrm{x}))/\delta t\}\,.$$

THEOREM. *Let there exist the VLF $\upsilon : \mathcal{H} \to \overline{\Re}_{+}^{k}$, such that:*

1^0 $(\exists \mathrm{a} \in \mathbb{R}_{+}^{l})$ $(\forall \mathrm{x} \in \mathcal{H})$ $\quad \rho\,(\mathrm{x}) \leq \mathrm{a} \cdot \overline{\upsilon}(\mathrm{x})$;

2^0 $(\exists \mathrm{b} \in \mathbb{R}_{+}^{l_0})$ $(\forall \mathrm{x}_0 \in \mathcal{H}_0)$ $\|\upsilon(\mathrm{x}_0)\| \leq \langle \mathrm{b}\,, \rho^{0}(\mathrm{x}_0)\rangle$;

3^0 $(\exists \gamma_c \in \overline{\mathbb{R}}_{+}^{k})$ *and a function* $\varphi_\gamma(\cdot) = \{\varphi_\gamma^{s}(\cdot)\}$ *exists such that* $\gamma_c \leq \varphi_\gamma(\mathrm{a}, \gamma)$;

4^0 $(\forall (t, \mathrm{x}) \in (\mathrm{T}_{t_0} \times \mathcal{H})$ *the conditions are satisfied:*

a) $\upsilon_\gamma'(\mathrm{x}) \;\dot{\leq}\; \mathrm{f}_c\,(t, \upsilon_\gamma(\mathrm{x})) \equiv \mathrm{P} \cdot \upsilon_\gamma(\mathrm{x}) + \tilde{\mathrm{f}}_c\,(t, \upsilon_\gamma(\mathrm{x}))$ *where* $\upsilon_\gamma(\mathrm{x}) \equiv \upsilon(\mathrm{x}) - \gamma_c$;

b) *Hurwitz condition* $(\mathrm{Re}\,\lambda_s(\mathrm{P}) < 0, s = 1 : k)$ *for the positive* $(k \times k)$*-matrix* P;

c) *Ważewski condition on the quasi-monotonicity for the vector-function* $\tilde{\mathrm{f}}_c(t, \mathrm{y})$*:* $\tilde{f}_{cs}(t, \mathrm{y}') \leq \tilde{f}_{cs}\,(t, \mathrm{y}), s = 1 : k$ *for* $y_s{}' = y_s$ *and* $y_j{}' \leq y_j$, $j \neq s$;

d) *Carateodory condition for the vector-function* $\tilde{\mathrm{f}}_c(t, \mathrm{y})$*, bounded in each domain:* $\Omega_c^r = (\mathrm{T}_{t_0} \times S_c^r)$*, where* $r > 0$ *and* $S_c^r = \{\mathrm{y} \in \mathbb{R}^k : \|\mathrm{y}\|_{\mathrm{E}} < r\}$;

e) $\left\{\tilde{\mathrm{f}}_c(t, \mathrm{y})/\|\mathrm{y}\|\right\} \overset{t \in \mathrm{T}_{t_0}}{\Longrightarrow} 0$ *for* $\mathrm{y} \to 0$ *uniformly with respect to time* $t \in \mathrm{T}_{t_0}$.

Then the system (10) is $\rho\rho^0$-exponential invariant and the matrix $\mathcal{B}$ in the definition (11) has the form $\mathcal{B} = \mathrm{ab}^{\mathrm{t}}$ with a parameter $c \in \mathbb{R}_{+}$.

PROOF. In the traditional scheme of the *comparison principle* [8], the basis of inequality for vector norm $\rho\,(x(t))$ in (11) is attained [13, 14, 15, 16] using the *maximum right-sided* solution $\overline{x}_c(t) \equiv \overline{\mathrm{x}}_c(t_0, \mathrm{x}_{c0}; t)$ of the *comparison system* (CS):

$$D^{+}\mathrm{x}_c(t) \dot{=} \mathrm{f}_c(t, \mathrm{x}_c(t) - \gamma_c) \equiv \mathrm{P}(\mathrm{x}_c(t) - \gamma_c) + \tilde{\mathrm{f}}_c(t, \mathrm{x}_c(t) - \gamma_c); \mathrm{x}_c(t_0) \equiv \mathrm{x}_{c0} = \upsilon(\mathrm{x}_0). \tag{12}$$

For the conditions (4^0) according to the comparison principle under the condition

$$\xi_c \equiv \max\{v^s(\mathrm{x}_0) - \gamma_c^s, s = 1:k\} \equiv \max\{\mathrm{x}_{c0} - \gamma_c, s = 1:k\} \le \delta_c; \;\; \delta_c \in \mathbb{R}_+,$$

the CS (12) has the dynamic property *upper exponential semi-invariance*, R.I. Kozlov [17, Ch. III, §1]: $\exists \alpha_c \in \mathbb{R}_+$ and $\exists \beta_c \in \mathbb{R}_+^k$ such that the vector inequality

$$v(x(t)) \le \overline{x}_c(t) \le \gamma_c + \overline{\beta}_c(\xi_c) \exp(-\alpha_c(t - t_0)) \; \forall t \in \mathrm{T}_{t_0};$$

$$\overline{\beta}_c(\xi_c) \equiv \begin{cases} 0 & v(\mathrm{x}_0) \le \gamma_c; \\ \beta_c \cdot \xi_c & v(\mathrm{x}_0) \nleq \gamma_c, \end{cases}$$

is satisfied. Moreover, according to the conditions (1^0) – (3^0) the *upper bound* of the vector norm $\rho\,(x(t))$ is valid:

$$\begin{aligned} \rho\,(x(t)) &\le \mathrm{a} \cdot \max\{\gamma_c^s + \overline{\beta}_c^s(\xi_c) \exp(-\alpha_c(t - t_0), s = 1:l_k, 1 \le l_k \le k\} \\ &\le \mathrm{a} \cdot \max\{\gamma_c^s, s = 1:l_k\} + \max\{\beta_c^s, s = 1:l_k\}\; \mathrm{a} \cdot \|v(\mathrm{x}_0)\| \; \exp(-\alpha_c(t - t_0)) \\ &\le \gamma + c \cdot \mathcal{B}\, \rho^0\,(\mathrm{x}_0) \cdot \exp(-\alpha(t - t_0)) \quad \forall t \in \mathrm{T}_{t_0}, \end{aligned}$$

where $\gamma = \mathrm{a} \cdot \max\{\varphi_\gamma^s(\mathrm{a}, \gamma), s = 1:l_k\}$, $c = \max\{\beta_c^s, s = 1:l_k\}$, and $\alpha = \alpha_c - \varepsilon$ with $\varepsilon \ge 0$. ■

As is well-known [18], a lot of similar theorems [17] are *quite useless for a control engineering practice* until we have solved an important problem: by what approach is it possible to create *constructive* techniques for constructing the vector Lyapunov function $v(\mathrm{x})$ and *simultaneous* synthesis of a nonlinear control law $\mathrm{u} = \mathcal{U}(\mathrm{x})$ for the closed-loop system (10) with given vector norms $\rho(\mathrm{x})$ and $\rho^0(\mathrm{x}_0)$?

During the last 25 years we have persistently searched for a general approach to this basic problem and carefully varied our results with application to the gyromoment ACSs of information spacecraft in Kazan [19, 20, 21, 22], since 1984 in Irkutsk [13, 14, 23, 24, 25], and since 1997 in Moscow [15, 16, 26, 27, 28]. As a result, today a pithy technique on constructing VLF for the synthesis of nonlinear control law has been elaborated. This method is based on a *nonlinear transformation* of the nonlinear control system's model to a canonical representation by solving the problem in the two stages.

In stage 1, the right side $\mathcal{F}(\cdot)$ in the model (10) is transformed as

$$\mathcal{F}(\mathrm{x}(t), \mathrm{u}, \mathrm{p}^f(t, \mathrm{x})) = \mathrm{f}(\mathrm{x}) + \mathrm{G}(\mathrm{x})\,\mathrm{u} + \tilde{\mathcal{F}}(t, \mathrm{x}(t), \mathrm{u})$$

and for some *principal variables* in a state vector $\mathrm{x} \in \tilde{\mathcal{H}} \subset \mathbb{R}^{\tilde{n}} \subseteq \mathbb{R}^n$ with $\tilde{n} \le n$, $\mathrm{x}_0 \in \tilde{\mathcal{H}}_0 \subseteq \tilde{\mathcal{H}}$, a *simplified nonlinear model* of a nonlinear controlled object is presented in the form of an affine *quite smooth* nonlinear system

$$\dot{\mathrm{x}} = \mathrm{F}(\mathrm{x}, \mathrm{u}) \equiv \mathrm{f}(\mathrm{x}) + \mathrm{G}(\mathrm{x})\mathrm{u} = \mathrm{f}(\mathrm{x}) + \sum \mathrm{g}_j(\mathrm{x}) \cdot u_j, \tag{13}$$

which is *structurally* synthesized by the *exact feedback linearization* technique. In this aspect, based on the structural analysis of *given* vector norms $\rho(\mathrm{x})$ and $\rho^0(\mathrm{x})$, and also vector-functions $\mathrm{f}(\mathrm{x})$ and $\mathrm{g}_j(\mathrm{x})$, the *output vector-function* $\mathrm{h}(\mathrm{x}) = \{\mathrm{h}_i(\mathrm{x})\}$ is carefully selected. Futhermore, the nonlinear invertible (one-to-one) coordinate transformation $\mathrm{z} = \Phi(\mathrm{x})\; \forall \mathrm{x} \in \mathcal{S}_\mathrm{h} \subseteq \tilde{\mathcal{H}}$ with $\Phi(0) = 0$ is analytically obtained according to

$$\begin{aligned} &\mathrm{z}(\mathrm{x}) = \{\mathrm{z}_i(\mathrm{x}), i = 1:r\},\; \mathrm{z}_i = \{z_i^k, k = 1:\tilde{n}_i\}; \quad z_i^k(\mathrm{x}) = L_f^{k-1}\mathrm{h}_i(\mathrm{x}); \\ &L_f^0\mathrm{h}_i(\mathrm{x}) \equiv L_f\mathrm{h}_i(\mathrm{x}) = \langle \partial(\mathrm{h}_i(\mathrm{x})/\partial\mathrm{x}, \mathrm{h}_i(\mathrm{x})\rangle; \;\; L_f^k\mathrm{h}_i(\mathrm{x}) \equiv L_f(L_f^{k-1}\mathrm{h}_i(\mathrm{x})), \end{aligned}$$

where $L_f \mathrm{h}_i(\mathrm{x})$ is the *Lie* derivative of the function $\mathrm{h}_i(\mathrm{x})$ with respect to the vector field $\mathrm{f}(\mathrm{x})$ on $\mathbb{R}^{\tilde{n}}$, and $(\tilde{n}_1, \cdots, \tilde{n}_r)$ is the *relative degree* of the nonlinear system (13) with respect to the vector-function $\mathrm{h}(\mathrm{x})$, which $\forall \mathrm{x} \in \mathcal{S}_{\mathrm{h}}$ is defined as

$$L_{g_j} L_f^k \mathrm{h}_i(\mathrm{x}) = 0 \ \forall k \ < \tilde{n}_i \ \& \ \forall \, i, j \in (1:r).$$

For some assumptions, including that matrix $\mathrm{A}_c(\mathrm{x}) = \{[L_{g_j} L_f^{\tilde{n}_i - 1} \mathrm{h}_i(\mathrm{x}), j = 1:r], i = 1:r\}$ is nonsingular $\forall \mathrm{x} \in \mathcal{S}_{\mathrm{h}}$, the system (13) is transformed to the canonical representation:

$$\dot{z}_i^k = z_i^{k+1}, k = 1:(\tilde{n}_i - 1), i = 1:r; \quad \dot{\mathrm{z}}_c \equiv \{\dot{z}_i^{\tilde{n}_i}\} = \mathrm{b}_c(\mathrm{x}) \ + \ \mathrm{A}_c(\mathrm{x}) \cdot \mathrm{u}, \qquad (14)$$

where the vector-function $\mathrm{b}_c(\mathrm{x}) = \{L_f^{\tilde{n}_i} \mathrm{h}_i(\mathrm{x}), i = 1:r\}$.

Now, if the control vector $\mathrm{u} = \mathcal{U}(\mathrm{x})$ is chosen in the form $\mathrm{u} = \mathrm{A}_c^{-1}(\mathrm{x})(-\mathrm{b}_c(\mathrm{x}) + \nu)$ with the *canonical control* $\nu \in \mathbb{R}^r$, then the system (14) can be written in the *Brunovsky* canonical form with the state vector z. Futhermore, for the desired spectrum $\mathrm{S}_c^* = \cup \mathrm{S}_{ci}^*$ of that closed-loop system, where

$$\mathrm{S}_{ci}^* = (\lambda_{c1}^{*i}, \ldots \lambda_{cn_i}^{*i}), \ i = 1:r; \ \lambda_{cs}^{*i} = -\alpha_{cs}^i \pm j\omega_{cs}^i, \ s = 1:\tilde{p}_i;$$

$$\lambda_{cs}^{*i} = -\alpha_{cs}^i, \ s = (\tilde{p}_i + 1) : \tilde{k}_i$$

with $\tilde{k}_i = \tilde{n}_i - \tilde{p}_i$ and $\alpha_{cs}^i > 0$, the *modal synthesis* of the *canonical control law* $\nu = \mathrm{K} \cdot \mathrm{z} \equiv \mathrm{K} \cdot \Phi(\mathrm{x})$ is executed, with *simultaneous construction* of the VLF [27]:

$$\upsilon(\mathrm{x}) = \{\{\upsilon_i^s(\mathrm{x}), s = 1:\tilde{k}_i\}, i = 1:r\}; \upsilon_i^s(\mathrm{x}) = |\mathrm{w}_i^s(\mathrm{x})|; \mathrm{w}(\mathrm{x}) \equiv \mathrm{V}_c^{-1}(\mathrm{S}_c^*)\Phi(\mathrm{x}), \qquad (15)$$

with dimension $\tilde{k} = \sum \tilde{k}_i$, where $\mathrm{V}_c(\mathrm{S}_c^*)$ is the block-diagonal *Vandermonde* matrix, which is analytically formed and inverted [27].

The *linear canonical control law* $\nu = \mathrm{K} \cdot \mathrm{z}$ provides the exponential stability for the *linear* system (14) in an equilibrium position $\mathrm{z} = \Phi(\mathrm{x}) = 0 \Longleftrightarrow \mathrm{x} = 0$ for the *nonlinear* system (13) with *nonlinear control law* $\mathrm{u}(\mathrm{x}) = \mathrm{A}_c^{-1}(\mathrm{x})(-\mathrm{b}_c(\mathrm{x}) + \mathrm{K} \cdot \Phi(\mathrm{x}))$. Moreover, the matrix P in the comparison system (12) for the VLF (15) has the elementary diagonal form. Finally, bilateral componentwise inequalities for the vectors $\mathrm{x}, \mathrm{z}, \upsilon(\mathrm{x}), \rho(\mathrm{x})$, and $\rho^0(\mathrm{x}_0)$ are derived. It is most desirable to obtain the *explicit* form for the nonlinear transformation $\mathrm{x} = \Psi(\mathrm{z})$, inverted with respect to $\mathrm{z} = \Phi(\mathrm{x})$. The *aggregation procedure* for the VLF is carried out with an analysis of the proximity for the *singular* of directions in the *Jacobian* $[\partial \mathrm{F}(\mathrm{x}, \mathcal{U}(\mathrm{x}))/\partial \mathrm{x}]$ [27].

Since the forming control u_k is discrete on a period T_u and a measurement of the model (13) coordinate's state is discrete and incomplete, we obtained a simplified discrete object's model (9) from the continuous model (13) by means of the *Taylor-Lie* series [29], and then we form a nonlinear discrete observer in the nonlinear control law (8) by so-called $k(x)$-duality of the nonlinear control systems [30] and carry out its modal synthesis in canonical variables with parameterization of its desired *discrete* spectrum Z_s^*, and also simultaneously analytically [27] construct a *discrete subvector* Lyapunov function $\upsilon_s(\hat{\mathrm{x}}_{ek})$ for the estimate $\hat{\mathrm{x}}_{ek}$ of the principal variables' vector x_{ek}.

In stage 2, with the help of a VLF with the above structure for the *principal variables*, the problem of nonlinear control law synthesis for the *complete model* of the nonlinear control system (10), taking rejected coordinates and nonlinearities, and also the restriction on

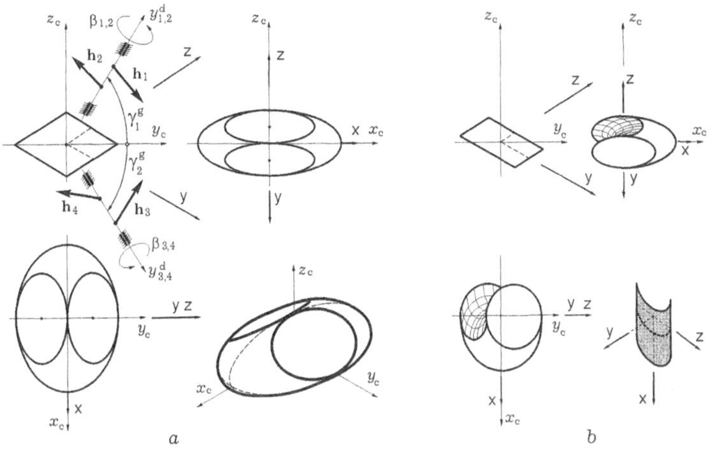

Figure 1 The minimum-excessive fault-tolerant scheme of MGC and envelope its AM.

control, into account is solved. That is, fulfilled as a multistage numerical process of *parametric* optimization of the *nonlinear* comparison system (12) with a view to selecting a spectrum S_c^* (or *discrete* spectra Z_c^* and Z_s^*) [23] as well as some other parameters for the additionally considered coordinates and nonlinearities in the complete model (10), which provide the $\rho\rho^0$-*exponential invariance* for closed-loop nonlinear control system and optimize its main quality criterion.

3 MATHEMATICAL MODELS

Any model of a moving information spacecraft takes into account:

- spatial angular motion of a spacecraft bogy;
- movements of flexible contructional units (SAPs, RTAs, etc.) about the spacecraft bogy by single- or two-gimbal gear-stepping drives (GSDs);
- the external torque's vector $\mathbf{M}_d^o$ as the vector sum of gravity-gradient $\mathbf{M}_g^o$, magnetic $\mathbf{M}_m^o$ (both disturbing $\mathbf{M}_{md}^o$ and control $\mathbf{M}_{mc}^o$) torques, and also that of solar pressure forces $\mathbf{M}_s^o$ for communication spacecraft and of aerodynamic forces $\mathbf{M}_a^o$ for the remote-sensing spacecraft;
- operation of the MGC in the form of minimum-excessive scheme "2-SPEED"-type [31], see Fig. 1, where $\mathbf{h}_p, p = 1:4$, is the angular momentum (AM) unit vector;
- model of each gyrodine (GD), which describes
  - nonlinear rotation dynamics for gyrorotor with regard to its imbalance;
  - flexibility of gyroshell's and gyrorotor's preloaded ball bearings;

- nonlinear dynamics of the moment gearless drive (MGD) on the GD precession axis for communication spacecraft as well as flexibility, a dead band, and kinematic defects in the GSD for remote-sensing spacecraft;
- nonlinear dynamics of stepping motor and *electromagnetic damper* on the GD precession axis taking into account the *dry friction* torque;

- models of desaturation loops of the accumulated AM at the expense of low-draft reaction thrusters (RTs) and/or a magnetic drive (MD), and also of aerodynamic forces from the ongoing atmospheric flow for the LEO remote-sensing spacecraft;
- models of the system's meter elements take into account the proper dynamics of these devices, nonlinearities, digital forming of output signals, discrete noise influences, and delay with respect to the main cycle of the onboard computer operation.

It is very important to provide the opportunity for reconfiguration of the gyrocomplex structure and control algorithms for 1 to 2 abrupt possible faults in any GD of the MGC, since in this case the gyrocomplex control characteristics are essentially changed — see the AM envelope for the MGC without GD-4 in Fig. 1*b*.

In communication spacecraft 4 GDs are also used, but with MGD under bounded precession angles with respect to their neutral positions, see Fig. 2*a*, where $\gamma_1^{\mathrm{g}} = \pi/2$ and $\gamma_1^{\mathrm{g}} = 0$; furthermore for the main mode of the spacecraft attitude control in each MGC configuration only 3 GDs are used — the fourth GD is in the "old" reserve, and all GDs' electric components are duplicated.

For the MGC's Z-arrangement on the spacecraft body, when the axis $\mathrm{O}_g x_c$ is the same as the axis $\mathrm{O}z$ of the body's reference frame (BRF), for $\sigma = \pi/6$ and $\beta_p \in [-\pi/2, \pi/2]$, the following 4 efficient (for 3-axis spacecraft attitude control) MGC configurations are possible on the basis of only 3 active GDs — configurations Z-I, $\mathrm{I} = 1 : 4$ — the MGC without GD-I, represented at the nominal state in Fig. 2*b* (Z-4 or Z-3) and in Fig. 2*c* (Z-2 or Z-1). Therefore, the MGC scheme in Fig. 2*a* is really fault-tolerant under *diagnostics* of the faulted GD and *reconfiguration* of the used GDs by passages between configurations Z-I with *special logic conditions*.

The BRF's attitude with respect to the inertial reference frame (IRF) is defined by the quaternion $\mathbf{\Lambda} = (\lambda_0, \boldsymbol{\lambda}), \boldsymbol{\lambda} = (\lambda_1, \lambda_2, \lambda_3)$. Assume that $\mathbf{\Lambda}^p(t)$ is a quaternion, and $\boldsymbol{\omega}^p(t) \equiv \{\omega_i^p(t)\}$ and $\dot{\boldsymbol{\omega}}^p(t)$ are angular rate and acceleration vectors of the programmed spacecraft body's motion in the IRF.

The spacecraft attitude error quaternion is $\mathbf{E} \equiv (e_0, \mathbf{e}) = \tilde{\mathbf{\Lambda}}^p(t) \circ \mathbf{\Lambda}$, the *Euler* parameters' vector is $\boldsymbol{\mathcal{E}} \equiv \{e_0, \mathbf{e}\}$, and the attitude error's matrix is $\mathbf{C}_e \equiv \mathbf{C}(\boldsymbol{\mathcal{E}}) = \mathbf{I}_3 - 2[\mathbf{e}\times]\mathbf{Q}_e^{\mathrm{t}}$, where $\mathbf{Q}_e \equiv \mathbf{Q}(\boldsymbol{\mathcal{E}}) = \mathbf{I}_3 e_0 + [\mathbf{e}\times]$ with $\det(\mathbf{Q}_e) = e_0$. Here $\mathbf{I}_n$ is n-dimensional identity matrix, symbols $\langle\cdot,\cdot\rangle$, $\times$, $\{\cdot\}$, $[\cdot]$ for vectors and $[\mathbf{a}\times]$, $(\cdot)^{\mathrm{t}}$ for matrixes are conventional denotations.

The communication spacecraft attitude with respect to the orbital reference frame (ORF) $\mathrm{O}x^o y^o z^o$, see Fig. 2*d*, is defined by quaternion $\mathbf{E}^o \equiv (e_0^o, \mathbf{e}^o) = \tilde{\mathbf{\Lambda}}_o(t) \circ \mathbf{\Lambda}$, where $\mathbf{\Lambda}_o$ is a known quaternion of the ORF's attitude with respect to the IRF, by angles of yaw ψ, roll φ, and pitch θ for the rotational sequence $\{1\text{-}3\text{-}2\}$, and also by the matrix of direction cosines $\mathbf{C}_e^o \equiv [\varphi]_2\,[\theta]_3\,[\psi]_1$, where $[\alpha]_i$ is the matrix of elementary rotation, and by vector of *Euler's* parameters $\boldsymbol{\mathcal{E}}^o = \{e_0^o, \mathbf{e}^o\}$; moreover $\mathbf{C}_e^o = \mathbf{C}(\boldsymbol{\mathcal{E}}^o)$. For a fixed position of onboard equipment's *flexible* structures on a spacecraft body with some simplifying assumptions, the motion model of a spacecraft ACS equipped with the MGC and the moment

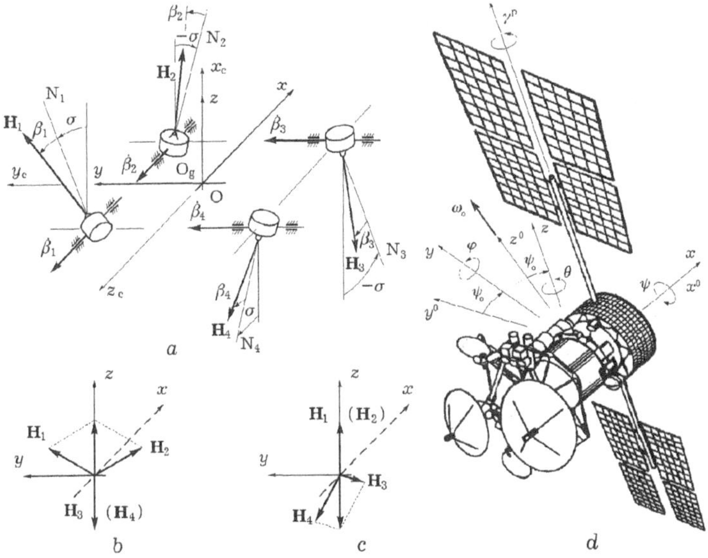

Figure 2 The fault-tolerant scheme of MGC and the spacecraft attitude angles.

gearless drives on the GD precession axes appears as:

$$\dot{\Lambda} = \Lambda \circ \omega \,/\, 2; \qquad \mathbf{A}^o \cdot \{\dot{\omega}, \ddot{\mathbf{q}}, \ddot{\beta}, \dot{\Omega}^h\} = \{\mathbf{F}^\omega, \mathbf{F}^q, \mathbf{F}^\beta, \mathbf{F}^h\}, \tag{16}$$

where

$$\mathbf{F}^\omega = \mathbf{M}^{\mathrm{g}} - \omega \times \mathbf{G} + \mathbf{M}^o_d + \mathbf{Q}^o; \quad \mathbf{M}^{\mathrm{g}} = -\dot{\mathcal{H}} = -\mathbf{A}_{\mathrm{h}}\,\dot{\beta};$$

$$\mathbf{F}^q = \{-a^q_{jj}((\delta^q/\pi)\,\Omega^q_j \cdot \dot{q}_j \; + \; (\Omega^q_j)^2 \cdot q_j) + \mathbf{Q}^q_j(\omega, \dot{q}_j, q_j)\};$$

$$\mathbf{F}^\beta = \mathbf{A}^{\mathrm{t}}_{\mathrm{h}}\omega + \mathbf{M}^g_c + \mathbf{M}^g_d + \mathbf{M}^g_b + \mathbf{M}^g_f + \mathbf{Q}^g; \; \mathbf{M}^g_c = \mathbf{M}^g + \mathbf{M}^{gd} + \mathbf{M}^{ga};$$

$$\mathbf{F}^h = \mathbf{M}^h_c + \mathbf{M}^h_d + \mathbf{M}^h_f + \mathbf{Q}^h(\Omega, \beta, \dot{\beta}, \omega); \quad \mathbf{M}^h_c = \mathbf{M}^h + \mathbf{M}^{ha};$$

$$\mathbf{A}^o = \begin{bmatrix} \mathbf{J}^o & \mathbf{D}_q & \mathbf{D}_g & \mathbf{D}_h \\ \mathbf{D}^{\mathrm{t}}_q & \mathbf{A}^q & 0 & 0 \\ \mathbf{D}^{\mathrm{t}}_g & 0 & \mathbf{A}^g & 0 \\ \mathbf{D}^{\mathrm{t}}_h & 0 & 0 & \mathbf{A}^h \end{bmatrix}; \qquad \begin{array}{l} \mathbf{H} = \{\mathrm{H}_p\}; \; \mathcal{H}(\beta) = \sum \mathbf{H}_p(\beta_p); \\ \mathbf{G} = \mathbf{G}^o + \mathbf{D}_q\dot{\mathbf{q}} + \mathbf{D}_g\dot{\beta}; \\ \mathbf{G}^o = \mathbf{J}^o\,\omega + \mathcal{H}(\beta); \\ \mathbf{A}_{\mathrm{h}}(\mathbf{H}, \beta) \; = \; [\partial\mathcal{H}(\beta) \,/\, \partial\beta]; \end{array}$$

- the inertia matrices $\mathbf{J}^o$, $\mathbf{A}^q$, $\mathbf{D}_q$, $\mathbf{A}^g$, $\mathbf{A}^h$ and $\mathbf{D}_g$ are constant in the BRF, whereas the matrix $\mathbf{D}_h$ can be found from the relation $\mathcal{H}(\beta) = \mathbf{D}_h(\beta)\Omega^h$;
- $\omega = \{\omega_i\}$ is a spacecraft body angular rate vector;
- $\mathrm{q} = \{q_j\}$ is a coordinate vector by *flexible oscillations* of its construction;

- $\boldsymbol{\beta} = \{\beta_p\}$, $\boldsymbol{\Omega}^h = \{\Omega^h_p\}$ and $\mathbf{H} \equiv \{\mathrm{H}_p\} = \mathbf{A}^h\ \boldsymbol{\Omega}^h$ are vectors of the precession angles, the gyrorotor's angular rates and the proper AMs of the GDs in the MGC;
- $\mathbf{G}$ is the AM vector of a spacecraft together with the MGC and the vector $\mathbf{G}^o = \mathbf{J}^o\boldsymbol{\omega} + \boldsymbol{\mathcal{H}}(\boldsymbol{\beta})$ is its main component;
- $\mathbf{Q}^o(\boldsymbol{\omega}, \dot{\mathbf{q}})$ and $\mathrm{Q}^q_j = \mathrm{Q}^q_j(\boldsymbol{\omega}, \dot{q}_j)$ are nonlinear continuously differentiable functions, as well as the vector-functions $\mathbf{Q}^g(\boldsymbol{\Omega}, \boldsymbol{\beta}, \dot{\boldsymbol{\beta}}, \boldsymbol{\omega})$ and $\mathbf{Q}^h(\boldsymbol{\Omega}, \boldsymbol{\beta}, \dot{\boldsymbol{\beta}}, \boldsymbol{\omega})$;
- the torques $\mathbf{M}^g_b(\boldsymbol{\beta}, \dot{\boldsymbol{\beta}})$, describing the influence of limiting supports on the gyrodine precession axes, as well as the torque's vectors of rolling friction forces in the bearings on the gyrodine's rotor axes $\mathbf{M}^h_f(\boldsymbol{\Omega}, \boldsymbol{\beta}, \dot{\boldsymbol{\beta}}, \boldsymbol{\omega})$ and in the bearings on gyrodine's precession axes $\mathbf{M}^g_f(\boldsymbol{\Omega}, \boldsymbol{\beta}, \dot{\boldsymbol{\beta}}, \ddot{\boldsymbol{\beta}}, \boldsymbol{\omega})$ are nonlinear *discontinuous* functions; moreover the last function vs. $\ddot{\boldsymbol{\beta}}$ for the *extended definition* of the gyrodine's motion under the conditions $\dot{\beta}_p = 0$ for any $p = 1:4$;
- the torques of the physically implemented damping $\mathrm{M}^g_{dp}(\dot{\beta}_p)$ and $\mathrm{M}^h_{dp}(\Omega_p)$ are nonlinear continuous functions.

The components of the gyrocomplex control vectors $\mathbf{M}^g_c$ and $\mathbf{M}^h_c$ with regard for the possible *faults* in electric circuits of the MGDs as well as electromagnetic dampers on the GD's precession axes, and also that of the electric drives on the gyrorotor axes and arresters (cages) are described by *hybrid functions*:

$$\mathrm{M}^x_p = \sum_{l=1}^{2} \gamma^{fxl}_p(\nu)\, \gamma^{rxl}_p(\mu)\, a^x_p\, i^{xl}_p, \quad x = g, gd, ga, h, ha, \tag{17}$$

where coordinates $\gamma^{yxl}_p, y = f, r$ are *logic* variables $\gamma^{yxl}_p \in \{0, 1\}$; $\gamma^{yx1}_p \wedge \gamma^{yx2}_p = 0$; $\gamma^{yx1}_p \vee \gamma^{yx2}_p = 1, p = 1:4$; i^{xl}_p are the control currents and currents at GD's electromagnetic arresters in main ($l = 1$) and in reserve ($l = 2$) circuits, and a^x_p are constants. The functions $\gamma^{fxl}_p(\nu)$ are outputs of an ALA $\boldsymbol{\mathcal{A}}^f$ with memory used for representing fault occurrences and damage development. Functions $\gamma^{rxl}_k(\mu)$ are outputs of an SLA $\boldsymbol{\mathcal{A}}^r$, also with memory, for description of damage or fault blockkeeping and the reconfiguration sequence. The currents in GD's control circuits $i^{gl}_p(t)$ for $\gamma^{rgl}_p = 1$ and $i^{hl}_p(t)$ for $\gamma^{rhl}_p = 1$ are proportional to GD's digital control voltages:

$$u^x_p(t) = \mathrm{Zh}[\mathrm{Sat}(\mathrm{Qntr}(u^x_{pk}, b^x_u), B^x_u), \mathrm{T}_u], \quad x = g, h, \tag{18}$$

where $u^x_{pk}, x = g, h$ are the outputs of NCLs on the GD's precession and gyrototor's axes, and functions $\mathrm{Sat}(x, a)$ and $\mathrm{Qntr}(x, a)$ are general-usage ones, while the holder model with the period T_u is of the type: $y(t) = \mathrm{Zh}[x_k, \mathrm{T}_u] = x_k \quad \forall t \in [t_k, t_{k+1})$.

4 THE MGC MOMENTUM DISTRIBUTION

For any gyrodine at gyrocomplex in Fig. 1 equipped with the GSDs on the GD precession axes, we investigated the influence of such factors as flexibility, dead band, kinematic defects in the gear, quantization of the control signal, limitation on the stepping motor current, etc. We revealed [32] amplitude of ranges for the GD's precession rate and acceleration, in which the command and true precession rates are close in spite of the existence of large gyroscopic torques on the GD's precession axes by virtue of the spacecraft's rapid SRM. This means that for selected GD parameters, the assumptions for the *precession theory* of the CMGs are satisfied for indicated ranges of the input command signals, and for $\mathrm{H}_p = h_g = \mathrm{const}\ \forall p = 1:4$ the vector of the MGC's output control torque $\mathbf{M}^{\mathrm{g}}$ can be presented by

$$\mathbf{M}^{\mathrm{g}}(\boldsymbol{\beta}, \mathbf{u}) = -\dot{\boldsymbol{\mathcal{H}}}(\boldsymbol{\beta}, \mathbf{u}) = -h_g \mathbf{A}_\gamma \mathbf{A}_h(\boldsymbol{\beta}) \cdot \mathbf{u};\ \dot{\boldsymbol{\beta}} = \mathbf{u};\ \ \mathbf{A}_h(\boldsymbol{\beta}) = [\partial \mathbf{h}(\boldsymbol{\beta})/\partial \boldsymbol{\beta}], \quad (19)$$

where $\boldsymbol{\mathcal{H}}(\boldsymbol{\beta}) = h_g \mathbf{h}^{\mathrm{c}};\ \mathbf{h}^{\mathrm{c}} = \{x_c, y_c, z_c\} = \mathbf{A}_\gamma \mathbf{h};\ \mathbf{h}(\boldsymbol{\beta}) = \{\mathsf{x}, \mathsf{y}, \mathsf{z}\} = \sum \mathbf{h}_p(\beta_p)$;
$\gamma = \gamma_1^g = \gamma_2^g$; $\mathsf{x} = x_{12} + x_{34}$; $\mathsf{y} = y_1 + y_2$; $\mathsf{z} = -(z_3 + z_4)$; $x_p = c_{\beta_p}$; $y_p = s_{\beta_p}$; $z_p = s_{\beta_p}$; $x_{12} = x_1 + x_2$; $x_{34} = x_3 + x_4$; $\mathbf{A}_\gamma = [\{1, 0, 0\}, \{0, s_\gamma, -c_\gamma\}, \{0, s_\gamma, c_\gamma\}]$, and $c_\alpha \equiv \cos\alpha$; $s_\alpha \equiv \sin\alpha$.

The exact distribution law (DL) [33] of the normed AM $\mathbf{h}(\boldsymbol{\beta}) = \mathbf{A}_\gamma^{-1}\boldsymbol{\mathcal{H}}(\boldsymbol{\beta})/h_g$ between the GD pairs:

$$f_\rho(\boldsymbol{\beta}) = (\tilde{x}_1 - \tilde{x}_2) + \rho\,(\tilde{x}_1\,\tilde{x}_2 - 1) = 0; \quad (20)$$

$$\tilde{x}_1 \equiv x_{12}/q_y;\ \tilde{x}_2 \equiv x_{34}/q_z;\ q_s \equiv \sqrt{4 - s^2}, s = y, z,$$

with $\rho = 2\sqrt{6}/5$, which differs from that proposed by Crenshaw [31], ensures *global maximum* of the *Grame*'s determinant $\det(\mathbf{A}_h \mathbf{A}_h^{\mathrm{t}}) = 64/27$ and the maximum module of the warranted control torque's vector $\mathbf{M}^{\mathrm{g}}$ in an arbitrary direction for the MGC "park" state $\mathbf{h}(\boldsymbol{\beta}) = \mathbf{0}$, as well as large singularities within the central part of the MGC AM's variation domain:

$$\mathbf{S} \equiv \{x^2 + y^2 + z^2 - 2q_y q_z < 8; |y| < 2; |z| < 2\},$$

see Fig. 1*a*, and analytically described discrete curves in the set of smoothly passed MGC's *internal singularities* $\mathbf{Q}_{yz}(\boldsymbol{\beta}) = \mathbf{Q}_y^p \cup \mathbf{Q}_z^p$ with subsets in a peripheral part of domain $\mathbf{S}$ and $\mathbf{Q}_s^p = \mathbf{Q}_s^* \cap \mathbf{S}_s^*; \mathbf{S}_s^* = \{s = 0; |s_1| = |s_2| = 1, s = y, z\}$, and subsets $\mathbf{Q}_s^*$ are described as

$$\mathbf{Q}_y^* = \{(x_{34}/(2\rho))^2 + (z/2)^2 = 1;\ x_{34} < 0\};$$
$$\mathbf{Q}_z^* = \{(x_{12}/(2\rho))^2 + (y/2)^2 = 1;\ x_{12} > 0\}.$$

For the DL (20) the "right-sided differential relay-hysteresis" tuning [33]:

$$D^+ f_\rho(\boldsymbol{\beta}) = \Phi_\rho(f_\rho(\boldsymbol{\beta}), \mathbf{h}(\boldsymbol{\beta})); \quad (21)$$

$$\Phi_\rho(\cdot,\cdot) \triangleq \begin{cases} -\mathrm{Sat}(\phi_\rho, \mu_\rho \cdot f_\rho(\boldsymbol{\beta})), & \mathbf{h}(\boldsymbol{\beta}) \in \mathbf{S} \setminus \mathbf{Q}_{yz}; \\ \phi_\rho \mathrm{Relh}(a_s, l_\rho, r_s), & \mathbf{h}(\boldsymbol{\beta}) \in \mathbf{Q}_s^p, s = y, z, \end{cases}$$

where ϕ_ρ, μ_ρ and l_ρ are positive constants; nonlinear function $\mathrm{Sat}(a, x)$ is well-known and relay hysteresis function

$$\mathrm{Relh}(a, l_\rho, x) = (1,\ \text{if}\ x > -l_\rho) \vee (-1,\ \text{if}\ x < l_\rho),$$

with

$$\mathrm{Relh}(a_s, l_\rho, r_s(\boldsymbol{\beta}(t_0))) = a_s \in \{-1; 1\}, s = y, z;\quad r_y = M_\pi(\beta_1 - \beta_2 - \pi);$$
$$r_z = M_\pi(\beta_3 - \beta_4 - \pi);\ \ M_\pi(\alpha) = (\alpha,\ \text{if}\ |\alpha| \le \pi) \vee (\alpha - 2\pi\ \mathrm{Sign}(\alpha),\ \text{if}\ |\alpha| > \pi),$$

for MGC's AM vector ensures its inclusion in the set $\mathbf{Q}_{yz}(\boldsymbol{\beta})$ of singular states only at *separate* time moments, *bijectively* connects the vectors $\mathbf{M}^{\mathrm{g}}$ and $\boldsymbol{\beta}$, $\dot{\boldsymbol{\beta}}$.

5 OPTIMIZATION OF PROGRAMMED CONTROL

During the recent 25 years we have persistently researched optimizing the programmed attitude motions of information spacecraft [28, 34, 35, 36] and control algorithms for their precise stabilization based on the methods of Lyapunov functions and vector functions.

For the simplified controlled object model (16), (19) within the CMG precession theory under the spacecraft model as a free rigid body,

$$\dot{\mathbf{\Lambda}} = \mathbf{\Lambda} \circ \boldsymbol{\omega}/2; \quad \mathbf{J}^o\dot{\boldsymbol{\omega}} + [\boldsymbol{\omega}\times]\,\mathbf{G}^o = \mathbf{M}^{\mathrm{g}}(\boldsymbol{\beta},\mathbf{u}); \quad \dot{\boldsymbol{\beta}} = \mathbf{u}, \tag{22}$$

the computer-aided algorithm for synthesis of strict optimal control $\mathbf{u} = \mathbf{u}^p(t)$, $t \in \mathrm{T}_r = [t_i, t_f]$ for the AM's DL of the MGC during the spacecraft SRM with general *boundary conditions* on the initial and final states $\mathbf{\Lambda}^s = \mathbf{\Lambda}(t_s); \boldsymbol{\omega}^s = \boldsymbol{\omega}(t_s), s = i, f; \boldsymbol{\beta}^i = \boldsymbol{\beta}(t_i)$ with $\mathbf{h}(\boldsymbol{\beta}^i) \subseteq \mathbf{S}^o \subset \mathbf{S}$ in the *optimization* problem of the energy-type index:

$$\mathrm{I}^o = \frac{1}{2}\int_{t_i}^{t_f} \langle \mathbf{u}^p(\tau), \mathbf{u}^p(\tau)\rangle d\tau \Longrightarrow \min,$$

with preassigned time $T_r = t_f - t_i$ for $\mathbf{\Lambda}^i, \boldsymbol{\omega}^i \Rightarrow \mathbf{\Lambda}^f, \boldsymbol{\omega}^f$; $\mathbf{h}(\boldsymbol{\beta}(t)) \subseteq \mathbf{S}^o \; \forall\, t \in \mathrm{T}_r \subset \mathrm{T}_{t_0}$, has been created. The condition that $\mathbf{S}^o \subset \mathbf{S}$ and the used AM's DL (20) are inexplicit restrictions on the programmed control $\mathbf{u}^p(t)$.

The algorithm is based on numerical solving the *two-point boundary-value* problem by *Newton's* iteration method for differential equations of *Pontryagin's* maximum principle. For ensuring numerical convergence for this method, the technique of descending along the parameter from the analytically resolvable *starting* problem, differing from that by Junkins and Turner [2], to the initially stated problem was used.

We elaborated [33, 37] the fast approximate algorithms for parametric optimization of the programmed control $\mathbf{u}^p(t)$ with restrictions to $\boldsymbol{\omega}(t), \dot{\boldsymbol{\beta}}(t), \ddot{\boldsymbol{\beta}}(t)$ in a class of spacecraft angular motions. Therewith we used conditions of solving *Darboux's* kinematic problem; the spacecraft angular motion is represented in the form of *three simultaneous* rotations, each executed about the axis *motionless* with respect to the preceding coordinate frame. By exact DL (20) of a *given* normed AM $\mathbf{h}(\boldsymbol{\beta})$ there is derived a main relation

$$\delta = d(1 - (1 - 2ac\rho - e\rho^2)^{1/2})/\rho,$$

where $a = x/d; b = q_y q_z/d^2; c = (q_y - q_z)/d; d = q_y + q_z; e = 4b - a^2$, and due to explicit analytical formulae $x_s = ((2 - r_s)\,(2 + r_s))^{1/2}/r_s; r_s = (a_s^2 + b_s^2)^{1/2}$, $s = y, z$, where $a_y = (x + \delta)/2$; $b_y = y$; $a_z = (x - \delta)/2$; $b_z = -z$, results in $x_{1,2}(t) = (a_y \mp b_y\, x_y)/2$; $y_{1,2}(t) = (b_y \pm a_y\, x_y)/2$ for the first GD's pair and similarly for the second pair.

6 NONLINEAR FEEDBACK SPACECRAFT CONTROL

If the error $\tilde{\boldsymbol{\omega}}$ in the angular rate vector $\boldsymbol{\omega}$ is defined as $\tilde{\boldsymbol{\omega}} = \boldsymbol{\omega} - \mathbf{C}_e\boldsymbol{\omega}^p(t)$, and the MGC's required control torque vector $\mathbf{M}^{\mathrm{g}}$ (19) for the spacecraft model (22) is formed as $\mathbf{M}^{\mathrm{g}}(\cdot) = \boldsymbol{\omega}\times\mathbf{G}^o + \mathbf{J}^o(\mathbf{C}_e\dot{\boldsymbol{\omega}}^p(t) - [\boldsymbol{\omega}\times]\mathbf{C}_e\boldsymbol{\omega}^p(t) + \tilde{\mathbf{m}})$, then the simplest normed nonlinear model of spacecraft attitude error kinematics and dynamics is as follows:

$$\dot{e}_0 = -\langle \mathbf{e}, \tilde{\boldsymbol{\omega}}\rangle / 2; \quad \dot{\mathbf{e}} = \mathbf{Q}_e\,\tilde{\boldsymbol{\omega}}/2; \quad \dot{\tilde{\boldsymbol{\omega}}} = \tilde{\mathbf{m}}. \tag{23}$$

By analytical transformations we derived [14, 15] the following relations: $\mathbf{Q}_e^{-1}\mathbf{Q}_e^{\mathrm{t}} = \mathbf{C}_e$; $\mathbf{Q}_e^{-1} = \mathbf{Q}_e^{\mathrm{t}} + \mathbf{e}\cdot\mathbf{e}^{\mathrm{t}}/e_0$; $\mathbf{Q}_e^{-1}\mathbf{e} = \mathbf{e}/e_0$; $\mathbf{I}_3 - e_0\mathbf{Q}_e^{-1} = \mathbf{Q}_e^{\mathrm{t}}[\mathbf{e}\times]$, which are used for $e_0 \neq 0$.

On stage 1, for model (23) let $\mathbf{x} = \{\mathbf{e}, \tilde{\boldsymbol{\omega}}\}$, $e_0^2(\mathbf{x}) = 1 - \mathbf{e}^2$; $\mathbf{f}(\mathbf{x}) = \{\mathbf{Q}_e\tilde{\boldsymbol{\omega}}/2, \mathbf{0}\}$. Now, choosing $\mathbf{h}(\mathbf{x}) = \mathbf{z}^1 = -2\mathbf{e}$ we can define a nonlocal nonlinear coordinate transformation and use the EFL analytical synthesis. So, we derive

$$\dot{\mathbf{z}}^1 = \mathbf{z}^2 = \mathbf{z}_c = -\mathbf{Q}_e\tilde{\boldsymbol{\omega}};\ \dot{\mathbf{z}}_c = \mathbf{b}_c(\mathbf{x}) + \mathbf{A}_c(\mathbf{x})\tilde{\mathbf{m}};$$

$$\mathbf{A}_c(\mathbf{x}) = -\mathbf{Q}_e;\ \ \mathbf{b}_c(\mathbf{x}) = \mathbf{P}_{e\omega}(\mathcal{E}, \tilde{\boldsymbol{\omega}})\tilde{\boldsymbol{\omega}} + \tilde{\boldsymbol{\omega}}^2\mathbf{e}/2,$$

with matrix $\mathbf{P}_{e\omega} = \langle \mathbf{e}, \tilde{\boldsymbol{\omega}}\rangle(\mathbf{Q}_e - e_0\mathbf{I}_3)/(2e_0)$. If we choose

$$\tilde{\mathbf{m}} = \mathbf{A}_c^{-1}(-\mathbf{b}_c + \boldsymbol{\nu}) = \mathbf{R}_{e\omega}(\mathcal{E}, \tilde{\boldsymbol{\omega}})\cdot\tilde{\boldsymbol{\omega}} + (\tilde{\boldsymbol{\omega}}^2/e_0)\cdot\mathbf{e}/2 - \mathbf{Q}_e^{-1}\boldsymbol{\nu},$$

where $\mathbf{R}_{e\omega} = \langle \mathbf{e}, \tilde{\boldsymbol{\omega}}\rangle\mathbf{Q}_e[\mathbf{e}\times]/(2e_0)$, and $\boldsymbol{\nu} = -a_0^*\,\mathbf{z}^1 - a_1^*\,\mathbf{z}^2$ with constants a_0^* and a_1^*, which are analytically calculated by *Vieta's* formulas on spectrum $\mathrm{S}_{ci}^* = -\alpha_c \pm j\omega_c$, this results in the normed nonlinear control law (NCL):

$$\tilde{\mathbf{m}}(\boldsymbol{\mathcal{E}}, \tilde{\boldsymbol{\omega}}) = -\mathbf{A}_0(\boldsymbol{\mathcal{E}}, \tilde{\boldsymbol{\omega}})\cdot\mathbf{e}\cdot\mathrm{Sgn}(e_0) - \mathbf{A}_1(\boldsymbol{\mathcal{E}}, \tilde{\boldsymbol{\omega}})\cdot\tilde{\boldsymbol{\omega}}, \tag{24}$$

$\mathbf{A}_0(\cdot) = ((2a_0^* - \tilde{\omega}^2/2)/e_0)\mathbf{I}_3$; $\mathbf{A}_1(\cdot) = a_1^*\mathbf{I}_3 - \mathbf{R}_{e\omega}$; $\mathrm{Sgn}(e_0) = (1, \text{if } e_0 \geq 0)\vee(-1, \text{if } e_0 < 0)$. Moreover, the closed-loop nonlinear system (23), (24), is transformed to the simplest *linear* system for the spacecraft *spatial* attitude control. *Simultaneously* constructing the VLF's components using the *Vandermonde* matrix is analytically executed by relation (15), the VLF's structure $\boldsymbol{v}(\mathbf{x}) = \boldsymbol{v}(\boldsymbol{\mathcal{E}}, \tilde{\boldsymbol{\omega}})$ has become known; in this case $\boldsymbol{v}(\mathbf{x}) \in \overline{\mathbb{R}}_+^3$, and for obtaining the inverse nonlinear transformation $\mathbf{x} = \boldsymbol{\Psi}(\mathbf{z})$ in the explicit form it is sufficient to use only the analytic inverse of matrix $\mathbf{Q}_e$ [14]:

$$\mathbf{Q}_e^{-1} = e_0^{-1}\cdot\begin{bmatrix} p_{11}^o & q_{03}^o + q_{12}^o & -q_{02}^o + q_{13}^o \\ -q_{03}^o + q_{12}^o & p_{22}^o & q_{01}^0 + q_{23}^o \\ q_{02}^o + q_{13}^o & -q_{01}^o + q_{23}^o & p_{33}^0 \end{bmatrix},$$

where $p_{11}^o = 1 - q_{22}^o - q_{33}^o$ (123) with the standard chain (123); $q_{ij}^o = e_i\cdot e_j$, $i, j \in \{0:3\}$.

Without restrictions on the moment gyrocomplex control $\mathbf{u} \in \mathbb{R}^4$, the NCL $\mathbf{u}(t, \boldsymbol{\Lambda}, \boldsymbol{\omega}, \boldsymbol{\beta})$ is evaluated analytically using the vector relations

$$\{\mathbf{A}_h(\boldsymbol{\beta}), [\partial f_\rho(\boldsymbol{\beta})/\partial\boldsymbol{\beta}]\}\cdot\mathbf{u}(\cdot) = \{\mathbf{v}_u(\cdot), \Phi_\rho(f_\rho(\boldsymbol{\beta}), \mathbf{h}(\boldsymbol{\beta}))\};\ \ \mathbf{v}_u(\cdot) = -\mathbf{A}_\gamma^{-1}\mathbf{M}^{\mathrm{g}}(\cdot)/h_g.$$

This results in the solution $\mathbf{u}(\cdot) = \lceil s_p^o(\boldsymbol{\beta})\rfloor^{-1}\mathbf{q}_g(\cdot)$, where

$$\sin(\beta_1 - \beta_2) \equiv s_{1,2}^o = y_1x_2 - y_2x_1;\ \ \sin(\beta_3 - \beta_4) \equiv s_{3,4}^o = z_3x_4 - y_4x_3,$$

vectors

$$\begin{aligned} \mathbf{q}_g(\cdot) &= \tilde{\mathbf{D}}(\boldsymbol{\beta})\cdot\mathbf{v}_u + \tilde{\mathbf{d}}_f(\boldsymbol{\beta})\cdot\Phi_\rho(f_\rho(\boldsymbol{\beta}), \mathbf{h}(\boldsymbol{\beta})); \\ \tilde{\mathbf{d}}_f(\boldsymbol{\beta}) &= (q_yq_z/d_{yz})\cdot\{-x_2, x_1, x_4, -x_3\}, \end{aligned} \tag{25}$$

matrix

$$\tilde{\mathbf{D}}(\boldsymbol{\beta}) = \begin{bmatrix} -x_2p_{yx} & -y_2 + x_2r_{yx} & -x_2r_{zx} \\ x_1p_{yx} & y_1 - x_1r_{yx} & x_1r_{zx} \\ -x_4p_{zx} & -x_4r_{yx} & z_4 + x_4r_{zx} \\ x_3p_{zx} & x_3r_{yx} & -z_3 - x_3r_{zx} \end{bmatrix},$$

and also

$$\begin{aligned} d_{yz} &= d_{yx} + d_{zx}; \quad d_{yx} = q_y - \rho\, x_{12}; \quad d_{zx} = q_z + \rho\, x_{34}; \\ r_{yx} &= x_{12}\, y\, p_{zx}/q_y^2;\ r_{zx} = x_{34}\, z\, p_{yx}/q_z^2;\ p_s = d_s/d_{yz},\ s = yx, zx; \end{aligned}$$

moreover the identity $\mathbf{A}_h(\boldsymbol{\beta})\tilde{\mathbf{D}}(\boldsymbol{\beta}) \cdot \tilde{\mathbf{d}}_f(\boldsymbol{\beta}) \equiv \mathbf{0}$ is valid for any vector $\boldsymbol{\beta}$.

Taking into account *restrictions* with respect to the gyrocomplex control in the form $|u_p(t)| \le \overline{u}\ \forall p \in \{1:4\}$ and $\forall t \in \mathrm{T}_{t_0}$ for a *given* constant $\overline{u} \in \mathbb{R}_+$, special nonlinear *scalar* and *vector* functions of the type *"division of variables with scaling"* are *introduced* [15, 33]. The *scalar* function is defined as follows:

$$d^s(x, s, a, m) \triangleq \begin{cases} 0 & x = 0; \\ mx/s & (x \neq 0)\ \&\ (m|x| < a|s|); \\ a\, m\, \mathrm{Sign}(xs) & m|x| > a|s|, \end{cases} \tag{26}$$

for the scalar arguments $x, s \in \mathbb{R}; a, m \in \mathbb{R}_+$. The *vector* function $\mathbf{y} = \mathbf{Md}_4^s(\mathbf{x}, \mathbf{s}, \overline{u})$ has the vector arguments $\mathbf{s}, \mathbf{x} \in \mathbb{R}^4$ and the internal working constant $u^* \gg \overline{u}$. Its components $y_p, p = 1:4$, are defined by the algorithm using the scalar function (26):

$$\begin{array}{ll} \text{for } p = 1:4 \text{ do} & r_p := d^s(x_p, s_p, u^*, 1); \\ & r := \max\{|r_p|,\ p = 1:4\}; \\ \text{for } p = 1:4 \text{ do} & \text{if } r > 0 \text{ then } m_p := |r_p|/r \text{ else } m_p := 1; \\ \text{for } p = 1:4 \text{ do} & y_p := \mathrm{d}^s(x_p, s_p, \overline{u}, m_p). \end{array} \tag{27}$$

As a result, for the vector $\mathbf{s}^o(\cdot) = \{s_p^o(\cdot)\}$ the discrete NCL obtained and applied has the form

$$\mathbf{u}_k^g \equiv \{u_{pk}^g\} = \mathbf{u}^g(\cdot) = \mathbf{Md}_4^s(\mathbf{q}_g(\cdot), \mathbf{s}^o(\cdot), \overline{u}), \tag{28}$$

using the algorithm (27) and the vector function $\mathbf{q}_g(\cdot)$ (25) together with the discrete form of tuning (21) for the distribution law (20).

On stage 2, the problems of this nonlinear control law synthesis are solved for the gyromoment ACS model of the LEO remote-sensing spacecraft (16), (19), (24) with the GD's GSDs [38]. Moreover, the selection of only 3 parameters $\alpha_c, \omega_c, \mu_\rho$ in the nonlinear control law structure (28) and gain k_d^g in GDs' electomagnetic damper, which optimize the main quality criterion and provide the $\rho\rho^0$-*exponential invariance* for given restrictions, is fulfilled on the basis of multistage numerical analysis and optimization of *nonlinear* CS for VLF having derived the above structure for the *principle error coordinates* $\boldsymbol{\mathcal{E}}, \tilde{\boldsymbol{\omega}}$ and other VLF components' structure in the form of *sublinear norms* for vector variables $\dot{\boldsymbol{\beta}}(t)$ and $\mathbf{q}(t)$, $\dot{\mathbf{q}}(t)$, vs. the vector $\boldsymbol{\beta}(t)$.

7 NONLINEAR CONTROL OF COMMUNICATION SPACECRAFT

An arbitrary behavior of gyromoment ACS under requirements (i) and (ii) for communication spacecraft is essentially nonlinear under the gyrodine faults. The nonlinear dynamic problems of the fault-tolerant ACS are quite real for the *nonstoping* communication sessions and spacecraft spatial angular movements:

- the spacecraft body orbital stabilization, when the given state in the ORF is $\mathcal{E}_c^o = \{1, \mathbf{0}\}$, i.e., the attitude angles $\psi = \varphi = \theta = 0$, with reconfiguration of moment gyrocomplex, the indirect SAPs orienting and tracing the Sun by the gear-stepping drives digital control at the angles' vector $\gamma^p = \{\gamma_1^p, \gamma_2^p\}$ on two-gimbal axes for aligning the normal-unit $\mathbf{e}_n^p$ to the panels' surface with respect to the Sun direction unit $\mathbf{e}_n^s$, for all that the matrixes $\mathbf{J}^o(\gamma^p(t)), \mathbf{D}_q(\gamma^p(t))$ and the vector $\mathbf{M}_p^o(\gamma^p, \dot{\gamma}^p, \ddot{\gamma}^p, \boldsymbol{\omega})$ is added to the vector $\mathbf{F}^\omega$ in (16);

- the precision spacecraft stabilization in the Sun-Earth reference frame at $\varphi = \theta = 0$ and $\psi = \psi_0(t)$, with reconfiguration of the MGC, the indirect SAPs guidance into the Sun by means of both the *simultaneous* attitude control of the spacecraft body with respect to the yaw angle ψ and the digital control of the single-gimbal gear-stepping drive on angle γ^p for the panels, see Fig. 2*d*;

- the slow angular reorientation of the spacecraft body into a preassigned state $\mathcal{E}_c^o(t)$ in the ORF or in the Sun-Earth reference frame both before and after the correction of the spacecraft orbit and/or in the passage of some singular kinematic conditions, with reconfiguration of the MGC and the indirect SAPs guidance;

- the precision spacecraft angular stabilization of the given commanded state $\mathcal{E}_c^o(t)$ under correction of the spacecraft orbit and movements of the correcting RTs by digital control of the GSDs on the thruster 2-gimbal axes, and also with reconfiguration of the MGC and the indirect SAPs guidance.

By VLF-method we have also carried out research on nonlinear dynamics for fault-tolerant gyromoment ACSs of communication spacecraft under their substantial *parametric uncertainty*, incomplete discrete measurement of the state and resource restrictions on digital control for small-mass gyrocomplexes of various types, including the mimimum-excessive scheme in Fig. 2*a*. For this case the kinematic relations in (16) have the form

$$\dot{e}_0^o = -\langle \mathbf{e}^o, \tilde{\boldsymbol{\omega}} \rangle / 2; \quad \dot{\mathbf{e}}^o = \mathbf{Q}_e^o \, \tilde{\boldsymbol{\omega}} / 2, \tag{29}$$

where $\mathbf{Q}_e^o \equiv \mathbf{Q}(\mathcal{E}^o)$; $\tilde{\boldsymbol{\omega}} = \boldsymbol{\omega} - \mathbf{C}_e^o \boldsymbol{\omega}_o$; $\boldsymbol{\omega}_o = \{0, 0, \omega_o\}$ is the orbital rate vector, defined in the ORF, see Fig. 2*d*, the matrix

$$\mathbf{A}_h \equiv \mathbf{A}_h(\mathbf{H}, \boldsymbol{\beta}) = \left[\frac{\partial \boldsymbol{\mathcal{H}}(\boldsymbol{\beta})}{\partial \boldsymbol{\beta}} \right] = \begin{bmatrix} -C_{h3}^\sigma & -C_{h4}^\sigma & 0 & 0 \\ 0 & 0 & C_{h1}^\sigma & C_{h2}^\sigma \\ -S_{h1}^\sigma & -S_{h2}^\sigma & S_{h3}^\sigma & S_{h4}^\sigma \end{bmatrix},$$

where $X_{12}^\sigma = X_{h1}^\sigma + X_{h2}^\sigma$; $X_{34}^\sigma = X_{h3}^\sigma + X_{h4}^\sigma$, $X = S, C$, and

$$\delta C_g^\sigma \equiv C_{12}^\sigma - C_{34}^\sigma; \;\; S_{hp}^\sigma \equiv \mathrm{H}_p \sin\beta_p^\sigma; \;\; C_{hp}^\sigma \equiv \mathrm{H}_p \cos\beta_p^\sigma; \;\; \beta_p^\sigma = \beta_p + (-1)^{p-1}\sigma, /p = 1:4,$$

and summary angular momentum of MGC is

$$\boldsymbol{\mathcal{H}}(\boldsymbol{\beta}) = \sum \mathbf{H}_p(\beta_p) = \{-S_{34}^\sigma, S_{12}^\sigma, \delta C_g^\sigma\}; \qquad \mathbf{H}_p(\beta_p) = \mathrm{H}_p \cdot \mathbf{h}_p(\beta_p).$$

Due to accessible information:

- the command order vector of *Euler*'s parameters $\mathcal{E}_c^o(t)$,

- the measured attitude angles $\psi(t), \varphi(t), \theta(t)$ with the sampling period T_q and a vector $\boldsymbol{\mathcal{E}}_k^{\mathrm{of}}$ of *Euler*'s parameters, which is filtered out at the time $t = t_k$,

- the measured and filtered-out gyrodine vectors $\boldsymbol{\beta}_k^{\mathrm{f}}$ of their precession angles and $\boldsymbol{\Omega}_k^{h\mathrm{f}}$ of the gyrorotor's angular rates,

the vector $\delta\mathbf{H}_k$ of error on GDs angular momentums and the *Euler*'s vector $\boldsymbol{\mathcal{E}}_k^{\varepsilon} = \{\varepsilon_{0k}, \boldsymbol{\varepsilon}_k\}$ of the spacecraft attitude error are reprsented as

$$\mathbf{H}_k^{\mathrm{f}} = \mathbf{A}^h \boldsymbol{\Omega}_k^{h\mathrm{f}}; \ \ \delta\mathbf{H}_k = \mathrm{h}_g \cdot \mathbf{I}_4 - \mathbf{H}_k^{\mathrm{f}};$$
$$\varepsilon_{0k} = e_{0k}^{\mathrm{f}} e_{0k}^{\mathrm{c}} + \langle \mathbf{e}_k^{\mathrm{f}}, \mathbf{e}_k^{\mathrm{c}} \rangle; \ \ \boldsymbol{\varepsilon}_k = \mathbf{Q}^{\mathrm{t}}(\boldsymbol{\mathcal{E}}_k^{\mathrm{c}})\, \mathbf{e}_k^{\mathrm{f}} - e_{0k}^{\mathrm{f}}\, \mathbf{e}_k^{\mathrm{c}},$$

where $\mathrm{h}_g = \mathrm{const}$ is a nominal value of angular momentum for each gyrodine. As a result of the nonlinear control law synthesis for such ACS, the gyrocomplex control vectors $\mathbf{u}_k^g$ and $\mathbf{u}_k^h$ are computed onboard by the relations

$$\boldsymbol{\epsilon}_k = -2\boldsymbol{\varepsilon}_k; \quad \mathbf{w}_k^g = \mathbf{w}_{k-1}^g + \mathrm{k}_g(\boldsymbol{\epsilon}_k + \mathrm{a}_g \boldsymbol{\epsilon}_{k-1});$$
$$\mathbf{u}_k^g = -\mathbf{A}_h^{\mathrm{t}}(\mathbf{H}_k^{\mathrm{f}}, \boldsymbol{\beta}_k^{\mathrm{f}})\,(\mathbf{w}_k^g + \boldsymbol{\omega}_{\mathrm{o}k}^{\mathrm{ef}}); \ \ \mathbf{u}_k^h = -\mathrm{k}_h \cdot \delta\mathbf{H}_k, \ \ k \in \mathrm{N}_0, \tag{30}$$

where $\boldsymbol{\omega}_{\mathrm{o}k}^{\mathrm{ef}} = \mathbf{C}(\boldsymbol{\mathcal{E}}_k^{\mathrm{f}})\boldsymbol{\omega}_{\mathrm{o}}$, and $\mathrm{k}_h, \mathrm{k}_g, \mathrm{a}_g$ are the constant scalar parameters.

The synthesized digital nonlinear control law (30) is universal for the given gyrocomplex type; it provides the precision 3-axis attitude control of the spacecraft body in the process of the gyrocomplex reconfiguration and is robust with respect to the accumulated angular momentum. Furthermore, the digital information on only the attitude of the spacecraft body and the gyrodine positions are used for forming the GDs' control. Moreover, the flexible oscillations of spacecraft structures (the SAPs, the high-gain RTAs, etc.) are damped perfectly well. Operability of this digital nonlinear control law under the large torque perturbations arising in the process of the spacecraft orbit correction, as well as under slow large-angle spatial reorientations of spacecraft body has been investigated.

Digital nonlinear control laws for 2-gimbal gear-stepping drivers of solar array panels have been developed with the aid of the *Lyapunov functions method*. These control laws provide asymptotic stability of indirect guidance for the unit $\mathbf{e}_n^p$ to the Sun's direction unit $\mathbf{e}^s$ under arbitrary, but measured, angular motion of the spacecraft body. Moreover, the *Lyapunov function*, $v = 1 - \langle \mathbf{e}^s, \mathbf{e}_n^p \rangle \geq 0$, is a natural measure of closeness for the vectors $\mathbf{e}^s$ and $\mathbf{e}_n^p$ with $v = 0$ iff $\mathbf{e}^s = \mathbf{e}_n^p$. The nonlinear dynamics of both such indirect SAPs orienting and simultaneous spacecraft attitude control in the Sun-Earth reference frame with respect to yawing and by digital control of the SAPs with 1-gimbal gear-stepping driver, as well as tracing the Sun, has been investigated, and problems of passage through singular states in the process of panel and spacecraft attitude control have also been considered. Digital control algorithms for the gear-stepping driver precision guidance of high-gain antennas have also been synthesized.

8 PROVISION OF FAULT TOLERANCE

Onboard ACSs of the communication spacecraft use the two-level logic-digital system of the fault diagnosis:

- on the lower level — an integral local SLA $\mathcal{A}_d^d$ with memory for automatic discrete monitoring of the status of all relevant devices with a sampling period T_q for measurement of available physical variables (currents, movements, rates, etc.)
- on the higher "system" level — an SLA $\mathcal{A}^d$, also with memory, bound up with the sampling period T_u for the functional diagnostics of the main control loop by comparison of outputs for normal and emergency models of the ACS operation.

The high fail-safe operation of gyromoment ACS has been achieved with the use of the onboard digital algorithms, in accordance with the SLAs $\mathcal{A}_d^d$, $\mathcal{A}^d$, and $\mathcal{A}^r$, allowing application of all the reverse complete sets of onboard devices or their electric circuits [39].

The *verbal* description of the provision of fault tolerance of a spacecraft ACS with a gyrocomplex in Fig. 2*a*, for its initial configuration Z-4, when $\mathrm{H}_p = \mathrm{h}_g, p = 1:3$ and

$$\gamma_p^{fxl} = \gamma_p^{rx1} = 1, x = \mathrm{g}, gd, \mathrm{h};\ \gamma_p^{rx1} = 0, x = ga, ha,$$

and GD-4 in the stopping state

$$\mathrm{H}_4 = \beta_4 = 0;\ \gamma_4^{rxl} = 0, x = \mathrm{g}, gd, \mathrm{h};\ \gamma_4^{rxl} = 1, x = ga, ha,$$

see (17), is as follows. In the normal mode, the magnetic desaturation loop ensures the condition $\mathbf{G}^o \approx \mathbf{0}$ under formation of the magnetic control torque vector:

$$\mathbf{M}_{mc}^o = \mathbf{L}_m(t) \times \mathbf{B}_\oplus;\ \mathbf{L}_m(t) = \mathbf{Zh}[\mathbf{L}_{mk}, \mathrm{T}_u],$$

where $\mathbf{B}_\oplus$ is a magnetic displacement vector of geomagnetic field and

$$\mathbf{L}_{mk} = -l_m^o \phi_m^o(\mathrm{R}_0, \lambda_m, b_m, \mathrm{R}_k)\, \mathbf{e}_{mk};\ \mathbf{e}_{mk} = \mathbf{c}_k / \mathrm{c}_k;$$

$$\mathbf{c}_k = \mathbf{R}_k \times \mathbf{B}_{\oplus k}^{\mathrm{f}};\ \mathbf{R}_k = \mathbf{J}^o(\gamma_k^p)\boldsymbol{\omega}_{ok}^{\mathrm{ef}} + \mathcal{H}(\boldsymbol{\beta}_k^{\mathrm{f}}),$$

where a scalar relay-hysteresis function:

$$\phi_m^o(a, \lambda_m, b_m, x) \equiv \{(1\,\forall x > \lambda_m b_m) \vee (0\,\forall x < b_m)\};$$

moreover $\phi_m^o(a, \lambda_m, b_m, a) = a, a \in \{0, 1\}$, with threshold of operation b_m and coefficient of return $0 < \lambda_m < 1$, and l_m^o is the modulus of the magnetic driver dipole torque.

Let the fault of the torque gearless driver current circuit in GD-3 occur at any time moment $t = t_\nu^f \in [t_{k_*-1}, t_{k_*})$; $\nu = 1$, $\gamma_3^{fg1}(1) = 0$. Then by SLAs $\mathcal{A}_{\mathrm{GD-3}}^d$ or $\mathcal{A}^d$, and by SLA $\mathcal{A}^r$ as the result of circuits switching ($\gamma_3^{rg1} = 0; \gamma_3^{rg2} = 1$) is guaranteed for the discrete time $k = k^* = k_*$ or $k = k^* = k_* + 1$, respectively. Moreover, the intensity of dynamic processes for the attitude control channels is essentially dependent not only on the time interval duration $\delta t_{k_*}^f = t_{k^*} - t_\nu^f$, when there is no control, but also on the potentialities of the gyrodines, which remained operable in the aspect of compensation of disturbing influence of the angular rate vector $\boldsymbol{\omega}_{\mathrm{o}}$ because of the spacecraft orbital motion. After such an isolation of the fault, the scheduled reconfiguration of Z-4 $\Rightarrow$ Z-3 process starts:

- $\gamma_4^{rha1} = 0$ and $\gamma_4^{rh1} = 1$ with speeding up from the rest state of GD-4 rotor,
- the magnetic driver desaturation loop operates with $\mathbf{R}_k = \mathcal{H}(\boldsymbol{\beta}_k^{\mathrm{f}})$,

- at achieving of a small neighborhood for the gyrocomplex "park" state $\mathrm{H}_p = \mathrm{h}_g$; $\beta_p = 0$, $p = 1:4$, there takes place *simultaneously*: the GD-3 caging ($\gamma_3^{rga1} = 1$), and GD-4 uncaging ($\gamma_4^{rga1} = 0$) and switching ($\gamma_4^{rg1} = 1$) on-line the control closed-loop.

In the final stage of this process: the magnetic desaturation loop is returned into the nominal mode; the SLA's $\mathcal{A}^r$ output $\gamma_3^{rh1} = 0$, and the GD-3 rotor is speeding down to the rest state; finally, after reaching the condition $\mathrm{H}_3 \approx 0$, the GD-3 is caged ($\gamma_3^{rha1} = 1$). Thus, the moment gyrocomplex restores its excessiveness with respect to control circuits of torque gearless drivers for the on-line gyrodines, and it is prepared for the rapid isolation of any new gyrodine fault and for new reconfiguration.

9 SOFTWARE AND RESULTS

The developed methods for modeling and dynamic research of the fault-tolerant ACS have been implemented in the form of the software system DYNAMICS [40, 41, 42]. The general technology for usage of this software system consists of obtaining models of separate components of ACS, their *automatic* transforming with *automatic* assembling into an integrated model, and then a subsequent dialogue investigation based on both known and unique methods [43, 44, 45, 46] implemented in the software system.

As discussed above, the intensity of dynamic processes is essentially dependent on the potentialities of the gyrodines that remained operable in the aspect of compensation of the spacecraft orbital motion. This fact is illustrated in Fig. 3, where the sampling period values $\mathrm{T}_q = 0.25\,s$ and $\mathrm{T}_u = 4\,s$. Such processes are presented with respect to the pitch channel of ACS under the orbital stabilization for configurations Y-4, Fig. 3a and Z-4, Fig. 3b, when the GD-3 fault takes place under $t = 300.1\,s$.

In the Y-4 case, there are no operable gyrodines needed to create control torques along the axis Oz, so despite the "fast" fault diagnostics by the SLA $\mathcal{A}^d_{\mathrm{GD-3}}$ and switching the torque gearless driver's reserve circuit into the control closed-loop on-line at time $t = t_{k^*} = 304\,s$, there take place substantial overshoots of attitude errors.

Such overshoots are absent for a similar fault in GD-3 within the gyrocomplex according to the configuration Z-4 see Fig. 3a, since GD-1 and GD-2 in this case remain operable to create control torques along the axis Oz, see Fig. 3b. So, even for the "slow" GD-3 fault diagnosis with the aid of SLA $\mathcal{A}^d$ and switching the torque gearless driver's reserve circuit in GD-3 by time $t = t_{k^*} = 308\,s$, the precision angular stabilization with respect to pitch remains the same.

10 CONCLUSIONS

With the aid of the logic-dynamical models, nonlinear methods, and software developed, we have conducted multilateral dynamic investigations of robust fault-tolerant gyromoment ACSs on an order from the State Research and Production Association of the Russian Space Agency — *TsSKB-Progress* (Samara) [47, 48] and *NPO PM* (Krasnoyarsk) [49, 50], for spacecraft both manufactured and presently under design, including those being developed for international projects: the maneuvering vehicle *Ikar* [51] in cooperation with

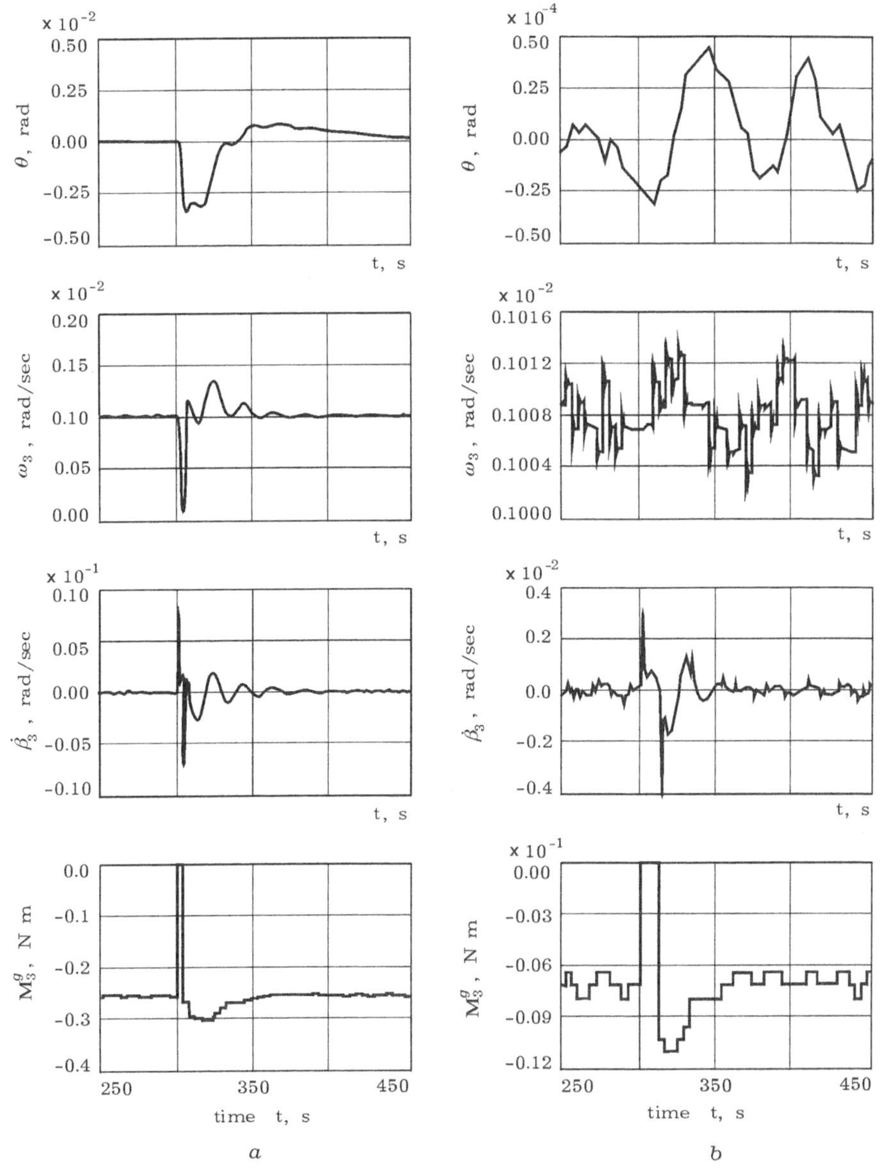

Figure 3 The processes for fault in the MGD's control current circuit of GD-3.

Aerospatiale (France) and communication satellite *Sesat* [52, 53, 54] in cooperation with *Alcatel Space Industries* (France) on the *Eutelsat* order.

Flight exploitation of designed spacecraft ACSs have proved the high efficiency of principles and methods employed for providing the fault-tolerant operation.

Acknowledgments

This work was supported by the Russian Space Agency and by the Russian Foundation for Basic Research (RFBR), grant 01-00-293.

REFERENCES

1. B.V. Raushenbakh and Ye.N. Tokar, *Spacecraft Attitude Control.* Nauka, Moscow, 1974.
2. J.L. Junkins and J.D. Turner, *Optimal Spacecraft Rotational Maneuvers.* Elsevier, Oxford, 1986.
3. B.R. Hoelscher and S.R. Vadali, "Optimal open-loop and feedback control using single gimbal control moment gyroscopes," *Journal of the Astronautical Sciences* 42(2):189–206, 1994.
4. A. Isidori, *Nonlinear Control Systems*, Springer-Verlag, New York, 2nd edition, 1989.
5. S.N. Singh and T. Bossart. "Exact feedback linearization and control of space station using a control moment gyros," *IEEE Trans. on Automatic Control* AC-38(1):184–187, 1993.
6. P.M. Frank, "Fault diagnosis in dynamic systems using analytical and knowledge-based redundancy — a survey and some new results," *Automatica* 26(3):459–474, 1990.
7. P.M. Frank, "Enhancement of robustness in observer-based fault detection," *International Journal of Control* 59(5):955–981, 1994.
8. V.M. Matrosov, "Vector Lyapunov functions in the analysis of nonlinear interconnected systems." In *Symposia Mathematica*, vol. 6, pp. 209–242, Academic Press, London, New York, 1971.
9. V.M. Matrosov, "Comparison method in system's dynamics." In *Equation Differentielles et Functionnelles Non Lineaires*, pp. 407–445, Herman, Paris, 1973.
10. N. Rouche, P. Habets, and M. Lalya, *Stability Theory by Lyapunov's Direct Method.* Springer-Verlag, Heidelberg, 1977.
11. V.M. Matrosov, "Method of vector Lyapunov functions in nonlinear mechanics," *Advances in Mechanics* 6(3):59–82, 1989.
12. V. Lakshmikamtham, V.M. Matrosov, and S. Sivasundaram, *Vector Lyapunov Functions and Stability Analysis for Nonlinear Systems*, Kluwer Academic Publishers, Dordrecht, Boston, London, 1991.
13. Ye.I. Somov, "On synthesis of nonlinear control systems using vector Lyapunov functions." In *Differential Equations and Numerical Methods*, pp. 72–89, Nauka, Novosibirsk, 1986.
14. Ye.I. Somov, "Feedback linearization and VLF techniques on the synthesis of spacecraft gyromoment attitude control systems." In *Proceedings of the IEEE International Conference on Systems, Man and Cybernetics*, vol. 4, pp. 2522–2527, Beijing, 1996.
15. Ye.I. Somov, "Nonlinear spacecraft gyromoment attitude control." In *Proceedings of 1st International Conference on "Nonlinear Problems in Aviation and Aerospace,"* pp. 625–630, Daytona Beach, FL, 1997. ERAU.
16. Ye.I. Somov, "Precision nonlinear gyromoment attitude control of the remote sensing spacecraft," *Actual Problems of Aviation and Aerospace Systems* 2(1):44–53, 1997.

17. V.M. Matrosov, S.N. Vassiljev, R.I. Kozlov, et al., *Algorithms for Theorem Derivation of Vector Lyapunov Functions Method*, Nauka, Novosibirsk, 1981.
18. Lj.T. Grujić, "Coordinated Lyapunov's methodology for stationary nonlinear systems," *Automation and Remote Control* (12):35–73, 1997.
19. A.S. Zemlyakov, Ye.I. Somov, and R.A. Sabitov, "Computer-aided synthesis of the hierarchical motion control systems on the basis of vector Lyapunov functions." In *Stability Theory and Its Applications* pp. 264–274, Nauka, Novosibirsk, 1979.
20. Ye.I. Somov, "On the synthesis of spacecraft attitude control systems." In *Computing and Controlling Systems of Flying Vehicles*, pp. 13–22, KAI, Kazan, 1981.
21. Ye.I. Somov, "Synthesis of invariant systems for spacecraft spatial gyromoment stabilization using the vector Lyapunov functions method." In *Invariance Theory, Sensitivity Theory and their Applications*, pp. 184–185, ICS, the USSR Academy of Sciences, Moscow, 1982.
22. Ye.I. Somov, "Synthesis of nonlinear control law for spacecraft spatial rotational maneuver." In *Optimizing Processes in Aviation Engineering*, pp. 66–70, KAI, Kazan, 1982.
23. V.M. Matrosov, R.I. Kozlov, Ye.I. Somov, et al., "Methods and software for computer-aided design of spacecraft attitude control systems." In *Spacecraft Dymamics and Control*, pp. 163–179, Nauka, Novosibirsk, 1992.
24. G.P. Anshakov, V.M. Matrosov, Ye.I. Somov, et al., "Gyromoment attitude control systems dynamics of rapid manoeuvring remote sensing spacecraft." In *Proceedings of 1st International Aerospace Congress*, vol. 2, pp. 129–132, Petrovka, Moscow, 1995.
25. Ye.I. Somov, "Nonlinear methods for synthesis of spacecraft fail-safe gyromoment attitude control systems." In *Proceedings of the 3rd International Workshop "New Computer Technologies in Control Systems,"* pp. 68–70, IPS of RAS, Pereslavl-Zalessky, 1996.
26. V.M. Matrosov, M.F. Reshetnev, V.A. Rayevsky, and Ye.I. Somov, "Nonlinear methods for dynamic synthesis of fault-tolerant spacecraft attitude control systems," *Journal of Computer and Systems Sciences International* (6):120–130, 1997.
27. Ye.I. Somov, "Constructing the vector Lyapunov functions at synthesis of linear control systems with a partial measurement of state," *Journal of Computer and Systems Sciences International* (3):73–86, 1997.
28. Ye.I. Somov, "Nonlinear dynamics and optimization of the spacecraft precision gyromoment attitude control systems." In *Proceedings of 2nd IMACS/IEEE Conference "Computational Engineering in Systems Application,"* vol. 2, pp. 72–77, Ecole Centrale, Lille, 1998.
29. N. Kazantzis and C. Kravaris, "System-theoretic properties of sampled-data repersentations of nonlinear systems obtained via Taylor-Lie series," *Intern. Journal of Control* 67(6):997–1020, 1997.
30. A.P. Krishenko and S.B. Tkachev, "Nonlinear $k(x)$-dual systems," *Automation and Remote Control* (2):21–34, 1995.
31. J.W. Crenshaw, "2-SPEED, a single-gimbal control moment gyro attitude control systems," *AIAA Paper* (73-895):1–10, 1973.
32. Ye.I. Somov, A.V. Sorokin, and O.A. Kondrat'yev, "Precession models of control moment gyroscopes on optimization of spacecraft singularly perturbed control systems." In *Proceedings of 3rd International Workshop, "Singular Solutions and Perturbations in Control Systems,"* pp. 99–102, IPS of RAS, Pereslavl-Zalessky, 1997.

33. Ye.I. Somov, Ye.A. Bondarenko, and N.B. Kapitonova, "Synthesis of a gyromoment spatial stabilization system on the basis of vector Lyapunov functions and parametric optimization." In *Problems of Analytic Mechanics, Stability and Control*, pp. 257–264, Nauka, Novosibirsk, 1991.
34. Ye.I. Somov and E.F. Fatkhullin, "Optimal control of a gyrostat programmed motion." In *Memoirs of Kazan Aviation Institute*, vol. 180, pp. 37–40, KAI, Kazan, 1975.
35. Ye.I. Somov, "Optimization of extensive control for spacecraft re-orientation by control moment gyroscopes." In *Optimizing Processes in Aviation Engineering*, pp. 78–82, KAI, Kazan, 1981.
36. Ye.I. Somov, "Stabilization of the extensive re-orientation optimal behavior for a spacecraft with control moment gyroscopes." In *Stability and Control*, pp. 55–59, KAI, Kazan, 1983.
37. Ye.I. Somov and I.A. Gerasin, "The rotational maneuver realizability estimation for spacecraft controlled by an excessive gyrodine system." In *Proceedings of the Tsiolkovsky Cosmonautics Academy*, pp. 138–143, Russian Academy of Cosmonautics, Samara, 1998.
38. Ye.I. Somov, V.M. Matrosov, G.P. Anshakov, et al., "Ultra-precision attitude control of the low-orbital remote sensing spacecraft." In *Proceedings of the 14th IFAC Symposium on Automatic Control in Aerospace*, pp. 45–50, Elsevier Science, Oxford, 1999.
39. V.A. Rayevsky, V.V. Monakhov, and A.F. Glazunov, "Digital orientation and stabilizaton system for geostationary satellite-retransmitter *Lutch*." In *Spacecraft Dynamics and Control*, pp. 40–46, Nauka, Novosibirsk, 1992.
40. Ye.I. Somov, S.A. Butyrin, et al., "The software system dynamics for computer-aided design of spacecraft attitude control systems." In *Proceedings of the All-Russian Scientific School "Computer Logic, Algebra and Intelligent Control,"* vol. 3, pp 401–431, ICC, the RAS, Irkutsk, 1994.
41. Ye.I. Somov, S.A. Butyrin, et al., "Software tool DYNAMICS in simulation of fault-tolerant spacecraft attitude control systems," *Gyroscopy and Navigation* 2(25):92–107, 1999.
42. Ye.I. Somov, S.A. Butyrin, V.A. Rayevsky, et al., "Computer aided design and flight support of spacecraft control systems." In *Proceedings of the 14th IFAC Congress*, vol. L, pp. 445–450, Elsevier Science, Oxford, 1999.
43. Ye.I. Somov and Ye.A. Bondarenko, "Stability analysis of multiply discrete stabilization system of the flexible flying vehicle," *Soviet Aeronautics* (4):26–30, 1988.
44. Ye.I. Somov, "Dynamics of the multiply digital system for spatial gyromoment stabilization of flexible spacecraft." In *Spacecraft Dymamics and Control*, pp. 46–76, Nauka, Novosibirsk, 1992.
45. Ye.I. Somov and I.A. Gerasin, "Stability of linear continuous-discrete multiply control systems with a delay." In *Proceedings of the Tsiolkovsky Cosmonautics Academy*, pp. 133–137, Russian Academy of Cosmonautics, Samara, 1998.
46. Ye.I. Somov, "Robust stabilization of a flexible spacecraft at partial discrete measurement and a delay in forming control," *Journal of Computer and Systems Sciences International* 40(2):287–307, 2001.
47. D.I. Kozlov, G.P. Anshakov, Yu.G. Antonov, Ye.I. Somov, et al., "Precision control systems of motion on Russian spacecraft for ecological remote sensing." In *Proceedings of the 14th IFAC Symposium on Automatic Control in Aerospace*, pp. 27–38, Elsevier Science, Oxford, 1999.

48. Ye.I. Somov, "Nonlinear synthesis, optimization and design of the spacecraft gyromoment attitude control systems." In *Proceedings of 11th IFAC Workshop "Control Applications of Optimization,"* vol. 1, pp. 327–332, Elsevier Science, Oxford, 2000.
49. V.A. Bartenev, V.A. Rayevsky, Ye.I. Somov, et al., "Orbit and attitude control systems of Russian communication, geogetic and navigation spacecraft." In *Proceedings of the 14th IFAC Symposium on Automatic Control in Aerospace*, pp. 15–26, Elsevier Science, Oxford, 1999.
50. Ye.I. Somov, V.M. Matrosov, V.A. Rayevsky, et al., "Fault-tolerant attitude control systems of the communication and navigation spacecraft." In *Proceedings of the 14th IFAC Symposium on Automatic Control in Aerospace*, pp. 128–133, Elsevier Science, Oxford, 1999.
51. Ye.I. Somov, S.A. Butyrin, G.P. Anshakov, Yu.G. Antonov, et al., "Dynamics of the maneuvering vehicle *Ikar* control system by orbital placement of *Globalstar* satellites." In *Preprints of 15th IFAC Symposium on Automatic Control in Aerospace*, pp. 372–377, University of Bologna, Bologna, 2001.
52. V.A. Bartenev, G.P. Titov, V.A. Rayevsky, Ye.I. Somov, et al., "The attitude-orbit control subsystem and the antenna pointing perfermance of the *Sesat* spacecraft." In *Proceedings of 4th ESA International Conference on Spacecraft Guidance, Navigation and Control Systems*, pp. 35–48, ESTEC, Noordwjik, the Netherlands, 2000.
53. G.P. Titov, Ye.I. Somov, V.A. Rayevsky, et al., "Nonlinear dynamics of gyromoment attitude control and precise antenna pointing by the *Sesat* spacecraft." In *Preprints of 15th IFAC Symposium on Automatic Control in Aerospace*, pp. 73–78, University of Bologna, Bologna, 2001.
54. Ye.I. Somov, S.A. Butyrin, V.A. Rayevsky, G.P. Titov, et al., "Nonlinear dynamics of gyromoment attitude control system at communication satellite *Sesat*." In *Preprints of 5th IFAC Symposium "Nonlinear Control Systems,"* vol. 5, pp. 1481–1486, IPME of RAS, St. Petersburg, 2001.

SUBJECT INDEX